Prentice-Hall Computer Applications in Electrical Engineering Series

FRANKLIN F. KUO, *Editor*

NETWORK ANALYSIS
Theory and Computer Methods

RANDALL W. JENSEN

Department of Computer Science and Applied Statistics
Utah State University

Senior Staff Engineer
Hughes Aircraft Company

BRUCE O. WATKINS

Department of Electrical Engineering
Utah State University

PRENTICE-HALL, INC.

Englewood Cliffs, New Jersey

Library of Congress Cataloging in Publication Data

JENSEN, RANDALL W.
    Network Analysis.

    Includes bibliographies.
        1. Electric networks.   2. Electronic data
processing—Electric networks.   3. Electric engineering—
Mathematics.   I. Watkins, Bruce O., joint author.
II. Title.
TK454.2.J46   621.319′2   72-11750
ISBN 0-13-611061-4

PRENTICE-HALL INTERNATIONAL, INC., *London*
PRENTICE-HALL OF AUSTRALIA, PTY. LTD., *Sydney*
PRENTICE-HALL OF CANADA, LTD., *Toronto*
PRENTICE-HALL OF INDIA PRIVATE LIMITED, *New Delhi*
PRENTICE-HALL OF JAPAN, INC., *Tokyo*

To Our Families

CONTENTS

PREFACE

The application of the digital computer to electrical engineering has profoundly affected many areas, particularly network analysis and design. A new discipline, referred to as computer-aided design (CAD), has not only become accepted by industry as an important field, but now receives considerable support for further development. Some examples of the enthusiastic acceptance of computer-aided analysis methods are (1) the widespread use of commercial programs such as ECAP and SCEPTRE, (2) the large number of new analysis and design programs introduced each year, and (3) the increased emphasis on CAD in the IEEE publications.

Up to a few years ago, networks of more than three loops or nodes led to theoretically exciting ideas seldom consummated. In network analysis courses the network complexity rarely exceeded two nodes or loops simply because the perspiration expended in obtaining the networks' response to a small class of inputs far exceeded the increase in knowledge derived. This was particularly true in undergraduate classes. Now the computer has made it possible for the student to easily analyze the response of large networks driven by complex input signals. This new power frees the student from most of the traditional "pick and shovel" work so that he may devote his energy to network theory rather than elementary algebra. The student can also examine a broader scope of network topics where he was previously limited to only a few.

A risk involved in the indiscriminate use of the computer in network analysis consists of the danger that the student may lose sight of the fundamental basis of network analysis and rely on the computer to do all of the work. (One wonders what happens if the computer breaks down.) Many concepts exist in network design which still require mastery. These concepts, such as Kirchhoff's Laws, remain every bit as important as the computer and the electrical engineer is useless without them.

It is the purpose of this text to provide the traditional introductory electrical engineering network theory normally given at about the sophomore or junior level, but at the same time demonstrate the use of the computer in applying the theory. The text also introduces several areas of network theory which are basic to an understanding of computer methods of analysis. Although the text provides a number of programs developed by the authors and some programming problems, we do not intend to develop computer programmers. The student should know how to develop and how to analyze programs, but we prefer that he become an engineer.

We originally intended to include nonlinear analysis, but we soon found the subject of linear networks to be more than enough for one text. In studying so much linear analysis the student faces the hazard of accepting linearity as the norm of things; on the other hand a thorough basis in this theory is vital to any understanding of nonlinear phenomena.

The instructor will wish to emphasize certain portions of the text at the expense of other portions. While orienting the book toward the computer, efforts have been made to maintain a fair amount of mathematical rigor. However, the approach is that of the engineer, and not the mathematician. Some theorems (Tellegens' theorem, for example) are omitted. Space limitations did not allow treatment of n-port theory, or of many special aspects. Chapters 7, 11, and 12 are primarily devoted to the construction of automatic network analysis programs, although other chapters rely heavily on the computer for problem solutions. There is no claim that the programs shown are the most efficient that could be devised, but we have attempted to make them easy to read. The text is expected to occupy one academic year.

We have attempted to keep errors to a minimum, but will welcome your notice of these and also your criticism. Duplicates of the text programs punched in cards are available at nominal cost from Bruce O. Watkins, Electrical Engineering Dept., Utah State University, Logan, Utah 84321.

RANDALL W. JENSEN
BRUCE O. WATKINS

Logan, Utah

NETWORK ANALYSIS
Theory and Computer Methods

Chapter 1

INTRODUCTION

When I use a word, Humpty-Dumpty said, it means just what I choose it to mean— neither more nor less.

LEWIS CARROLL, *Through the Looking Glass and What Alice Found There.*

1.1 Introduction

During the 1960s, the digital computer became a widely accepted tool in the design of electrical networks. The advent of computer methods in the electrical network field proved to be a valuable extension to the network analyst's capabilities. Before the computer (BC), the analyst generally had to reduce the network to a very simple approximation to even begin to write the network equations and obtain a solution. Frequently, the network was oversimplified and yielded a poor prediction of the network's true response or characteristics. After constructing a better approximation of the network, the network equations became unwieldy at best and frequently so cumbersome that much of the network was laboratory designed. The cost of performing a worst-case network analysis in the laboratory is extremely high because (1) worst-case components, such as transistors, with simultaneous minimum and maximum parameters in the same device are hard to find, and (2) experimental procedures to find the best combination of extreme values for a large network are very time consuming. The difficulty and cost of worst-case analysis prohibited its use in many designs. The application of the digital computer now makes it possible to analyze networks containing several hundred nodes and branches in both the frequency and the time domain under both nominal and worst-case conditions at realistic cost and effort.

The introduction of the computer has brought to the analyst many additional methods, of which some are new while others were considered only to be of academic interest in the years BC. Many of the topological methods described in this text, such as the use of the incidence matrix (Chapter 7), have been known since the early 1950s but were too cumbersome to be of any practical use for hand calculations. These techniques are well suited to the use of the digital computer and allow the computer to formulate as well as solve the network equations.

1.2 Experiments and Mathematical Models

Most likely you have connected a circuit such as shown in Fig. 1.1, which diagramatically shows a dc battery, a resistance R, a variable resistance R_h, an ammeter, and a voltmeter. Varying R_h changes the voltmeter reading. You find that as the reading of the voltmeter changes, the reading of the ammeter simultaneously changes. Recording the ammeter and voltmeter readings results in a table of values corresponding to each setting of R_h. If we call the voltmeter reading v and the ammeter

2

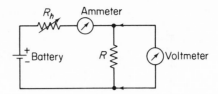

Figure 1.1 Simple nonlinear resistance network.

R_h Setting	i Amps	v Volts	$R = \dfrac{v}{i}$
1	1.10	11.0	10.00
2	1.82	18.1	9.95
3	2.41	24.2	10.02
4	3.06	30.6	10.00
5	3.98	39.9	10.01
6	4.52	45.1	9.98
7	6.85	70.1	10.23
8	8.24	90.6	10.99
9	9.47	117.0	12.37
10	10.30	138.0	13.40

Figure 1.2 Resistance data table.

reading i, we obtain a set of tabulated numbers such as shown in Fig. 1.2 (we assume each meter can be read to three significant figures).

The table of Fig. 1.2 establishes a functional relationship between v and i, and if we postulate the current as the independent variable, we say that v is a function of i, or in algebraic shorthand,

$$v = f(i) \tag{1.1}$$

In Eq. (1.1) the functional relation f is established by the table of Fig. 1.2, which shows the relationship of only 10 discrete points, or we have a discrete function.

If we plot the data of the table on rectangular graph paper, we obtain points such as shown by the small circles of Fig. 1.3. By disconnecting the battery of Fig. 1.1, we find that when $i = 0$, $v = 0$. Hence the origin in Fig. 1.3 is an additional point. The points on Fig. 1.3 give values of the function f for discrete values of i in Eq. (1.1), but by drawing in a smooth curve which comes as close as possible to the circles, we obtain the curve in Fig. 1.3, which then predicts the values of v which might occur for values of i other than those used in the experiment. In some cases the curve "misses" the points slightly. We could justify these differences by a "misreading" of the instruments, an error in plotting, an instrument error, or an inaccuracy in the curve. Where the points of the table of Fig. 1.2 give the functional relation of v to i for only discrete values of i, the curve in Fig. 1.3 gives v as a continuous function of i, that is, for all values of i within a specified set of currents. The curve may now be used to find a continuous functional relationship $v = f_1(i)$ for any value of i within the set.

Although tables express a type of functional relationship and curves another,

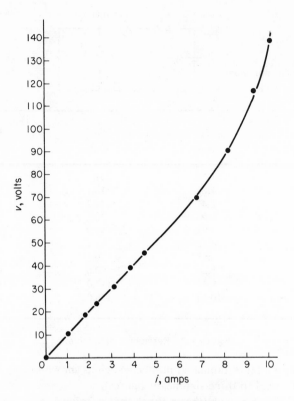

Figure 1.3 Nonlinear resistance characteristic.

we know that for analytic purposes it is even better if we can establish an algebraic function. To try to do so, we define the resistance R as the relationship between v and i for any two corresponding values; thus, by definition,

$$R = \frac{v}{i} \qquad \text{ohms\dag}$$

(1.2)

Equation (1.2) becomes quite revealing, for if we look at the curve in Fig. 1.3 we see that up to a value of i of about 6 amps (A), the value of R is constant; that is, the curve in Fig. 1.3 is a straight line. For i greater than 6 A, we find that R is not constant, and the curve is not a straight line. We might think of the resistance R as a small *network*, which operating on i produces v, as in Fig. 1.4. This can be expressed algebraically by showing R as a general network operator. Thus,

$$v = R(i)$$

(1.3)

Figure 1.4 Resistance network "black-box".

\dag The symbol for ohms is Ω.

where R now operates on i to produce v. For values of current i up to $i = 6$ A, the network is said to be linear. For values of i beyond $i = 6$ A, the network is said to be nonlinear.† In our immediate discussion, we see that in the linear region R itself is a constant, or within experimental error,

$$R = 10 \, \Omega \qquad 0 \le i \le 6 \, A \tag{1.4}$$

To obtain an algebraic relationship for R for $i > 6$ A we could resort to a process called *curve fitting*. One method for this assumes that the function $v = f_1(i)$ is a polynomial. That is,

$$v = ai + bi^2 + ci^3 + di^4 + \ldots \tag{1.5}$$

or from Eq. (1.2),

$$R = a + bi + ci^2 + di^3 + \ldots \tag{1.6}$$

For an approximation over a small region, we can terminate Eq. (1.5) after two or three terms. Thus, if R is considered a quadratic,

$$R = a + bi + ci^2 \tag{1.7}$$

To find a, b, and c, we need three values of R from Fig. 1.2 or Fig. 1.3, and, further, we need to define the set of i for which Eq. (1.7) may be true. Thus, for the region $6 \, A \le i \le 10 \, A$, we might select values of i at 6, 8, and 10 A. Then we obtain the equations

$$10 = a + b(6) + c(6)^2 \tag{1.8a}$$

$$\frac{90.6}{8.24} = a + b(8) + c(8)^2 \tag{1.8b}$$

$$\frac{138}{10.3} = a + b(10) + c(10)^2 \tag{1.8c}$$

We can solve Eqs. (1.8) for a, b, and c and obtain one possible algebraic functional relationship for R within a limited region of i [or for v in terms of i in Eq. (1.5)]. A polynomial *fit* for R may not necessarily be the best type nor the described method of obtaining the coefficients the best to find a functional relationship for R in the nonlinear region. There are many other functions and techniques, but this illustration demonstrates a general possible approach. At any rate, some sort of algebraic relationship becomes desirable for the solution of problems involving resistances similar to R, for then we may use these relationships in analysis and design.

When we have a mathematical relationship for the actual device or network, we often say that we have a *model* for the device or network.

The experiment described in this section illustrates several aspects of science. These aspects are:

1. Gather data by experimental means.

2. Collect the data in logical order.

---

†The terms "linear" and "nonlinear" will be more precisely defined subsequently.

3. Form a hypothesis which explains the experimental results.

4. Replace the actual experimental object by an abstraction or model, using mathematical relations if possible.

5. Test the hypothesis by repeating the experiment or by using alternative experiments.

6. Challenge others to test the hypothesis.

All these aspects become a part of a scientific theory which may become useful, and if so, the engineer may use the results of the scientific investigation in the analysis and design of devices and networks.

In network terms, the engineer often faces a problem of the type illustrated by Fig. 1.5, where Fig. 1.4 is altered slightly to show i and v as functions of another

Figure 1.5 General network operator.

$i(t)$ N $v(t)$

independent variable time, t. N is a general network which operates on $i(t)$ to produce some $v(t)$. The engineering problem usually reduces to one of the following three:

1. Given $i(t)$ and N, find $v(t)$.

2. Given $v(t)$ and N, find $i(t)$. Problems 1 and 2 involve analysis and assume that N has been described adequately by theory.

3. Given $i(t)$ and $v(t)$, find N. This is a problem in synthesis, or design, and for many networks may be much more difficult than the other situations, since the engineer must postulate the structure or the model for N. In other words, there may be many N which give the required results, and the engineer may then have to select the best of these on the basis of some other criteria.

We will illustrate these ideas in the next sections.

1.3   Analysis and Algorithms

Figure 1.6 shows a circuit with one nonlinear resistance R_2 and one resistor R_1 which operates in its linear region and hence is designated a linear resistance. The current i measured by the ammeter I is common to R_1 and R_2 if we make the voltmeter

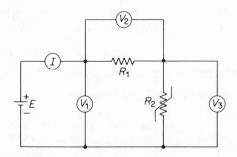

Figure 1.6 Nonlinear resistance network.

input resistances extremely high (approaching infinity) compared to R_1 and R_2.

Assume that we know the value of R_1 and have a relation for the value of v_3 (as measured by V_3) as a function of i. We wish to predict the value of i for any given value of v_1 (as measured by V_1). This is a typical problem in analysis. Let the R_2 characteristic be represented by the equation

$$v_3 = f(i) \tag{1.9}$$

First, we observe (or learn from theoretical developments) that $v_1 = v_2 + v_3$ and that i is the common current. Hence, since $R_1 = v_2/i$ and $R_2 = v_3/i$, then

$$R_T = R_1 + R_2 \tag{1.10}$$

where $R_T = v_1/i$ is the total resistance. In this situation the series resistance problem can be solved graphically by *load-line analysis*. We plot the function of Eq. (1.9) (it is convenient to use v_3 as the independent variable) as in Fig. 1.7. Since

$$v_1 = v_2 + v_3$$

and

$$v_2 = R_1 i$$

then

$$R_1 i = -v_3 + v_1$$

or

$$i = -\frac{v_3}{R_1} + \frac{v_1}{R_1} \tag{1.11}$$

Equation (1.11) is the equation of a straight line with slope $= -1/R_1$ if $v_1 = K_1$, a constant. Two convenient points to determine this line are with $v_3 = 0$, for which case $i = v_1/R_1$, and with $i = 0$, for which case $v_3 = v_1$. These two points are shown

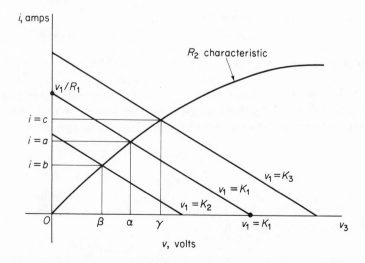

Figure 1.7 "Load line" analysis of nonlinear resistance network.

in Fig. 1.7, with the straight line drawn between them and for $v_1 = K_1$, a constant. The intersection of the line and the curve for R_2 gives the common current $i = a$ for this particular v_1, and the voltage $v_3 = \alpha$ across R_2. As a check, we require that $\alpha + aR_1 = K_1$. As v_1 varies, the straight line moves parallel to itself as shown in Fig. 1.7, since R_1, a constant, governs the slope. By taking a few values of v_1, such as K_1, K_2, and K_3, we may then prepare a table of i as a function of v_1 (or vice versa) as previously.

The graphical technique of Fig. 1.7 has particular appeal when the function of Eq. (1.9) is given in tabular or graphical form (such as transistor characteristics), v_1 is known, and R_1 is to be found to achieve a desired i. For more general use, we might apply Eq. (1.10) and plot the curve of i versus v_1 by graphically adding the two characteristics as in Fig. 1.8. We may immediately extend this to two or more nonlinear resistances.

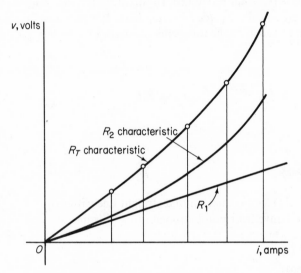

Figure 1.8 "Load line" analysis of two-resistance network.

The graphical techniques suggested by Figs. 1.7 and 1.8 may not be very exciting, since they involve tedious plotting and are not particularly accurate. Since this text purports to be computer oriented, let us introduce the computer and see how it can do the "pick-and-shovel" work. Assume that we have two nonlinear resistances R_1 and R_2 in series and that their respective characteristics are given by the algebraic equations

$$v = f_1(i) \tag{1.12a}$$

$$v = f_2(i) \tag{1.12b}$$

We wish to find i for any given v_1 placed across the series combination. Thus, we need to solve an equation of the form

$$v_1 = f_1(i) + f_2(i)$$

or

$$f_3(i) - v_1 = 0 \qquad (1.13)$$

where

$$f_3(i) = f_1(i) + f_2(i)$$

[We have already briefly discussed how we might obtain $f_1(i)$ or $f_2(i)$ if these functions are originally given in tabular or graphical form.] The problem then reduces to finding the roots of Eq. (1.13). One method of solving Eq. (1.13) is to get it into the form $f(i) = i$. A very simple-minded idea is to add (or subtract) i from each side. Thus, in Eq. (1.13),

$$f_3(i) + i - v_1 = i$$

or

$$f_4(i) = i \qquad (1.14)$$

where

$$f_4(i) = f_3(i) + i - v_1$$

Now let

$$y_1 = f_4(i) \qquad (1.15)$$

and

$$y_2 = i \qquad (1.16)$$

Obviously from Fig. 1.9, the solution lies at the intersection of y_1 and y_2. To find this solution using the computer, we assume an initial value of $i = i_{(1)}$. We then solve Eq. (1.15) for $y_{1(1)}$ corresponding to $i_{(1)}$. In the next step we let $y_{2(2)} = y_{1(1)}$, and since in general $y_2 = i$, we obtain a new value of $i = i_{(2)}$. With this value $i_{(2)}$, we start a new iteration by again solving Eq. (1.15) to obtain $y_{1(2)}$. From Fig. 1.9 it seems probable that if we continue this process, the intersection will be approached rather rapidly. The method used here is termed *iteration*, the subscript in parentheses

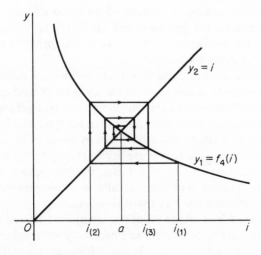

Figure 1.9 Iterative solution of nonlinear network operation.

indicating the iteration number. The general process utilizes what is known as an *algorithm*.

An algorithm may be described as a process, method, technique, routine, procedure, or recipe for solving a segment of a problem. The word "recipe" brings to mind the mixing at the proper times of the proper amounts of the correct ingredients to produce gastronomic joy. Somewhat similarly, a good algorithm produces joy and satisfaction in obtaining desired solutions to a given problem. A good algorithm must (1) terminate in not too many steps, (2) consist of precise steps, (3) have input and output quantities, and (4) be effective. An algorithm precisely describes the steps necessary to be taken whether we solve the problem by hand or by computer. A program is a collection of algorithms, each one solving a segment of a broader and more difficult problem. The word "program" usually implies that we intend the solution to occur on a computer of a particular kind. Thus, a program utilizes in logical order one or more algorithms and is written in a specific form called a *language*, usually intended for a specific type of computer. Although computer-oriented languages are becoming more universal, unfortunately a program written in Fortran, say, will not usually run on any machine but will require some revision for each. We assume that you are familiar with the rudiments of digital-computer operations. In this text we will use Fortran IV as a programming language because of its universality as a scientific language rather than the newer, more powerful PL/1.

Algorithms frequently use iterative techniques such as the one previously described. The one under consideration may be placed in diagramatic form, as in Fig. 1.10, which is called a *flow diagram* or *flowchart*. The general relationship used in this example consists of the iteration

$$i_{(n)} = f_4(i_{(n-1)}) \qquad n = 2, 3, \ldots \tag{1.17}$$

with $f_4(i_{(1)})$ computed from an assumed starting value $i_{(1)}$.

At first glance, the algorithmic approach seems to offer little. However, you must recall that the iterations occur at tremendous speed and that the solutions are computed in microseconds. The precision and speed of the digital computer bring into play many possible mathematical methods which, prior to the advent of the computer, were only of academic interest.

The branch of mathematics used to explore techniques particularly adaptable for digital-computer solution is called *numerical analysis*. Numerical analysis, as the name implies, gives results in numerical form (as opposed to algebraic or other forms). The stress is on specific numerical results rather than analytic solutions. The computer numbers may be used to form graphical displays by using other peripheral devices, but the basic data are digital. The results are accurate only to within some tolerance, which depend on the program and the basic machine characteristics. Furthermore, in many cases, correct numerical results may be obtained when existence theorems which guarantee the results cannot be found.

Proceeding through Fig. 1.10, we input the functions $f_1(i)$ and $f_2(i)$. It is possible to store tabular values on the computer, and necessary intermediate values are then found at the current iteration by interpolation. However, normal memory space is

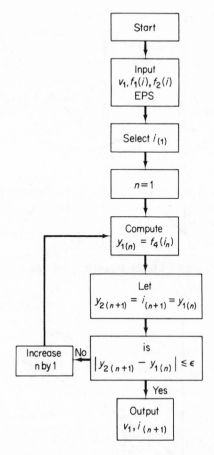

Figure 1.10 Iterative solution
                flowchart.

too limited for this method, and the functions, if in tabular form, are approximated by some sort of curve-fitting algorithm in a preliminary operation [see Eq. (1.7), for example]. In many cases the functions are already known in algebraic form. We then input v_1 and proceed as shown in Fig. 1.10. After the computation of $f_4(i_{(n)})$, we need to make a test to see if the algorithm has reached a solution. Many possible tests may prove satisfactory. In Fig. 1.10 we compare the magnitude of the difference between y_1 and y_2 [Eqs. (1.15) and (1.16)] with some test-value ϵ which must be input by the programmer or furnished by the program. Other more sophisticated tests may be used. If the test does not result in a sufficiently close agreement between y_1 and y_2, we then proceed with the algorithm; if it does result in close agreement, the solution (to within some tolerance) is found.

We may now ask a reasonable question: will the algorithm converge to a single solution? By sketching various $f_4(i)$ curves (see Figs. 1.9, 1.11, and 1.12) it appears that convergence is likely to occur if the magnitude of the slope of y_1 is less than the magnitude of the slope of y_2 in the region of interest, or, since the slope of $y_2 = 1$, we require that $|y_1'| = |f_4'(i)| \leq 1$ over the interval $i_{(1)}, i_{(2)}, \ldots, i_{(n)}$.

This may be proved as follows: Let $i = a$ be the correct solution for Eq. (1.14)

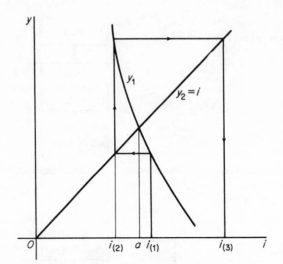

Figure 1.11 Divergent iterative solution for $|y_1'| > |y_2'|$.

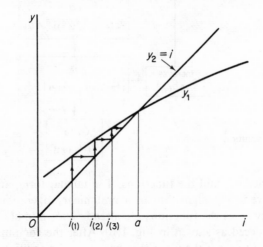

Figure 1.12 Convergent iterative solution for $|y_1'| < |y_2'|$.

and let $f_4(i) = f(i)$ for convenience; then

$$a = f(a) \tag{1.18}$$

Assume the algorithm described in Eq. (1.17), or

$$i_{(n)} = f(i_{(n-1)})$$

Then

$$i_{(n)} - a = f(i_{(n-1)}) - f(a) \tag{1.19}$$

Multiply the right side of Eq. (1.19) by $(i_{(n-1)} - a)/(i_{(n-1)} - a)$ and apply the mean-value theorem[†] to get

$$i_{(n)} - a = f'(\xi)(i_{(n-1)} - a) \tag{1.20}$$

where ξ lies somewhere between $i_{(n-1)}$ and a and $f'(\xi)$ is the derivative with respect to i.

Let k be the maximum value of $|f'(i)|$ over the interval considered. Then

$$|i_{(n)} - a| \leq k|i_{(n-1)} - a|$$

and

$$|i_{(n-1)} - a| \leq k|i_{(n-2)} - a|$$

Thus

$$|i_{(n)} - a| \leq k^2|i_{(n-2)} - a|$$

and so on, or finally

$$|i_{(n)} - a| \leq k^\gamma|i_{(1)} - a| \tag{1.21}$$

where $\gamma > 1$. Then in Eq. (1.21), if $k < 1$ over the interval, the initial choice of $i_{(1)}$ will cause the algorithm described in Eq. (1.17) to converge to the proper solution.

More extensive analysis has established the following theorem[‡]: Given the equation

$$x = f(x) \tag{1.22}$$

where f is continuous on the closed and bounded interval I and has its values in I. Then for any two points x_1 and x_2 in I if

$$|f(x_1) - f(x_2)| \leq L|x_1 - x_2| \tag{1.23}$$

where $0 \leq L < 1$, then the sequence

$$\xi_{(n)} = f(\xi_{(n-1)}), \qquad n = 2, 3, \ldots \tag{1.24}$$

converges to a solution of Eq. (1.22) for any $\xi_{(1)}$ in I, and this is the only solution in I. The condition, Eq. (1.23), is termed the *Lipschitz condition*. The theorem does not utilize or require information on slopes. Practically, this means that it is not necessary to know or calculate the derivative of $f(x)$ to prove convergence.

The Lipschitz condition may require difficult or time-consuming testing in a program, and we may try to find a solution using Fig. 1.10 by any arbitrary choice of $i_{(1)}$. Moreover, since this Lipschitz condition is sufficient, but not necessary, it is possible that convergence may occur even if it is violated. In such a case, however, we need to provide a way of stopping the iterations if the solution does not converge.

[†]*Theorem of mean value:* If a single-valued function $f(x)$ and its derivative $f'(x)$ are continuous in the interval $a \leq x \leq b$, there exists at least one point x_1 between a and b such that

$$\frac{f(b) - f(a)}{b - a} = f'(x_1)$$

[‡]Peter Henrici, *Discrete Variable Methods in Ordinary Differential Equations*, John Wiley & Sons, Inc., New York, 1962, pp. 15–16.

One method would be to check $y_{2(n)} - y_{1(n)}$ with $y_{2(n-1)} - y_{1(n-1)}$ to make sure that the difference does become smaller as n grows. If it does not, we might try another $i_{(1)}$. We may also wish to place a limit on the maximum number of iterations and the overall number of initial conditions, since if divergence occurs, we can spend much expensive computer time uselessly.

Again look at Eq. (1.17), which we can rewrite as

$$i_{(n+1)} = f(i_{(n)}) \qquad n = 1, 2, \dots$$

or

$$i_{(n+1)} = i_{(n)} + [f(i_{(n)}) - i_{(n)}]$$

and thus

$$i_{(n+1)} = i_{(n)} + \Delta i_{(n)} \tag{1.25}$$

We might consider multiplying the second term by a quantity which would hasten convergence (overshoot i). Thus make

$$i_{(n+1)} = i_{(n)} + B[f(i_{(n)}) - i_{(n)}] \tag{1.26}$$

If we let

$$B = \frac{1}{1 - f'(i_{(n)})} \tag{1.27}$$

then

$$i_{(n+1)} = i_{(n)} + \frac{f(i_{(n)}) - i_{(n)}}{1 - f'(i_{(n)})}$$

Now let

$$F(i_{(n)}) = f(i_{(n)}) - i_{(n)}$$

Then

$$F'(i_{(n)}) = -[1 - f'(i_{(n)})]$$

and Eq. (1.26) becomes

$$i_{(n+1)} = i_{(n)} - \frac{F(i_{(n)})}{F'(i_{(n)})} \tag{1.28}$$

where the derivative F' is taken with respect to $i_{(n)}$.

Equation (1.28) is the famous Newton–Raphson algorithm, which converges much faster than the previous method described by Eq. (1.17). (It is said to have quadratic convergence, as opposed to linear convergence.) Furthermore, $F(i)$ is our original function [for example, $F(i) = f_3(i) - v_1$ in Eq. (1.13), rather than $f_3(i) - v_1 + i$ as in Eq. (1.14)]. However, Eq. (1.28) actually converges only if $i_{(1)}$ is sufficiently close to a root of $F(i) = 0$, $F'(i)$ is not too small, and $F''(i)$ is not too large in the region of interest. In other words, compared to Eq. (1.17), we have lost our precise test for convergence, and now derivative information becomes necessary. The assumption given in Eq. (1.26) causes us to proceed along a tangent to $F(i_{(n)})$ at $i_{(n)}$ to find $i_{(n+1)}$. Figure 1.13 shows a Newton–Raphson solution. Usually in a well-behaved function, the Newton–Raphson method works quite well if $i_{(1)}$ is reasonably close to a and all roots are reasonably separated.

We can solve networks with parallel nonlinear resistances in a manner analogous to that used for series resistances. In this case the voltage across the resistances is

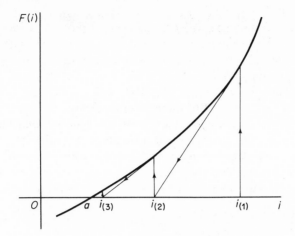

Figure 1.13 Newton–Raphson convergent solution.

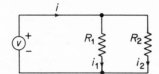

Figure 1.14 Nonlinear parallel resistance network.

common. Thus, in Fig. 1.14 we may have R_1 and R_2, respectively, described by the equations

$$i_1 = f_1(v) \tag{1.29a}$$

$$i_2 = f_2(v) \tag{1.29b}$$

We desire to find v for any given current i flowing into the network. Then

$$i = f_1(v) + f_2(v)$$

or

$$f_3(v) - i = 0 \tag{1.30}$$

where

$$f_3(v) = f_1(v) + f_2(v)$$

We now find the roots of Eq. (1.30) for various values of i using one of the algorithms previously described. Thus, root-finding algorithms have wide application and are quite generally useful. Since $f_3(i)$ in Eq. (1.13) and $f_3(v)$ in Eq. (1.28) may be quite complicated functions, it should not surprise you that in many cases we may arrive at more than one solution, because the function curves cross more than once, which means that we may have to supply many starting values. Preliminary sketching may reveal the regions of interest and eliminate unapplicable roots. (Some roots may be complex, which might have no physical significance in certain situations.)

This brief discussion of nonlinear networks shows how much more complicated they become than simple linear networks, in which only one solution exists. Before we continue with these ideas, however, it behooves us to define some terms and review some principles that will be needed in future work.

1.4 Terms and Definitions

Before we proceed, we need to more precisely define some terms we have used up to now. To this onerous but necessary task we now proceed. Generally the engineer deals with input and output quantities which are functions of the independent variable time. Thus, we may designate an input m to a device or network as $m(t)$ and the output y as $y(t)$. For now, assume that these functions are continuous functions of time, as in Fig. 1.15; that is, values are known for any time t lying within some given set.

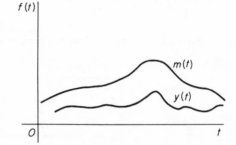

Figure 1.15 Input $m(t)$ and output $y(t)$ of network operator P.

These time functions often are called *signals*. A signal is merely a general name for voltage, current, force, velocity, or other quantity that appears in a network. Often it becomes convenient to consider a signal as entering a network as an input, then proceeding through the network, and exiting as an output, the network altering or reshaping the signal and often changing its units as it proceeds. We now define a network by using an operator P in the equation

$$y(t) = P(m(t)) \tag{1.31}$$

where the equation refers to Fig. 1.16. In this figure P is an operator such that the output $y(t)$ results from the input $m(t)$. In general, the operator may be such that $y(t)$ depends not only on present values of $m(t)$, but also on past (and perhaps even

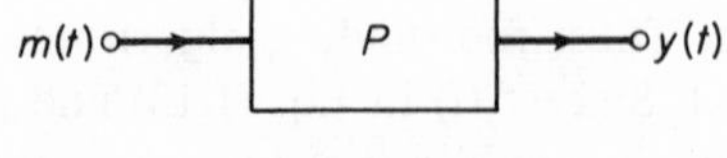

Figure 1.16 Network function P.

$m(t) \circ \!\!\!\longrightarrow \boxed{\qquad P \qquad} \longrightarrow \!\!\! \circ y(t)$

future) values of $m(t)$. You have already seen rather simple operators in the previous sections; that is, resistance may be considered an operator such that current becomes altered to voltage, or

$$v(t) = Ri(t) \tag{1.32}$$

Here the operator P specified the simple product of R and i. Another example might be that of a capacitance. Here

$$v(t) = \frac{1}{C} \int_{-\infty}^{t} i(\tau) \, d\tau \qquad (1.33)$$

and the operator specifies an integration process, or $v(t)$ depends on values of $i(t)$ from the infinite past to the time t considered. Thus, the operator shown in Eq. (1.31) corresponds to such operators as d/dt log and $\int$ and is not an entity in itself.

If we have many inputs and outputs occurring at different points and networks as in Fig. 1.17, we may consider these to be ordered arrays. That is,

$$\mathbf{m} = \begin{bmatrix} m_1 \\ m_2 \\ \cdots \\ m_n \end{bmatrix} \qquad (1.34)$$

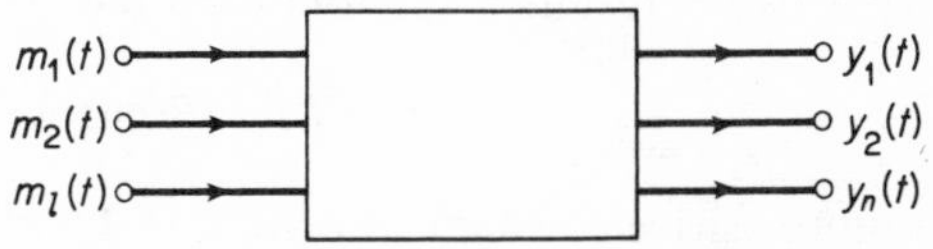

Figure 1.17 Multiple input–output network function.

where $m_1, m_2, \ldots, m_n$ are inputs. The ordered array on the right side of Eq. (1.34) is a matrix (another name for an ordered array) of special characteristics, an $n \times 1$ matrix. Such a matrix is usually termed a *vector*. Now Eq. (1.31) may be written

$$\mathbf{y}(t) = \mathbf{P}(\mathbf{m}(t)) \qquad (1.35)$$

where $\mathbf{P}$ becomes another matrix if the network is linear.

What do we mean by linear? Going back to a single input–output network for simplicity, we call a network *linear* if the principle of superposition applies. This principle states that if the output $y_k(t)$ can be found for any given input $m_k(t)$, then the output to simultaneous inputs $m_1(t), m_2(t), \ldots, m_k(t), \ldots, m_n(t)$ will be the sum of the outputs resulting from each input acting alone. Here the subscripts 1, 2, k, and n refer to different forms of inputs applied at the same point. In equation form, given a network where

$$y(t) = P(m(t))$$

then if the network is linear,

$$P(Am_1(t) + Bm_2(t)) = AP(m_1(t)) + BP(m_2(t)) \qquad (1.36)$$

where A and B are constants. That is, the linear operator acting on two simultaneous inputs (1) gives the same quantity as the sum of the results given by the operator acting on each individually, and (2) a constant input multiplier carries over unchanged into the output.

Linear networks may be treated by mathematical relations in a much simplier fashion than nonlinear networks. For example, Laplace, Fourier, and Z transform methods apply to linear networks. In many cases linear theory provides solutions which are sufficiently good for design purposes, and in other cases linear theory can be extended by making an average linear approximation, or using linear approximations over limited operating regions. Thus we find that most theory presented to the undergraduate is of the linear variety. However, bear in mind that natural phenomena are usually nonlinear. With the advent of the digital and analog computers, we may treat more nonlinear cases than was done in the past, as shown by the examples of Section 1.2. Even so, however, linearity gives such simplifications that nonlinear operators should be used only if necessity makes such a demand.

Either a linear or nonlinear network may be time varying; loosely the operator is a function of the independent variable time. A non-time-varying network is termed *fixed* or *stationary*. Thus, for a network

$$y(t) = P(m(t))$$

a stationary network results if the following relation holds:

$$y(t - T) = P(m(t - T)) \tag{1.37}$$

where T is any positive constant. Equation (1.37) indicates that the shape or form of the response is independent of the time of application of the input. Thus, we see the same trace on an oscilloscope for the output whether we apply the input at time t or at time $(t - T)$.

For continuous networks, we will be dealing primarily with operators which are described by an ordinary differential equation of the form

$$K_n \frac{d^n x}{dt^n} + K_{n-1} \frac{d^{n-1} x}{dt^{n-1}} + \ldots + K_0 = f(t) \tag{1.38}$$

where

$$x = x(t)$$

Rather loosely (exceptions can be found), if the coefficients K_i are not functions of the independent variable t, the network is stationary, and if the coefficients K_i are not functions of the dependent variable x, the network is linear. The use of ordinary differential equations indicates description of a lumped parameter network, as opposed to a distributed parameter network, which must be described by partial differential equations. Most networks are also causal and dynamic. A *dynamic network* is one in which past events influence the present, or we say the network has a *memory*. If the output depends only on the present, it has no memory. For example, in the network described by

$$v(t) = Ri(t) \tag{1.39}$$

$v(t_1)$ depends only on the present value $i(t_1)$, but in the network described by

$$v(t) = \frac{1}{C} \int_{-\infty}^{t} i(\tau) \, d\tau \tag{1.40}$$

the value of $v(t)$ depends on all past values of $i(t)$. In nonrigorous language, the response of a *causal* (or *physical*) *network* depends only on past events (or does not depend on future events). A network made up of physical elements must be causal, whereas a hypothetical network described by mathematical equations need not be causal.

A network may be described by lumped parameters if the wavelength $\lambda = v/f$ is large compared to the physical components of the network, where v is the velocity and f the frequency, as defined above. In an electrical network $v \approx 10^8$ meters per second, the velocity of light. Hence lumped parameter networks are usually adequate for physically small networks for frequencies from zero to the megahertz range. A transmission line whose length is comparable to λ must be described by distributed parameters. In this text we consider only linear, fixed, lumped, causal networks.

1.5 Signals

We now need to discuss signals we repeatedly encounter. Signals may be classified as deterministic or stochastic. A *deterministic signal* is one which we can predict with small uncertainty, and write an equation which gives values at any time. For example,

$$m(t) = 60 \sin(5t + 30°) \qquad t \geq 0 \qquad (1.41)$$

describes a signal which will have a calculated value of 30 at $t = 0$, a value of 60 at $t = 12$, and so on. In any case, the actual value of the signal will be within a given tolerance of the calculated value, but this tolerance reflects accuracies in measurement or production and can be specified. A deterministic signal is thus predictable and repeatable (to within specified tolerance limits). A signal not so predictable is called *random* or *stochastic*. In working with random signals we have little idea as to its value at any particular time, and must therefore adopt some kind of statistical analysis. In this book we deal with deterministic signals, which are much more amenable to mathematical treatment than random signals.

Signals may also be classified as to whether or not the independent variable time has continuity. If the signal is a function of the continuous variable t, it is termed *continuous*. Note that the signal itself, or the dependent variable, may have discontinuities. For example, Fig. 1.18 shows a signal $m(t)$ which is a function of the continu-

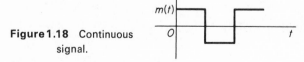

Figure 1.18 Continuous signal.

ous variable time and is known at any value of t (except at the dependent variable discontinuities).

If a signal is a function of time at particular instants and is not known for intervals between these points in time, it is a function of a discrete variable; a network utilizing such signals is said to be a *discrete time network*. Figure 1.19 shows a signal $m(t)$ which is known only at time $t_1, t_2, t_3, \ldots, t_n$, but not at values of t intermediate in the interval $t_1, t_2, \ldots$. Such a signal may be written

$$m = m(t_k)$$

or

$$m = m(k)$$

where k specifies the instant at which the signal is known.

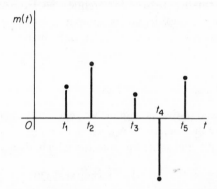

Figure 1.19   Discrete signal.

A large class of discrete time networks now exist, and in particular digital computers always handle discrete time signals because of their nature. Frequently the time intervals between known values are constant. We indicate that a continuous network may be characterized by a differential equation such as

$$\frac{dy(t)}{dt} = f(y(t), m(t))$$

[see Eq. (1.38), for example], while a discrete network may be depicted by the corresponding difference equation

$$y(k + 1) = f(y(k), m(k)) \tag{1.42}$$

The concepts of linearity, stationarity, and so on, discussed previously for continuous time networks, carry over with only slight revisions into discrete time networks. We now present some continuous time signals which are frequently used in engineering work.

One of the oldest signals used in analysis is the sine-wave signal. We define this signal by the equation

$$m(t) = A_k \sin(\omega_k t + \phi_k) \tag{1.43}$$

where $\omega_k = 2\pi f_k$, and f_k is the frequency in hertz, A_k is the peak value of the signal and ϕ_k is the phase angle. One cycle of this signal is shown in Fig. 1.20 for $0 < \phi_k < 90°$. The sine function repeats every 360° (or 2π radians) and the time T for one such

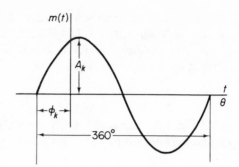

Figure 1.20 Sinusoidal continuous signal.

cycle is $T = 1/f_k$. Hence the independent variable may also be an angle θ. In electrical degrees, $\theta = 360\,t/T$.

Several reasons exist for the popularity of the sine-function signal. First, simple but reliable generators exist for producing the signal, and for accurately measuring ω_k, A_k, and ϕ_k. Second, after transients have vanished, the output of a linear network with sine-wave input is another sine wave, with altered magnitude and phase angle. The ratio of input and output magnitudes and the difference of phase angles leads to the definition of a characteristic of the network termed *impedance*.

Finally, and perhaps of most importance, any signal periodic in time may be represented by a Fourier series, which is an infinite sum of terms of the type given by Eq. (1.43). Thus, if any $m(t)$ is periodic, then

$$m(t) = \sum_{k=0}^{\infty} A_k \sin(k\omega_k + \phi_k) \tag{1.44}$$

where k takes on all positive integral values. This implies that if we can analyze a device or network for the single general signal of Eq. (1.43) we have essentially solved it for all periodic signals by merely using a summation of the type given in Eq. (1.44). Alternatively, if we know the network response both as to gain (ratio of output magnitude to input magnitude) and phase (difference between output and input phase angles), we know a great deal about the network. We explore this further subsequently.

The step function perhaps follows the sine function in importance. It is shown in Fig. 1.21 and is described mathematically by the relations

$$m(t) = \begin{cases} 0 & t < 0 \\ A & t > 0 \end{cases} \tag{1.45}$$

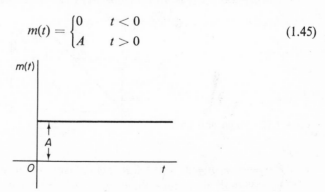

Figure 1.21 Step function.

At $t = 0$, $m(t)$ may be defined in various ways, depending on the situation.† The unit step function, which exists when $A = 1$, has sufficient use that we give it the special designation $u(t)$. A time function $m(t) = 0$, $t < t_1$ and $m(t) = A$, $t > t_1$ ($t_1 > 0$) can thus be described by the equation

$$m(t) = Au(t - t_1) \tag{1.46}$$

which is the step of Fig. 1.21 displaced t_1 units along the t-axis. The step function is generated easily, and network output with this input becomes a measure of network performance.

The step function is a special case of the general polynomial function

$$m(t) = \begin{cases} 0 & t < 0 \\ m_0 + m_1 t + m_2 t^2 + \dots, & t > 0 \end{cases} \tag{1.47}$$

where $m_0, m_1, m_2, \dots$ are constant. The second term comprises the ramp function, the third term the parabolic function, and so on. For "unit" functions we may obtain each successive term by integrating the preceding term. Thus, the unit ramp function is

$$m(t) = 0 \qquad t < 0$$
$$m(t) = \int_0^t u(\tau) \, d\tau \tag{1.48}$$
$$= t \qquad t \geq 0$$

and the unit parabolic function is

$$m(t) = 0 \qquad t < 0$$
$$m = \int_0^t \tau u(\tau) \, d\tau \tag{1.49}$$
$$= \tfrac{1}{2} t^2 \qquad t \geq 0$$

These functions are shown in Fig. 1.22.

An integration process which obtains higher-order unit terms tempts us to proceed the other way, that is, to differentiate. If we differentiate the unit step function

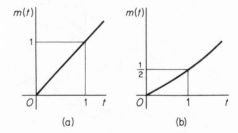

Figure 1.22 Ramp and parabolic signals.

(a)   (b)

---

†For example, $m(0) = (0 + A)/2 = A/2$, or the average of 0 and A at $t = 0$ using the inverse Laplace transform. Or $m(0_+) = A$ if 0_+ implies a time slightly greater than $t = 0$.

with respect to time, we obtain a new signal, called the *unit impulse*, $\delta(t)$, that is

$$\delta(t) = \frac{du(t)}{dt} \tag{1.50}$$

Equation (1.50) must be considered in a special way, since if we tried to differentiate the signal of Fig. 1.21 (with $A = 1$), we would run into some problems of limits. To thoroughly understand the impulse $\delta(t)$ one needs to study the theory of distributions, introduced by Schwartz.† The impulse function $\delta(t)$ is a special form of distribution which may be obtained as a generalized limit of a series of more ordinary functions.‡ One of these functions may be represented by the equation

$$f(t) = \lim_{\lambda \to 0} \frac{1}{\lambda}\left[u\left(t + \frac{\lambda}{2}\right) - u\left(t - \frac{\lambda}{2}\right)\right] \tag{1.51}$$

where "lim" implies the limit of the series of functions represented by the right-hand side of Eq. (1.51) as λ becomes smaller. One function in the series is depicted by the rectangle shown in Fig. 1.23. The area of the rectangle is $(\lambda)1/\lambda = 1$. Then Eq. (1.51) is sometimes described as allowing λ to approach zero in Fig. 1.23 while the area

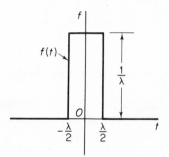

Figure 1.23 Unit impulse approximation.

remains constant. As we follow this to the limit, we have some visualization difficulty, which emphasizes the fact that $\delta(t)$ cannot be classed as an ordinary function. Two possible (among many) other impulse signals are shown in the following equations:

$$\delta(t) = \lim_{\omega \to \infty} \frac{\sin \omega t}{\pi t} \tag{1.52}$$

$$\delta(t) = \lim_{\lambda \to 0} \frac{e^{-t^2/\lambda}}{\sqrt{\lambda \pi}} \tag{1.53}$$

where again both limits are to be taken in the sense of the limit of a series of the functions indicated.

---

†See E. F. Beckenbach, *Modern Mathematics for the Engineer*, McGraw-Hill Book Company, New York, 1961, pp. 5–67.

‡A distribution assigns a number $N[\phi(t)]$ to a test function $\phi(t)$. The test function must meet certain requirements.

Rather than defining $\delta(t)$ by Eqs. (1.51), (1.52), or (1.53), and without going into the Schwartz theory here, we adopt the simple and yet understandable definitions of the impulse given by the following three relations:

$$\int_{-\infty}^{\infty} \delta(t - t_1) f(t)\, dt = f(t_1) \qquad (1.54a)$$

$$\delta(t - t_1) = 0 \qquad t \neq t_1 \qquad (1.54b)$$

$$\delta(t - t_1) = \delta(t_1 - t) \qquad (1.54c)$$

where $f(t)$ is continuous at $t = t_1$.

The last two relations say that the impulse signal is an even function and that it is zero everywhere except at the time t_1 when it occurs. Equation (1.54a) may be visualized by imagining $\delta(t - t_1)$ as a very long spike at t_1, as in Fig. 1.24. (Consider Fig.

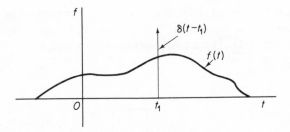

Figure 1.24 Signal and superimposed sifting function.

1.23 as $\lambda \rightarrow 0$.) Now look at the product inside the integral of Eq. (1.54a), considering $\delta(t - t_1)$ as a function which weights (or influences) $f(t)$. Clearly from Fig. 1.24, the only value of $f(t)$ not weighted to zero is at t_1. At this point $f(t) = f(t_1)$, a constant which can be brought outside the integral. Hence, in Eq. (1.54a),

$$\int_{-\infty}^{\infty} \delta(t - t_1) f(t)\, dt = f(t_1) \int_{-\infty}^{\infty} \delta(t - t_1)\, dt \qquad (1.55)$$

If we meet the definition given by Eq. (1.54a), then from Eq. (1.55),

$$\int_{-\infty}^{\infty} \delta(t - t_1)\, dt = 1 \qquad (1.56)$$

Thus, the area of the impulse is unity. The impulse $\delta(t)$ occurs when $t_1 = 0$, or $\delta(t - t_1)$ is merely the unit impulse displaced t_1 units along the positive t-axis, as in Fig. 1.24. In some situations we also need the relation

$$f(t)\delta(t) = f(0)\delta(t) \qquad (1.57)$$

if $f(t)$ is continuous at $t = 0$.

Equation (1.54) defines the impulse by the results it produces rather than as an entity in itself. It is a signal of great theoretical value, as we shall see. We will not do so here, but it may be shown that the previous signals described by Eqs. (1.51), (1.52), and (1.53) satisfy the requirements of our definition. [You are invited to show that Eq. (1.56) is satisfied.]

The impulse may not seem a very practical signal, as it requires the achievement of infinite magnitude in zero time. However, we may closely approximate an impulse by a signal which has a time duration that is short compared to time constants of the network to which it is applied. We give the following example:

EXAMPLE 1.1

In Fig. 1.25a find $v_c(t)$ if $v_c = 0$ just before $t = 0$, e is the pulse signal shown in part b, where $\tau = RC$ is the time constant of the network, and α is a positive constant. Vary α, find $v_c(\tau/\alpha)$, and compare this with the response v_c at $t = \tau/\alpha$ when e is an impulse $\delta(t)$. Both R and C are constant (the network is linear).

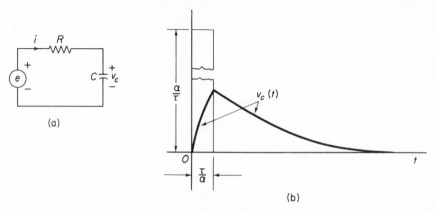

Figure 1.25 Impulse testing example.

Solution. As we will verify in Chapter 5, the differential equation for Fig. 1.25a with R and C constant is

$$RC\frac{dv_c}{dt} + v_c = e \tag{1.58}$$

From time $t = 0$ to $t = \tau/\alpha$, the transient (homogeneous) solution is

$$v_c = K_1 e^{-(1/RC)t} \tag{1.59}$$

and the steady state (particular integral) is

$$v_c = \frac{\alpha}{\tau}$$
$$= \frac{\alpha}{RC} \tag{1.60}$$

The complete solution therefore becomes

$$v_c = K_1 e^{-(1/RC)t} + \frac{\alpha}{\tau}$$

Since with a finite source it takes a finite time to charge the capacitance C, and thus to change the voltage across it (as will be discussed in more detail subsequently), if

$v_c = 0$ just before $t = 0$, then $v_c = 0$ just after $t = 0$, or v_c is continuous. Hence, $K_1 = -\alpha/\tau$ or

$$v_c = \frac{\alpha}{\tau}(1 - e^{-(1/RC)t}) \qquad 0 \leq t \leq \infty \tag{1.61}$$

Now when

$$t = \frac{\tau}{\alpha} = \frac{RC}{\alpha}$$

$$v_c\left(\frac{\tau}{\alpha}\right) = \frac{\alpha}{\tau}(1 - e^{-1/\alpha}) \qquad \text{(pulse input)} \tag{1.62}$$

From time $t = \tau/\alpha$ to infinity, $v_c(t)$ obviously then decays exponentially, or

$$v_c(t) = v_c\left(\frac{\tau}{\alpha}\right)e^{-(1/\tau)(t-\tau/\alpha)} \qquad \frac{\tau}{\alpha} \leq t \leq \alpha \tag{1.63}$$

The solution to Eq. (1.58) when $e = \delta(t)$ will be shown in Chapter 5 to consist of the equation

$$v_c(t) = \frac{1}{\tau}e^{-t/\tau} \qquad 0 \leq t \leq \infty \tag{1.64}$$

The value of Eq. (1.64) when $t = \tau/\alpha = RC/\alpha$ is thus

$$v_c\left(\frac{\tau}{\alpha}\right) = \frac{1}{\tau}e^{-1/\alpha} \qquad \text{(impulse input)} \tag{1.65}$$

The difference between Eqs. (1.62) and (1.65) is

$$\text{dif} = \frac{1}{\tau}[\alpha - e^{-1/\alpha}(\alpha + 1)] \qquad \alpha > 0\dagger \tag{1.66}$$

The maximum value of Eq. (1.64) is

$$v_c(\text{max}) = \frac{1}{\tau} \qquad \text{for the impulse input}$$

then

$$\% \, \text{dif} = 100[\alpha - e^{-1/\alpha}(\alpha + 1)] \tag{1.67}$$

where $v_c(\text{max})$ represents the 100 percent value, or we let $\tau = 1$ for comparative purposes. Then if $\tau/\alpha = 0.1$, $\alpha = 10$ and $\%$ dif $= 4.68$. If $\tau/\alpha = 0.01$, $\%$ dif $= 0.495$. Applying

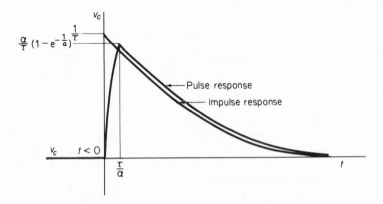

Figure 1.26 Impulse response approximations.

$\dagger$Problem note: Does Eq. (1.66) ever become negative for $\alpha > 0$?

L'Hospital's rule to Eq. (1.62), the maximum of the pulse response approaches $1/\tau$ as α gets large, or the same as the maximum for the impulse response. Finally, since Eq. (1.63) shows the same exponential decay as Eq. (1.64) when t becomes large, we conclude that the pulse response will be substantially the same as the impulse response if the pulse width is made sufficiently short compared to the network time constant (or to all time constants if more than one), and the pulse height correspondingly adjusted. Figure 1.26 compares the two responses with a magnified time scale for small t. Note that the impulse response v_c violates the premise that v_c has time continuity at $t = 0$. This emphasizes the fact that the impulse is an unusual signal, able to inject finite energy into the network in zero time.

Although the input pulse of Fig. 1.25 is not the even function shown by Fig. 1.23, the analysis using the pulse of Fig. 1.23 would proceed in much the same manner as Example 1.1 and give the same conclusion. In fact, from Eqs. (1.52) and (1.53), the shape of the pulse is not important. Similar analysis may be applied to more complicated networks. Of course, in some cases it might be difficult to apply a pulse of limited duration. (For example, take a large airplane.) We may then have to resort to other types of signals for test purposes.

Just as we continued integrating the unit step function, we may continue to differentiate (in the distribution sense). The second differentiation gives the unit doublet, the third the unit triplet, and so on. These signals are not too meaningful in the time domain but become more useful when we use the Laplace transform (Chapter 8).

The signals previously discussed usually suffice for most situations. When we discuss the superposition integral, we will observe that linear network responses to almost any type of signal may be put in terms of a response to an impulse or a step function. Alternatively, linear network design may be based on the response to general sine function with frequency as a variable, as previously mentioned. Nonlinear devices or networks do not offer such advantages, and thus are much more difficult. We will concentrate on linear networks but point out situations in which nonlinearities may seriously alter linear solutions.

1.6 Voltage, Current, and Power

We quickly review elementary definitions for electrical terms, using the meter-kilogram-second-coulomb (MKSQ) system. These four units are considered basic and are defined by international agreement. As we know, a force of 1 newton is defined as a force producing an acceleration of 1 meter per second per second, and 1 joule of work or energy consists of 1 newton of force operating parallel to itself over a distance of 1 meter. A coulomb may be considered as a unit quantity (or charge) of electricity. (We now know that a coulomb constitutes the absence of an experimentally determined number of electrons.) We designate charge by the symbol q.

If 1 joule of work is done in moving 1 coulomb of electricity from point B to point A, the difference of potential between A and B is said to be 1 volt. If it requires work from an external agent to accomplish this task, then A is at a higher potential than B, and this may be indicated by placing a $+$ sign at A and a $-$ sign at B, as shown in Fig. 1.27. Showing both signs gives redundant information, so some texts show the + sign only. The voltage polarity may also be designated by an arrow, with the head at the higher potential and the tail at the lower potential, also as in Fig. 1.27. Some texts use this arrow convention; others show the arrow with the head at the − point. Whatever the convention, it must be understood. We will use the (+ −) convention or the arrow convention shown in Fig. 1.27 (and sometimes both). Note that points A and B are any two points in space and are not necessarily connected by any device. We designate voltage by the symbol v.

Figure 1.27 Positive voltage convention representation.

If the rate of charge flow from one region of space to another is 1 coulomb per second, a current of 1 ampere is said to flow. If we eliminate sources from consideration, the current direction is from a region of high potential to one of lower potential. For example, if we connect a resistance R between the points A and B of Fig. 1.27, then the current is in the direction shown by the arrow designated in Fig. 1.28. (Electrons flow in the opposite direction, but current, a mathematical abstraction, was

Figure 1.28 Positive current convention representation.

defined long before electrons were known.) However, note that a conductor need not be present for a current to exist. For example, the electron beam in an oscilloscope or television tube constitutes a current. Thus,

$$i = \frac{dq}{dt} \quad \text{amperes} \tag{1.68}$$

Instantaneous power is the product of voltage and current, or

$$p = vi \quad \text{watts} \tag{1.69}$$

where v is in volts and i in amperes. The units of Eq. (1.69) are watts, since a watt is 1 joule per second.

Finally the energy E expended in the system from t_0 to t_1 is, from Eq. (1.69),

$$E = \int_{t_0}^{t_1} p(t) \, dt \quad \text{joules} \tag{1.70}$$

or

$$E = \int_{t_0}^{t_1} v(t)i(t)\, dt \qquad \text{joules} \qquad (1.71)$$

1.7 Errors Using a Digital Computer

The appearance of six- or eight-digit answers printed out by a computer lead the neophite, and sometimes the experienced user, to blithely assume that these answers are correct. Of such an assumption the student should rapidly and completely disabuse himself.

First, the computer will, of course, usually print out answers as long as the program compiles and executes. The source program may have outright statement errors, but some kind of answers often result. For example, if the programmer writes

$$A = B/C + D$$

when he means

$$A = B/(C + D)$$

an incorrect result for A occurs, but the result is stored in six or more decimal digits nevertheless. Such errors are easy to make, and can usually be eliminated only by a hand computation of representative solutions. Slide-rule calculations often produce answers that detect gross errors, but a desk calculator may be necessary to find more subtle errors of this type.

Assuming that the program source statements are correct, the nature of the calculation may result in large errors. For example, if we compute the sine of a large angle by using the formula

$$\sin x = x - \frac{x^3}{3!} + \frac{x^5}{5!} - \frac{x^7}{7!} + \cdots$$

the results will be incomprehensible. McCracken computed sin 1470° using eight-digit arithmetic and using enough terms of the series so that the terms differed by less than 10^{-8}. The result was 24.45401855, which obviously has no meaning.[†] This problem can be eliminated by limiting the angle x to less than $\pi/2$ radians (90°). In general, use of series solutions requires great caution, and regions of convergence must be carefully observed.

Another example of computation limitations is the calculation of the roots of a quadratic equation

$$x^2 + ax + b = 0$$

by the usual formula,

$$x = \frac{-b \pm \sqrt{b^2 - 4ac}}{2a}$$

---

†D. D. McCracken and W. S. Dorn, *Numerical Methods and Fortran Programming*, John Wiley & Sons, Inc., New York, 1964, p. 44.

If two real roots to the equation exist, one large and one small, the small root may be vastly in error (see the Problems). A preferable method for finding the small root consists of first finding the large root and then locating the other root by knowing that the root product is b. Thus, if the quadratic is represented as

$$(x + r_1)(x + r_2) = 0$$

then

$$a = r_1 + r_2 \quad \text{and} \quad b = r_1 r_2$$

where $-r_1$ and $-r_2$ are the roots. In general, the roots of a polynomial are very sensitive to variations in the coefficients. Thus, given the polynomial equation

$$a_n x^n + a_{n-1} x^{n-1} + a_{n-2} x^{n-2} + \cdots + a_k + \cdots + a_0 x = 0$$

slight changes in any a_k will generally cause large variations in some roots. Since the original data making up the coefficients are often accurate to three significant figures at best, in many polynomials some roots may not be accurate to one significant figure, even though the computer prints eight digits. One check on this consists of computing the roots from largest to smallest first and then reversing the process. Another involves the multiplication of calculated factors and comparison with the original coefficients.

Even though the programmer anticipates errors of the sort previously mentioned, often sizable errors occur as a result of the nature of the computer itself. Decimal numbers are stored in binary form (or some variant of this form). Thus, 0.1 might be stored as 0.000110011001100. . . , a nonterminating binary fraction of which only a finite number of digits can be stored. Hence, if we add up 100 numbers each 0.1 in value, the result is not likely to be 10.000. . . .

A more subtle error occurs in computer calculation roundoff. Since the computer operates with a finite number of digits, certain computations may result in large error. For example, if two numbers of almost the same size are subtracted, the result will have a large roundoff error. Suppose that two numbers x and y are exactly known to four significant figures, or

$$x = 0.6528 \times 10^4$$
$$y = 0.6531 \times 10^4$$

then

$$y - x = 0.0003 \times 10^4$$

and we know the result to only one significant figure! If $y = (y - x)(0.7529 \times 10^4)$, then assuming that all digits beyond four are chopped off,

$$z = 0.2258 \times 10^5$$

in this case, but in this result only one digit is significant.

Each time the computer makes a computation, some roundoff error occurs, and although the error is usually not as serious as that illustrated, many hundreds of operations produce sizable errors, and in fact some computations result in "garbage" simply as a result of this type of error. Some of this error may be reduced by double-precision computations, at the cost of longer program execution times.

We hope that we have said enough to make the reader wary. The computer user must always adopt a skeptical attitude and remind himself that six-digit printouts may have little significance.

Problems

1.1. Given the linear equations from Eqs. (1.8),

$$a + 6b + 36c = 10 \tag{1}$$

$$a + 8b + 64c = \frac{90.6}{8.24} \tag{2}$$

$$a + 10b + 100c = \frac{138}{10.3} \tag{3}$$

find a, b, and c, and hence an equation for R in Eq. (1.7). One method is Cramer's rule (see Chapter 3). Another method is to subtract equation (1) from equations (2) and (3); then multiply the new equation, (2), by 4/2 and subtract from the new equation, (3). The result provides a value for c which we may then substitute into (2) and (1).

1.2. Choose voltage values in Fig. 1.2 corresponding to current values of 7, 9, and 10 A and repeat Problem 1.1. Using the results, predict the voltage for current values of 8.5, 9.5, and 11 A.

1.3. Assuming R in Fig. 1.2 is to be represented by a cubic equation, $R = a + bi + ci^2 + di^3$, select current values of 6, 8, 9, and 10 A; calculate the coefficients a, b, c, and d. Compare the coefficients with those obtained in Problem 1.1.

1.4. The resistance R in Fig. 1.2 might be represented by an exponential-type equation such as $R = a + bi^c$, where c is not necessarily an integer. Suggest a method for finding a, b, and c. Is this equation reasonable?

1.5. Assume that we have two resistances, R_1 and R_2, in series. R_1 is a linear resistance with a value of 5000 Ω. R_2 is a nonlinear resistance which has the characteristic $i = 2 \times 10^{-3} (1 - e^{-0.1v})$, where $e = 2.718. \ldots$ If a voltage of 5 V is applied across the combination, find the current through the resistance using the load-line analysis of Eq. (1.11). Repeat for voltages of 1, 3, 7, 9, and 11; using these results, plot a curve showing the voltage–current relationship for the series combination.

1.6. Find the series combination volt-amp characteristic for the resistances of Problem 1.5 by adding the characteristics directly, as shown in Fig. 1.8. Compare the results with those obtained in Problem 1.5.

1.7. Find the real root of $F(i) = i^3 + 3i^2 + 6i + 12 = 0$ using Eq. (1.28) to four significant figures. As a first approximate value for the root, try $i = -\frac{12}{6} = -2$. (Lin's method takes the negative of the ratio of the last two coefficients as the approximate real root of a polynomial of odd degree.) By dividing out the real root, the resulting quadratic may be solved to find all roots of the cubic.

1.8.   Write computer programs to calculate a real root of third-degree and fifth-degree polynomials with given constant coefficients using the Newton–Raphson method. Stop the iteration when successive values differ by less than 0.001 percent.

1.9.   The positive square root of a positive number N may be found by solving the equation $x^2 - N = 0$. Using the Newton–Raphson formula Eq. (1.28), if $R_{(n)}$ is the first try, then

$$R_{(n+1)} = R_{(n)} - \frac{R_{(n)}^2 - N}{2R_{(n)}}$$

$$= \tfrac{1}{2}\left(R_{(n)} + \frac{N}{R_{(n)}}\right)$$

Find $\sqrt{10}$ to four places using this method with $R_{(1)} = 3$.

1.10. Using the ideas presented in Problem 1.9, devise a formula to find the cube root of N.

1.11. If resistance $R_1 = 3i^2 + 2i + 3$ is in series with resistance $R_2 = 2i^2 + 3i + 4$, find the current which flows through the combination if a voltage of 5 V is applied. Use the Newton–Raphson method and write a computer program to perform the calculation.

1.12. Write a computer program to obtain the summation indicated by Eq. (1.44). The sum is to be terminated with a finite number of terms N; that is, sum from $k = 0$ to $k = N$, $N < \infty$. Use time t as the basis of iteration. For the purpose of this problem, let the fundamental angular velocity $\omega = 2\pi$. The fundamental period T [the time for periodic repetition of $m(t)$] is then $T = 2\pi/\omega = 1$. Let t go from 0 to T in increments of $\Delta t = 0.02$ so that there will be 50 values of $m(t)$, $0 \leq t \leq 1$, or 50 iterations of the algorithm. Find the series sum for $k = N - 1$ and $k = N$ so that you may evaluate the effect of the last term. Plot $m(t)$ and predict its shape as $k \rightarrow \infty$ for the following series (you may increase N if desired):

(a) $m(t) = \dfrac{4}{\pi}[1 \sin \omega t + \tfrac{1}{3} \sin 3 \omega t$

$+ \tfrac{1}{5} \sin 5 \omega t + \tfrac{1}{7} \sin 7 \omega t$

$+ \tfrac{1}{9} \sin 9 \omega t + \tfrac{1}{11} \sin 11 \omega t + \cdots]$

(b) $m(t) = \dfrac{2}{\pi}[1 \sin \omega t - \tfrac{1}{2} \sin 2 \omega t + \tfrac{1}{3} \sin 3 \omega t$

$- \tfrac{1}{4} \sin 4 \omega t + \tfrac{1}{5} \sin 5 \omega t - \tfrac{1}{6} \sin 6 \omega t$

$+ \tfrac{1}{7} \sin 7 \omega t - \tfrac{1}{8} \sin 8 \omega t + \tfrac{1}{9} \sin 9 \omega t$

$- \tfrac{1}{10} \sin 10 \omega t + \tfrac{1}{11} \sin 11 \omega t + \cdots]$

(c) $m(t) = \dfrac{1}{\pi} + \tfrac{1}{2} \sin \omega t - \dfrac{2}{\pi}\left[\tfrac{1}{3} \sin(2 \omega t + 90°)\right.$

$+ \tfrac{1}{15} \sin(4 \omega t + 90°) + \tfrac{1}{35} \sin(6 \omega t + 90°) + \cdots\Big]$

Note: See *Standard Mathematical Tables*, Chemical Rubber Company, Cleveland.

1.13. Write a computer program to evaluate Eq. (1.66). In that equation adopt τ as a reference value of unity and find the per unit difference by varying α from 1 to 1000 in steps of 20. What value of τ/α would you adopt as reasonable for hand calculations using a pulse to approximate an impulse? What value might be necessary for computer calculations?

1.14. Write a computer program to solve a quadratic equation. First, obtain the roots by simply applying the quadratic formula. Second, obtain the roots by first finding the larger root and then the smaller, as suggested in the text. Compare the results of the two methods for the quadratic equation $x^2 + x + 1 \times 10^{-8} = 0$.

1.15. Write a computer program to compare the results of $\sin^2 x/x$ with $(1 - \cos^2 x)/x$ for increasingly smaller values of x. Compare single-precision results with double-precision results.

Chapter 2

NETWORK ELEMENTS

The end justifies the means.

Practice is the best of all instructors.

PUBLILIUS SYRUS, 42 B.C.

2.1 Models

Engineers construct electric circuits with physical components such as resistors, capacitors, diodes, and transistors. Each of these components makes use of some primary or *first-order* physical property or properties. For example, the simple resistor is ideally a two-terminal device with finite conductivity which exhibits a voltage drop across its terminals proportional to the current through it. However, the essential first-order properties of the component are always accompanied by other inherent properties, which, more often than not, are unwanted and degrade performance. In the case of the resistor, a magnetic field exists when a current is present; consequently, energy becomes stored in this magnetic field, and since a potential difference occurs, energy is also stored in an electrostatic field. Thus, both electrostatic and magnetic coupling effects appear because of the physical construction of the resistor and the proximity of the resistor to other components in the circuit. In many cases, particularly in a preliminary design, the engineer neglects the undesirable *second-order* effects, but some designs require that higher-order effects be considered.

In an approximate sense, we can think of the physical resistor in terms of a lumped-circuit model, the complexity of the model depending on the mode of operation of the model. The resistor model used under dc conditions is quite different from the model used for analysis at gigahertz frequencies. We can construct equivalent circuits or models of the first-order physical functions of many physical components with great accuracy. For example, a mylar capacitor can almost always be modeled as an ideal linear capacitance without including the dielectric leakage resistance which exists between the capacitor plates. Tantalum capacitors can also be modeled as ideal capacitances at low frequencies and low current levels. A junction diode at low frequencies can be treated as a nonlinear resistance, which in turn can be modeled as an ideal diode characteristic in series with a linear resistance. To include second-order effects, we revise the model, making it more complicated. In any case, however, the model is only a representation of the physical device and cannot express all the characteristics of the device.

To efficiently produce a design, the model used must be as simple as possible. On the other hand, the model must include effects which seriously influence circuit performance. An important task of the circuit designer is the selection of the proper circuit model for the intended operating conditions.† In constructing component models we must, therefore, understand the conditions under which its equivalent circuit is valid and under which conditions the model must be modified.

---

†A circuit model is a collection of component models interconnected to form a circuit.

35

As the goal of this text consists of developing methods of network analysis of circuits made up of equivalent circuits or models, the student should realize that engineering is not an exact science, and the results achieved are only valid insofar as the circuit model is valid. In the following sections of this chapter, we describe a set of network elements from which circuit models can be constructed. The models will vary from that of a simple two-terminal device, such as a resistor or capacitor, to that of a complex electronic circuit. The elements provide for construction of nearly exact models of first-order circuit characteristics, and also provide for the inclusion of many second-order characteristics, such as parasitic capacitance and inductance and high-frequency effects in semiconductor devices. As partially demonstrated in Chapter 1, linear elements result in simpler mathematics than do nonlinear elements. Fortunately, most analysis applications allow the use of linear elements, so we will not use nonlinear models unless circumstances direct.

In the later model development in this chapter we will use two basic laws, known as *Kirchhoff's laws*, which we discuss in great detail in Chapter 4. Here we simply state these laws as the Kirchhoff voltage law (KVL) and the Kirchhoff current law (KCL). The voltage law states that the summation of voltages around a closed loop is zero; the current law states that the summation of currents into a junction is zero. We apply these laws to build some of the subsequent models but do not justify the laws or describe them in detail at this point.

2.2 Two-Terminal Devices

Resistance

An ideal resistance is a two-terminal network element whose voltage $v_r(t)$ is defined by the relationship

$$v_r(t) = Ri_r(t) \tag{2.1}$$

where R is the resistance value in ohms and $i_r(t)$ is the current in amperes, as shown in Fig. 2.1. R is a constant in the linear case.

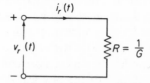

Figure 2.1 Symbolic representation for a resistance (conductance).

Notice that the resistance model defines only a relationship between the voltage and the current and also that the value of resistance can be positive, zero, or negative, which provides some latitude in using the resistance element in circuit models. The normal physical two-terminal resistor is limited to only positive resistance values. If $R = 0$, the terminal voltage across the resistance is zero and independent of the applied current. In this case the terminals of the resistor are *short-circuited*.

Equation (2.1) can be rewritten as

$$i_r(t) = Gv_r(t) \tag{2.2}$$

where $G = 1/R$ is the conductance of the element in mhos ($\mho$). Here, if $G = 0$, the terminal current through the resistance is zero for all values of applied voltage, and the resistor terminals are *open-circuited*. The instantaneous electrical power $p_r(t)$ dissipated in the resistance is

$$p_r(t) = v_r(t)i_r(t) \qquad \text{watts} \qquad (2.3\text{a})$$

$$= i_r(t)^2 R \qquad (2.3\text{b})$$

$$= \frac{v_r(t)^2}{R} \qquad (2.3\text{c})$$

The energy dissipated (turned into heat) in the resistance in a given period from time t_0 to time t is

$$W_r(t) = \int_{t_0}^{t} p_r(\tau)\,d\tau \qquad \text{joules} \qquad (2.4)$$

A resistance with positive R is termed a *passive* device, as it absorbs power from the circuit to which it is connected.

For many applications a linear, time-invariant resistance is an accurate approximation to the physical resistor. There are conditions, however, under which the physical properties of the resistor may change. For example, a large applied voltage may cause the component temperature to change, producing a change in the resistance characteristic. At higher voltages or ambient temperatures, the physical resistor might fail completely. Under conditions such as these, the element shown is not a satisfactory model of the resistor with R constant, and we must resort to other models, such as the equations discussed in Section 1.2.

Independent Voltage Sources

The independent voltage source is the two-terminal network element shown in Fig. 2.2. The independent voltage ideal source maintains the source voltage E_s with the polarity indicated by the plus and minus signs outside the source terminals independent of the current through the source. Thus, the terminal voltage is given by

$$v = E_s \qquad (2.5)$$

Figure 2.2 Ideal independent voltage source.

where v is the terminal voltage and E_s is the source voltage in volts.

The source voltage E_s may be either a constant (battery) or a function of time (signal generator).

For most practical voltage sources a more realistic model of the source contains resistance in series with the terminals, as shown in Fig. 2.3. The terminal voltage for the network in Fig. 2.3 is given by

$$v = E_s - iR_g \tag{2.6}$$

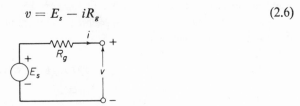

Figure 2.3 Practical voltage
source model.

The value of the series resistance R_g is about 0.05 Ω in an automobile battery and
between about 50 and 2000 Ω in a high-frequency signal generator.

Independent Current Sources

The independent current source is a two-terminal network element as shown in
Fig. 2.4. The independent current source (ideal) maintains the terminal source current

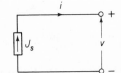

Figure 2.4 Ideal independent
current source.

J_s, with the polarity indicated by the arrow in the rectangular block designating the
source, independent of the terminal voltage. Thus, the terminal current is given by

$$i = J_s \tag{2.7}$$

where i is the terminal current and J_s is the source current in amperes. The source
current may be either a constant value or a function of time (signal generator).

Practical independent current source models contain a resistance across the source
terminals to represent the internal resistance of the source, as shown in Fig. 2.5.
The terminal current for the network in Fig. 2.5 is given by

$$i = J_s - \frac{v}{R_g} \tag{2.8}$$

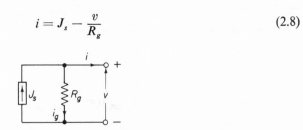

Figure 2.5 Practical current
source model.

Capacitance

The capacitance element shown schematically in Fig. 2.6 is the next passive
element we consider. The voltage polarity and current direction for the element are
chosen so that the current enters the positive terminal.

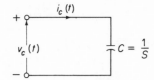

Figure 2.6 Symbolic representation for a capacitance (elastance).

The capacitance is defined in terms of the charge q stored in the capacitance and the voltage v_c across the terminals. The relationship is given by

$$q(t) = Cv_c(t) \qquad (2.9a)$$

$$= \frac{1}{S}v_c(t) \qquad (2.9b)$$

where C is the capacitance in farads (coulombs/volt) and S is the elastance in darafs. The capacitance has a constant value for the linear time-invariant case.

Physically, capacitance consists of parallel conducting plates, separated by an insulator. When the capacitance is connected to a voltage, charge q flows into one plate, and an equal amount of charge flows out of the other plate. Since the current $i = dq/dt$, the current flowing into and out of the capacitance depends on the rate of change of charge. We thus refer to current "through" the capacitance, even though, strictly speaking, current in the form of electron flow does not occur between the plates. The capacitance charge $q(t)$ is positive on the plate into which the arrow representing $i(t)$ is directed. The voltage–current relationship for a linear time-invariant capacitance is given by

$$i_c(t) = \frac{dq(t)}{dt}$$

Thus

$$i_c(t) = \frac{d(Cv_c(t))}{dt}$$

$$= C\frac{dv_c(t)}{dt} \qquad (2.10a)$$

$$= \frac{1}{S}\frac{dv_c(t)}{dt} \qquad (2.10b)$$

In terms of the current through a capacitance, the voltage across the element is given by

$$v_c(t) = \frac{1}{C}\int_{0_+}^{t} i_c(\tau)\,d\tau + v_c(0_+) \qquad (2.11)$$

where $v_c(0_+)$ is the terminal voltage across the capacitance at time $t = 0_+$. Time $t = 0_+$ is defined as a time just slightly greater than $t = 0$.

The initial voltage $v_c(0_+)$ in Eq. (2.11) appears as a dc voltage in series with an initially uncharged capacitance. Since $v_c(0_+)$ is a voltage, we can represent it as an equivalent source, as demonstrated by the equivalent circuit shown in Fig. 2.7.

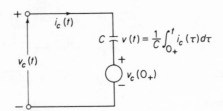

Figure 2.7 Equivalent circuit for initial voltage on capacitances.

We justify briefly the validity of the equivalent circuit in Fig. 2.7 by observing the response of the equivalent circuit shunted by a resistance R. The voltage across a capacitance initially charged to $v_c(0_+)$ and then shunted by a resistance is given by

$$v_c(t) = v_c(0_+)e^{-t/RC}$$

As $t \rightarrow \infty$, the terminal voltage tends to zero. In the equivalent circuit of Fig. 2.7 the voltage of the capacitance reaches a value which is equal and opposite to the dc source. By applying KVL we see that the terminal voltage $v_c(t)$ also tends to zero as proposed. This equivalent circuit may seem strange at this point, but it will be a powerful aid when we begin to write network equations.

The nonlinear capacitance exists in a large number of devices, such as the semiconductor diode, in which the capacitance value is a function of the voltage applied across the component terminals. The q versus v_c curve for the diode and a constant-value capacitance is shown in Fig. 2.8.

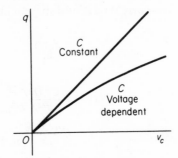

Figure 2.8 Nonlinear voltage-dependent capacitance characteristic.

In the case in which $C = f(v_c)$, the terminal current becomes

$$i_c(t) = \frac{d(Cv_c)}{dt}$$

$$= C\frac{dv_c(t)}{dt} + v_c\frac{dC}{dv_c}\frac{dv_c(t)}{dt} \tag{2.12a}$$

$$= f(v_c)\frac{dv_c(t)}{dt} + v_c\frac{df(v_c)}{dv_c}\frac{dv_c(t)}{dt} \tag{2.12b}$$

which reduces to Eq. (2.10) if the q versus v_c curve is linear, or $dC/dv_c = 0$ and $f(v_c)$

$= C$. The energy stored in a capacitance of constant value C during time t is

$$E_c = \int_{-\infty}^{t} v_c(\tau) i_c(\tau) \, d\tau$$

$$= \int_{-\infty}^{\tau} v_c(\tau) \frac{dq}{d\tau} \, d\tau$$

$$= \int_{-\infty}^{q(t)} v_c(q') \, dq'$$

$$= \frac{1}{C} \int_{-\infty}^{q(t)} q' \, dq'$$

$$= \frac{1}{2} \frac{q^2(t)}{C} \tag{2.13a}$$

$$= \frac{1}{2} C v_c^2(t) \tag{2.13b}$$

where $q(-\infty)$ is assumed to be zero.

Before we leave the capacitance element, we described some second-order characteristics of the linear, time-invariant physical capacitor.[†] When a current flows in a capacitor lead a voltage is induced in the lead proportional to its resistance R_L. In addition, the dielectric material used to separate the capacitor plates has a high resistivity and a large cross-sectional area, which reduces the total dielectric resistance to a value of the order of $10^7 \, \Omega$. This dielectric resistance R_d appears between capacitor plates as shown in Fig. 2.9. In nost applications both the series resistance R_L and the

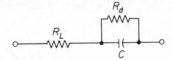

Figure 2.9 A more exact model for a physical capacitor.

dielectric resistance R_d can be neglected. At very high frequencies, the lead inductance effect, which is not included in Fig. 2.9, may also be appreciable.

Inductance

The inductance element, shown schematically in Fig. 2.10, is the last two-terminal passive element we describe. The voltage polarity and current direction for the element are chosen so that the current enters the positive terminal.

Before defining the inductance network element L, we need to define magnetic flux ϕ and flux linkages λ. As to magnetic flux, ϕ webers link a one-turn coil if the voltage v at the coil terminals is given by the relation $v(t) = d\phi(t)/dt$ volts. Figure 2.11

[†]We use the terms "capacitor," "resistor," and "inductor" to refer to physical components and the terms "capacitance," "resistance," and "inductance" to refer to ideal network elements.

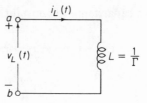

Figure 2.10 Symbolic representation for an inductance element.

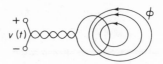

Figure 2.11 Flux and coil in linkage relationship.

indicates that by *linking* we refer to the relation between the links of a chain; the coil represents one link and the flux another. If the coil has more than one turn, then obviously the terminal voltage will increase, and we define the total flux linkages λ as the sum of the individual flux links, or $\lambda = \sum_{i=1}^{N} \phi_i$, where N is the total number of turns and ϕ_i the flux linking the ith turn. If all flux ϕ linked all N turns, then $\lambda = N\phi$, but usually some flux (termed *leakage flux*) does not link with all turns. The voltage is then given by the equation

$$v(t) = \frac{d\lambda(t)}{dt} \tag{2.14}$$

The coil–flux combination is termed a self-inductance if the current producing the flux ϕ flows through the coil itself. If there is a linear relationship between current i and the flux ϕ produced by it, as in Fig. 2.12 (the slope of the characteristic is con-

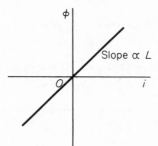

Figure 2.12 Flux–current relationship in linear inductance.

stant), then the self-inductance L in henries is given by the definition

$$\lambda(t) = Li_L(t) \tag{2.15}$$

From Eqs. (2.14) and (2.15), the relationship between the voltage and the terminals of the coil and the current through it, if L is linear and stationary, is

$$v_L(t) = L\frac{di_L(t)}{dt} \qquad \text{volts} \tag{2.16}$$

Figure 2.10, symbolizing inductance L, shows that the polarity of $v_L(t)$ is such that if it were considered a source with terminals ab connected to an external resistance, it would cause a current to flow in L such as to oppose the current $i(t)$ producing the changing flux $\phi(t)$. We know this as *Lenz's law*. If the current is unchanging, then $v_L = 0$.

In terms of the voltage across the inductance, the current through the element is given by

$$i_L(t) = \frac{1}{L} \int_{0_+}^{t} v_L(\tau)\, d\tau + i_L(0_+) \qquad (2.17\text{a})$$

$$= \Gamma \int_{0_+}^{t} v_L(\tau)\, d\tau + i_L(0_+) \qquad (2.17\text{b})$$

where $\Gamma = 1/L$ and $i_L(0_+)$ is the terminal current through the inductance at time $t = 0_+$.

Since $i_L(0_+)$ is a current, we can represent it as an equivalent source, as demonstrated by the equivalent circuit shown in Fig. 2.13. The use of the dc source to repre-

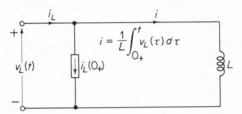

Figure 2.13 Equivalent circuit for initial current in an inductance.

sent initial conditions on the inductance can be justified in the same way that we justified the dc voltage source to represent initial conditions on the capacitance. The verification is left to the reader as an exercise. We recall in this connection that only the terminal characteristics are important in a model or equivalent circuit.

The energy E_L stored in an inductor of constant value L at any time t is

$$\begin{aligned}
E_L &= \int_{-\infty}^{t} v_L(\tau) i_L(\tau)\, d\tau \\
&= \int_{-\infty}^{t} i_L(\tau) \frac{d\lambda}{d\tau}\, d\tau \\
&= \int_{-\infty}^{\lambda(t)} i_L(\lambda')\, d\lambda' \\
&= \int_{-\infty}^{\lambda(t)} \frac{\lambda'}{L}\, d\lambda' \\
&= \frac{1}{2} \frac{\lambda^2(t)}{L} \\
&= \frac{1}{2} L i_L^2(t) \qquad \text{joules}
\end{aligned} \qquad (2.18)$$

where $\lambda(-\infty)$ is assumed to be zero.

Most physical inductors are nonlinear, owing to the fact that ferromagnetic cores are normally used which have nonlinear ϕ versus i characteristics. One of the few exceptions of this statement is the air-core inductor, which possesses the characteristic shown in Fig. 2.12. Since this text is primarily concerned with the analysis of linear networks, we will restrict our discussion to cases where the magnetic flux–current characteristic is essentially linear.

In the preceding discussion covering first-order inductance properties, we ignored the fact that the inductance element is made from a coil of wire which has a finite resistance value. A more precise model of a physical inductance should then include a small series resistance R_L to represent the voltage drop in the wire. Also, the coil is constructed of several turns of wire, one above the other, with a voltage between them, much like a series of small capacitances. Thus, we must also add a small capacitance C_L, as shown in Fig. 2.14, in parallel with the inductance, to provide a better

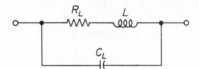

Figure 2.14 A more exact model of a physical inductor.

model of a physical inductor. In many situations, both R_L and C_L may be neglected, at least as a first approximation.

2.3 Four-Terminal Devices

Dependent Sources

Many electronic devices, such as transistors and transformers, have more than two terminals. For example, the transistor, which possesses three distinct terminals, is sometimes represented as a four-terminal (two terminal pairs) device. The transformer has two terminal pairs, or four terminals.

In these devices the voltage (or current) at the output terminal pair is a function of the voltage (or current) at the input terminal pair. Most electronic devices fall into the one- or two-terminal-pair classification. Resistors, capacitors, inductors, and independent sources are one-terminal-pair devices. Other devices fall into the two-pair category.

Transistors and similar devices cannot be modeled using the passive two-terminal elements we have described so far in the discussion of network elements. To model these devices we must introduce a *dependent source* class of network elements. There are four members of this element class, each of which has four terminals (two terminal pairs). These elements are defined by two equations, which relate the input voltage and/or current to the output voltage and/or current. The names for the symbols are chosen to indicate that coupling exists *from* one branch (or element) *to* another; that is, the *cause*, or controlling variable, is in the "from" branch and the effect is in the "to" branch.

Voltage-Controlled Voltage Source
(Transpotential)

The symbol for the voltage-controlled voltage source is shown in Fig. 2.15. The input voltage v_1 and current i_1 are shown at the left-hand terminals of the schematic,

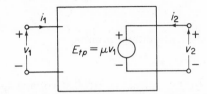

Figure 2.15 Schematic representation of a voltage-controlled voltage source (transpotential).

and the output voltage v_2 and current i_2 are shown at the right-hand terminals. The voltage-controlled voltage source is defined by the set of equations

$$i_1 = 0 \tag{2.19a}$$

$$E_{tp} = \mu v_1 \tag{2.19b}$$

$$v_2 = E_{tp} \tag{2.19c}$$

where the *transpotential*, or voltage-amplification factor, μ is a dimensionless constant. The input voltage v_1 is referred to as the "control" voltage since it determines the value of the "controlled" voltage source E_{tp}. The characteristics of the controlled source are the same as those for the independent voltage source in the sense that the voltage at the terminals of the source is maintained constant independent of all loading effects. However, the controlled source is dependent upon (completely specified by) the conditions at the input terminals of the element. The second equation given by Eq. (2.19) is referred to as the *control equation* for the dependent source since it relates the source voltage to the input, or control voltage.

Voltage-Controlled Current Source
(Transconductance)

The symbol for the voltage-controlled current source is shown in Fig. 2.16. The input voltage v_1 in this element controls the output current i_2 as defined by

$$i_1 = 0 \tag{2.20a}$$

$$J_{tc} = g_m v_1 \tag{2.20b}$$

$$i_2 = J_{tc} \tag{2.20c}$$

where the constant g_m is the *transconductance*, or mutual conductance, specified in units of amperes per volt. The control voltage v_1 determines the current source current J_{tc} independent of the voltage v_2. Hence, the output current i_2 is determined solely by voltage v_1, the control voltage.

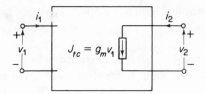

Figure 2.16 Schematic representation of a voltage-controlled current source (transconductance).

Current-Controlled Voltage Source (Transresistance)

The symbol for the current-controlled voltage source is shown in Fig. 2.17. The output voltage v_2 in this element is controlled by the input current i_1 according to the relationships

$$v_1 = 0 \tag{2.21a}$$

$$E_{tr} = r_m i_1 \tag{2.21b}$$

$$v_2 = E_{tr} \tag{2.21c}$$

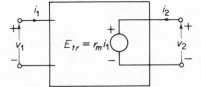

Figure 2.17 Schematic representation of a current-controlled voltage source (transresistance).

where r_m is the *transresistance*, or mutual resistance, and has the dimensions of volts per ampere. Because the value of the voltage is maintained at $r_m i_1$ volts independent of the output current i_2, i_1 is called the *control current*. Equation (2.21b) is the control equation for the dependent voltage source.

Current-Controlled Current Source (Transfluence)

The symbol for the current-controlled current source is shown in Fig. 2.18. The output current i_2 in this element is controlled by the input current i_1 as defined by

$$v_1 = 0 \tag{2.22a}$$

$$J_{tf} = \alpha i_1 \tag{2.22b}$$

$$i_2 = J_{tf} \tag{2.22c}$$

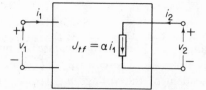

Figure 2.18 Schematic representation of a current-controlled current source (transfluence).

where the constant α is dimensionless and called the current gain, or *transfluence*, of the element. The control current i_1 determines the dependent current source current J_{tf} regardless of the output voltage v_2. The control equation for this element is given by Eq. (2.22b).

The four dependent sources defined in this discussion are both linear and time invariant. They will be used throughout the text in developing the equivalent networks for a wide range of active and passive electronic devices. Some simple examples of the dependent sources are the vacuum triode model (voltage-controlled voltage source), the FET model (voltage-controlled current source), and the bipolar transistor (current-controlled current source). These may be combined with other elements as previously defined to more exactly depict device characteristics, as previously illustrated with the capacitor and the inductor.

Coupled Inductances

In some circuits we may have an inductance whose magnetic flux couples or links with other inductances. The model of such a device consists of two-terminal pairs if two inductances are coupled, and more than two pairs if additional inductances enter into the coupling. The coupling occurs because one coil has close proximity to another, such as in Fig. 2.19. In this figure, if we apply a voltage source $v_1(t)$ to

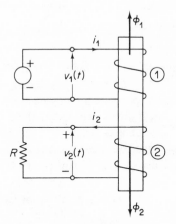

Figure 2.19 Two coupled inductances.

inductance 1 with the polarity shown, current i_1 causes an increasing flux ϕ_1, which has the direction indicated. (If the thumb of the right hand grasping a wire points in the direction of the current, conventionally the flux is in the direction of the fingers.) Some of this flux links inductance 2, and the increasing flux linkages in inductance 2 produce a voltage v_2 across the terminals. By Lenz's law this voltage has a polarity which causes current i_2 to flow in a direction such as to set up a flux ϕ_2 opposing the original flux ϕ_1. The polarity of v_2 is, therefore, positive on the top terminal and negative on the bottom in Fig. 2.19. The voltage v_2 exists even though $i_2 = 0$ (we may visualize $R \longrightarrow \infty$). If inductance 2 were wound in the opposite direction, its terminal polarity would be reversed.

If we assume that the flux linking two coupled inductances increases proportionately, as in Fig. 2.12 (linear conditions), then the respective flux linkages are given by

$$\lambda_1 = L_1 i_1 \pm M_{12} i_2 \tag{2.23a}$$

$$\lambda_2 = \pm M_{21} i_1 + L_2 i_2 \tag{2.23b}$$

where M_{12} and M_{21} (in henries) are constants. From energy considerations, $M_{12} = M_{21}$, and we label this M (mutual inductance), a positive constant for any particular coupling. Since from Farady's law, voltage $v(t) = d\lambda(t)/dt$, we may depict coupled inductances by the two-terminal-pair model of Fig. 2.20, where

$$v_1(t) = L_1 \frac{di_1}{dt} \pm M \frac{di_2}{dt} \tag{2.24a}$$

$$v_2(t) = \pm M \frac{di_1}{dt} + L_2 \frac{di_2}{dt} \tag{2.24b}$$

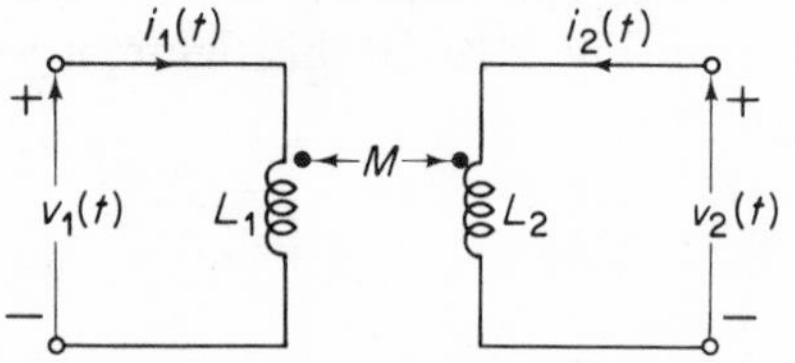

Figure 2.20 Coupled inductance schematic illustrating dot convention.

In Eqs. (2.24a) and (2.24b) we indicate that the signs of the mutually induced voltages $M(di_2/dt)$ and $M(di_1/dt)$ may be positive or negative depending on the relative coil winding directions. The proper sign may be determined by the *dot convention*. The dot convention states that if the current enters the dot on one inductance the induced voltage in the other inductance is positive at the other dot. Thus, with the dots as shown in Fig. 2.20, the positive sign is selected for the mutual voltages $M(di_2/dt)$ and $M(di_1/dt)$. For the case of Fig. 2.21, the correct equations are

$$v_1(t) = L_1 \frac{di_1}{dt} - M \frac{di_2}{dt} \tag{2.24a}$$

$$v_2(t) = -M \frac{di_1}{dt} + L_2 \frac{di_2}{dt} \tag{2.25b}$$

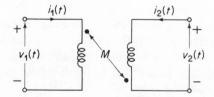

Figure 2.21 Alternate dot location.

In Fig. 2.21 the relative coil windings are such that the current in inductance 2 reduces the flux due to inductance 1, rather than increasing it, as in the previous case for Fig. 2.20.

The dot convention may also be stated as follows: If the currents both enter or leave the dots, the mutual voltage term is positive; otherwise it is negative. Note that in both statements of the dot-convention law, the chosen voltage polarities and current directions must be such that the device is passive; that is, it absorbs power from a connecting circuit. Thus, if the $v(t)$ polarity signs are reversed in Fig. 2.20 or 2.21, i_1 must be reversed to use the associated equations and the dot-convention rules.

The dots are usually indicated on the terminals of coupled inductance devices by the manufacturer. If they are not, the first rule indicates a test procedure. We connect a dc source through a switch to one inductance and a dc voltmeter to the other, as in Fig. 2.22. Place a dot on terminal 1. If, on closing the switch, the voltmeter

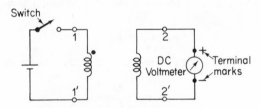

Figure 2.22 DC location test circuit.

deflects upscale, terminal 2 must be dotted; if it deflects downscale, 2′ must have a dot. The test may be hazardous, so care should be taken to avoid contact with the terminals.

If the winding directions are visible we may use Lenz's law to place the dots. Thus, in Fig. 2.19, following the discussion given in connection with that figure, the top terminals of inductances 1 and 2 should have dots. If inductance 2 is wound in the opposite direction, then its lower terminal should have the dot.

We conclude this discussion by calculating the total inductance L_T of two mutually coupled coils, as in Fig. 2.23. First, applying the previous arguments used in relation to Fig. 2.19, the top terminals of each coil should have dots. Thus, we are able to

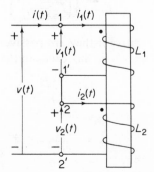

Figure 2.23 Two mutually coupled coils connected.

apply Eqs. (2.24) with the $+$ sign. But $i_2 = i_1$, and $v(t) = v_1(t) + v_2(t)$, so

$$v(t) = L_1 \frac{di_1}{dt} + M \frac{di_1}{dt} + L_2 \frac{di_1}{dt} + M \frac{di_1}{dt}$$

$$= (L_1 + L_2 + 2M) \frac{di_1}{dt}$$

Thus,

$$L_T = L_1 + L_2 + 2M \tag{2.26}$$

Looking at the problem from a flux standpoint, we observe that

$$\lambda_1 = L_1 i_1 + M i_2$$
$$\lambda_2 = M i_1 + L_2 i_2$$

and since $i_1 = i_2$, the total flux is increased. The total inductance L_T is larger than $L_1 + L_2$. If the winding of L_2 were reversed, or if terminal $1'$ were connected to $2'$, then

$$L_T = L_1 + L_2 - 2M \tag{2.27}$$

A coefficient of coupling for two inductances may now be defined as

$$k = \frac{M}{\sqrt{L_1 L_2}} \tag{2.28}$$

By proceeding as in the case of the inductance, we can show that the energy stored in the mutual inductance is

$$E_M = M i_1 i_2 \quad \text{joules} \tag{2.29}$$

Hence the total energy E_T stored in two coupled coils is

$$E_T = \tfrac{1}{2} L_1 i_1^2 + M i_1 i_2 + \tfrac{1}{2} L_2 i_2^2$$

$$= \tfrac{1}{2} L_1 \left(i_1 + \frac{M}{L_1} i_2 \right)^2 + \tfrac{1}{2} \left(L_2 - \frac{M^2}{L_1} \right) i_2^2$$

If we choose $i_1 = -(M/L_1) i_2$, then

$$E_T = \tfrac{1}{2} \left(L_2 - \frac{M^2}{L_1} \right) i_2^2$$

However, L_1 and L_2 are positive if they are passive (do not deliver power), and E_T must always be positive if magnetic flux exists in the core. Hence

$$L_2 - \frac{M^2}{L_1} \geq 0$$

and

$$k = \frac{M}{\sqrt{L_1 L_2}} \leq 1 \tag{2.30}$$

If $k = 1$, physically this means that all the flux produced by i_1 links with L_2, and vice versa, or we have a *tight coupling*. Nonlinking flux is called *leakage flux*.

Two or more coupled inductances are often referred to as a *transformer*. Most transformers with core materials other than air have relatively high coefficients of coupling. Some specially designed transformers have values of k approaching 0.99999. Transformer models also contain resistances, as a result of the physical conductors required for the windings. To achieve a transformer with a high primary (input) inductance value many turns are necessary, which leads us to the observation that the primary resistance R_p is directly proportional to L_1.

Assuming that we could achieve a transformer design without winding resistance and with $k = 1$ (the total flux ϕ links all coils), we would have a *perfect transformer*. The perfect transformer should not be confused with *ideal transformer*, which is a model used by network analysts and synthesists to fit mathematical requirements. In addition to the attributes of a perfect transformer, an ideal transformer has infinite primary and secondary inductances. We summarize by noting that the practical transformer is characterized by its winding resistances, self-inductances, and mutual inductance. In a model of a perfect transformer we neglect the winding resistance and any two inductances must satisfy the relationship $M = \sqrt{L_1 L_2}$. Finally, we characterize the ideal transformer of two windings by the relation $v_1(t)/v_2(t) = N_1/N_2 = n$, the turns ratio. Later we will consider the ideal transformer further.

Some texts consider the mutual M as carrying a sign rather than the mutual voltage terms as requiring a sign determination, as we have done here. This has advantages in some equivalent circuit representations. The sign of M is determined by arguments as previously; in any case, you should be aware of the possibility of a different approach. Mutual couplings may of course exist between more than two inductances, in which case they are considered in pairs, as previously.

We now formulate a current–voltage relationship for the coupled circuit similar to the one derived for the isolated inductance in Eq. (2.17). First, we solve Eq. (2.24) for di_1/dt and di_2/dt and obtain

$$\frac{di_1}{dt} = \frac{L_2}{L_1 L_2 - M^2} v_1 \mp \frac{M}{L_1 L_2 - M^2} v_2 \qquad (2.31a)$$

$$\frac{di_2}{dt} = \mp \frac{M}{L_1 L_2 - M^2} v_1 + \frac{L_1}{L_1 L_2 - M^2} v_2 \qquad (2.31b)$$

Since $k^2 = M^2/L_1 L_2$, we can reduce Eqs. (2.31) to

$$\frac{di_1}{dt} = \frac{1}{L_{e1}} v_1 \mp \frac{M}{L_1 L_{e2}} v_2 \qquad (2.32a)$$

$$\frac{di_2}{dt} = \mp \frac{M}{L_2 L_{e1}} v_1 + \frac{1}{L_{e2}} v_2 \qquad (2.32b)$$

where $L_{e1} = L_1(1 - k^2)$ and $L_{e2} = L_2(1 - k^2)$. Therefore, the current–voltage relations for the circuit of Fig. 2.20 are

$$i_1(t) = \frac{1}{L_{e1}} \int_{0_+}^{t} v_1(\tau)\, d\tau \mp \frac{M}{L_1 L_{e2}} \int_{0_+}^{t} v_2(\tau)\, d\tau + i_1(0_+) \tag{2.33a}$$

$$i_2(t) = \mp \frac{M}{L_2 L_{e1}} \int_{0_+}^{t} v_1(\tau)\, d\tau + \frac{1}{L_{e2}} \int_{0_+}^{t} v_2(\tau)\, d\tau + i_2(0_+) \tag{2.33b}$$

where the sign is selected as − if the mutual voltage term of Eqs. (2.24) is + (as in Fig. 2.20), and vice versa otherwise.

Using Eqs. (2.33), an equivalent circuit for coupled inductances can be constructed using only the inductance elements and transfluences (current-controlled current sources), as shown in Fig. 2.24. The equivalent circuit in this figure satisfies the

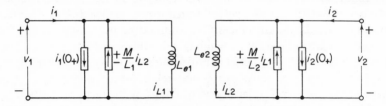

Figure 2.24 Equivalent circuit for coupled inductances defined by Eqs. (2.33a) and (2.33b).

defining equations and provides the isolation between the primary and secondary windings which is usually present in transformers. A simpler, but less powerful, equivalent circuit which satisfies Eqs. (2.23a) and (2.23b) with zero initial conditions and no winding isolation is shown in Fig. 2.25.

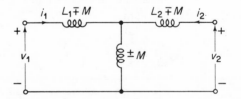

Figure 2.25 Three-terminal equivalent circuit for the coupled inductances defined by Eqs. (2.33a) and (2.33b).

Note that the coil resistances and capacitances, as well as intercoil capacitance, have been neglected in these models.

Ideal Transformer

An ideal transformer (Fig. 2.26) is a device that consists of a group of two or more inductances which are perfectly coupled; that is $k_{ij} = 1$ (unity coefficient of coupling). In addition, all inductances are infinite. The following voltage and current

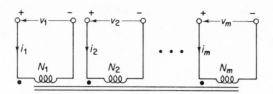

Figure 2.26 Ideal transformer.

relationships define the ideal transformer:

$$\sum_{i=1}^{m} N_i \cdot i_i = 0 \tag{2.34}$$

$$\frac{v_1}{N_1} = \frac{v_2}{N_2} = \ldots \frac{v_m}{N_m} \tag{2.35}$$

where N_i = number of turns of the ith winding

$\qquad i_i$ = current flowing in the ith winding

$\qquad v_i$ = voltage developed across the terminals of the ith winding

For simplicity, the ideal-transformer-model development will be limited to the two-winding transformer. In this case the basic equations, (2.34) and (2.35), reduce to

$$N_i i_1 + N_2 i_2 = 0 \tag{2.36}$$

$$\frac{v_1}{N_1} = \frac{v_2}{N_2} \tag{2.37}$$

These equations provide the information necessary to construct a model of an ideal two-winding transformer.

Consider the transformer in Fig. 2.27. Resistance R_1 in series with winding N_1 is necessary to relate the current i_1 and i_2 to voltages v_1 and v_2. The need for this resistance will become more apparent as the development progresses.

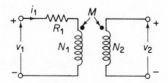

Figure 2.27 Modified ideal transformer.

From Kirchhoff's laws and Eqs. (2.36) and (2.37), the equations for the circuit (Fig. 2.27) are

$$v_1 = i_1 R_1 + v_2 \frac{N_1}{N_2} \tag{2.38}$$

$$i_2 = -\frac{N_1}{N_2} i_1 \tag{2.39}$$

If R_1 is kept small, the equations are a good approximation of the ideal transformer.

By rearranging Eqs. (2.38) and (2.39) and solving for currents i_1 and i_2, the expressions become

$$i_1 = \frac{v_1}{R_1} - \left[\frac{v_2}{R_1(N_2/N_1)^2} \cdot \frac{N_2}{N_1} \right] \tag{2.40a}$$

and

$$i_2 = -\frac{v_1}{R_1}\frac{N_1}{N_2} + \frac{v_2}{R_1(N_2/N_1)^2} \tag{2.40b}$$

The equations are written in this form to isolate the terms involving the primary and secondary voltages and to establish the relationships between the terms of the primary and secondary currents.

The equations are obviously of the forms

$$i_1 = A - \frac{1}{n}B \tag{2.41a}$$

$$i_2 = -nA + B \tag{2.41b}$$

where

$$A = \frac{v_1}{R_1}$$

$$B = \frac{v_2}{R_1}\left(\frac{N_2}{N_1}\right)^2$$

$$n = \frac{N_1}{N_2}$$

Utilizing the relationship established in Eqs. (2.40), the equivalent circuit (Fig. 2.28) can be constructed. The circuit shown possesses the same black box characteristics as the transformer in Fig. 2.27. Note the dot placement assumed in Fig. 2.27. For the

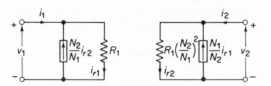

Figure 2.28 Ideal transformer equivalent circuit.

opposite placement, reverse the polarity of both dependent current sources in Fig. 2.28.

Additional Reading

DESOER, C. A., and E. S. KUH, *Basic Circuit Theory*, McGraw-Hill Book Company, New York, 1969.

JENSEN, R. W., and M. D. LIEBERMAN, *IBM Electronic Circuit Analysis Program: Techniques and Applications*, Prentice-Hall, Inc., Englewood Cliffs, N.J., 1968.

KUO, B. C., *Linear Networks and Systems*, McGraw-Hill Book Company, New York, 1967.

LEON, B. J., and P. A. WINTZ, *Basic Linear Networks For Electrical and Electronics Engineers*, Holt, Rinehart and Winston, Inc., New York, 1970.

SANFORD, R. S., *Physical Networks*, Prentice-Hall, Inc., Englewood Cliffs, N.J., 1965.

SESHU, S., and N. BALABANIAN, *Linear Network Analysis*, John Wiley & Sons, Inc., New York, 1959.

Problems

2.1. The electrical power dissipated in a resistor is 10 W when the terminal voltage is 10 V. Find the resistance in ohms and the current in amperes.

2.2. A water heater with a constant resistance of 50 Ω and powered by a dc source is immersed in 0.6 kilogram (kg) of water at a temperature of 30°C. After 30 minutes the water temperature is 40°C. If the water container is well insulated, determine the current through the resistance. 1 kilogram-calorie (kg-cal) of heat increases the temperature of 1 kg of water 1°C. 1 kg-cal is also equivalent to 4200 joules.

2.3. If a current through a constant resistance of 5 Ω has the sawtoothed shape of Fig. P2.1, find the energy changed to heat in the resistance over a period of 0.1 second (s), and from this find the energy over a time of 10 s.

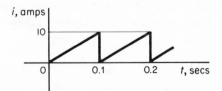

Figure P2.1

2.4. If a certain R varies with time such that $R = 5 \sin t$ while the current through it is $i = 10 \sin 5t$, find the energy turned into heat in 10 s.

2.5. Using the fact that current i through two series linear resistances R_1 and R_2 is common, while the resistor voltage v_1 and v_2 add such that $v_1 + v_2 = v_t$, the source voltage, show that the equivalent resistance R_t as viewed from the source is $R_t = v_t/i = R_1 + R_2$. Find R_t in terms of G_1 and G_2, where $G = 1/R$.

2.6. Using the fact that the voltage v across two parallel linear resistances R_1 and R_2 is common, while the currents i_1 and i_2 add such that $i_1 + i_2 = i_t$, the source current, show that $R_t = v/i_t = R_1R_2/(R_1 + R_2)$. Find G_T in terms of G_1 and G_2 and R_1 and R_2.

2.7. Proceeding similarly to Problems 2.5 and 2.6, show that for two linear inductances the total inductance $L_t = L_1$ and L_2 if they are in series and $L_t = L_1L_2/(L_1 + L_2)$ if they are in parallel. Also show that for two linear capacitances in parallel, the total capacitance is $C_t = C_1 + C_2$, while for two capacitances in series $C_t = C_1C_2/(C_1 + C_2)$. Develop procedures to find the equivalents for the case of three or more inductances and three or more capacitances. *Hint:* With capacitance C, use the concept of charge q.

2.8. Using the results of Problem 2.6, find the resistance $R_t = v/i$ for the circuit of Fig. P2.2. The linear resistance values shown are in ohms.

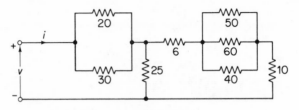

Figure P2.2

2.9. A nonlinear resistance R has a resistance which varies as the two-thirds power of the current i through it. The value of R when $i = 40$ milliamps (mA) is 20 kilohms (kΩ). What is R when $i = 95$ mA?

2.10. A voltmeter shows that a dry cell has a terminal voltage of 45 V with only the voltmeter attached, and a terminal voltage of 43 V when a 5000-Ω load is placed across the cell using the same voltmeter. The voltmeter resistance is 10,000 Ω. Find the internal resistance of the cell; take into account the voltmeter resistance.

2.11. The output of a power amplifier may be modeled by the box of Fig. P2.3 with terminals 1-2. If we place a resistance $R_1 = 600 \Omega$ across the terminals, the voltmeter reads 100 V. If we remove R_1, V reads 150 V directly across terminals 1-2. If the voltmeter resistance is very high (about $10^6 \Omega$), what is the amplifier resistance R? What is V?

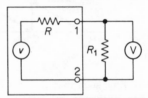

Figure P2.3

2.12. Find the energy in joules stored in a 500-μF capacitor with a voltage of 500 V. How long must a dc current of 10 mA flow through a 2000-Ω resistor to produce the same energy?

2.13. Find the energy in joules stored in an inductance of 50 millihenries (mH) with a current of 2 mA flowing through it. What is the energy if the current is 20 mA?

2.14. Find the equivalent effective inductance seen at the terminals 1-2 of the circuit of Fig. P2.4. All values shown are in henries.

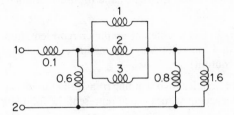

Figure P2.4

2.15. Find the equivalent effective capacitance seen at the terminals 1-2 of the circuit of Fig. P2.5. All values shown are in microfarads (μF).

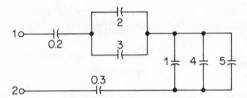

Figure P2.5

2.16. The current through an inductance of 0.2 H is given by the term $0.030 \sin 500t$. Find the voltage across the inductance, and plot current and voltage as a function of time.

2.17. The current through a capacitance of 50 μF is given by the term $0.030 \sin 500t$. Find the voltage across the capacitance, and plot current and voltage as a function of time. Compare the capacitance voltage with the inductance voltage of Problem 2.16. How large would the capacitance have to be to make these voltages exactly equal in magnitude? If this capacitance and the inductance were now connected in series, what would be the total voltage across the series combination as a function of time?

2.18. In Fig. P2.6 what is the voltage v_t? All resistances are in ohms, all voltages in volts, and all currents in amps.

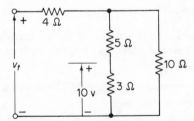

Figure P2.6

2.19. A current $i = 2t$ flows through a linear inductor of 0.2 H. Find the voltage across it. Repeat if $i = 3e^{-2t}$.

2.20. A current of $i = 2t$ flows through a capacitance of 20 μF. Find an expression for the charge $q(t)$ if $q(0) = 0$. Find an expression for the voltage. Repeat if $i = 3e^{-2t}$. Compare with Problem 2.19.

2.21. Assume Fig. 2.15 represents an amplifier and we have a load resistance of 1000 Ω connected across the right-hand (output) terminals with voltage v_2. If $v_1 = 1$ V and $\mu = 20$, find i_2.

2.22. Figure 2.16 is a different amplifier representation. If $v_1 = 1$ V and $g_m = 1 \times 10^{-4}$ A/V, find the voltage v_2 if a resistance of 100,000 Ω is connected across the right-hand (output) terminals.

2.23. For the practical two-winding transformer model in Fig. 2.24, calculate the equivalent-circuit-element values which represent the following transformer specifications:

$$L_1 = 1 \text{ mH}$$
$$L_2 = 4 \text{ mH}$$
$$k = 0.5$$

2.24. For the ideal two-winding transformer model in Fig. 2.28, calculate equivalent-circuit-element values which represent the following transformer specifications:

$$n = 2 \qquad \text{where } n = \frac{N_1}{N_2}$$
$$R = 20 \, \Omega$$

2.25. Derive a general equivalent circuit for an ideal transformer with three windings. Consider two of the three windings to be driving windings with winding resistances R_1 and R_2, respectively. The third winding is a driven winding.

2.26. Derive a general equivalent circuit for an ideal transformer with four windings. *Hint:* Consider three of the windings to be driving windings with associated winding resistances.

2.27. Derive a general equivalent circuit for a practical three-winding transformer with coupling coefficients k_{ij} less than 0.99995. Include winding resistances and isolation between the three windings.

2.28. Use the model derived in Problem 2.27 to obtain a model for a transformer with the following parameters:

$$R_1 = R_2 = R_3 = 0$$
$$L_1 = L_2 = L_3 = 1 \text{ mH}$$
$$k_{12} = k_{13} = k_{23} = 0.75$$

2.29. Derive a general equivalent circuit for a practical four-winding transformer with coupling coefficients k_{ij} less than 0.99995. Include winding resistances and isolation between the four windings.

2.30. Use the model derived in Problem 2.29 to obtain a model for a transformer with the following parameters:

$$R_1 = R_2 = R_3 = R_4 = 0$$
$$L_1 = L_2 = L_3 = L_4 = 1 \text{ mH}$$
$$k_{12} = k_{13} = k_{14} = k_{23} = k_{24} = k_{34} = 0.75$$

2.31. Show that a perfect transformer becomes an ideal transformer under the condition that the winding inductances approach infinity.

2.32. Write a computer program which will compute parameters for practical transformer models of the type shown in Fig. 2.24 and Problems 2.27 and 2.29 from the winding inductances L_i and coupling coefficients k_{ij}. The program should be able to compute parameters for two-, three-, and four-winding models. The input data should include the number of windings, the winding inductances, and the interwinding coupling coefficients.

2.33. If instantaneous power $p(t)$ is defined as the product of the voltage and current, or $p(t) = e(t)i(t)$, write a program to find $p(t)$ and the energy $E_L(t)$ over a complete period if the current through an inductor L of 2 H is $i_L(t) = 5 \sin 4\pi t$. Take about 25 iterations to cover the period ($\Delta t = 0.02$).

Chapter 3

MATRIX ALGEBRA

I have hardly ever known a mathematician who was capable of reasoning.

PLATO, *The Republic, Book VII*

3.1 Introduction

In network analysis we frequently deal with large systems of linear algebraic or differential equations. For linear electrical networks these equations generally result from the application of Kirchhoff's laws to the network (see Chapter 4). To save both time and space, it is often desirable to use appropriate shorthand methods to write the equations in a compact, convenient form. These shorthand methods will not simplify the actual numerical calculations, but the methods will make it easier for us to manipulate complex systems of equations symbolically and comprehend the manipulations.

As a simple illustration, consider a system of equations of the form

$$
\begin{aligned}
a_{11}x_1 + a_{12}x_2 + \cdots + a_{1n}x_n &= b_1 \\
a_{21}x_1 + a_{22}x_2 + \cdots + a_{2n}x_n &= b_2 \\
\cdots \cdots \cdots \cdots \cdots \cdots \cdots \cdots \\
a_{n1}x_1 + a_{n2}x_2 + \cdots + a_{nn}x_n &= b_n
\end{aligned}
\tag{3.1}
$$

In matrix form, this set of equations is written as

$$
\begin{bmatrix}
a_{11} & a_{12} & \cdots & a_{1n} \\
a_{21} & a_{22} & \cdots & a_{2n} \\
\cdots & \cdots & \cdots & \cdots \\
a_{n1} & a_{n2} & \cdots & a_{nn}
\end{bmatrix}
\begin{bmatrix}
x_1 \\ x_2 \\ \cdots \\ x_n
\end{bmatrix}
=
\begin{bmatrix}
b_1 \\ b_2 \\ \cdots \\ b_n
\end{bmatrix}
\tag{3.2}
$$

or, more simply, as

$$
\mathbf{Ax} = \mathbf{b} \tag{3.3}
$$

where

$$
\mathbf{A} =
\begin{bmatrix}
a_{11} & a_{12} & \cdots & a_{1n} \\
a_{21} & a_{22} & \cdots & a_{2n} \\
\cdots & \cdots & \cdots & \cdots \\
a_{n1} & a_{n2} & \cdots & a_{nn}
\end{bmatrix}
\qquad
\mathbf{x} =
\begin{bmatrix}
x_1 \\ x_2 \\ \cdots \\ x_n
\end{bmatrix}
\qquad
\mathbf{b} =
\begin{bmatrix}
b_1 \\ b_2 \\ \cdots \\ b_n
\end{bmatrix}
\tag{3.4}
$$

The symbols $\mathbf{A}$, $\mathbf{x}$, and $\mathbf{b}$ are called *matrices*; they possess specific relations with the coefficients and variables of the original equations. In Eq. (3.3) the matrix $\mathbf{A}$ may be said to transform the vector $\mathbf{x}$ into the vector $\mathbf{b}$. The transformation is referred to as a *linear transformation*. Note that the matrix $\mathbf{A}$ is the array of coefficients a_{ij} in the same order relative to their positions in rows and columns as they appear in the defining equations, Eq. (3.1). That is, the element a_{ij}, called the (i, j)th element of the matrix **A**, appears in the ith row and the jth column. Since we do not plan to write a book on matrix algebra, we do not attempt to provide a comprehensive coverage of matrix theory, only a brief outline of the essential operations. Some excellent texts on the subject are listed under Additional Reading at the end of the chapter.

3.2 Definitions

Matrix. An $n \times m$ matrix is a collection of numbers or elements arranged in a rectangular array, containing n rows and m columns. An $n \times m$ matrix is said to be of order $n \times m$.

A matrix is represented in several ways. The matrix can be enclosed in brackets as in Eq. (3.4) or it can be written simply as

$$\mathbf{A} = [a_{ij}]_{n,m} \tag{3.5}$$

which implies the previous definitions of matrix elements and order.

Square Matrix. A square matrix of order n is one that has n rows and n columns. Examples of a square matrix are

$$\mathbf{A} = [a_{ij}]_{3,3} = \begin{bmatrix} 2 & -4 & 0 \\ 0 & -1 & 2 \\ 1 & 0 & 3 \end{bmatrix}$$

$$\mathbf{B} = [b_{ij}]_{2,2} = \begin{bmatrix} 2 & -0.5 \\ 3 & 7 \end{bmatrix}$$

In a square matrix, the elements $a_{11}, a_{22}, \ldots, a_{nn}$ are called *diagonal* elements.

Column Matrix. A column matrix (or vector) is a matrix containing only one column. The order of the matrix is $n \times 1$, where $n \neq 1$, thus:

$$\mathbf{a} = \begin{bmatrix} a_{11} \\ \cdots \\ a_{n1} \end{bmatrix} \qquad \mathbf{x} = \begin{bmatrix} x_1 \\ \cdots \\ x_n \end{bmatrix} \tag{3.6}$$

The $(n \times 1)$-order column matrix corresponds to an n-dimensional vector. The elements $x_1, x_2, \ldots, x_n$ of **x** in Eq. (3.6) may be considered as components of a vector (or directed line segment) in an n-dimensional space. Figure 3.1 shows a vector **x** of two components, x_1 and x_2, in a two-dimensional space. Obviously, it would be difficult to depict a vector in more than three dimensions, but the concept becomes clarified by this figure. The vector components are assumed to be linearly independent; that is, no two of the n-component vectors are in the same direction.

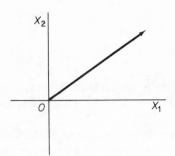

Figure 3.1

Row Matrix. A row matrix (or vector) is a matrix containing only one row. The order of the matrix is $1 \times m$, where $m \neq 1$, thus:

$$\mathbf{a} = [a_{11} \dots a_{1m}]$$

Occasionally a column matrix may be written column $[a_{11} \dots a_{1m}]$ to save space, but we will avoid this form.

Real Matrix. A real matrix is a matrix in which all the elements are real.

Complex Matrix. A complex matrix is a matrix in which some or all of the elements are complex.

Diagonal Matrix. A diagonal matrix is an $n \times n$ matrix in which $a_{ij} = 0$ for all $i \neq j$. Examples of diagonal matrices are

$$\mathbf{A} = \begin{bmatrix} a_{11} & 0 & 0 \\ 0 & a_{22} & 0 \\ 0 & 0 & a_{33} \end{bmatrix} \quad \mathbf{B} = \begin{bmatrix} 3 & 0 \\ 0 & 3 \end{bmatrix} \quad \mathbf{I} = \begin{bmatrix} 1 & 0 \\ 0 & 1 \end{bmatrix}$$

There are two important special cases of the diagonal matrix. The first of these is the unit (or identity) matrix $\mathbf{I}$, in which $a_{ij} = 1$ for all $i = j$ as in the third matrix above. The second special case is the scalar matrix, in which $a_{ij} = k$ for all $i = j$, as in matrix $\mathbf{B}$ above.

Null Matrix. A null matrix is a matrix in which $a_{ij} = 0$ for all i and j.

Transposed Matrix. The transpose of a matrix $\mathbf{A}$ is a new matrix $\mathbf{A}^{\mathsf{T}}$ obtained by interchanging the corresponding rows and columns of $\mathbf{A}$. If we let $\mathbf{A} = [a_{ij}]_{n,m}$, the transpose of $\mathbf{A}$, designated by $\mathbf{A}^{\mathsf{T}}$, is given by

$$\mathbf{A}^{\mathsf{T}} = [a_{ji}]_{m,n} \tag{3.7}$$

As an example of a transposed matrix, assume that

$$\mathbf{A} = \begin{bmatrix} 3 & 4 \\ 1 & 7 \\ 2 & 1 \end{bmatrix}$$

Then

$$\mathbf{A}^\mathsf{T} = \begin{bmatrix} 3 & 1 & 2 \\ 4 & 7 & 1 \end{bmatrix}$$

Symmetrical Matrix. A symmetrical matrix is a square matrix which remains unaltered in transposition, so $\mathbf{A} = \mathbf{A}^\mathsf{T}$. This characteristic implies that $a_{ij} = a_{ji}$ for all i and j.

We will observe later that all passive networks without coupling between elements or dependent sources have symmetrical admittance and impedance matrices.

Skew-Symmetrical Matrix. A skew-symmetric matrix $\mathbf{A}$ is defined by the condition $a_{ij} = -a_{ji}$, or $\mathbf{A}^\mathsf{T} = -\mathbf{A}$. For example,

$$\begin{bmatrix} 1 & 2 & 3 & 4 \\ 2 & 4 & 5 & 9 \\ 3 & 5 & 6 & 1 \\ 4 & 9 & 1 & 7 \end{bmatrix}$$

is symmetrical, but

$$\begin{bmatrix} 0 & 2 & 3 & 4 \\ -2 & 0 & 5 & 9 \\ -3 & -5 & 0 & 1 \\ -4 & -9 & -1 & 0 \end{bmatrix}$$

is skew symmetric. (Note that the diagonal elements are all zero.)

Determinant of a Matrix. The determinant of a square matrix $\mathbf{A}$ is a number associated with $\mathbf{A}$ and is designated by

$$\Delta_A = \det \mathbf{A} = |\mathbf{A}| = \begin{vmatrix} a_{11} & a_{12} & \cdots & a_{1n} \\ a_{21} & a_{22} & \cdots & a_{2n} \\ \cdot & \cdot & \cdots & \cdot \\ a_{n1} & a_{n2} & \cdots & a_{nn} \end{vmatrix}$$

There are several methods available for evaluating the determinant of a matrix. We limit our present discussion to the cofactor-expansion method.

One form of the cofactor-expansion method expands the determinant about a row of the matrix (row expansion). This is given by

$$\Delta_A = \sum_{k=1}^{n} a_{ik} C_{ik} \tag{3.8}$$

where i is any integer from 1 to n and C_{ik} represents the cofactor of the element a_{ik}. The cofactor C_{ik} is the determinant obtained by omitting the ith row and kth column of the $\mathbf{A}$ matrix and multiplying by $(-1)^{i+k}$. The cofactor without the sign term is called a *minor*. A determinant may also be expanded about any column. Thus,

$$\Delta_A = \sum_{k=1}^{n} a_{kj}C_{kj} \tag{3.9}$$

where j is any integer from 1 to n and C_{kj} is the cofactor of the element a_{kj}.

EXAMPLE 3.1

Find $\Delta_A = \det \mathbf{A}$, where

$$\mathbf{A} = \begin{bmatrix} 1 & 0 & 2 \\ 3 & 4 & 5 \\ 1 & 1 & 1 \end{bmatrix}$$

Solution. Expanding about the first row [$i = 1$ in Eq. (3.8)],

$$\Delta_A = (-1)^2(1)\begin{vmatrix} 4 & 5 \\ 1 & 1 \end{vmatrix} + (-1)^3(0)\begin{vmatrix} 3 & 5 \\ 1 & 1 \end{vmatrix} + (-1)^4(2)\begin{vmatrix} 3 & 4 \\ 1 & 1 \end{vmatrix}$$

$$= \begin{vmatrix} 4 & 5 \\ 1 & 1 \end{vmatrix} + 2\begin{vmatrix} 3 & 4 \\ 1 & 1 \end{vmatrix}$$

Since

$$\begin{vmatrix} a_{11} & a_{12} \\ a_{21} & a_{22} \end{vmatrix} = a_{11}a_{22} - a_{12}a_{21}$$

$$\Delta_A = (4 - 5) + 2(3 - 4)$$

or

$$\Delta_A = -3$$

To verify the calculated value, we expand the determinant about the second column ($j = 2$) using Eq. (3.9). Then

$$\Delta_A = (-1)^3(0)\begin{vmatrix} 3 & 5 \\ 1 & 1 \end{vmatrix} + (-4)^4(4)\begin{vmatrix} 1 & 2 \\ 1 & 1 \end{vmatrix} + (-1)^5(1)\begin{vmatrix} 1 & 2 \\ 3 & 5 \end{vmatrix}$$

$$= 0 + 4(1 - 2) - 1(5 - 6)$$

or

$$\Delta_A = -3$$

Singular Matrix. A singular matrix is a square matrix whose determinant is zero. This condition indicates that one or more of the rows (or of the columns) of the matrix, when considered as vectors, are dependent on another row (or column). For example, look at the following set of equations:

$$x_1 + x_2 + x_3 = 0$$
$$2x_1 - x_2 + x_3 = 0$$
$$3x_1 \qquad + 2x_3 = 0$$

These equations can be written in matrix form as

$$\mathbf{Ax} = 0$$

where

$$\mathbf{A} = \begin{bmatrix} 1 & 1 & 1 \\ 2 & -1 & 1 \\ 3 & 0 & 2 \end{bmatrix} \qquad \mathbf{x} = \begin{bmatrix} x_1 \\ x_2 \\ x_3 \end{bmatrix} \qquad \mathbf{0} = \begin{bmatrix} 0 \\ 0 \\ 0 \end{bmatrix}$$

Since $\mathbf{A}$ is a square matrix, a determinant Δ_A exists which is equal to

$$\begin{aligned} \Delta_A &= (-1)^3 \begin{vmatrix} 2 & 1 \\ 3 & 2 \end{vmatrix} + (-1)^4(-1) \begin{vmatrix} 1 & 1 \\ 3 & 2 \end{vmatrix} \\ &= (-1)(4 - 3) - 1(2 - 3) \\ &= 0 \end{aligned}$$

We can observe by careful inspection that the three original equations are not independent since the third equation is equal to the sum of the first two equations.

Adjoint Matrix. The adjoint matrix of the square matrix $\mathbf{A}$, denoted by adj $\mathbf{A}$, is defined as

$$\text{adj } \mathbf{A} = \text{adj}[a_{ij}]_{n,n} = [C_{ij}]_{n,n}^{\mathsf{T}} \tag{3.10}$$

where C_{ij} is the cofactor of the element a_{ij} of the matrix $\mathbf{A}$ as defined in Eq. (3.8).

EXAMPLE 3.2

The adjoint matrix of $\mathbf{A} = [a_{ij}]_{3,3}$ is given by

$$\begin{aligned} \text{adj } \mathbf{A} &= \begin{bmatrix} C_{11} & C_{12} & C_{13} \\ C_{21} & C_{22} & C_{23} \\ C_{31} & C_{32} & C_{33} \end{bmatrix}^{\mathsf{T}} = \begin{bmatrix} C_{11} & C_{21} & C_{31} \\ C_{12} & C_{22} & C_{32} \\ C_{13} & C_{23} & C_{33} \end{bmatrix} \\[2mm] &= \begin{bmatrix} a_{22}a_{33} - a_{23}a_{32} & -(a_{12}a_{33} - a_{13}a_{32}) & a_{12}a_{23} - a_{13}a_{22} \\ -(a_{21}a_{33} - a_{23}a_{31}) & a_{11}a_{33} - a_{13}a_{31} & -(a_{11}a_{23} - a_{13}a_{21}) \\ a_{21}a_{32} - a_{22}a_{31} & -(a_{11}a_{32} - a_{12}a_{31}) & a_{11}a_{22} - a_{12}a_{21} \end{bmatrix} \end{aligned}$$

3.3 Matrix Algebra

Before we can manipulate and solve equations in matrix form we must define certain basic matrix operations, such as equality, addition, subtraction, multiplication, division, and exponentiation. Some of these operations have rules which are identical to the rules of algebra for ordinary real numbers. However, some of the rules have entirely different meanings.

Equality

Two matrices **A** and **B** are equal if and only if

1. They are of the same order.
2. $a_{ij} = b_{ij}$ for all i and j.

Addition

The sum $\mathbf{A} + \mathbf{B}$ of two matrices $\mathbf{A} = [a_{ij}]_{n,m}$ and $\mathbf{B} = [b_{ij}]_{n,m}$ is defined only when the matrices are of the same order. The sum is then given by

$$[a_{ij}]_{n,m} + [b_{ij}]_{n,m} = [a_{ij} + b_{ij}]_{n,m} = [c_{ij}]_{n,m} \qquad (3.11)$$

The order of the resulting matric $\mathbf{C} = [C_{ij}]_{n,m}$ is unchanged by the operation.

The process of matrix addition is described in Fortran IV by the block of statements

```
C    MATRIX ADDITION ALGORITHM
     DO 10 I = 1,N
     DO 10 J = 1,M
10 C(I,J) = A(I,J) + B(I,J)
```

where the variables A, B, C, I, J, N, and M are defined in Eq. (3.11).

EXAMPLE 3.3

Find $\mathbf{C} = \mathbf{A} + \mathbf{B}$ for the matrices

$$\mathbf{A} = \begin{bmatrix} 1 & 3 & 6 \\ 5 & 4 & 2 \end{bmatrix} \qquad \mathbf{B} = \begin{bmatrix} 2 & 6 & 3 \\ 4 & 7 & 1 \end{bmatrix}$$

$$\mathbf{A} + \mathbf{B} = \begin{bmatrix} 2+1 & 3+6 & 6+3 \\ 5+4 & 4+7 & 2+1 \end{bmatrix} = \begin{bmatrix} 3 & 9 & 9 \\ 9 & 11 & 3 \end{bmatrix}$$

Subtraction

The difference $\mathbf{A} - \mathbf{B}$ between two matrices $\mathbf{A} = [a_{ij}]_{n,m}$ and $\mathbf{B} = [b_{ij}]_{n,m}$ is defined only when the matrices are of the same order. The difference is then given by

$$[a_{ij}]_{n,m} - [b_{ij}]_{n,m} = [a_{ij} - b_{ij}]_{n,m} = [c_{ij}]_{n,m} \qquad (3.12)$$

The order of the resulting matrix, $\mathbf{C} = [c_{ij}]_{n,m}$, is unchanged by the operation.

Multiplication

The product **AB** of two matrices $\mathbf{A} = [a_{ij}]_{n,m}$ and $\mathbf{B} = [b_{ij}]_{r,k}$ is defined only when the matrices are conformable; that is, the number of columns of **A** must equal the number of rows of **B**. Using the definitions of matrices **A** and **B**, the matrices are conformable, to form the product

$$\mathbf{AB} = [a_{ij}]_{n,m} \times [b_{ij}]_{r,k}$$
$$= [c_{ij}]_{n,k} = \mathbf{C} \tag{3.13}$$

if and only if $m = r$. The resulting product matrix **C** will have the same number of rows as **A** and the same number of columns as **B**. Notice that if **A** and **B** are conformable to form the product **AB**, they will not be conformable to form **BA** unless in Eq. (3.13) $n = k$. The commutative law for matrix mutiplication is not valid in general. It does hold for a special class of square matrices in which either **B** or **A** is an identity matrix or a scaler matrix or **B** is the inverse of **A**. To differentiate between the products **AB** and **BA**, we will refer to the product **AB** as either **B** *premultiplied* by **A** or **A** *postmultiplied* by **B**.

The formal definition of matrix multiplication is given by Eq. (3.13), where

$$c_{ij} = \sum_{s=1}^{m} a_{is}b_{sj} \tag{3.14}$$

EXAMPLE 3.4

Given $\mathbf{A} = [a_{ij}]_{2,3}$ and $\mathbf{B} = [b_{ij}]_{3,2}$, then

$$\mathbf{AB} = \begin{bmatrix} a_{11} & a_{12} & a_{13} \\ a_{21} & a_{22} & a_{23} \end{bmatrix} \begin{bmatrix} b_{11} & b_{12} \\ b_{21} & b_{22} \\ b_{31} & b_{32} \end{bmatrix}$$

$$= \begin{bmatrix} a_{11}b_{11} + a_{12}b_{21} + a_{13}b_{31} & a_{11}b_{12} + a_{12}b_{22} + a_{13}b_{32} \\ a_{21}b_{11} + a_{22}b_{21} + a_{23}b_{31} & a_{21}b_{12} + a_{22}b_{22} + a_{23}b_{32} \end{bmatrix}$$

$$= \begin{bmatrix} c_{11} & c_{12} \\ c_{21} & c_{22} \end{bmatrix}$$

In Example 3.4 the products **AB** and **BA** are both conformable and **C** is square. However, $\mathbf{AB} = [c_{ij}]_{2,2}$, $\mathbf{BA} = [c_{ij}]_{3,3}$, and $\mathbf{AB} \neq \mathbf{BA}$.

EXAMPLE 3.5

Given

$$\mathbf{A} = \begin{bmatrix} 1 & 3 \\ 5 & 4 \end{bmatrix} \quad \text{and} \quad \mathbf{B} = \begin{bmatrix} 2 & 6 \\ 4 & 7 \end{bmatrix}$$

then

$$\mathbf{AB} = \begin{bmatrix} 1(2) + 3(4) & 1(6) + 3(7) \\ 5(2) + 4(4) & 5(6) + 4(7) \end{bmatrix}$$

$$= \begin{bmatrix} 14 & 27 \\ 26 & 58 \end{bmatrix}$$

Similarly,

$$\mathbf{BA} = \begin{bmatrix} 32 & 30 \\ 39 & 40 \end{bmatrix}$$

Therefore $\mathbf{AB} \neq \mathbf{BA}$, although both products are conformable.

EXAMPLE 3.6

Given

$$\mathbf{A} = \begin{bmatrix} 1 & 3 \\ 5 & 4 \end{bmatrix} \qquad \mathbf{I} = \begin{bmatrix} 1 & 0 \\ 0 & 1 \end{bmatrix}$$

then

$$\mathbf{AI} = \mathbf{IA} = \mathbf{A} = \begin{bmatrix} 1 & 3 \\ 5 & 4 \end{bmatrix}$$

The identity matrix $\mathbf{I}$ thus plays the same role as the number 1 in ordinary algebra. The process of matrix multiplication is described in Fortran IV by the statements

```
C       MATRIX MULTIPLICATION ALGORITHM  AB·A is n × m, B is m × k
        DO 10 I = 1, N
        DO 10 J = 1, K
        C(I,J) = 0.0
        DO 10 IND = 1, M
     10 C(I,J) = C(I,J) + A(I,IND) * B(IND,J)
```

Power of a Matrix

The matrix product $\mathbf{AA} = \mathbf{A}^2$ exists only if $\mathbf{A}$ is square. The product $\mathbf{A}^3 = \mathbf{A}^2\mathbf{A}$, and by induction we define $\mathbf{A}^k$ as the kth continued product $\mathbf{A}^{k-1}\mathbf{A} = \mathbf{A}^k$ or the kth power of $\mathbf{A}$. In particular, when $k = 0$, $\mathbf{A}^0$ is defined as the unit matrix $\mathbf{I}$.

Associative, Commutative, and Distributive Laws

The associative law for real numbers is also valid for addition, subtraction, and multiplication in matrix operations. The commutative law is valid only for the addition and subtraction operations. The laws for matrix algebra are summarized in Table 3.1.

TABLE 3.1 Associative, Commutative, and Distributive Laws
for Matrix Algbra

	Addition and Subtraction	Multiplication
Associative law	$(\mathbf{A} + \mathbf{B}) + \mathbf{C} = \mathbf{A} + (\mathbf{B} + \mathbf{C})$	$(\mathbf{AB})\mathbf{C} = \mathbf{A}(\mathbf{BC})$
Commutative law	$\mathbf{A} + \mathbf{B} + \mathbf{C} = \mathbf{B} + \mathbf{C} + \mathbf{A}$	$\mathbf{ABC} \neq \mathbf{BCA} \neq \mathbf{CAB}$
	$= \mathbf{C} + \mathbf{A} + \mathbf{B}$	
Distributive law		$\mathbf{A}(\mathbf{B} + \mathbf{C}) = \mathbf{AB} + \mathbf{AC}$

Multiplication by a Scalar

The product of a matrix $\mathbf{A} = [a_{ij}]_{n,m}$ by a (scalar) quantity k is the matrix

$$kA = [ka_{ij}]_{n,m}$$

That is, each element of $\mathbf{A}$ is multiplied by k. The quantity k is referred to as a *scalar* and the process is called *scalar multiplication*.

Division (Matrix Inversion)

Division is not defined in matrix algebra. In the algebra for real numbers the single equation $ax = b$ can be solved as follows:

$$ax = b \qquad a \neq 0$$
$$a^{-1}ax = a^{-1}b$$
$$x = a^{-1}b$$

where a^{-1} denotes the *inverse* of a and is a quantity such that $a^{-1}a = 1$.

A similar relation exists in matrix algebra using the inverse of the matrix $\mathbf{A}$. Thus, define the inverse of $\mathbf{A}$ as $\mathbf{A}^{-1}$ such that

$$\mathbf{A}^{-1}\mathbf{A} = \mathbf{I} \tag{3.15}$$

Then given the equation

$$\mathbf{A}\mathbf{x} = \mathbf{b}$$

we can premultiply both sides by $\mathbf{A}^{-1}$ to get

$$\mathbf{A}^{-1}\mathbf{A}\mathbf{x} = \mathbf{A}^{-1}\mathbf{b}$$

or

$$\mathbf{I}\mathbf{x} = \mathbf{A}^{-1}\mathbf{b}$$
$$\mathbf{x} = \mathbf{A}^{-1}\mathbf{b} \tag{3.16}$$

To find $\mathbf{A}^{-1}$ let us look at the set of linear equations

$$a_{11}x_1 + a_{12}x_2 + \cdots + a_{1n}x_n = b_1$$
$$a_{21}x_1 + a_{22}x_2 + \cdots + a_{2n}x_n = b_2$$
$$\cdots \cdots \cdots \cdots \cdots \cdots \cdots \tag{3.17}$$
$$a_{m1}x_1 + a_{m2}x_2 + \cdots + a_{mn}x_n = b_n$$

It is well known that in order to obtain unique solutions for $x_1, x_2, \ldots, x_n$, we need to have as many independent equations as unknowns; that is, m must equal n.

In terms of the equivalent matrix equation,

$$\mathbf{Ax} = \mathbf{b} \tag{3.18}$$

where

$$
\mathbf{A} = \begin{bmatrix} a_{11} & a_{12} & \cdots & a_{1n} \\ a_{21} & a_{22} & \cdots & a_{2n} \\ \cdot & \cdot & \cdot & \cdot \\ a_{n1} & a_{n2} & \cdots & a_{nn} \end{bmatrix} \qquad
\mathbf{x} = \begin{bmatrix} x_1 \\ x_2 \\ \cdots \\ x_n \end{bmatrix} \qquad
\mathbf{b} = \begin{bmatrix} b_1 \\ b_2 \\ \cdots \\ b_n \end{bmatrix}
$$

Thus, $\mathbf{A}$ must be square and ($n \times n$) if $\mathbf{b}$ is of dimension n.

One way to solve Eq. (3.17) is through *Cramer's rule*, which states that

$$
x_i = \frac{\begin{vmatrix} a_{11} & \cdots & a_{1,i-1} & b_1 & a_{1,i+1} & \cdots & a_{1n} \\ a_{21} & \cdots & a_{2,i-1} & b_2 & a_{2,i+1} & \cdots & a_{2n} \\ \cdot & \cdot & \cdot & \cdot & \cdot & \cdot & \cdot \\ a_{n1} & \cdots & a_{n,i-1} & b_n & a_{n,i+1} & \cdots & a_{nn} \end{vmatrix}}{\det \mathbf{A}} \tag{3.19}
$$

The numerator on the right-hand side of Eq. (3.19) is $\det \mathbf{A}$ except that the ith column of $\mathbf{A}$ has been replaced by the elements of the vector $\mathbf{b}$ of Eq. (3.17). For x_1, Eq. (3.19) becomes

$$
x_1 = \frac{\begin{vmatrix} b_1 & a_{12} & \cdots & a_{1n} \\ b_2 & a_{22} & \cdots & a_{2n} \\ \cdot & \cdot & \cdot & \cdot \\ b_n & a_{n2} & \cdots & a_{nn} \end{vmatrix}}{\det \mathbf{A}}
$$

and if we expand this columnwise as per Eq. (3.9), then

$$x_1 = \frac{1}{\det \mathbf{A}}[b_1 C_{11} + b_2 C_{21} + \cdots + b_n C_{n1}] \tag{3.20}$$

where $C_{11}, C_{21}, \ldots$, and C_{n1} are cofactors as previously defined. Similarly,

$$
x_2 = \frac{\begin{vmatrix} a_{11} & b_1 & a_{13} & \cdots & a_{1n} \\ a_{21} & b_2 & a_{23} & \cdots & a_{2n} \\ \cdot & \cdot & \cdot & \cdot & \cdot \\ a_{n1} & b_n & a_{n3} & \cdots & a_{nn} \end{vmatrix}}{\det \mathbf{A}}
$$

or

$$x_2 = \frac{1}{\det \mathbf{A}}[b_1 C_{12} + b_2 C_{22} + \cdots + b_n C_{n2}] \tag{3.21}$$

Continuing, it is clear that the solutions for all x_i could be expressed by the compact matrix equation

$$
\begin{bmatrix} x_1 \\ x_2 \\ \cdots \\ x_n \end{bmatrix} = \frac{1}{\det \mathbf{A}} \begin{bmatrix} C_{11} & C_{21} & \cdots & C_{n1} \\ C_{12} & C_{22} & \cdots & C_{n2} \\ \cdots & \cdots & \cdots & \cdots \\ C_{1n} & C_{2n} & \cdots & C_{nn} \end{bmatrix} \begin{bmatrix} b_1 \\ b_2 \\ \cdots \\ b_n \end{bmatrix} \tag{3.22}
$$

or

$$
x = \frac{\mathrm{adj}\,\mathbf{A}}{\det \mathbf{A}} \mathbf{b} \tag{3.23}
$$

where

$$
\mathrm{adj}\,\mathbf{A} = \begin{bmatrix} C_{11} & C_{21} & \cdots & C_{n1} \\ C_{12} & C_{22} & \cdots & C_{n2} \\ \cdots & \cdots & \cdots & \cdots \\ C_{1n} & C_{2n} & \cdots & C_{nn} \end{bmatrix}
$$

as previously defined. Comparing Eq. (3.16) with Eq. (3.23) you see that

$$
\mathbf{A}^{-1} = \frac{\mathrm{adj}\,\mathbf{A}}{\det \mathbf{A}} \qquad \det \mathbf{A} \neq 0 \tag{3.24}
$$

EXAMPLE 3.7

If

$$
\mathbf{A} = \begin{bmatrix} 1 & 0 & 2 \\ 3 & 4 & 5 \\ 1 & 1 & 1 \end{bmatrix}
$$

then

$$
\mathbf{A}^{-1} = \frac{\mathrm{adj}\,\mathbf{A}}{\Delta_A} = \frac{1}{-3} \begin{bmatrix} -1 & 2 & -8 \\ 2 & -1 & 1 \\ -1 & -1 & 4 \end{bmatrix}
$$

To verify the calculation,

$$
\mathbf{A}\mathbf{A}^{-1} = \frac{1}{-3} \begin{bmatrix} 1 & 0 & 2 \\ 3 & 4 & 5 \\ 1 & 1 & 1 \end{bmatrix} \begin{bmatrix} -1 & 2 & -8 \\ 2 & -1 & 1 \\ -1 & -1 & 4 \end{bmatrix} = \begin{bmatrix} 1 & 0 & 0 \\ 0 & 1 & 0 \\ 0 & 0 & 1 \end{bmatrix} = \mathbf{I}
$$

There are many methods of obtaining $\mathbf{A}^{-1}$ for a given $\mathbf{A}$, some of which are superior to the row or column-expansion method previously shown. By superior, we mean less computation time and greater computational accuracy when using a digital computer. It is rather clear from the example that a computer becomes almost mandatory if the order of the matrix exceeds three. (Try inverting a 4×4 matrix.) Later we will present an efficient commonly used matrix-inversion method,

Matrix Partitioning

It is often possible in the matrix analysis of networks to reduce the number and complexity of computations by dividing a given matrix into a set of two or more submatrices. The partitioning of a matrix is the process of subdividing the matrix into smaller arrays so that each array can be treated as a new matrix. For example, we can arbitrarily partition the matrix $\mathbf{A}$ as

$$\mathbf{A} = \begin{bmatrix} a_{11} & a_{12} & \cdot & a_{13} \\ a_{21} & a_{22} & \cdot & a_{23} \\ \cdot & \cdot & \cdot & \cdot & \cdot & \cdot \\ a_{31} & a_{32} & \cdot & a_{33} \end{bmatrix} = \begin{bmatrix} \mathbf{A}_{11} & \cdot & \mathbf{A}_{12} \\ \cdot & \cdot & \cdot & \cdot & \cdot \\ \mathbf{A}_{21} & \cdot & \mathbf{A}_{22} \end{bmatrix} \tag{3.25}$$

where

$$\mathbf{A}_{11} = \begin{bmatrix} a_{11} & a_{12} \\ a_{21} & a_{22} \end{bmatrix} \qquad \mathbf{A}_{12} = \begin{bmatrix} a_{13} \\ a_{23} \end{bmatrix} \qquad \mathbf{A}_{21} = [a_{31} \quad a_{32}] \qquad \mathbf{A}_{22} = [a_{33}]$$

are submatrices of $\mathbf{A}$. The dotted lines in Eq. (3.25) indicate the partitions in the matrix $\mathbf{A}$. A *row partition* separates the rows of a matrix into subgroups such as the partition between row two and row three in Eq. (3.25). Similarly, *column partitions* separate columns of a matrix into subgroups.

If the matrix is associated with other matrices in a matrix equation, restrictions in the allowable ways a matrix can be partitioned become apparent. The operations, such as equality, addition, subtraction, and multiplication, limit the number of allowable partition configurations, as we now show.

Equality, Addition, and Subtraction. The matrix operations

$$\mathbf{A} = \mathbf{B} \quad \text{or} \quad \mathbf{A} + \mathbf{B} = \mathbf{C} \quad \text{or} \quad \mathbf{A} - \mathbf{B} = \mathbf{C}$$

require that the row partitioning of $\mathbf{A}$ and $\mathbf{B}$ be identical and that the column partitioning of $\mathbf{A}$ and $\mathbf{B}$ be identical. That is, matrices $\mathbf{A}$ and $\mathbf{B}$ must be partitioned in the same way.

EXAMPLE 3.8

Given

$$\mathbf{A} = \begin{bmatrix} a_{11} & a_{12} & a_{13} \\ b_{21} & b_{22} & b_{23} \end{bmatrix} \quad \text{and} \quad \mathbf{B} = \begin{bmatrix} b_{11} & b_{12} & b_{13} \\ b_{21} & b_{22} & b_{23} \end{bmatrix}$$

If we partition $\mathbf{A}$ arbitrarily as

$$\mathbf{A} = \begin{bmatrix} a_{11} & a_{12} & \cdot & a_{13} \\ \cdot & \cdot & \cdot & \cdot & \cdot & \cdot \\ a_{21} & a_{22} & \cdot & a_{23} \end{bmatrix} = \begin{bmatrix} \mathbf{A}_{11} & \cdot & \mathbf{A}_{12} \\ \cdot & \cdot & \cdot & \cdot \\ \mathbf{A}_{21} & \cdot & \mathbf{A}_{22} \end{bmatrix}$$

then the partitions of **B** must be identical to the partitions of **A**, so

$$
\mathbf{A} \pm \mathbf{B} = \begin{bmatrix} \mathbf{A}_{11} & \vdots & \mathbf{A}_{12} \\ \cdots & \cdots & \cdots \\ \mathbf{A}_{21} & \vdots & \mathbf{A}_{22} \end{bmatrix} \pm \begin{bmatrix} \mathbf{B}_{11} & \vdots & \mathbf{B}_{12} \\ \cdots & \cdots & \cdots \\ \mathbf{B}_{21} & \vdots & \mathbf{B}_{22} \end{bmatrix}
$$

$$
= \begin{bmatrix} \mathbf{A}_{11} \pm \mathbf{B}_{11} & \vdots & \mathbf{A}_{12} \pm \mathbf{B}_{12} \\ \cdots & \cdots & \cdots \\ \mathbf{A}_{21} \pm \mathbf{B}_{21} & \vdots & \mathbf{A}_{22} \pm \mathbf{B}_{22} \end{bmatrix}
$$

or

$$
\mathbf{A} \pm \mathbf{B} = \begin{bmatrix} a_{11} \pm b_{11} & a_{12} \pm b_{12} & \vdots & a_{13} \pm b_{13} \\ \cdots & \cdots & \cdots & \cdots & \cdots \\ a_{21} \pm b_{21} & a_{22} \pm b_{22} & \vdots & a_{23} \pm b_{23} \end{bmatrix} \tag{3.26}
$$

Multiplication. The following rules for partitioning matrices **A** and **B** to form the product **AB** = **C** must be observed:

1. The rows of **A** and the columns of **B** may be partitioned arbitrarily.

2. The column partitioning of **A** must be identical to the row partitioning of **B**.

The submatrices of **A** must still be conformable to the submatrices of **B** after partitioning to form the product **AB**.

EXAMPLE 3.9

Given the matrices

$$
\mathbf{A} = \begin{bmatrix} a_{11} & a_{12} & a_{13} \\ a_{21} & a_{22} & a_{23} \end{bmatrix} \quad \text{and} \quad \mathbf{B} = \begin{bmatrix} b_{11} & b_{12} \\ b_{21} & b_{22} \\ b_{31} & b_{32} \end{bmatrix}
$$

which are conformable to form the product **AB** and

$$
\mathbf{AB} = \begin{bmatrix} a_{11}b_{11} + a_{12}b_{21} + a_{13}b_{31} & a_{11}b_{12} + a_{12}b_{22} + a_{13}b_{32} \\ a_{21}b_{11} + a_{22}b_{21} + a_{23}b_{31} & a_{21}b_{12} + a_{22}b_{22} + a_{23}b_{32} \end{bmatrix} \tag{3.27}
$$

we can arbitrarily partition **A** as

$$
\mathbf{A} = \begin{bmatrix} a_{11} & \vdots & a_{12} & a_{13} \\ \cdots & \cdots & \cdots & \cdots \\ a_{21} & \vdots & a_{22} & a_{23} \end{bmatrix} = \begin{bmatrix} \mathbf{A}_{11} & \vdots & \mathbf{A}_{12} \\ \cdots & \cdots & \cdots \\ \mathbf{A}_{21} & \vdots & \mathbf{A}_{22} \end{bmatrix}
$$

Then, in order that **A** be conformable to **B**, we may partition **B** as

$$
\mathbf{B} = \begin{bmatrix} b_{11} & \vdots & b_{12} \\ \cdots & \cdots & \cdots \\ b_{21} & \vdots & b_{22} \\ b_{31} & \vdots & b_{32} \end{bmatrix} = \begin{bmatrix} \mathbf{B}_{11} & \vdots & \mathbf{B}_{12} \\ \cdots & \cdots & \cdots \\ \mathbf{B}_{21} & \vdots & \mathbf{B}_{22} \end{bmatrix}
$$

The product **AB** is then

$$\mathbf{AB} = \begin{bmatrix} \mathbf{A}_{11} & \vdots & \mathbf{A}_{12} \\ \cdots & \vdots & \cdots \\ \mathbf{A}_{21} & \vdots & \mathbf{A}_{22} \end{bmatrix} \begin{bmatrix} \mathbf{B}_{11} & \vdots & \mathbf{B}_{12} \\ \cdots & \vdots & \cdots \\ \mathbf{B}_{21} & \vdots & \mathbf{B}_{22} \end{bmatrix}$$

$$= \begin{bmatrix} \mathbf{A}_{11}\mathbf{B}_{11} + \mathbf{A}_{12}\mathbf{B}_{21} & \vdots & \mathbf{A}_{11}\mathbf{B}_{12} + \mathbf{A}_{12}\mathbf{B}_{22} \\ \cdots\cdots\cdots\cdots\cdots\cdots & \vdots & \cdots\cdots\cdots\cdots\cdots\cdots \\ \mathbf{A}_{21}\mathbf{B}_{11} + \mathbf{A}_{22}\mathbf{B}_{21} & \vdots & \mathbf{A}_{21}\mathbf{B}_{12} + \mathbf{A}_{22}\mathbf{B}_{22} \end{bmatrix}$$

$$= \begin{bmatrix} a_{11}b_{11} + \begin{bmatrix} a_{12} & a_{13} \end{bmatrix}\begin{bmatrix} b_{21} \\ b_{31} \end{bmatrix} & \vdots & a_{11}b_{12} + \begin{bmatrix} a_{12} & a_{13} \end{bmatrix}\begin{bmatrix} b_{22} \\ b_{32} \end{bmatrix} \\ \cdots\cdots\cdots\cdots\cdots\cdots\cdots\cdots\cdots\cdots\cdots\cdots\cdots\cdots\cdots \\ a_{21}b_{11} + \begin{bmatrix} a_{22} & a_{33} \end{bmatrix}\begin{bmatrix} b_{21} \\ b_{31} \end{bmatrix} & \vdots & a_{21}b_{12} + \begin{bmatrix} a_{22} & a_{23} \end{bmatrix}\begin{bmatrix} b_{22} \\ b_{32} \end{bmatrix} \end{bmatrix}$$

which is identical to the result obtained in Eq. (3.27).

We could have partitioned **B** as

$$\mathbf{B} = \begin{bmatrix} b_{11} & b_{12} \\ \cdots & \cdots \\ b_{21} & b_{22} \\ b_{31} & b_{32} \end{bmatrix} = \begin{bmatrix} \mathbf{B}_{11} \\ \cdots \\ \mathbf{B}_{21} \end{bmatrix}$$

which also satisfies the rule for multiplication partitioning. In this case

$$\mathbf{AB} = \begin{bmatrix} \mathbf{A}_{11} & \mathbf{A}_{12} \\ \mathbf{A}_{21} & \mathbf{A}_{22} \end{bmatrix} \begin{bmatrix} \mathbf{B}_{11} \\ \mathbf{B}_{21} \end{bmatrix}$$

$$= \begin{bmatrix} \mathbf{A}_{11}\mathbf{B}_{11} + \mathbf{A}_{12}\mathbf{B}_{21} \\ \cdots\cdots\cdots\cdots \\ \mathbf{A}_{21}\mathbf{B}_{11} + \mathbf{A}_{21}\mathbf{B}_{21} \end{bmatrix}$$

$$= \begin{bmatrix} a_{11}\begin{bmatrix} b_{11} & b_{12} \end{bmatrix} + \begin{bmatrix} a_{12} & a_{13} \end{bmatrix}\begin{bmatrix} b_{21} & b_{22} \\ b_{31} & b_{31} \end{bmatrix} \\ \cdots\cdots\cdots\cdots\cdots\cdots\cdots\cdots\cdots\cdots \\ a_{21}\begin{bmatrix} b_{11} & b_{12} \end{bmatrix} + \begin{bmatrix} a_{22} & a_{23} \end{bmatrix}\begin{bmatrix} b_{21} & b_{22} \\ b_{31} & b_{32} \end{bmatrix} \end{bmatrix}$$

$$= \begin{bmatrix} a_{11}b_{11} + a_{12}b_{21} + a_{13}b_{31} & \vdots & a_{11}b_{12} + a_{12}b_{22} + a_{13}b_{32} \\ \cdots\cdots\cdots\cdots\cdots\cdots\cdots\cdots & \vdots & \cdots\cdots\cdots\cdots\cdots\cdots\cdots\cdots \\ a_{21}b_{11} + a_{22}b_{21} + a_{23}b_{31} & \vdots & a_{21}b_{12} + a_{22}b_{22} + a_{23}b_{32} \end{bmatrix}$$

which is also identical to Eq. (3.27).

3.4 Solution of Systems of Linear Equations

In engineering, many problems have as the final computational step a system of linear equations of the form

$$\mathbf{A}\mathbf{x} = \mathbf{b} \tag{3.28}$$

where the $\mathbf{A}$ matrix elements (a_{ij}'s and b_i's) are real numbers and the x_j's are unknown real numbers. The n equations in Eq. (3.28) are referred to as a *system of linear equations*. We indicated one method of solution by using Cramer's rule in the previous section.

In this section we develop two methods for determining the unknown numbers $x_1, x_2, x_3, \ldots, x_n$ in Eq. (3.28), which are easily implemented on the digital computer.

The first method, known as the *Gauss-elimination method*, determines $\mathbf{x}$ by converting Eq. (3.28) into upper-triangular form and solving for each x_i by back substitution. The second method determines $\mathbf{x}$ from the equation

$$\mathbf{x} = \mathbf{A}^{-1}\mathbf{b} \tag{3.29}$$

where $\mathbf{A}^{-1}$ is, as before, the inverse of $\mathbf{A}$.

Basic studies of algebra have found certain transformations which can be made on an equation, such as multiplying both sides by a nonzero constant without changing the roots of the equation. We call the second equation an *equivalent equation*. In dealing with a linear system of equations, we will use three types of elementary transformations, each of which results in an equivalent system of equations. The transformations are (1) multiplying an equation by a nonzero constant, (2) interchanging the order of two equations, and (3) adding a multiple of one equation to another equation.

Suppose we have a set of n linear equations

$$
\begin{aligned}
a_{11}x_1 + a_{12}x_2 + \cdots + a_{1n}x_n &= b_1 \\
a_{21}x_1 + a_{22}x_2 + \cdots + a_{2n}x_n &= b_2 \\
a_{31}x_1 + a_{32}x_2 + \cdots + a_{3n}x_n &= b_3 \\
\cdot\ \cdot\ \cdot\ \cdot\ \cdot\ \cdot\ &\cdot\ \cdot\ \cdot\ \cdot\ \cdot \\
a_{k1}x_1 + a_{k2}x_2 + \cdots + a_{kn}x_n &= b_k \\
\cdot\ \cdot\ \cdot\ \cdot\ \cdot\ \cdot\ &\cdot\ \cdot\ \cdot\ \cdot\ \cdot \\
a_{n1}x_1 + a_{n2}x_2 + \cdots + a_{nn}x_n &= b_n
\end{aligned}
\tag{3.30}
$$

We multiply row 1 of Eq. (3.30) by a_{21}/a_{11} and subtract the resulting row from row 2, multiply row 1 by a_{31}/a_{11} and subtract from row 3, and in general, multiply row 1 by a_{k1}/a_{11} and subtract from row k. Continuing for all $n-1$ rows, the result will be the equivalent system

$$a_{11}x_1 + a_{12}x_2 + a_{13}x_3 + \cdots + a_{1n}x_x = b_1$$
$$a'_{22}x_2 + a'_{23}x_3 + \cdots + a'_{2n}x_n = b'_2$$
$$a'_{32}x_2 + a'_{33}x_3 + \cdots + a'_{2n}x_n = b'_3$$
$$\cdot \quad \cdot \quad \cdot \quad \cdot \quad \cdot \quad \cdot \quad \cdot \quad \cdot \quad \cdot \quad \cdot \quad \cdot \quad \cdot \quad \cdot \quad \cdot$$
$$\tag{3.31}$$
$$a'_{k2}x_2 + a'_{k3}x_3 + \cdots + a'_{kn}x_n = b'_k$$
$$\cdot \quad \cdot \quad \cdot \quad \cdot \quad \cdot \quad \cdot \quad \cdot \quad \cdot \quad \cdot \quad \cdot \quad \cdot \quad \cdot \quad \cdot \quad \cdot$$
$$a'_{n2}x_2 + a'_{n3}x_3 + \cdots + a'_{nn}x_n = b'_n$$

where $a'_{22} = a_{22} - a_{12}(a_{21}/a_{11})$, $a'_{23} = a_{23} - a_{13}(a_{21}/a_{11})$, $b'_2 = b_2 - b_1(a_{21}/a_{11})$, $a'_{kj} = a_{kj} - a_{k-1j}(a_{k1}/a_{11})$, and so on. (Note that a_{11} must not be zero.)

We then apply the same algorithm to the $(n-1) \times (n-1)$ system formed when the first equation in (3.31) is deleted to eliminate all a'_{k2}, $k = 3, 4, \ldots, n$. Continuing this scheme recursively until we only have x_n with a nonzero coefficient in the last equation (all other coefficients in the last equation equal zero) gives an algorithm for the *upper triangulation* of a linear system.

Thus, an upper-triangulation algorithm for the computer consists of two steps. First, an equation must be located with a nonzero leading coefficient. Once the equation is found, if it is not the first equation, it must be interchanged with the first equation. The nonzero leading coefficient is called the *pivot element*. As long as a nonzero pivot element can be found, the algorithm can be completed. Otherwise, the coefficient matrix **A** is singular. Better computational results are achieved by using the leading coefficient with the largest absolute value as the pivot element. We will interchange rows such that this occurs.

The second step of the algorithm is to subtract appropriate multiples of the resulting first equation from each of the other equations of the system to produce the equivalent system in Eq. (3.31). The first equation is then disregarded and the same process duplicated on the remaining equations. Repeating this entire recursion a total of $n - 1$ times, we obtain an *upper-triangular form*.

At the completion of the upper-triangulation algorithm, the resulting equivalent system is of the form

$$c_{11}x_1 + c_{12}x_2 + \cdots + c_{1,n-1}x_{n-1} + c_{1n}x_n = d_1$$
$$c_{22}x_2 + \cdots + c_{2,n-1}x_{n-1} + c_{2n}x_n = d_2$$
$$\cdot \quad \cdot \quad \cdot \quad \cdot \quad \cdot \quad \cdot \quad \cdot \quad \cdot \quad \cdot \quad \cdot \quad \cdot \quad \cdot \quad \cdot$$
$$\tag{3.32}$$
$$c_{n-1,n-1}x_{n-1} + c_{n-1,n}x_n = d_{n-1}$$
$$c_{nn}x_n = d_n$$

Matrix $\mathbf{C} = [c_{ij}]_{n,n}$ is called *upper triangular* since all elements below the diagonal are zero.

Since we intend to implement the upper-triangulation algorithm in Fortran, it

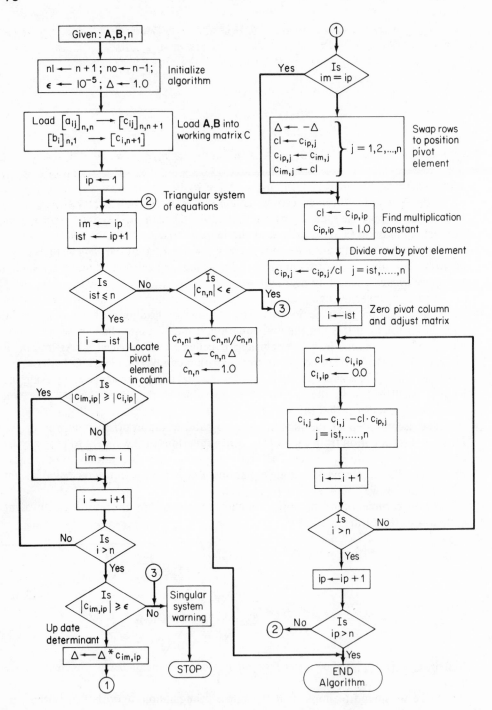

Figure 3.2 Upper-triangulation algorithm
flowchart (Gauss-elimination).

will be convenient to make a slight change in notation. The elements of the right-hand side of the equation, $b_1, b_2, \ldots, b_n$, can be stored as $a_{1,n+1}, a_{2,n+1}, \ldots, a_{n,n+1}$ in the $(n + 1)$st column of an augmented matrix A_1. Thus, the system becomes

$$a_{11}x_1 + a_{12}x_2 + \cdots + a_{1n}x_n = a_{1,n+1}$$
$$\cdot \quad \cdot \quad \cdot \quad \cdot \quad \cdot \quad \cdot \quad \cdot \quad \cdot \quad \cdot \quad \cdot \quad \cdot \quad \cdot \quad \cdot \quad \cdot \quad \cdot \quad \cdot \tag{3.33}$$
$$a_{n1}x_1 + a_{n2}x_2 + \cdots + a_{nn}x_n = a_{n,n+1}$$

This change in notation allows us to automatically manipulate the $(n + 1)$st index and do row operations on the vector b while performing the required operations on the rows of the matrix A.

The algorithm for upper triangulation is presented in Fig. 3.2. This leads us to an algorithm for solving upper-triangular systems of linear equations. This algorithm is commonly called *solution by back substitution*.

If we assume that the linear system is of the form defined in Eq. (3.32), where $c_{ii} \neq 0$ for $i = 1, 2, \ldots, n$, the back-substitution algorithm proceeds as follows:

1. Let $x_n = d_n/c_{nn}$.

2. Let $i = 1$.

3. $x_{n-i} = (d_{n-i} - \sum_{j=0}^{n-i} c_{n-i,n-j}x_{n-j})/c_{n-i,n-i}$.

4. Increase i by 1.

5. If $i = n$, go to step 6. If $i < n$, go to step 3.

6. The algorithm is complete.

This amounts to first solving for x_n in the last equation, (3.32), then using this value to solve for x_{n-1} in the next-to-last equation, and so on, until all x values are found.

The flowchart for the back-substitution algorithm is presented in Fig. 3.3.

EXAMPLE 3.10

Find the solution to

$$x_1 + 2x_2 + x_3 = 3 \tag{1}$$
$$x_1 + x_2 + 2x_3 = 2 \tag{2}$$
$$2x_1 + 3x_2 + x_3 = 1 \tag{3}$$

or

$$\begin{bmatrix} 1 & 2 & 1 \\ 1 & 1 & 2 \\ 2 & 3 & 1 \end{bmatrix} \begin{bmatrix} x_1 \\ x_2 \\ x_3 \end{bmatrix} = \begin{bmatrix} 3 \\ 2 \\ 1 \end{bmatrix}$$

Solution

$$a_{21} = 1 \quad a_{11} = 1 \quad \text{or} \quad \frac{a_{21}}{a_{11}} = 1$$

$$a_{31} = 2 \quad a_{11} = 1 \quad \text{or} \quad \frac{a_{31}}{a_{11}} = 2$$

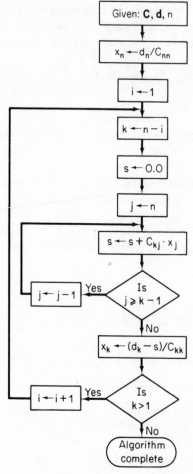

Figure 3.3 Back substitution
algorithm   flowchart   (Gauss-
elimination).

$$\frac{a_{21}}{a_{11}} \times \text{row 1 gives } x_1 + 2x_2 + x_3 = 3 \tag{2a}$$

$$\frac{a_{31}}{a_{11}} \times \text{row 1 gives } 2x_1 + 4x_2 + 2x_3 = 6 \tag{3a}$$

Subtracting (2a) from (2) and (3a) from (3) gives

$$x_1 + 2x_2 + x_3 = 3 \tag{1}$$
$$0 - x_3 + x_2 = -1 \tag{2$'$}$$
$$0 - x_2 - x_3 = -5 \tag{3$'$}$$

Now $a'_{32}/a_{22} \times$ row 2$'$ gives

$$-x_2 + x_3 = -1 \tag{2$'$a}$$

Subtracting row (2$'$a) from row (3$'$) we get

$$x_1 + 2x_2 + x_3 = 3 \tag{1}$$
$$0 - x_2 + x_3 = -1 \tag{2$'$}$$
$$0 \quad 0 - 2x_3 = -4 \tag{3$''$}$$

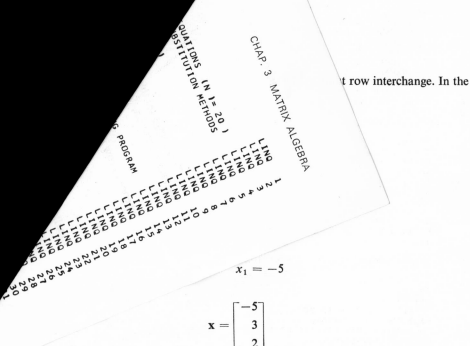

t row interchange. In the

$$x_1 = -5$$

$$\mathbf{x} = \begin{bmatrix} -5 \\ 3 \\ 2 \end{bmatrix}$$

Figure 3.4 presents a Fortran subroutine LINQ that uses the Gauss-elimination procedure to solve linear systems of equations of order less than or equal to 20. The order can be increased by changing the dimensions of the $\mathbf{A}$, $\mathbf{B}$, $\mathbf{C}$, and $\mathbf{D}$ arrays in the DIMENSION statement.

The subroutine LINQ also provides the determinant Δ_A of the matrix $\mathbf{A}$. It is obvious that the determinant Δ_{AU} of an upper-triangular system in Eq. (3.32) is given by

$$\Delta_{AU} = \Delta_A = (-1)^k \prod_{i=1}^{n} c_{ii} \tag{3.34}$$

where k is the number of row interchanges required to obtain the upper-triangular system.

The Gauss-elimination method of solving a system of linear algebraic equations can be extended by forcing zero coefficients both above and below the main diagonal. This method, called the Gauss-Jordan method, reduces the system defined in Eq. (3.30) to the form

$$
\begin{aligned}
x_1 \quad\quad\quad &= g_1 \\
x_2 \quad\quad &= g_2 \\
\cdots \cdots \cdots \\
x_n &= g_n
\end{aligned}
\tag{3.35}
$$

```
      SUBROUTINE LINQ(A,B,D,N,DET)
C
C     PURPOSE
C     SOLUTION OF AN NTH ORDER LINEAR SYSTEM OF
C     USING GAUSS' UPPER TRIANGULATION AND BACK S
C  DEFINITION OF VARIABLES
C     A - NXN COEFFICIENT MATRIX (DOUBLE SUBSCRIPTED
C     B - RIGHT HAND VECTOR FROM AX=B
C     C - NXN+1 WORKING MATRIX
C     D - SOLUTION VECTOR X
C     N - NUMBER OF EQUATIONS IN SYSTEM
C     DET - DETERMINANT OF MATRIX A
C     EPS - ZERO TOLERANCE
C  NOTE ..... VARIABLES A,B,D MUST BE DIMENSIONED IN CALLI
C
      DIMENSION A(20,20),B(20),C(20,21),D(20)
C     INITIALIZE SUBROUTINE PARAMETERS
      N1=N+1
      NO=N-1
      EPS=1.0E-5
      DET=1.0
C     LOAD A AND B INTO WORKING MATRIX C
      DO 10 I=1,N
      C(I,N1)=B(I)
      DO 10 J=1,N
   10 C(I,J)=A(I,J)                                           L
C                                                             LI
C     TRIANGULARIZE SYSTEM OF EQUATIONS                       LIN
C                                                             LINQ
      DO 110 IP=1,N                                           LINQ
      IM=IP                                                   LINQ
      IST=IP+1                                                LINQ   32
      IF (IST.LE.N) GO TO 20                                  LINQ   33
C     LAST ROW CALCULATIONS                                   LINQ   34
      IF (ABS(C(N,N)).LT.EPS) GO TO 40                        LINQ   35
      C(N,N1)=C(N,N1)/C(N,N)                                  LINQ   36
      DET=C(N,N)*DET                                          LINQ   37
      C(N,N)=1.0                                              LINQ   38
      GO TO 120                                               LINQ   39
C     DETERMINE PIVOT ELEMENT AND UPDATE DETERMINANT          LINQ   40
   20 DO 30 I=IST,N                                           LINQ   41
      IF (ABS(C(IM,IP)).GE.ABS(C(I,IP))) GO TO 30             LINQ   42
      IM=I                                                    LINQ   43
   30 CONTINUE                                                LINQ   44
      IF (ABS(C(IM,IP)).GE.EPS) GO TO 70                      LINQ   45
C     ERROR OUTPUT SECTION                                    LINQ   46
   40 WRITE (6,50) IP,C(IM,IP)                                LINQ   47
   50 FORMAT (1H ,///10X,23HPIVOT ELEMENT IN COLUMN,I3,2H =,1PE12.5)  LINQ   48
      IF (C(IM,IP).EQ.0.0) WRITE (6,60)                       LINQ   49
   60 FORMAT (1H0,9X,46HSOLUTION IMPOSSIBLE, DIAGONAL ELEMENT IS ZERO.)  LINQ   50
      IF (C(IM,IP).EQ.0.0) RETURN                             LINQ   51
   70 DET=C(IM,IP)*DET                                        LINQ   52
C     TEST FOR DIAGONAL ELEMENT AND SWAP ROWS                 LINQ   53
      IF (IM.EQ.IP) GO TO 90                                  LINQ   54
C     SWAP ROWS TO LOCATE PIVOT ELEMENT                       LINQ   55
      DET=-DET                                                LINQ   56
      DO 80 J=IP,N1                                           LINQ   57
      CL=C(IP,J)                                              LINQ   58
      C(IP,J)=C(IM,J)                                         LINQ   59
   80 C(IM,J)=CL                                              LINQ   60
C     FIND MULTIPLICATION CONSTANT                            LINQ   61
   90 CL=C(IP,IP)                                             LINQ   62
      C(IP,IP)=1.0                                            LINQ   63
```

Figure 3.4 A Fortran subroutine for solving an
nth-order linear system of algebraic equations.

```
C      DIVIDE ROW BY PIVOT ELEMENT                         LINQ   64
       DO 80 J=IST,N1                                      LINQ   65
       C(I,IP)=0.0                                         LINQ   66
       DO 100 J=IST,N1                                     LINQ   67
  100 C(I,J)=C(I,J)-CL*C(IP,J)                             LINQ   68
  110 CONTINUE                                             LINQ   69
C      DETERMINE SOLUTION BY BACK SUBSTITUTION             LINQ   70
  120 DO 140 I=1,NO                                        LINQ   71
       K=N-I                                               LINQ   72
       SUM=0.0                                             LINQ   73
       DO 130 L=1,I                                        LINQ   74
       J=N1-L                                              LINQ   75
  130 SUM=SUM+C(K,J)*C(J,N1)                               LINQ   76
  140 C(K,N1)=C(K,N1)-SUM                                  LINQ   77
C      LOAD D VECTOR = SOLUTION VECTOR X                   LINQ   78
       DO 150 I=1,N                                        LINQ   79
  150 D(I)=C(I,N1)                                         LINQ   80
       RETURN                                              LINQ   81
       END                                                 LINQ   82
```

Figure 3.4—*Cont.*

or

$$\mathbf{I}\mathbf{x} = \mathbf{g} \qquad\qquad (3.36)$$

As a consequence, the values of $x_1, x_2, \ldots, x_n$ can be found directly without using the back-substitution algorithm. This form may be obtained from the upper-triangular form of Eq. (3.32) by proceeding with Gauss-elimination, using the trailing coefficients as the pivot elements, in the reverse direction (no row interchange). The implementation in Fortran of the Gauss–Jordan method is left to the reader as an exercise.

3.5 Matrix Inversion

Since the solution of a set of linear equations essentially involves a matrix inversion, as already shown, matrix inversion may be accomplished by Gaussian elimination. Thus,

$$\mathbf{A}\mathbf{A}^{-1} = \mathbf{I}$$

Let

$$\mathbf{A}^{-1} = [\mathbf{e}_1 \quad \mathbf{e}_2 \quad \cdots \quad \mathbf{e}_n] \qquad\qquad (3.37)$$

where the e_i are column vectors. Then

$$\mathbf{A}[\mathbf{e}_1 \quad \mathbf{e}_2 \quad \cdots \quad \mathbf{e}_n] = \begin{bmatrix} 1 & 0 & 0 & \cdots & 0 \\ 0 & 1 & 0 & \cdots & 0 \\ \cdot & \cdot & \cdot & \cdot & \cdot \\ 0 & 0 & 0 & \cdots & 1 \end{bmatrix}$$

or

$$\mathbf{A}\mathbf{e}_1 = \begin{bmatrix} 1 \\ 0 \\ \cdots \\ 0 \end{bmatrix} \qquad \mathbf{A}\mathbf{e}_2 = \begin{bmatrix} 0 \\ 1 \\ \cdots \\ 0 \end{bmatrix} \qquad \text{etc.}$$

EXAMPLE 3.11

Find A^{-1}, where

$$A = \begin{bmatrix} 1 & 2 & 1 \\ 1 & 1 & 2 \\ 2 & 3 & 1 \end{bmatrix}$$

Solution

$$\begin{bmatrix} 1 & 2 & 1 \\ 1 & 1 & 2 \\ 2 & 3 & 1 \end{bmatrix} \begin{bmatrix} e_{11} \\ e_{12} \\ e_{13} \end{bmatrix} = \begin{bmatrix} 1 \\ 0 \\ 0 \end{bmatrix}$$

Proceeding with Gaussian elimination, A reduces, as before, to the upper-triangular matrix found in Example 3.10. The right side of the matrix equation converts to $[1 - 1 - 1]^{\tau}$, or we get

$$\begin{bmatrix} 1 & 2 & 1 \\ 0 & -1 & 1 \\ 0 & 0 & -2 \end{bmatrix} \begin{bmatrix} e_{11} \\ e_{12} \\ e_{13} \end{bmatrix} = \begin{bmatrix} 1 \\ -1 \\ -1 \end{bmatrix}$$

whence $e_{13} = \frac{1}{2}$, $e_{12} = \frac{3}{2}$, $e_{11} = -\frac{5}{2}$. In the second round,

$$\begin{bmatrix} 1 & 2 & 1 \\ 0 & -1 & 1 \\ 0 & 0 & -2 \end{bmatrix} \begin{bmatrix} e_{21} \\ e_{22} \\ e_{23} \end{bmatrix} = \begin{bmatrix} 0 \\ 1 \\ -1 \end{bmatrix}$$

whence $e_{23} = \frac{1}{2}$, $e_{22} = -\frac{1}{2}$, $e_{21} = \frac{1}{2}$. Finally,

$$\begin{bmatrix} 1 & 2 & 1 \\ 1 & 1 & 2 \\ 2 & 3 & 1 \end{bmatrix} \begin{bmatrix} e_{31} \\ e_{32} \\ e_{33} \end{bmatrix} = \begin{bmatrix} 0 \\ 0 \\ 1 \end{bmatrix}$$

which reduces to

$$\begin{bmatrix} 1 & 2 & 1 \\ 0 & -1 & 1 \\ 0 & 0 & -2 \end{bmatrix} \begin{bmatrix} e_{31} \\ e_{32} \\ e_{33} \end{bmatrix} = \begin{bmatrix} 0 \\ 0 \\ 1 \end{bmatrix}$$

or $e_{33} = -\frac{1}{2}$, $e_{32} = -\frac{1}{2}$, $e_{31} = \frac{3}{2}$. Then

$$A^{-1} = [e_1 \quad e_2 \quad e_3] = \begin{bmatrix} e_{11} & e_{21} & e_{31} \\ e_{12} & e_{22} & e_{32} \\ e_{13} & e_{23} & e_{33} \end{bmatrix}$$

$$= \frac{1}{2} \begin{bmatrix} -5 & 1 & 3 \\ 3 & -1 & -1 \\ 1 & 1 & -1 \end{bmatrix}$$

The advantage of finding A^{-1} is that now you can easily solve

$$Ax = b$$

for any given **b**, since

$$\mathbf{x} = \mathbf{A}^{-1}\mathbf{b}$$

An alternative method for determining the inverse of a matrix uses a variation of the Gauss–Jordan reduction method described earlier. The method is based on the property that

$$\mathbf{A}\mathbf{A}^{-1} = \mathbf{I}$$

We begin by forming the augmented matrix **B**, where

$$\mathbf{B} = [\mathbf{A} \colon \mathbf{I}] \tag{3.38}$$

and the augmented vector **y** equal to

$$\mathbf{y} = \begin{bmatrix} \mathbf{x} \\ \cdots \\ -\mathbf{b} \end{bmatrix} \tag{3.39}$$

so that

$$\mathbf{B}\mathbf{y} = [\mathbf{A} \colon \mathbf{I}] \begin{bmatrix} \mathbf{x} \\ \cdots \\ -\mathbf{b} \end{bmatrix} = 0 \tag{3.40}$$

The ith row of Eq. (3.40) may then be associated with the equation

$$\sum_{k=1}^{n} a_{ik} x_k = b_i \tag{3.41}$$

and all rows represent the set of equations

$$\sum_{k=1}^{n} a_{ik} x_k = b_i \qquad i = 1, 2, \ldots, n \tag{3.42}$$

Suppose we alter **B** by elementary transformations such that we obtain the matrix

$$[\mathbf{I} \colon \mathbf{C}]$$

Similarly, this represents the set of equations

$$x_i = \sum_{k=1}^{n} c_{ix} b_k \qquad i = 1, 2, \ldots, n \tag{3.43}$$

Equation (3.43) is now a solution to Eq. (3.41) and hence **C** is the inverse of **A**, or

$$\mathbf{A}^{-1} = \mathbf{C}$$

The elementary transformations allowed consist of

1. Multiplying any row of **B** by a constant.
2. Adding or subtracting a multiple of any row to any other row.

EXAMPLE 3.12

Let

$$\mathbf{A} = \begin{bmatrix} a_{11} & a_{12} \\ a_{21} & a_{22} \end{bmatrix}$$

Then

$$\mathbf{B} = \begin{bmatrix} a_{11} & a_{12} & \vdots & 1 & 0 \\ a_{21} & a_{22} & \vdots & 0 & 1 \end{bmatrix}$$

1. Divide the first row by a_{11} to get

$$\begin{bmatrix} 1 & \dfrac{a_{12}}{a_{11}} & \vdots & \dfrac{1}{a_{11}} & 0 \\ a_{21} & a_{22} & \vdots & 0 & 1 \end{bmatrix}$$

2. Multiply the first row by a_{21} and subtract from the second row to find

$$\begin{bmatrix} 1 & \dfrac{a_{12}}{a_{11}} & \vdots & \dfrac{1}{a_{11}} & 0 \\ 0 & \dfrac{a_{21}a_{22} - a_{12}a_{21}}{a_{11}} & \vdots & -\dfrac{a_{21}}{a_{11}} & 1 \end{bmatrix}$$

```
      SUBROUTINE INVERT(A,X,N)                                          NVRT    1
C                                                                       NVRT    2
C     PURPOSE                                                           NVRT    3
C        COMPUTE THE INVERSE OF A MATRIX OF ORDER N (N )= 20)           NVRT    4
C     DEFINITION OF VARIABLES                                          NVRT    5
C        A - INPUT NXN MATRIX                                           NVRT    6
C        X - NXN INVERSE OF MATRIX A                                    NVRT    7
C        N - ORDER OF MATRIX A                                          NVRT    8
C        C - NX(2N) WORKING MATRIX                                      NVRT    9
C        EPS - ZERO TOLERANCE                                           NVRT   10
C     NOTE ..... VARIABLES A AND X MUST BE DIMENSIONED IN CALLING PRCGRAM NVRT 11
C                                                                       NVRT   12
      DIMENSION A(20,20),X(20,20),C(20,40)                              NVRT   13
C     INITIALIZE SUBROUTINE VARIABLES                                   NVRT   14
      DATA EPS/1.0E-6/                                                  NVRT   15
      N1=2*N                                                            NVRT   16
C     LOAD A MATRIX AND CONSTRUCT IDENTITY MATRIX                       NVRT   17
      DO 10 I=1,N                                                       NVRT   18
      DO 10 J=1,N                                                       NVRT   19
   10 C(I,J)=A(I,J)                                                     NVRT   20
      DO 30 I=1,N                                                       NVRT   21
      DO 30 J=1,N                                                       NVRT   22
      IF (I.EQ.J) GO TO 20                                              NVRT   23
      C(I,J+N)=0.0                                                      NVRT   24
      GO TO 30                                                          NVRT   25
   20 C(I,J+N)=1.0                                                      NVRT   26
   30 CONTINUE                                                          NVRT   27
C                                                                       NVRT   28
C     INVERT MATRIX                                                     NVRT   29
C                                                                       NVRT   30
      DO 130 IP=1,N                                                     NVRT   31
C     FIND PIVOT ELEMENT IN COLUMN IP                                   NVRT   32
      IM=IP                                                             NVRT   33
      IF (IP.GE.N) GO TO 50                                             NVRT   34
      IST=IP+1                                                          NVRT   35
      DO 40 I=IST,N                                                     NVRT   36
```

Figure 3.5 Gauss–Jordan reduction method
for matrix inversion.

```
      IF (ABS(C(IM,IP)).GE.ABS(C(I,IP))) GO TO 40            NVRT  37
      IM=I                                                    NVRT  38
   40 CONTINUE                                                NVRT  39
   50 IF (ABS(C(IM,IP)).GE.EPS) GO TO 70                      NVRT  40
C     NEAR ZERO DIAGONAL ELEMENT FLAG                         NVRT  41
      WRITE (6,60) IP,C(IM,IP)                                NVRT  42
   60 FORMAT (1H0,17HDIAGONAL ELEMENT ,I2,2H =,1PE13.5)       NVRT  43
      IF (C(IM,IP).EQ.0.0) RETURN                             NVRT  44
   70 IF (IM.EQ.IP) GO TO 90                                  NVRT  45
C     INTERCHANGE ROWS TO POSITION PIVOT ELEMENT              NVRT  46
      DO 80 J=IP,N1                                           NVRT  47
      CL=C(IP,J)                                              NVRT  48
      C(IP,J)=C(IM,J)                                         NVRT  49
   80 C(IM,J)=CL                                              NVRT  50
C     FIND MULTIPLICATION CONSTANT, SET C(I,I)=1              NVRT  51
   90 CL=C(IP,IP)                                             NVRT  52
      C(IP,IP)=1.0                                            NVRT  53
C     DIVIDE ELEMENT IN ROW BY PIVOT ELEMENT                  NVRT  54
      DO 100 J=IST,N1                                         NVRT  55
  100 C(IP,J)=C(IP,J)/CL                                      NVRT  56
C     ZERO COLUMN OF PIVOT ELEMENT                            NVRT  57
      DO 120 I=1,N                                            NVRT  58
      IF (I.EQ.IP) GO TO 120                                  NVRT  59
      IP1=IP+1                                                NVRT  60
      CL=C(I,IP)                                              NVRT  61
      C(I,IP)=0.0                                             NVRT  62
      DO 110 J=IP1,N1                                         NVRT  63
  110 C(I,J)=C(I,J)-CL*C(IP,J)                                NVRT  64
  120 CONTINUE                                                NVRT  65
  130 CONTINUE                                                NVRT  66
C     LOAD INVERSE MATRIX INTO X                              NVRT  67
      M=N+1                                                   NVRT  68
      DO 140 I=1,N                                            NVRT  69
      DO 140 J=M,N1                                           NVRT  70
      K=J-N                                                   NVRT  71
  140 X(I,K)=C(I,J)                                           NVRT  72
      RETURN                                                  NVRT  73
      END                                                     NVRT  74
```

Figure 3.5—Cont.

3. Divide the second row by $(a_{11}a_{22} - a_{12}a_{21})/a_{11}$ to get

$$\begin{bmatrix} 1 & \dfrac{a_{12}}{a_{11}} & \cdot & \dfrac{1}{a_{11}} & 0 \\[2ex] & & \cdot & & \\ 0 & 1 & \cdot & \dfrac{-a_{21}}{a_{11}a_{22} - a_{12}a_{21}} & \dfrac{a_{11}}{a_{11}a_{22} - a_{12}a_{21}} \end{bmatrix}$$

4. Multiply the second row by a_{12}/a_{11} and subtract from the first row to obtain

$$\begin{bmatrix} 1 & 0 & \cdot & \dfrac{a_{22}}{a_{11}a_{22} - a_{12}a_{21}} & \dfrac{-a_{12}}{a_{11}a_{22} - a_{12}a_{21}} \\[2ex] & & \cdot & & \\ 0 & 1 & \cdot & \dfrac{-a_{21}}{a_{11}a_{22} - a_{12}a_{21}} & \dfrac{a_{11}}{a_{11}a_{22} - a_{12}a_{21}} \end{bmatrix}$$

We may easily verify that the right square matrix in the partitioned matrix is $\mathbf{A}^{-1}$.

There are many other matrix-inversion techniques, such as pivotal condensation,[†] Faddeeva's method,[‡] and diagonalization. The Gauss–Jordan reduction

---

[†] P. M. DeRusso, R. J. Roy, and C. M. Close, *State Variables for Engineers*, John Wiley & Sons, Inc., New York, 1965.

[‡] B. O. Watkins, *Introduction to Control Systems*, The Macmillan Company, New York, 1969.

method is perhaps used more than the others. Figure 3.5 presents a program for the Gauss–Jordan reduction method of inversion.

3.6 Vector Spaces, Eigenvalues, and Quadratic Forms

We have implied that the **x** as represented by

$$\mathbf{x} = \begin{bmatrix} x_1 \\ x_2 \\ \cdots \\ x_n \end{bmatrix} \tag{3.44}$$

may be considered a vector in an n-dimensional space, with components $x_1, x_2, \ldots, x_n$.

We may define a method of measuring distances in such a space by extending the idea of length in the three-dimensional case. Thus, if $||\mathbf{x}||$ means the vector "length", we write

$$||\mathbf{x}||^2 = [x_1 \quad x_2 \quad \cdots \quad x_n] \begin{bmatrix} x_1 \\ x_2 \\ \cdots \\ x_n \end{bmatrix} \tag{3.45}$$

$$= x_1^2 + x_2^2 + \cdots + x_n^2$$

or

$$||\mathbf{x}|| = (x_1^2 + x_2^2 + \cdots + x_n^2)^{1/2}$$

Similarly, we may define an *angle* as follows:

First, let

$$[x_1 \quad x_2 \quad \cdots \quad x_n] \begin{bmatrix} y_1 \\ y_2 \\ \cdots \\ y_n \end{bmatrix} = \langle \mathbf{x}, \mathbf{y} \rangle \tag{3.46}$$

for convenience, where we term $\langle \mathbf{x}, \mathbf{y} \rangle$ the *inner product* of **x** and **y**. Then we define the angle θ between **x** and **y** by the relation

$$\cos \theta = \frac{\langle \mathbf{x}, \mathbf{y} \rangle}{||\mathbf{x}|| \, ||\mathbf{y}||} \tag{3.47}$$

where $||\mathbf{x}|| = \langle \mathbf{x}, \mathbf{x} \rangle^{1/2}$ and $||\mathbf{y}|| = \langle \mathbf{y}, \mathbf{y} \rangle^{1/2}$. From (3.47), if $\langle \mathbf{x}, \mathbf{y} \rangle = 0$, **x** and **y** are said to be orthogonal. Thus, you now see a justification for calling the **x** of Eq. (3.44) a *vector* of n dimensions. The elements $x_1, x_2, \ldots, x_n$ may be thought of as the lengths of component vectors making up **x**. These component vectors can be considered as orthogonal in the sense of Eq. (3.47) without great loss in generality.

As already shown, we frequently deal with equations of the type

$$\mathbf{Ax} = \mathbf{y} \tag{3.48}$$

Equation (3.48) represents the transformation of the vector $\mathbf{x}$ into the vector $\mathbf{y}$ by a rotation and an expansion or shrinkage of components (or a deformation of the space). A space translation cannot be represented by a matrix (a translation usually results in products and nonlinear equations). The vector $\mathbf{x}$ is said to lie in (or span) a space A^n of n dimensions if it is composed of n linearly independent vector components. By *linearly independent* we imply that no two of the component vectors are in the same direction.

A vector $\mathbf{y}$ may generate or span a space B^k, where $k < n$, and B^k is called a *subspace* of A^k if B^k contains the null vector ($\mathbf{x}$ and $\mathbf{y}$ go through a common origin). In linear networks study, a vector of n dimensions defines the network, and we are concerned with the length and direction of this vector as time evolves. A matrix of transformation becomes useful in finding solutions to vector equations.

EXAMPLE 3.13

If

$$\mathbf{x} = \begin{bmatrix} 2 \\ 1 \end{bmatrix} \quad \text{and} \quad \mathbf{A} = \begin{bmatrix} 1 & 2 \\ -2 & 1 \end{bmatrix}$$

then

$$\mathbf{y} = \begin{bmatrix} 1 & 2 \\ -2 & 1 \end{bmatrix} \begin{bmatrix} 2 \\ 1 \end{bmatrix}$$
$$= \begin{bmatrix} 4 \\ -3 \end{bmatrix}$$

representing a transformation in the space as shown in Fig. 3.6.

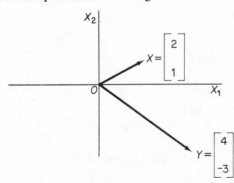

Figure 3.6 Vector transformation in space.

We now ask: Do any vectors $\mathbf{x}$ exist that have directions unchanged under the transformation $\mathbf{A}$? If so, a vector $\mathbf{x}$ of this set satisfies

$$\mathbf{Ax} = \lambda\mathbf{x} \tag{3.49}$$

where λ is a constant, indicating a change of length of **x**. Then

$$\mathbf{A}^2\mathbf{x} = \mathbf{A}(\mathbf{A}\mathbf{x}) = \mathbf{A}\lambda\mathbf{x} = \lambda\mathbf{A}\mathbf{x} = \lambda^2\mathbf{x}$$

and continuing,

$$\mathbf{A}^k\mathbf{x} = \lambda^k\mathbf{x} \tag{3.50}$$

In Eq. (3.49), if we bring the right side over to the left, we get

$$(\lambda\mathbf{I} - \mathbf{A})\mathbf{x} = \mathbf{0} \tag{3.51}$$

The **I** matrix must be introduced in Eq. (3.51) so as to make the quantity within the parentheses a matrix. If we write out Eq. (3.51) and change signs we obtain

$$
\begin{aligned}
(a_{11} - \lambda)x_1 + a_{12}x_2 + \cdots + a_{1n}x_n &= 0 \\
a_{21}x_1 + (a_{22} - \lambda)x_2 + \cdots + a_{2n}x_n &= 0 \\
\cdot \quad \cdot \quad \cdot \quad \cdot \quad \cdot \quad \cdot \quad \cdot \quad \cdot \quad \cdot \quad \cdot \quad \cdot \quad \cdot \quad \cdot \quad & \\
a_{n1}x_1 + a_{n2}x_2 + \cdots + (a_{nn} - \lambda)x_n &= 0
\end{aligned}
\tag{3.52}
$$

From Eq. (3.52) we can obtain a nonzero solution if and only if

$$\det[\lambda\mathbf{I} - \mathbf{A}] = 0 \tag{3.53}$$

Solving Eq. (3.53) results in a polynomial in λ, which, being set to zero, gives n values of λ which will satisfy Eq. (3.49). The λ values are called *eigenvalues*, *proper values*, or *characteristic values*, and may be real or complex. The $n\mathbf{x}$ vectors corresponding to the n eigenvalues are called *eigenvectors*. We will assume that the n values are all distinct, with none repeated. This assumption makes subsequent discussion simpler than otherwise and is not usually restrictive.

EXAMPLE 3.14

Find the eigenvalues and eigenvectors of the matrix

$$\mathbf{A} = \begin{bmatrix} 5 & \sqrt{2} \\ \sqrt{2} & 4 \end{bmatrix}$$

Solution

$$[\lambda\mathbf{I} - \mathbf{A}] = \begin{bmatrix} \lambda & 0 \\ 0 & \lambda \end{bmatrix} - \begin{bmatrix} 5 & \sqrt{2} \\ \sqrt{2} & 4 \end{bmatrix} = \begin{bmatrix} \lambda - 5 & -\sqrt{2} \\ -\sqrt{2} & \lambda - 4 \end{bmatrix}$$

Then

$$\det[\lambda\mathbf{I} - \mathbf{A}] = (\lambda - 5)(\lambda - 4) - (\sqrt{2}\sqrt{2}) = \lambda^2 - 9\lambda + 18$$

and so

$$\lambda^2 - 9\lambda + 18 = (\lambda - 6)(\lambda - 3) = 0$$

or

$$\lambda_1 = 3$$
$$\lambda_2 = 6$$

Substituting in Eq. (3.51) for $\lambda_2 = 6$,

$$\begin{bmatrix} 6-5 & -\sqrt{2} \\ -\sqrt{2} & 6-4 \end{bmatrix} \begin{bmatrix} x_1 \\ x_2 \end{bmatrix} = \begin{bmatrix} 0 \\ 0 \end{bmatrix}$$

or

$$x_1 - \sqrt{2}\,x_2 = 0 \quad \text{or} \quad x_1 = \sqrt{2}\,x_2$$
$$-\sqrt{2}\,x_1 + 2x_2 = 0 \quad \text{or} \quad x_1 = \sqrt{2}\,x_2$$

Both equations give the same result if no errors have occurred. Since

$$\mathbf{x} = \begin{bmatrix} x_1 \\ x_2 \end{bmatrix}$$

is specified only as to direction, we may select x_1 as any arbitrary length, and adjust x_2 accordingly, or one possible $\mathbf{x}$ could be

$$\mathbf{x}_2 = \begin{bmatrix} \sqrt{2} \\ 1 \end{bmatrix}$$

If we wish to make $\mathbf{x}_2$ of unit length, then $1 = x_1^2 + x_2^2$, or $1 = 2x_2^2 + x_2^2$, whence

$$\mathbf{x}_2 = \begin{bmatrix} \sqrt{\frac{2}{3}} \\ \frac{1}{\sqrt{3}} \end{bmatrix} \quad \text{where} \quad ||\mathbf{x}_2|| = 1$$

Similarly, using Eq. (3.51) for $\lambda_1 = 3$, we get

$$x_1 = -\frac{1}{\sqrt{2}}\,x_2$$

or

$$\mathbf{x}_1 = \begin{bmatrix} -\frac{1}{\sqrt{3}} \\ \sqrt{\frac{2}{3}} \end{bmatrix} \quad \text{where} \quad ||\mathbf{x}_1|| = 1$$

In this case ($\mathbf{A}$ symmetric), $\mathbf{x}_1$ and $\mathbf{x}_2$ are orthogonal; that is,

$$\langle \mathbf{x}_1, \mathbf{x}_2 \rangle = 0$$

If $\mathbf{A}$ is not symmetric, the eigenvectors are not necessarily orthogonal.

A matrix made up of eigenvectors as columns is called a *modal matrix*. Thus, a modal matrix connected with the $\mathbf{A}$ of Example 3.13 might be

$$\mathbf{T} = \begin{bmatrix} -1 & \sqrt{2} \\ \sqrt{2} & 1 \end{bmatrix}$$

where here **T** is not normalized. The importance of the modal matrix comes about through its ability to simplify matrix equations. It may be shown that any $n \times n$ matrix **A** with distinct eigenvalues $(\lambda_1, \lambda_2, \ldots, \lambda_n)$ can be diagonalized using the modal matrix **T**. Thus

$$\mathbf{T}^{-1}\mathbf{A}\mathbf{T} = \mathbf{\Lambda} = \begin{bmatrix} \lambda_1 & & & & 0 \\ & \lambda_2 & & & \\ & & \cdot & & \\ & & & \cdot & \\ 0 & & & & \lambda_n \end{bmatrix} \tag{3.54}$$

where $\mathbf{\Lambda}$ represents the diagonalized matrix shown. The eigenvalues used to find the first column (eigenvector) of **T** will appear in Eq. (3.54) as λ_1, the eigenvalue used to find the second column will appear as λ_2, and so on. To have some order it is often convenient to number the eigenvalues in some sequence, such that, for example, $|\lambda_1| < |\lambda_2| < \cdots < |\lambda_n|$. Now given the equation

$$\mathbf{A}\mathbf{x} = \mathbf{y} \tag{3.55}$$

if we let

$$\mathbf{x} = \mathbf{T}\mathbf{z}$$

then

$$\mathbf{A}\mathbf{T}\mathbf{z} = \mathbf{y}$$

and

$$\mathbf{T}^{-1}\mathbf{A}\mathbf{T}\mathbf{z} = \mathbf{T}^{-1}\mathbf{y}$$

or

$$\mathbf{\Lambda}\mathbf{z} = \mathbf{T}^{-1}\mathbf{y} = \mathbf{w} \tag{3.56}$$

The vector **z** may readily be found from Eq. (3.56) since, by expanding, we see that

$$\lambda_1 z_1 = w_1$$
$$\lambda_2 z_2 = w_2$$
$$\cdot \quad \cdot \quad \cdot \quad \cdot$$
$$\lambda_n z_n = w_n \tag{3.57}$$

where $w_1, w_2, \ldots, w_n$ are known from $\mathbf{w} = \mathbf{T}^{-1}\mathbf{y}$. Once **z** is known, then **x** may be found from the equation $\mathbf{x} = \mathbf{T}\mathbf{z}$.

EXAMPLE 3.15

$$\begin{bmatrix} 5 & \sqrt{2} \\ \sqrt{2} & 4 \end{bmatrix} \begin{bmatrix} x_1 \\ x_2 \end{bmatrix} = \begin{bmatrix} 6 \\ 3 \end{bmatrix}$$

Find x_1 and x_2.

Solution. From Example 3.14,

$$\mathbf{T} = \begin{bmatrix} -1 & \sqrt{2} \\ \sqrt{2} & 1 \end{bmatrix} \qquad \text{where } \lambda_1 = 3, \lambda_2 = 6$$

Then

$$T^{-1} = \begin{bmatrix} 1 & -\sqrt{2} \\ -\sqrt{2} & -1 \end{bmatrix} \frac{1}{-3}$$

$$T^{-1}AT = \Lambda = \begin{bmatrix} 3 & 0 \\ 0 & 6 \end{bmatrix}$$

$$T^{-1}y = -\frac{1}{3}\begin{bmatrix} 1 & -\sqrt{2} \\ -\sqrt{2} & -1 \end{bmatrix}\begin{bmatrix} 6 \\ 3 \end{bmatrix} = \begin{bmatrix} -2 & +\sqrt{2} \\ 2\sqrt{2} & +1 \end{bmatrix}$$

Then from Eq. (3.57),

$$3z_1 = -2 + \sqrt{2}$$
$$6z_2 = 2\sqrt{2} + 1$$

or

$$z = \frac{1}{6}\begin{bmatrix} -4 + 2\sqrt{2} \\ 2\sqrt{2} + 1 \end{bmatrix}$$

and

$$x = \frac{1}{6}\begin{bmatrix} -1 & \sqrt{2} \\ \sqrt{2} & 1 \end{bmatrix}\begin{bmatrix} -4 + 2\sqrt{2} \\ 2\sqrt{2} + 1 \end{bmatrix}$$

$$= \frac{1}{6}\begin{bmatrix} 8 - \sqrt{2} \\ -2\sqrt{2} + 5 \end{bmatrix}$$

This may be checked in the original equation.

Example 3.15 illustrates a technique (diagonalization) which is quite useful in theoretical developments and sometimes practical if A has particular forms. For example, if

$$A = \begin{bmatrix} 0 & 1 & 0 & 0 & \cdots & 0 \\ 0 & 0 & 1 & 0 & \cdots & 0 \\ \cdot & \cdot & \cdot & \cdot & \cdot & \cdot \\ 0 & 0 & 0 & 0 & \cdots & 1 \\ a_{n1} & a_{n2} & a_{n3} & a_{n4} & \cdots & a_{nn} \end{bmatrix}$$

and has distinct eigenvalues $\lambda_1, \lambda_2, \ldots \lambda_n$, then

$$T = \begin{bmatrix} 1 & 1 & \cdots & 1 \\ \lambda_1 & \lambda_2 & \cdots & \lambda_n \\ \lambda_1^2 & \lambda_2^2 & \cdots & \lambda_n^2 \\ \cdot & \cdot & \cdot & \cdot \\ \lambda_1^{n-1} & \lambda_2^{n-1} & \cdots & \lambda_n^{n-1} \end{bmatrix}$$

From the standpoint of the digital computer, eigenvalues are often difficult to accurately evaluate. Although Eq. (3.54) applies only to the case of distinct eigenvalues, a slight change in the coefficients of the polynomial of Eq. (3.53) will result in distinct roots if repeated roots otherwise occur. (This assumes floating-point coefficients. Integer coefficients often give repeated roots.)

A quadratic form Q may be defined as

$$Q = \mathbf{x}^T\mathbf{A}\mathbf{x} = \langle \mathbf{x}, \mathbf{A}\mathbf{x} \rangle \tag{3.58}$$

where $\mathbf{A}$ may be symmetric without loss of generality. Let $\mathbf{x} = \mathbf{T}\mathbf{y}$. Then

$$Q = (\mathbf{T}\mathbf{y})^T\mathbf{A}\mathbf{T}\mathbf{y}$$
$$= \mathbf{y}^T\mathbf{T}^T\mathbf{A}\mathbf{T}\mathbf{y}$$
$$= \mathbf{y}^T\mathbf{A}\mathbf{y}$$

since $\mathbf{T}^{-1} = \mathbf{T}^T$ if $\mathbf{A}$ is symmetric.

EXAMPLE 3.16

Given the quadratic form

$$\langle \mathbf{x}, \mathbf{A}\mathbf{x} \rangle = k$$

where

$$\mathbf{A} = \begin{bmatrix} 1 & -\sqrt{\frac{3}{4}} \\ -\sqrt{\frac{3}{4}} & 1 \end{bmatrix}$$

or

$$x_1^2 - \sqrt{\frac{3}{2}}\, x_1 x_2 + x_2^2 = k$$

find the curves in the $x_1 x_2$ space.

Solution

$$\mathbf{T} = \begin{bmatrix} \dfrac{1}{\sqrt{2}} & \dfrac{1}{\sqrt{2}} \\ -\dfrac{1}{\sqrt{2}} & \dfrac{1}{\sqrt{2}} \end{bmatrix}$$

and the eigenvalues are $\lambda_1 = (1 + \sqrt{\frac{3}{2}})$, $\lambda_2 = (1 - \sqrt{\frac{3}{2}})$. Then if

$$\mathbf{x} = \mathbf{T}\mathbf{y}$$
$$\langle \mathbf{y}, \mathbf{\Lambda}\mathbf{y} \rangle = (1 + \sqrt{\tfrac{3}{2}})y_1^2 + (1 - \sqrt{\tfrac{3}{2}})y_2^2 = k$$

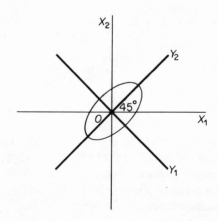

Figure 3.7 Spatial curves for
Example 3.15.

The result describes a family of elipses with major axis along y_2. The y_1, y_2 axis are rotated 45° with respect to the original x_1, x_2 axis as shown in Fig. 3.7.

As you will see, eigenvalues become very fundamental to a network, particularly a network described by linear differential equations. For example, in such a network, the eigenvalues must have negative real parts if the network is to be stable. In network analysis, eigenvalues are also termed *poles*. You will understand these terms and their usage subsequently.

Additional Reading

COMER, D. J., *Computer Analysis of Circuits*, International Textbook Company, Scranton, Pa., 1971.

DERUSSO, P. M., R. J. ROY, and C. M. CLOSE, *State Variables for Engineers*, John Wiley & Sons, Inc., New York, 1965.

FADDEEVA, V. N., *Computational Methods of Linear Algebra*, Dover Publications, Inc., New York, 1959.

FULLER, L. E., *Basic Matrix Theory*, Prentice-Hall, Inc., Englewood Cliffs, N.J., 1962.

GANTMACHER, F. R., *The Theory of Matrices*, Vols. I and II, Chelsea Publishing Company, New York, 1959.

LEY, B. J., *Computer Aided Analysis and Design for Electrical Engineers*, Holt, Rinehart and Winston, Inc., New York, 1970.

MOURSUND, D. G., and C. S. DURIS, *Elementary Theory and Application of Numerical Analysis*, McGraw-Hill Book Company, New York, 1967.

PIPES, L. A., *Matrix Methods for Engineering*, Prentice-Hall, Inc., Englewood Cliffs, N.J., 1963.

RALSTON, A., *A First Course in Numerical Analysis*, McGraw-Hill Book Company, New York, 1965.

TROPPER, A. M., *Matrix Theory for Electrical Engineers*, Addison-Wesley Publishing Company Inc., Reading Mass., 1962.

Problems

3.1. Given

$$
\mathbf{A} = \begin{bmatrix} 1 & 2 & -1 \\ 2 & -5 & 1 \\ 4 & 0 & 2 \end{bmatrix} \quad \mathbf{B} = \begin{bmatrix} 3 & -4 & 1 \\ 1 & 5 & 0 \\ 2 & -2 & 3 \end{bmatrix}
$$

determine

     (a) $\mathbf{A} + \mathbf{B}$
     (b) $\mathbf{A} - \mathbf{B}$
     (c) $\mathbf{AB}$
     (d) $\mathbf{BA}$
     (e) $\mathbf{A}^\mathsf{T}$
     (f) $\mathbf{A}^{-1}$
     (g) $\mathbf{B}^{-1}$

3.2. Find out if the following matrices are conformable for the products $\mathbf{ABC}, \mathbf{BAC}$, $\mathbf{BCA}, \mathbf{CAB}$, and $\mathbf{CBA}$. Find the valid products

$$\mathbf{A} = \begin{bmatrix} 1 & 1 \\ 2 & 5 \end{bmatrix} \qquad \mathbf{B} = \begin{bmatrix} 2 \\ 3 \end{bmatrix} \qquad \mathbf{C} = [6 \quad 1]$$

3.3. Prove that $(\mathbf{A} + \mathbf{B})^\mathsf{T} = \mathbf{A}^\mathsf{T} + \mathbf{B}^\mathsf{T}$.

3.4. Prove that $(\mathbf{AB})^\mathsf{T} = \mathbf{B}^\mathsf{T}\mathbf{A}^\mathsf{T}$.

3.5. Show why $(\mathbf{A} - \mathbf{B})^2 = \mathbf{A}^2 - 2\mathbf{AB} + \mathbf{B}^2$ is not necessarily true in matrix algebra. What is the correct matrix relation? Under what conditions is it true?

3.6. Prove that $(\mathbf{AB})^{-1} = \mathbf{B}^{-1}\mathbf{A}^{-1}$.

3.7. Compute $\mathbf{AB}$, where

$$\mathbf{A} = \begin{bmatrix} 1 & 2 & \vdots & 1 & 2 \\ 0 & 1 & \vdots & 2 & 1 \\ \cdots & \cdots & & \cdots & \cdots \\ 0 & 0 & \vdots & 0 & 1 \\ 0 & 0 & \vdots & 1 & 0 \end{bmatrix} \qquad \mathbf{B} = \begin{bmatrix} 1 & 2 & \vdots & 1 & 2 \\ 2 & 1 & \vdots & 0 & 1 \\ \cdots & \cdots & & \cdots & \cdots \\ 0 & 1 & \vdots & 0 & 0 \\ 1 & 0 & \vdots & 0 & 0 \end{bmatrix}$$

3.8. The partitions on the matrix $\mathbf{A}$ are given. Partition matrices $\mathbf{B}$ and $\mathbf{C}$ and carry out the required operations.

     (a) $\mathbf{A} + \mathbf{B} = \begin{bmatrix} 1 & 1 & -1 & \vdots & 0 \\ 1 & 1 & -1 & \vdots & 2 \\ 2 & 1 & 1 & \vdots & 2 \\ \cdots & \cdots & \cdots & & \cdots \\ 3 & -1 & 0 & \vdots & 1 \end{bmatrix} + \begin{bmatrix} 6 & 3 & 4 & 2 \\ 2 & 1 & 6 & 3 \\ 5 & 1 & 2 & 0 \\ 0 & 0 & 1 & 3 \end{bmatrix}$

     (b) $\mathbf{AB} = \begin{bmatrix} 1 & 2 & \vdots & -1 \\ \cdots & \cdots & & \cdots \\ 2 & -5 & \vdots & 1 \\ 4 & 0 & \vdots & 2 \end{bmatrix} \begin{bmatrix} 3 & -4 & 1 \\ 1 & 5 & 0 \\ 2 & -2 & 3 \end{bmatrix}$

     (c) $\mathbf{ABC} = \begin{bmatrix} 2 & \vdots & 1 & 3 \\ \cdots & & \cdots & \cdots \\ 4 & \vdots & -1 & 2 \end{bmatrix} \begin{bmatrix} -3 & 1 \\ 2 & 5 \\ 0 & 2 \end{bmatrix} \begin{bmatrix} -1 & 2 & 3 & 4 \\ 0 & -1 & 2 & 3 \end{bmatrix}$

3.9. Find the inverse (if they exist) of the following matrices. If the inverse does not exist, explain why.

     (a) $\mathbf{A} = \begin{bmatrix} 1 & 3 & -5 \\ 6 & 0 & 5 \\ 2 & 1 & -1 \end{bmatrix}$

(b) $\mathbf{A} = \begin{bmatrix} s(s+1) & 5 \\ s & 2 \end{bmatrix}$

(c) $\mathbf{A} = \begin{bmatrix} 7 & 3 \\ 5 & 1 \\ 6 & 2 \end{bmatrix}$

(d) $\mathbf{A} = \begin{bmatrix} 1 & 2 & 3 \\ 0 & 1 & 2 \\ 0 & 0 & 1 \end{bmatrix}$

(e) $\mathbf{A} = \begin{bmatrix} 1 & 1 & 1 & 0 \\ 2 & 3 & 3 & 2 \\ 1 & 3 & 3 & 2 \\ 4 & 6 & 7 & 4 \end{bmatrix}$

3.10. Using the Gauss-elimination method, hand-compute the value of the determinant of

$$\begin{bmatrix} 2 & 2 & 1 & 1 \\ 3 & 0 & 1 & 2 \\ 1 & 3 & 4 & 1 \\ 1 & 2 & 1 & 0 \end{bmatrix}$$

3.11. Determine if the set of equations represented in (a) and (b) are linearly independent.

(a) $x_1 + 2x_2 = 4 \ x_3 = 0$

$3x_1 - 4x_2 + 10x_3 = 0$

$2x_1 - x_2 + 3x_3 = 0$

(b) $\begin{bmatrix} 2 & -1 & 2 \\ 3 & 1 & -1 \\ 1 & 1 & 1 \end{bmatrix} \begin{bmatrix} x_1 \\ x_2 \\ x_3 \end{bmatrix} = \begin{bmatrix} 0 \\ 0 \\ 0 \end{bmatrix}$

3.12. Solve the following systems of equations for $\mathbf{x}$:

(a) $3x_1 - x_2 + 6x_3 = 1$

$x_1 + 2x_2 - 3x_3 = 0$

$2x_1 - 3x_2 - x_3 = -9$

(b) $\begin{bmatrix} 2 & -1 & -2 \\ 4 & 1 & 2 \\ 8 & -1 & 1 \end{bmatrix} \begin{bmatrix} x_1 \\ x_2 \\ x_3 \end{bmatrix} = \begin{bmatrix} 5 \\ 1 \\ 5 \end{bmatrix}$

(c) $x_1 - x_2 + 4x_3 - 2x_4 = 3$

$4x_4 - x_1 - x_2 - 2x_3 = 1$

$2x_2 - 4x_3 + 3x_4 + 5 = 0$

$3x_3 + x_4 - x_1 = 2$

(d) $x_1 - x_2 - x_3 - x_4 = 2$

$2x_1 + 4x_2 - 3x_3 = 6$

$$3x_2 - 4x_3 - 2x_4 + \quad 1 = 0$$

$$4x_3 + 3x_4 - 2x_1 + 3 \quad = 0$$

(e) $\quad x_1 + \quad x_2 + \quad x_3 + \quad x_4 = 4$

$$x_1 + 2x_2 + \quad x_3 - \quad x_4 = 0$$

$$2x_1 + \quad x_2 - \quad x_3 + 2x_4 = 0$$

$$3x_1 - 2x_2 + 4x_3 - \quad x_4 = 0$$

3.13. Solve for as many variables as possible in terms of the remaining variables.

(a) $\quad 2x_1 + \quad x_2 + \quad x_3 - \quad x_4 = 0$

$$x_1 + 2x_2 - \quad x_3 - 2x_4 = 0$$

$$-x_1 - \quad x_2 - 6x_3 + \quad x_4 = 0$$

$$2x_1 + 2x_2 - 6x_3 - 2x_4 = 0$$

(b) $\quad x_1 + 2x_2 + 9x_3 = 0$

$$2x_1 \qquad\quad + 2x_3 = 0$$

$$3x_1 - 2x_2 - 5x_3 = 0$$

(c) $\begin{bmatrix} 2 & 0 & 1 & -1 \\ 1 & 3 & 2 & 4 \\ 1 & 1 & 1 & 1 \end{bmatrix} \begin{bmatrix} x_1 \\ x_2 \\ x_3 \\ x_4 \end{bmatrix} = \begin{bmatrix} 0 \\ 0 \\ 0 \end{bmatrix}$

(d) $-4x_1 + 3x_2 = 2$

$$5x_1 - 4x_2 = 0$$

$$2x_1 - \quad x_2 = 5$$

(e) $4x_1 + 5x_3 = 6$

$$x_2 - 6x_3 = -2$$

$$3x_1 + 4x_3 = 3$$

3.14. Assuming that 100 microseconds (μs) are required for each computer multiplication operation and an addition operation takes 10 μs, estimate the time required to solve a set of 10 simultaneous equations using the Gauss-elimination method in Fig. 3.4.

3.15. Compute the number of arithmetic operations required to solve a set of N linear algebraic equations using the following methods:
(a) Gauss-elimination
(b) Gauss–Jordan

3.16. Write a Fortran subroutine that premultiplies a real matrix **A** by the transpose of another real matrix **B**.

3.17. Repeat Problem 3.16 for **A** and **B** with complex elements.

3.18. Modify the subroutine LINQ (Fig. 3.4) to obtain the solution to a system of linear equations by the Gauss–Jordan method.

3.19. Write a Fortran subroutine capable of solving a system of equations ($n \leq 20$) containing complex coefficients. The subroutine should be based upon either the Gauss–

elimination or Gauss–Jordan method and should produce the determinant of the coefficient matrix as a by-product. *Hint:* The subroutine LINQ can be easily converted to perform this function.

3.20. Modify the subroutine INVERT (Fig. 3.5) so that is capable of finding the inverse of a matrix that contains complex elements.

3.21. Find the eigenvalues and normalized eigenvectors of the following matrices:

(a) $\begin{bmatrix} 2 & -3 \\ -1 & 3 \end{bmatrix}$

(b) $\begin{bmatrix} 0 & 1 \\ -3 & -4 \end{bmatrix}$

(c) $\begin{bmatrix} 4 & 3 & -2 \\ -2 & -1 & 3 \\ 4 & 4 & -1 \end{bmatrix}$

3.22. Find a matrix $\mathbf{V}$ to diagonalize the matrix

$$\mathbf{A} = \begin{bmatrix} 0 & 1 & 0 \\ 0 & 0 & 1 \\ -6 & -11 & -6 \end{bmatrix}$$

and find $\mathbf{V}^{-1}$. Show that $\mathbf{V}^{-1}\mathbf{A}\mathbf{V} = \mathbf{\Lambda}$, where

$$\mathbf{\Lambda} = \begin{bmatrix} \lambda_1 & 0 & 0 \\ 0 & \lambda_2 & 0 \\ 0 & 0 & \lambda_3 \end{bmatrix}$$

Chapter 4

LOOP AND NODE
ANALYSIS

. . . Let him beware, however, of confusing simplicity with easiness, or complexity with difficulty. Reduction of anything, including circuits, to the bare essentials is seldom easy; whenever the process is trivial, the original network was probably trivial to begin with.

HAYT & HUGHES, *Introduction to Electrical Engineering*, 1968

4.1 Kirchhoff's Laws

Network analysis is based on fundamental physical laws which have been observed to be valid for the situation involved. Most of these laws relate to basic principles such as conservation of energy, conservation of matter, and conservation of momentum. In particular disciplines, some investigator has restated these basic principles in terms which are most easily applied to that discipline and for a particular set of conditions. As an example, we have D'Alembert's principle in macroscopic mechanics, which is a restatement of Newton's laws of motion, and in hydraulics we have Bernouli's principle. In 1847 G. Kirchhoff published a paper which stated several laws, of which two have been found to be very useful in the analysis of electrical networks with lumped parameters and driving functions at frequencies low enough to allow the lumped-parameter assumption. The first of these laws is known as the current law, the second as the voltage law.

Kirchhoff's current law (KCL) states that the summation of currents into a *node* (also called a *junction* or *vertex*) is equal to zero. A node is an intersection of two or more branches; a branch is any element (see Chapter 2) with two terminals. In Fig. 4.1 we show four elements connected at node *a*, which is depicted as a small

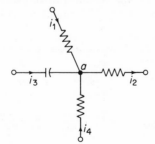

Figure 4.1 Four elements connected to a node.

circle (often filled in as a solid disk or heavy dot). We select the branch current directions arbitrarily and identify these currents as i_1, i_2, i_3, and i_4 in Fig. 4.1. These currents, once drawn on the diagram, become references for an algebraic equation based upon Kirchhoff's current law. From Kirchhoff's current law the net current flow into node *a* in Fig. 4.1 is

$$i_1 + i_3 + i_4 - i_2 = 0 \tag{4.1}$$

Note that in Eq. (4.1) we have defined currents oriented *toward* the node as positive and *away* from the node as negative. Kirchhoff states that currents *into* a node sum to zero. Hence currents *out* of a node have a sign opposite to those *into* a node, and thus i_2 has a negative sign. We could just as well have made the opposite choice, that is, defined currents toward the node as negative. Then the Kirchhoff equations would have been

$$-i_1 - i_3 - i_4 + i_2 = 0 \qquad (4.2)$$

If you compare Eq. (4.1) to Eq. (4.2) you see that they are algebraically identical. The choice of a "positive" or "negative" current direction with respect to a node is arbitrary, but once a choice is made, the definition must not be changed with respect to any particular node as you write the equation.

In some cases, a current may exist without an element, such as a beam of electrons in an oscilloscope tube, for example. Kirchhoff's law still holds with respect to the currents in such a case. Sometimes KCL is stated as: "The sum of currents directed out of a node must equal the current directed in." Application of this rule would merely move i_2 as a positive term to the right of Eq. (4.1). Perhaps in this form it is easier to see the basis of the current law, for if charge does not flow out of a node as fast as it flows into the node, a positive or negative charge (an excess or deficiency of electrons) will pile up on the node. If the charge is positive, then the potential of the node will increase (with respect to a zero reference potential). We exclude this possibility in network analysis by ruling out the presence of static potential charge at a node. Hence, the current law is based on the conservation of charge at a node. Charge may "pile up" on one plate of a capacitance but, charge is also drawn away from the other plate, and the total network charge is conserved. Capacitance plates are not considered nodes, although each plate may be connected to a node.

Kirchhoff's voltage law (KVL) states that the sum of the voltages around a closed loop (circuit) is zero. In Fig. 4.2, we show a simple, two-loop network, with element or branch voltages labeled $v_1, v_2, \ldots$, together with their defined polarities. The

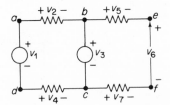

Figure 4.2 Two-loop network.

arbitrary voltage polarities are analogous to the current directions we defined in Fig. 4.1. Kirchhoff's voltage law states that if we proceed in a closed path around the loop *abcd*, taking a voltage as positive as we pass from a positive potential to a negative potential in each element, we will obtain the equation

$$v_2 + v_3 - v_4 - v_1 = 0 \qquad (4.3)$$

If we proceed around the loop in the opposite direction, or in the order *adcb*, we obtain

$$v_1 + v_4 - v_2 - v_3 = 0 \qquad (4.4)$$

which is algebraically the same as Eq. (4.4). In using the voltage law, there need not be elements all around the loop. For example, we could sum the voltages around the loop *abefcd* to obtain

$$v_2 + v_5 + v_6 - v_7 - v_4 - v_1 = 0 \qquad (4.5)$$

where v_6 will have a nonzero numerical value ($v_3 \neq 0$) even though no element exists between nodes e and f.

As we solve equations such as Eqs. (4.2) or (4.3), we find that currents have either positive or negative numerical values, and as a matter of fact, these values may change with time. Thus, in Fig. 4.2 we may get $v_4 = -3$ V at $t = 0$. (To take care of the matter of time variation, the figure may be considered to be drawn for a fixed time, either specified or understood.) These negative (or positive) numerical values are simply a result of the arbitrary selection of our reference voltage as placed on the diagram.

Kirchhoff's voltage law stems from the basic principle of conservation of energy. Since voltage is merely a measure of energy per unit charge, consequently the sum of voltages from a point around a loop back to the same point must be zero. (We assume that the voltages are constant or changing so slowly that energy does not become radiated. For rapidly changing currents and voltages, we must apply the more general Maxwell's laws.)

Kirchhoff's laws in themselves are not sufficient to analyze networks. In addition, we need the basic definition of electrical elements as shown in Chapter 2. The element definitions *and* the laws may now be used to interrelate currents and voltages and so solve the network equations.

4.2 Equivalent Networks

It is often possible and desirable to reduce a portion of a network into a simpler, more convenient form which is equivalent in some respects to the original network. This may occur when, in order to study the effects of a particular resistance value on a circuit's performance, we wish to reduce the circuit to a minimum number of components. Also, as we will show in the following sections, it is always easier to write the network equations in terms of node voltages if the independent and dependent sources in the network are current sources. Similarly, loop analysis becomes simpler if the sources are voltage sources. The "changes" in a network to transform it to an equivalent network of the desired form are made by applying the network theorems of Thévenin and Norton, and by making parallel–series element combinations.

An equivalent network can be defined as follows: Two networks are equivalent with respect to terminals $a_1, a_2, \ldots, a_n$ if they impose the same constraints on the voltages and currents associated with terminals $a_1, a_2, \ldots, a_n$.

EXAMPLE 4.1

Consider the resistive network in Fig. 4.3a. By Kirchhoff's current law (KCL) the total current i in the network equals the sum of the currents in resistances R_2 and

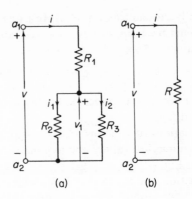

Figure 4.3 Resistive network
              equivalents.   (a)   (b)

R_3, or

$$i = i_1 + i_2 \tag{4.6}$$

Using Kirchhoff's voltage law (KVL) and the definition of resistance, the voltage drop v across the network is given by

$$v = R_1 i + R_2 i_1 \tag{4.7}$$

Since v_1 is common to R_2 and R_3, $R_2 i_1 = R_3 i_2$, and the total current i becomes

$$i = \left(1 + \frac{R_2}{R_3}\right) i_1 \tag{4.8}$$

Substituting Eq. (4.8) into Eq. (4.7) the former equation reduces to one involving only the terminal variables, v and i. The resulting equation is

$$
\begin{aligned}
v &= R_1 i + \frac{R_2 R_3}{R_2 + R_3} i \\
  &= \left(R_1 + \frac{R_2 R_3}{R_2 + R_3}\right) i
\end{aligned}
\tag{4.9}
$$

Thus Fig. 4.3b is equivalent to Fig. 4.3a if we make

$$R = R_1 + \frac{R_2 R_3}{R_2 + R_3} \tag{4.10}$$

The two networks are equivalent with respect to terminals a_1 and a_2 since we have identical voltage–current equations at these terminals.

Thévenin's Theorem

We can extend the idea of equivalent networks to include active networks containing dependent and independent sources by the application of a powerful analysis tool known as *Thévenin's theorem*. We will not prove the theorem; however, we will demonstrate its validity with specific examples.

Thévenin's theorem can be stated in two parts:

1. Any terminal pair in a network composed of active sources and linear passive elements can be represented relative to these terminals by an ideal voltage

source $E_T(t)$ in series with an impedance Z_T. The voltage source and impedance are equivalent to the original network with respect to any element or other network now connected to the two terminals.

2. The voltage source $E_T(t)$ is the open-circuit potential difference measured between the terminal pair mentioned above, and the impedance Z_T is the impedance measured "looking into" the pair of terminals in question, with all ideal *independent* sources inside the network set to zero. This is equivalent to replacing all nonideal independent sources with their internal impedances.† The impedance Z_T is defined as the *driving-point impedance*. Note that we do not set dependent sources to zero.

To provide later generality we use here the term *impedance*, which means the general relation between an element voltage and its current for time-varying conditions. For present illustrations we use constant currents and voltages, for which passive impedance becomes resistance.

EXAMPLE 4.2

Consider the active network in Fig. 4.4. The open-circuit voltage v_{oc} in part a of the figure is

$$v_{oc} = -JR_2$$
$$= -\alpha i R_2$$

or

$$v_{oc} = -\alpha R_2 \frac{E_s}{R_1} \tag{4.11}$$

Hence, equating the open-circuit voltage in Eq. (4.11) to the Thévenin voltage yields

$$E_T = -\frac{\alpha R_2}{R_1} E_s \tag{4.12}$$

for the voltage source in Fig. 4.4b.

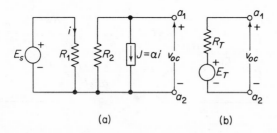

Figure 4.4 Thévenin's theorem example.

To obtain the Thévenin resistance R_T of the network in Fig. 4.4b, we first set the input source voltage E_s to zero as shown in Fig. 4.5 and apply an external voltage E_0 between terminals a_1 and a_2. The current i_0 and the source E_0 define the driving-point

---

†The impedance of an ideal voltage source is zero and of an ideal current source is infinite.

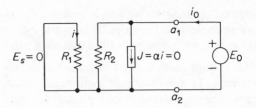

Figure 4.5 Thévenin resistance calculation.

resistance R_T of the network between terminals a_1 and a_2. Thus, from Fig. 4.4(a),

$$\frac{E_0}{i_0} = R_2 \tag{4.13}$$

The Thévenin resistance for the equivalent network in Fig. 4.4b is then

$$R_T = R_2 \tag{4.14}$$

Note that when applying this method, only the independent sources are set to zero. In the preceding example we did not arbitrarily set the current-controlled current source to zero as part of the procedure; it was coincidental in this particular example that the control current went to zero when $E_s = 0$ (For counterexamples see Problems 4.4, 4.5, and 4.8.)

At this point we have demonstrated that the networks at Figs. 4.4a and b are equivalent with the terminals a_1 and a_2 open-circuited and with $E_T = -\alpha R_2 E_s / R_1$ and $R_T = R_2$. More precisely, we have shown that they have the same open-circuit voltage and indirectly, the same short-circuit current between a_1 and a_2. We next demonstrate that they have the same terminal characteristics with respect to a network N connected across terminals a_1 and a_2. We term the network N a *load network*.

EXAMPLE 4.3

In Example 4.2 connect a network N consisting of a source E_1 in series with R_3 across terminals a_1 and a_2 as shown in Fig. 4.6b. We claim that the load current i_L and voltage v at terminals $a_1 a_2$ are equivalent to those in Fig. 4.6a (that is, i_L and v are equal in both figures).

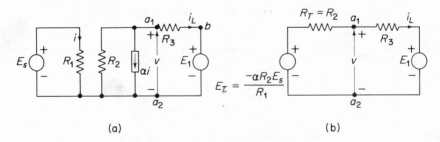

(a)   (b)

Figure 4.6 Thévenin's theorem application.

Proof. In Fig. 4.6a the current through the dependent source is

$$J = \alpha i$$

$$= \frac{\alpha E_s}{R_1}$$

Then using Kirchhoff's current law at node a_1 in Fig. 4.6a gives the current i_L out of terminal a_1 as

$$i_L = -\frac{v}{R_2} - \frac{\alpha E_s}{R_1} \tag{4.15}$$

Using Kirchhoff's voltage law around the loop $a_1 b a_2$ in the same figure results in the equation

$$i_L R_3 + E - v = 0$$

or

$$i_L = \frac{v - E_1}{R} \tag{4.16}$$

Setting the right-hand sides of Eqs. (4.15) and (4.16) equal and collecting terms we obtain

$$v\frac{R_2 + R_3}{R_2 R_3} = -\frac{\alpha E_s}{R_1} + \frac{E_1}{R_3} \tag{4.17}$$

Solving for v in Eq. (4.17) we have

$$v = \frac{-\alpha R_2 R_3}{R_1(R_2 + R_3)} E_s + \frac{R_2}{R_2 + R_3} E_1 \tag{4.18}$$

By Kirchhoff's voltage law in Fig. 4.6b, $i_L(R_T + R_3) - E_T + E_1 = 0$, or the loop current i_L is given by

$$i_L = \frac{E_T - E_1}{R_T + R_3} \tag{4.19}$$

From Kirchhoff's voltage law in Fig. 4.6b, $i_L R_3 + E_1 - v = 0$, or the terminal voltage in Fig. 4.6b can be expressed as

$$v = E_1 + i_L R_3$$

$$= E_1 + \frac{E_T - E_1}{R_2 + R_3} R_3$$

or

$$v = -\frac{\alpha R_2 R_3}{R_1(R_2 + R_3)} E_s + \frac{R_2}{R_2 + R_3} E_1$$

Furthermore, i_L from Fig. 4.6b is identical to Eq. (4.16) as obtained from Fig. 4.6a. Thus, the two networks are equivalent insofar as the current and voltage at the terminals $a_1 a_2$ and into the load are concerned. Note that the altered network in Figs. 4.6b to the left of the terminals $a_1 a_2$ is not equivalent to Fig. 4.6a; that is, source E_s carries current i, while source E_T carries current i_L ($\neq i$), and the power generated by E_s is not equal to the power generated by E_T. In addition, the current and power in the R_2 of Fig. 4.6a is not the same as in R_T in Fig. 4.6b.

From the example note that the Thévenin circuit is equivalent to the original network only insofar as the terminal conditions at the attached network (load) are concerned.

Norton's Theorem

Norton's theorem is essentially the dual of Thévenin's theorem and can be written by paraphrasing Thévenin's theorem:

1. Any terminal pair of a network composed of active sources and linear passive elements can be represented relative to these terminals by an ideal independent current source $J_N(t)$ in parallel with an impedance Z_N.

2. The magnitude of the current source $J_N(t)$ equals that of the current I which flows in a short circuit placed at the terminal pair, and the source direction is such as to cause a short-circuit current at the source terminals to flow in the same direction as the previous I. The impedance Z_N is the impedance measured looking into the pair of terminals with all ideal *independent* sources set to zero. This is equivalent to replacing all nonideal independent sources with their internal impedances. Dependent sources remain in the circuit.

We can demonstrate the validity of Norton's theorem as we did Thévenin's theorem, but our primary interest with the Norton equivalent circuit is its relationship to the Thévenin equivalent circuit. We leave the demonstration of the validity of Norton's theorem to the reader as an exercise.

Consider the Thévenin and Norton circuits shown in Fig. 4.7.† If we place a

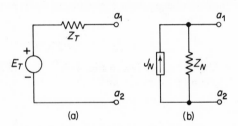

Figure 4.7 Thévenin and Norton equivalent circuits.

short-circuit across terminals a_1 and a_2 of the networks in Fig. 4.7, the current i_{sh} in the short circuit from a_1 to a_2 in Fig. 4.7a will be

$$i_{\text{sh}} = \frac{E_T}{Z_T}$$

while that in Fig. 4.7b will be

$$i_{\text{sh}} = J_N$$

Thus,

$$J_N = \frac{E_T}{Z_T} \tag{4.20}$$

The open-circuit voltage v_{oc} in Fig. 4.7b is

$$v_{\text{oc}} = J_N Z_N$$

and in Fig. 2.7a is

$$v_{\text{oc}} = E_T$$

Hence

$$E_T = J_N Z_N \tag{4.21}$$

---

†Z here stands for general impedance.

Comparing Eqs. (4.20) and (4.21), we see that

$$Z_N = Z_T \tag{4.22}$$

Let $Z_N = Z_T = Z$. Then the transformation between the Thévenin and Norton equivalent circuits is based upon the relationship

$$E_T = ZJ_N \tag{4.23a}$$

Also from Eq. (4.23a),

$$Z = \frac{E_T}{J_N} \tag{4.23b}$$

Equation (4.23b) may be useful in determining Z if dependent sources are present, since such sources may effect Z (see Problems 4.8 and 4.10).

The Thévenin network, incorporating a voltage source and series elements, is very useful when using the mesh (loop)-analysis methods described in the following sections. When using nodal-analysis methods it is generally advantageous to transform the network to a form containing only current sources and parallel elements.

EXAMPLE 4.4

In Fig. 4.8a find the Thévenin and Norton equivalent circuits of the resistive network N corresponding to the terminals a_1 and a_2 by experiment.

Solution. First measure the open-circuit voltage across terminals a_1 and a_2 with a high-resistance voltmeter or oscilloscope as in Fig. 4.8b. Call this voltage v_{oc}. Now place a known resistance R across the terminals as in Fig. 4.8c, and measure v. Then in the Thévenin equivalent circuit of Fig. 4.8d,

$$i = \frac{v}{R} = \frac{v_{oc}}{R + R_T}$$

Hence,

$$R_T = \frac{v_{oc} - v}{v} R$$

The Norton equivalent circuit is shown in Fig. 4.8e.
Note: Don't make R too small!

EXAMPLE 4.5

As an example of the transformation of an element in series with a voltage source to an element in parallel with a current source, consider the network shown in Fig. 4.9. The voltage v across the terminals a_1 and a_2 is

$$v(t) = \frac{1}{C_1} \int_0^t i\, dt + E_s(t) \tag{4.24}$$

where $v_{C_1} = 0$ when $t = 0$. Multiplying Eq. (4.24) by C_1 and differentiating with respect to time, we obtain

$$C_1 \frac{dv(t)}{dt} = i(t) + C_1 \frac{dE_s(t)}{dt} \tag{4.25}$$

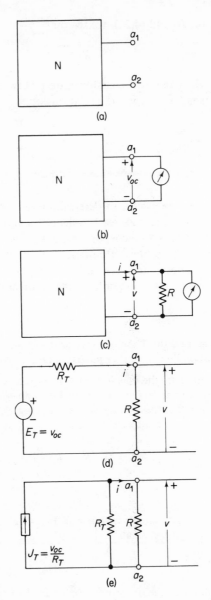

Figure 4.8 Thévenin and Norton circuit tests.

Solving Eq. (4.25) for $i(t)$,

$$i(t) = C_1 \frac{dv(t)}{dt} - C_1 \frac{dE_s(t)}{dt} \qquad (4.26)$$

The first term on the right-hand side of Eq. (4.26) corresponds to a capacitance across terminals a_1 and a_2. Thus,

$$C_2 = C_1 \qquad (4.27)$$

in Fig. 4.8b. Also the second term on the right-hand side of Eq. (4.26) corresponds to

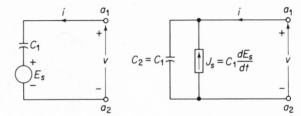

Figure 4.9 Series-parallel source transformation with capacitance element.

an independent current source oriented such that

$$J_s = C_1 \frac{dE_s(t)}{dt} \tag{4.28}$$

in Fig. 4.8b. Care must be taken that the reference current voltage and current directions are correct.

A similar transformation exists for a voltage source in series with an inductance element.

Source Transformations

In our discussions of methods of network analysis we assume that the number and location of independent and dependent sources is arbitrary as long as Kirchhoff's first and second laws are not violated (that is, as long as no groups of voltage sources form a loop and no node contains only current sources).

To aid in the formulation of the network equations by removing all branches containing only sources, it is useful to introduce two network transformations that allow us to relocate sources in the network without altering the analysis problem. These transformations are frequently referred to as the *E shift* and the *J shift*.

The *E* shift is illustrated in Fig. 4.10. The voltage source of interest is E_1 connected between nodes 1 and 2 of the network in Fig. 4.10a. If the current through the source E_1 is of no interest to us, we can replace the network with its equivalent in Fig. 4.10b. In the new circuit, the voltage source branch has been eliminated and a new node 1', which is a combination of nodes 1 and 2, is introduced. The network of Fig. 4.10b is constructed by essentially "shifting" the voltage source E_1 through node 2 into the branches containing R_2 and C_1.

It is not difficult to show that the transformation (*E* shift) is valid. We need only write the KVL equations for all the loops containing R_2 and all loops containing C_2 in both networks. The corresponding equations for both networks in Fig. 4.10 are the same. Also KCL applied to node 1' yields the same results as obtained at nodes 1 and 2 of the original network.

The *J* shift is illustrated in Fig. 4.11. The current source J_1 of interest is connected between nodes 1 and 3 of the network in Fig. 4.11. An equivalent network in which the original current source J_1 has been replaced by two current sources of the same value is shown in Fig. 4.11. We can verify that this transformation does not change the network by writing the KCL equations at each node for the networks of parts a

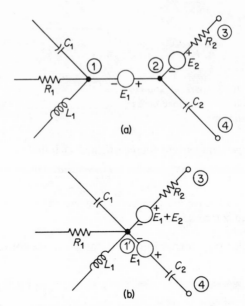

Figure 4.10 The E shift.

and b of the figure. Clearly, the corresponding equations are the same. An alternative transformation of the current source J_1 is presented in Fig. 4.11c to illustrate that the transformation in Fig. 4.11b is not unique. Again, the KCL equations are identical to the original set.

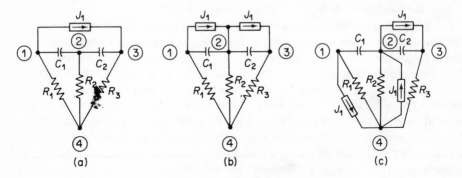

Figure 4.11 The J shift.

4.3 Nodal Analysis

In nodal analysis we apply the current law to determine the node voltages in the network. We illustrate by first using circuits containing only resistances with direct current (dc) voltages and currents. Figure 4.12 contains a four-node circuit, with the source current J and resistances R_1, R_2, R_3, and R_4 given. We connect one node (say node 3) to zero potential (commonly called *ground*) and refer all other node voltages to the ground voltage. Since nodes 3 and 4 are at the same potential, we arrive at the network in Fig. 4.13. We must then determine the voltages v_1 and v_2 of the remaining

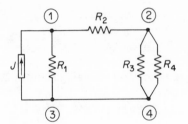

Figure 4.12 Four-node circuit.

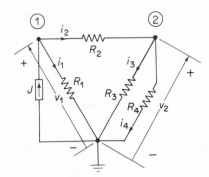

Figure 4.13 The circuit of Fig. 4.12 with nodes 3 and 4 grounded.

two nodes. The reference polarities of v_1 and v_2 are chosen arbitrarily; however, the selection of the voltages at nodes 1 and 2 as being *above* the ground node in potential simplifies the rules for the selection of matrix elements in later equations. We next define the branch currents i_1, i_2, i_3, and i_4. The directions of the branch currents satisfy the definition of resistance given in Section 2.2 namely, $R_1 = v_1/i_1$, $R_2 = (v_1 - v_2)/i_2$, $R_3 = v_2/i_3$, and $R_4 = v_2/i_4$. Then at node 1, we define those currents entering the node as positive to get

$$J - i_1 - i_2 = 0 \qquad (4.29a)$$

or

$$J = \frac{v_1}{R_1} - \frac{v_1 - v_2}{R_2} = 0 \qquad (4.29b)$$

Combining terms we then obtain

$$v_1 \left(\frac{1}{R_1} + \frac{1}{R_2} \right) - v_2 \frac{1}{R_2} = J \qquad (4.30)$$

Clearly we could also write Eq. (4.30) in terms of conductances as

$$v_1(G_1 + G_2) - v_2 G_2 = J \qquad (4.31)$$

where

$$G_1 = \frac{1}{R_1} \quad \text{and} \quad G_2 = \frac{1}{R_2}$$

Applying the same technique with node 2 we find

$$i_2 - i_3 - i_4 = 0 \qquad (4.32a)$$

or

$$\frac{v_1 - v_2}{R_2} - \frac{v_2}{R_3} - \frac{v_2}{R_4} = 0 \tag{4.32b}$$

and finally, combining terms and changing signs in Eq. (4.32b),

$$-v_1 G_2 + v_2(G_2 + G_3 + G_4) = 0 \tag{4.33}$$

where

$$G_2 = \frac{1}{R_2} \quad G_3 = \frac{1}{R_3} \quad \text{and} \quad G_4 = \frac{1}{R_4}$$

We now have two equations and two unknowns as follows:

$$v_1(G_1 + G_2) - v_2 G_2 = J \tag{4.34a}$$
$$-v_2(G_2) + v_2(G_2 + G_3 + G_4) = 0 \tag{4.34b}$$

or in matrix form,

$$\begin{bmatrix} G_1 + G_2 & -G_2 \\ -G_2 & G_2 + G_3 + G_4 \end{bmatrix} \begin{bmatrix} v_1 \\ v_2 \end{bmatrix} = \begin{bmatrix} J \\ 0 \end{bmatrix} \tag{4.35a}$$

or

$$\mathbf{G}\mathbf{v} = \mathbf{J} \tag{4.35b}$$

where $\mathbf{G} = [g_{ij}]$ is the nodal conductance matrix, $\mathbf{v}$ is the node voltage vector, and $\mathbf{J}$ is the equivalent current vector. The meaning of the "equivalent" current vector will become more apparent as the development progresses. We solve the system of equations in Eq. (4.35) either by an elimination method or by matrix inversion. Once v_1 and v_2 are known, all branch currents and power dissipations may be calculated.

EXAMPLE 4.6

Let $R_1 = 0.2$, $R_2 = 0.333$, $R_3 = 0.5$, $R_4 = 0.1$, and $J = 10$ in Fig. 4.12. Find all voltages, current, and element power dissipations.

Solution

$$G_1 = 5 \quad G_2 = 3 \quad G_3 = 2 \quad G_4 = 10$$

Equation (4.34) can be rewritten

$$v_1(8) - v_2(3) = 10 \tag{4.36a}$$
$$-v_1(3) + v_2(15) = 0 \tag{4.36b}$$

From Eq. (4.36b)

$$v_1 = v_2(\tfrac{15}{3})$$

$$v_1 = 5v_2$$

Substituting this result into Eq. (4.36a),

$$5v_2(8) - 3v_2 = 10$$

or

$$v_2 = \tfrac{10}{37}$$

$$v_1 = \tfrac{50}{37}$$

Then $i_1 = \frac{250}{37}$, $i_2 = \frac{40}{37}(3) = \frac{120}{37}$, $i_3 = \frac{20}{37}$, $i_4 = \frac{100}{37}$. The power in each resistor is easily calculated by Eq. (2.3). Note that the voltage across R_2 is $v_1 - v_2$. As a check, we apply KCL at node 1,

$$10 - \frac{250}{37} - \frac{120}{37} = \frac{370}{37} - \frac{370}{37} = 0$$

and at node 2,

$$\frac{120}{37} - \frac{20}{37} - \frac{100}{37} = 0$$

Using matrix methods, from Eq. (4.35) we obtain

$$\begin{bmatrix} 8 & -3 \\ -3 & 15 \end{bmatrix} \begin{bmatrix} v_1 \\ v_2 \end{bmatrix} = \begin{bmatrix} 10 \\ 0 \end{bmatrix}$$

Inverting the 2×2 conductance matrix and solving for v,

$$\begin{bmatrix} v_1 \\ v_2 \end{bmatrix} = \frac{1}{111} \begin{bmatrix} 15 & 3 \\ 3 & 8 \end{bmatrix} \begin{bmatrix} 10 \\ 0 \end{bmatrix}$$

or

$$v_1 = \frac{15(10)}{111} = \frac{150}{111} = \frac{50}{37}$$

$$v_2 = \frac{30}{111} = \frac{10}{37}$$

We can define the sum of all conductances attached to a node as the *self-conductance* of that node, and the sum of all conductance from the given node to another node as the *mutual conductance* between these nodes. Looking back at the nodal conductance matrix in Eq. (4.35), we see that the main diagonal elements g_{ii} are formed of the self-conductances at node i and the off-diagonal elements $g_{ij} = g_{ji}$ are formed with the mutual conductances (with negative sign) between nodes i and j while the equivalent current vector on the right side of the equation contains the current sources J_i *entering* node i (in order). We can organize the above procedure and establish the following recipe:

1. Select one node as the ground, zero potential, or reference node.

2. Assign sequential numbers to the remaining nodes in any arbitrary order.

3. Adopt the unknown voltages of the nodes of step 2 as a voltage vector $\mathbf{v} = [v_1, v_2, \ldots, v_n]^T$, where $v_1, v_2, \ldots, v_n$ are above the reference node in potential.

4. Write the matrix equation for the network as

$$
\begin{array}{c}
\begin{array}{cccc} 1 & 2 & \cdots & n \end{array} \\
\begin{array}{c} 1 \\ 2 \\ \cdots \\ n \end{array}
\begin{bmatrix} g_{11} & g_{12} & \cdots & g_{1n} \\ g_{21} & g_{22} & \cdots & g_{2n} \\ \cdot & \cdot & \cdots & \cdot \\ g_{n1} & g_{n2} & \cdots & g_{nn} \end{bmatrix}
\begin{bmatrix} v_1 \\ v_2 \\ \cdots \\ v_n \end{bmatrix}
=
\begin{bmatrix} J_1 \\ J_2 \\ \cdots \\ J_n \end{bmatrix}
\end{array}
\tag{4.37}
$$

or

$$\mathbf{Gv} = \mathbf{J} \tag{4.38}$$

where **G** is the nodal conductance matrix and the elements of the equivalent current vector on the right-hand side of Eq. (4.37) are the source currents entering the respective nodes. We have numbered the rows and columns of the **G** matrix to assist in locating elements.

5. The elements $g_{11}, g_{22}, g_{33}, \ldots, g_{nn}$ are self-conductances of nodes $1, 2, 3, \ldots, n$, respectively.

6. The elements $g_{ij} = g_{ji}$ are equal to the mutual conductance between node i and node j, but each has a negative sign. The **G** matrix of Eq. (4.38) is symmetric for a network containing no dependent source or coupled elements and most off-diagonal elements are equal to zero. In this case the matrix is said to be sparse. The signs of the matrix elements result from the selection of the potentials of the node voltages as being positive with respect to ground.

EXAMPLE 4.7

Given the network of Fig. 4.14, where the independent current sources are specified in amperes and the conductances are specified in mhos, write the network equations.

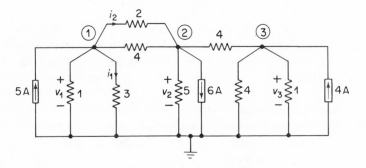

Figure 4.14 Example 4.7 resistive network.

Solution. We first ground the bottom node of Fig. 4.14 as shown. Then we write, by inspection, the matrix equation

$$
\begin{array}{ccc}
 & 1 & 2 & 3
\end{array}
$$

$$
\begin{array}{c}
1 \\ 2 \\ 3
\end{array}
\begin{bmatrix}
4+6 & -6 & 0 \\
-6 & 5+4+6 & -4 \\
0 & -4 & 4+5
\end{bmatrix}
\begin{bmatrix}
v_1 \\ v_2 \\ v_3
\end{bmatrix}
=
\begin{bmatrix}
5 \\ -6 \\ 4
\end{bmatrix}
\tag{4.39}
$$

where we have numbered the rows and columns along the matrix as our aid in locating the elements. In the source vector on the right-hand side of the equation, the second component is -6, since in this case the source current *leaves* node 2. There is no branch between nodes 1 and 3 and hence $g_{13} = g_{31} = 0$.

Suppose that we have a current source at node 3 of Fig. 4.14, with current proportional to the current i_1 in the 3-℧ conductance connected between nodes 1 and ground, or $J_3 = \alpha i_1$, where J_3 is a current into node 3, as shown in Fig. 4.15. Then the third component of the current vector contains the term αi_1. But $i_1 = v_1(3)$. Hence Eq.

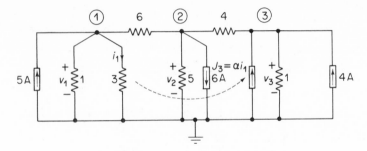

Figure 4.15 Active resistive network.

(4.39) becomes

$$\begin{bmatrix} 10 & -6 & 0 \\ -6 & 15 & -4 \\ 0 & -4 & 9 \end{bmatrix} \begin{bmatrix} v_1 \\ v_2 \\ v_3 \end{bmatrix} = \begin{bmatrix} 5 \\ -6 \\ 4 + 3\alpha v_1 \end{bmatrix} \qquad (4.40)$$

or

$$\begin{aligned} 10v_1 - 6v_2 + 0v_3 &= 5 \\ -6v_1 + 15v_2 - 4v_3 &= -6 \\ 0v_1 - 4v_2 + 9v_3 &= 4 + 3\alpha v_1 \end{aligned} \qquad (4.41)$$

In order to clear the right-hand side of Eq. (4.41), we rearrange it as

$$\begin{aligned} 10v_1 - 6v_2 + 0v_3 &= 5 \\ -6v_1 + 15v_2 - 4v_3 &= -6 \\ -3\alpha v_1 - 4v_2 + 9v_3 &= 4 \end{aligned}$$

or in matrix form,

$$\begin{array}{ccc} 1 & 2 & 3 \end{array}$$
$$\begin{array}{c} 1 \\ 2 \\ 3 \end{array} \begin{bmatrix} 10 & -6 & 0 \\ -6 & 15 & -4 \\ -3\alpha & -4 & 9 \end{bmatrix} \begin{bmatrix} v_1 \\ v_2 \\ v_3 \end{bmatrix} = \begin{bmatrix} 5 \\ -6 \\ 4 \end{bmatrix} \qquad (4.42)$$

We see from Eq. (4.42) that the recipe we established must be altered somewhat. First, only independent current sources may appear on the right-hand side of Eq. (4.42). Then a dependent current source into node j, which depends on the voltage at node i, brings about an addition or subtraction from element g_{ji} of the **G** matrix. If the source depends on a current in an element connected from node i to ground, then the dependency factor α becomes multiplied by the conductance of the element and the product is subtracted from g_{ji}. If the dependent current source depends on v_i directly as $J = g_m v_i$, then only the transconductance g_m is subtracted from g_{ji}.

We might have a current into node 3 dependent on current i_2 in the 2-℧ element of Fig. 4.14 as shown in Fig. 4.16. Rewriting Eqs. (4.41), including the transfluence dependent upon i_2, we obtain

$$\begin{aligned} 10v_1 - 6v_2 + 0v_3 &= 5 \\ -6v_1 + 15v_2 - 4v_3 &= -6 \\ 0v_1 - 4v_2 + 9v_3 &= 4 + 2\alpha(v_1 - v_2) \end{aligned} \qquad (4.43)$$

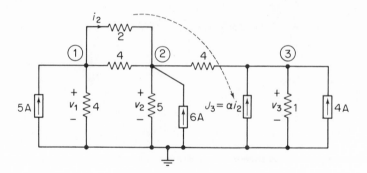

Figure 4.16 Modified active resistive network.

or

$$10v_1 - 6v_2 + 0v_3 = 5$$

$$-6v_1 + 15v_2 - 4v_3 = -6$$

$$-2\alpha v_1 + (-4 + 2\alpha)v_2 + 9v_3 = 4$$

In matrix form we obtain

$$
\begin{array}{ccc}
1 & 2 & 3
\end{array}
$$
$$
\begin{array}{c}
1 \\ 2 \\ 3
\end{array}
\begin{bmatrix}
10 & -6 & 0 \\
-6 & 15 & -4 \\
-2\alpha & -4+2\alpha & 9
\end{bmatrix}
\begin{bmatrix}
v_1 \\ v_2 \\ v_3
\end{bmatrix}
=
\begin{bmatrix}
5 \\ -6 \\ 4
\end{bmatrix}
$$

Now we see that a current source into node j which depends on the voltage at nodes i and k is involved in both g_{ji} and g_{jk}. If the current source depends on a current from node i to node k, then a subtraction is made from g_{ji} and an addition (of the same amount) to g_{jk}. Again the amount of change is the transfluence α times the conductance of the branch in which the current causing the dependent source current flows. If the current source depends on $(v_i - v_k)$, then only the transconductance g_m appears. Actually this rule covers the previous case also, since in that situation node k is ground, or does not appear in the **G** matrix. The transfluence α may be either positive or negative in value.

To prevent errors in sign and to select the correct conductance matrix elements to modify, it is usually preferable to first write the dependent sources as terms in the source vector, such as in Eqs. (4.40) or (4.43). Then as a second step these dependent source terms are transferred to the correct columns determined by the voltage subscript.

Independent voltage sources must be converted to equivalent current sources as discussed in Section 4.2 to correctly write node equations by inspection. If we had the network shown in Fig. 4.17 we would convert E_1 to an independent current source and E_4 to an independent current source as described in Section 4.2 to obtain the equivalent network in Fig. 4.18. The solution then proceeds as we have already described. Dependent voltage sources are handled in a similar fashion.

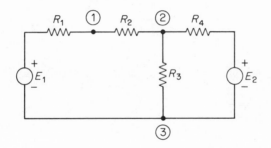

Figure 4.17 Resistive network with
independent voltage sources.

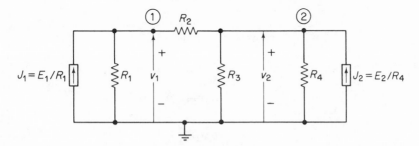

Figure 4.18 Network equivalent to resistive
network in Fig. 4.17.

EXAMPLE 4.8

Given the network shown in Fig. 4.19, in which all element values are specified in mhos. Find node voltages v_1, v_2, and v_3 for the network with respect to the reference potential at node 4.

Solution. We first convert the dependent voltage source to a dependent current source and obtain the network shown in Fig. 4.20. The matrix equation without the dependent

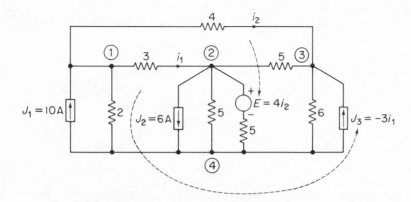

Figure 4.19 Network for Example 4.8.

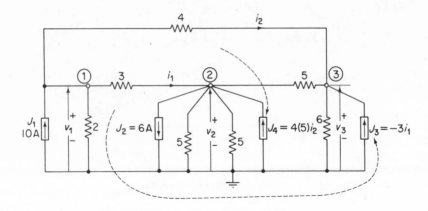

Figure 4.20 Nodal analysis equivalent
network for Example 4.8.

sources becomes

$$
\begin{array}{c}
\quad 1 \quad\ \ 2 \quad\ 3 \\
\begin{array}{c} 1 \\ 2 \\ 3 \end{array}
\left[\begin{array}{rrr}
9 & -3 & -4 \\
-3 & 18 & -5 \\
-4 & -5 & 15
\end{array}\right]
\left[\begin{array}{c} v_1 \\ v_2 \\ v_3 \end{array}\right]
=
\left[\begin{array}{r} 10 \\ -6 \\ 0 \end{array}\right]
\end{array}
\qquad (4.44)
$$

where **G** is symmetric but not sparse in this case.

There is a dependent source $J_3 = -3i_1 = -3(3)(v_1 - v_2)$ into node 3. Hence we modify Eq. (4.44) by adding these dependent sources to the equivalent current source vector, or

$$
\begin{array}{c}
\quad 1 \quad\ \ 2 \quad\ 3 \\
\begin{array}{c} 1 \\ 2 \\ 3 \end{array}
\left[\begin{array}{rrr}
9 & -3 & -4 \\
-3 & 18 & -5 \\
-4 & -5 & 15
\end{array}\right]
\left[\begin{array}{c} v_1 \\ v_2 \\ v_3 \end{array}\right]
=
\left[\begin{array}{c} 10 \\ -6 + 80(v_1 - v_3) \\ 0 - 9(v_1 - v_2) \end{array}\right]
\end{array}
$$

The first row of **G** is unaffected. In the second row the $80v_1$ in the source vector transfers into column 1 of the same row as -80, while the $-80v_3$ in the source vector transfers into column 3 of the same row as 80. Similarly, $+9$ is added to g_{31} and -9 to g_{32}. The resulting matrix equation which completely describes the network is

$$
\begin{array}{c}
\quad\quad 1 \qquad\quad 2 \qquad\quad 3 \\
\begin{array}{c} 1 \\ 2 \\ 3 \end{array}
\left[\begin{array}{ccc}
9 & -3 & -4 \\
-3 - 80 & 18 & -5 + 80 \\
-4 + 9 & -5 - 9 & 15
\end{array}\right]
\left[\begin{array}{c} v_1 \\ v_2 \\ v_3 \end{array}\right]
=
\left[\begin{array}{r} 10 \\ -6 \\ 0 \end{array}\right]
\end{array}
\qquad (4.45)
$$

or

$$
\left[\begin{array}{rrr}
9 & -3 & -4 \\
-83 & 18 & 75 \\
5 & -14 & 15
\end{array}\right]
\left[\begin{array}{c} v_1 \\ v_2 \\ v_3 \end{array}\right]
=
\left[\begin{array}{r} 10 \\ -6 \\ 0 \end{array}\right]
\qquad (4.46)
$$

In unusual cases, the dependent current source may depend on a current into the same node. In these cases, the elements on the main diagonal of **G** will be altered.

The digital computer becomes attractive as a method of obtaining the node volt-

age solutions when the number of nodes in the network increases beyond two or
three. For example, the solution of Eq. (4.46) is not trivial by hand. Using the methods
for solving systems of equations described in Chapter 3, we can obtain the node
voltages once we have written the equations in the form described in Eq. (4.37).
(In Chapter 7 we will discuss computer methods of constructing the nodal conductance matrix and equivalent current vector directly from the network description.)

Figure 4.21 contains a simple program using the subroutine LINQ described
in Section 3.4 to perform a dc network analysis. The program in Fig. 4.21 uses the
Gauss-elimination method to evaluate the unknown node voltages. These voltages
could have been evaluated using the INVERT subroutine (Section 3.5) and the method
described in Eq. (3.16). This procedure is left to the reader as an exercise.

```
C       PROGRAM TO EVALUATE NODE VOLTAGES FROM A GIVEN NODAL          DC01   1
C       CONDUCTANCE MATRIX AND EQUIVALENT CURRENT VECTOR              DC01   2
C                                                                     DC01   3
        DIMENSION G(20,20),XI(20),VLTGS(20)                          DC01   4
        READ (5,10) N                                                DC01   5
   10 FORMAT (I2)                                                     DC01   6
        DO 20 I=1,N                                                   DC01   7
   20 READ (5,30) (G(I,J),J=1,N)                                      DC01   8
   30 FORMAT (4F20.5)                                                 DC01   9
        READ (5,30) (XI(I),I=1,N)                                     DC01  10
C       SOLVE SYSTEM OF EQUATIONS                                     DC01  11
        CALL LINQ(G,XI,VLTGS,N,DET)                                   DC01  12
        IF (DET.EQ.0.0) GO TO 70                                      DC01  13
C       OUTPUT NODE VOLTAGES                                          DC01  14
        WRITE (6,40)                                                  DC01  15
   40 FORMAT (1H1)                                                    DC01  16
        DO 50 I=1,N                                                   DC01  17
   50 WRITE (6,60) I,VLTGS(I)                                         DC01  18
   60 FORMAT (5X,5HNODE ,I2,11H VOLTAGE = ,1PE12.5)                   DC01  19
        STOP                                                          DC01  20
C       ERROR OUTPUT                                                  DC01  21
   70 WRITE (6,80)                                                    DC01  22
   80 FORMAT (29H1 INVALID SYSTEM OF EQUATIONS)                       DC01  23
        STOP                                                          DC01  24
        END                                                           DC01  25
```

Figure 4.21 Node voltage calculation program
using nodal conductance matrix and equivalent
current vector.

Node Equations of *RLC* Networks

If the network contains energy-storing elements, application of Kirchhoff's
laws results in a set of simultaneous differential equations, as will be discussed in
detail in Chapter 5. These equations are linear and time-invariant if the elements are
also of this type. In order to demonstrate the method of writing the node equations
which describe an *RLC* network we will use the two-node network shown in Fig. 4.22.

The term conductance G is used in place of resistance R, and reciprocal inductance Γ is used in place of inductance L since it is more convenient to do so when
writing node equations. The current through a capacitance i_c and through an inductance i_L are defined by Eqs. (2.10a) and (2.17a), respectively. If we consider only the
total current from node 1 to ground (node 3) in Fig. 4.22 we see that this current,

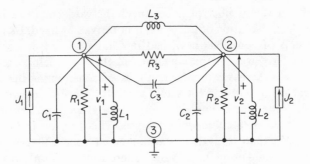

Figure 4.22 *RLC* nodal network analysis
example.

which we call i_a, is

$$i_a = \frac{1}{R_1}v_1 + C_1\frac{dv_1}{dt} + \frac{1}{L_1}\int v_1\,dt$$

$$= G_1 v_1 + C_1\frac{dv_1}{dt} + \Gamma_1 \int v_1\,dt \qquad (4.47)$$

At this point we adopt a special notation for equations of this type by defining the
following equation as equivalent to Eq. (4.47):

$$i_a = \left(G_1 + C_1\frac{d}{dt} + \Gamma_1 \int dt\right)v_1 \qquad (4.48)$$

This symbolism implies that the variable v_1 is operated on by multiplication by G_1,
multiplication by C_1 and differentiation, and finally multiplication by Γ_1 and integration. We must now consider all three operations of types of elements in writing node
equations.

We can proceed with the analysis now much as we did with the resistive networks.
Applying KCL to nodes 1 and 2 we arrive at the equations

$$\left[G_1 + G_3 + (C_1 + C_3)\frac{d}{dt} + (\Gamma_1 + \Gamma_3)\int dt\right]v_1$$

$$- \left[G_3 + C_3\frac{d}{dt} + \Gamma_3 \int dt\right]v_2 = J_1 \qquad (4.49a)$$

$$- \left[G_3 + C_3\frac{d}{dt} + \Gamma_3 \int dt\right]v_1$$

$$+ \left[G_2 + G_3 + (C_2 + C_3)\frac{d}{dt} + (\Gamma_2 + \Gamma_3)\int dt\right]v_2 = J_2 \qquad (4.49b)$$

which are of the form

$$\begin{bmatrix} a_{11} & -a_{12} \\ -a_{21} & a_{22} \end{bmatrix}\begin{bmatrix} v_1 \\ v_2 \end{bmatrix} = \begin{bmatrix} J_1 \\ J_2 \end{bmatrix} \qquad (4.50)$$

where

$$a_{11} = G_1 + G_3 + (C_1 + C_3)\frac{d}{dt} + (\Gamma_1 + \Gamma_3)\int dt$$

$$a_{12} = a_{21} = G_3 + C_3\frac{d}{dt} + \Gamma_3\int dt$$

$$a_{22} = G_2 + G_3 + (C_2 + C_3)\frac{d}{dt} + (\Gamma_2 + \Gamma_3)\int dt$$

Equations (4.49a) and (4.49b) are called *integrodifferential equations*. The equations imply that the voltages and currents vary with time; that is, $v_1 = v_1(t)$, $i_a = i_a(t)$, and so on.

EXAMPLE 4.9

Write the node equations for the network in Fig. 4.23.

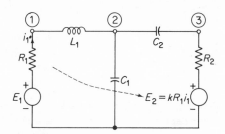

Figure 4.23 *RLC* active network.

Since the network is to be analyzed on the node basis with KCL, we must first manipulate the network sources so that only currents sources are present in the resulting network. Thus, by applying the Norton–Thévenin transformation, we reduce the network to the configuration in Fig. 4.24. At node 1,

$$J_1 = G_1 v_1 + \Gamma_1 \int (v_1 - v_2)\, dt$$

$$= \left(G_1 + \Gamma_1 \int dt\right)v_1 - \Gamma_1 \int dt\, v_2$$

(4.51a)

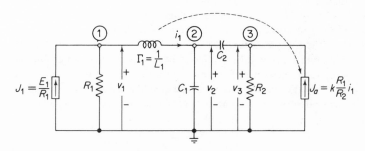

Figure 4.24 Nodal analysis form of active *RLC* network.

Similarly, at nodes 2 and 3,

$$
0 = -\Gamma_1 \int (v_1 - v_2)\, dt + C_1 \frac{dv_2}{dt} + C_2 \frac{d}{dt}(v_2 - v_3)
$$

$$
= -\Gamma_1 \int dt\, v_1 + \left[\Gamma_1 \int dt + (C_1 + C_2)\frac{d}{dt} \right] v_2 - C_2 \frac{dv_3}{dt} \tag{4.51b}
$$

$$
J_a = -C_2 \frac{dv_2}{dt} + \left(G_2 + C_2 \frac{d}{dt} \right) v_3
$$

However,

$$
J_a = \frac{kR_1}{R_2} i_1 = \frac{kR_1 \Gamma_1}{R_2} \int (v_1 - v_2)\, dt
$$

so at node 3 we obtain

$$
0 = -\frac{kR_1 \Gamma_1}{R_2} \int dt v_1 + \left(-C_2 \frac{d}{dt} + \frac{kR_1 \Gamma_1}{R_2} \int dt \right) v_2 + \left(G_2 + C_2 \frac{d}{dt} \right) v_3 \tag{4.51c}
$$

By inspection we could also write the node equations in matrix form as

$$
\begin{bmatrix} J_1 \\ 0 \\ J_a \end{bmatrix} = \begin{bmatrix} G_1 + \Gamma_1 \int dt & -\Gamma_1 \int dt & 0 \\ -\Gamma_1 \int dt & \Gamma_1 \int dt + (C_1 + C_2)\frac{d}{dt} & -C_2 \frac{d}{dt} \\ 0 & -C_2 \frac{d}{dt} & G_2 + C_2 \frac{t}{dt} \end{bmatrix} \begin{bmatrix} v_1 \\ v_2 \\ v_3 \end{bmatrix} \tag{4.52}
$$

Notice that in this case the dependent variable i_1 was not lost in the transformation of the voltage source E_a to the equivalent current source J_a. This was a stroke of luck since in most applications this variable would be lost in the transformation and would have had to be reconstructed. For example, we can reconstruct i_1 by observing that $i_1 = E_1 G_1 - G_1 v_1 = J_1 - G_1 v_1$ so that J_a is also defined by

$$
J_a = \frac{kR_1}{R_2} i_1
$$

$$
= k \frac{R_1}{R_2} \frac{1}{R_1} (E - v_1)
$$

$$
= \frac{k}{R_2} (E - v_1)
$$

The matrix in Eq. (4.52) possesses the same qualities as the conductance matrix described in Eq. (4.37). That is, the matrix is symmetric, the diagonal elements are all positive, and the off-diagonal elements are negative. Any diagonal element a_{ii} is in an abstract sense the *self-admittance* at node i. The direct relationship between the matrix elements and admittance will be explained in a subsequent chapter. Similarly, any element $a_{ij} = a_{ji}$ is the negative of the *mutual admittance* between nodes i and j.

The matrix equations for the node voltage calculation can be completed by translating the dependent source to the proper side of the equations. Since

$$
\begin{bmatrix} J_1 \\ 0 \\ J_a \end{bmatrix} = \begin{bmatrix} J_1 \\ 0 \\ \frac{kR_1 \Gamma_1}{R_2} \left[\int v_1\, dt - \int v_2\, dt \right] \end{bmatrix} \tag{4.53}
$$

the proper formulation of the network matrix equation is

$$\begin{bmatrix} J_1 \\ 0 \\ 0 \end{bmatrix} = \begin{bmatrix} G_1 + \Gamma_1 \int dt & -\Gamma_1 \int dt & 0 \\ -\Gamma_1 \int dt & \Gamma_1 \int dt + (C_1 + C_2)\dfrac{d}{dt} & -C_2\dfrac{d}{dt} \\ \dfrac{-kR_1\Gamma_1}{R_2} \int dt & -C_2\dfrac{d}{dt} + \dfrac{kR_1\Gamma_1}{R_2} \int dt & G_2 + C_2\dfrac{t}{dt} \end{bmatrix} \begin{bmatrix} v_1 \\ v_2 \\ v_3 \end{bmatrix} \quad (4.54)$$

Notice again that once the dependent sources are correctly inserted in the *admittance matrix*, the symmetry property is lost.

EXAMPLE 4.10

Write the node equations for the network in Fig. 4.25 which contains coupled inductances.

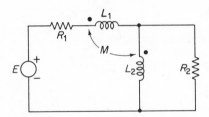

Figure 4.25 Inductively coupled network.

We begin by transforming the network in Fig. 4.25 into one containing only current sources and include the transformer equivalent circuit from Fig. 2.24 into the equivalent

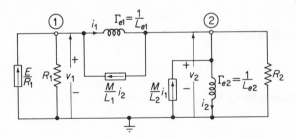

Figure 4.26 Nodal analysis form of inductively coupled circuit.

network in Fig. 4.26 before proceeding with the network equations. The node equations for the network in matrix form can be written directly from Fig. 4.26 as†

$$\begin{bmatrix} \dfrac{E}{R_1} + \dfrac{M}{L_1}i_2 \\ -\dfrac{M}{L_1}i_2 + \dfrac{M}{L_2}i_1 \end{bmatrix} = \begin{bmatrix} G_1 + \Gamma_{e1} \int dt & -\Gamma_{e1} \int dt \\ -\Gamma_{e1} \int dt & G_2 + (\Gamma_{e1} + \Gamma_{e2}) \int dt \end{bmatrix} \begin{bmatrix} v_1 \\ v_2 \end{bmatrix} \quad (4.55)$$

---

†$L_{e1} = L_1(1 - M^2/L_1L_2)$, $L_{e2} = L_2(1 - M^2/L_1L_2)$.

Since $i_1 = \Gamma_{e1} \int v_1 \, dt - \Gamma_{e1} \int v_2 \, dt$ and $i_2 = \Gamma_{e2} \int v_2 \, dt$,

$$
\begin{bmatrix} \dfrac{E}{R_1} + \dfrac{M}{L_1} i_2 \\[2ex] -\dfrac{M}{L_1} i_2 + \dfrac{M}{L_2} i_1 \end{bmatrix}
=
\begin{bmatrix} \dfrac{E}{R_1} + \dfrac{M\Gamma_{e2}}{L_1} \displaystyle\int v_2 \, dt \\[2ex] -\dfrac{M\Gamma_{e2}}{L_1} \displaystyle\int v_2 \, dt + \dfrac{M\Gamma_{e1}}{L_1} \Big[\displaystyle\int v_1 \, dt - \displaystyle\int v_2 \, dt \Big] \end{bmatrix}
\tag{4.56}
$$

Translating the terms due to the dependent sources from the source vector in Eq. (4.55) to the other side of the equation, we obtain the following proper matrix equation defining network in Fig. 4.24:

$$
\begin{bmatrix} \dfrac{E}{R_1} \\[2ex] 0 \end{bmatrix}
=
\begin{bmatrix} G_1 + \Gamma_{e1} \displaystyle\int dt & -\Big(\Gamma_{e1} + \dfrac{M}{L_1}\Gamma_{e2}\Big) \displaystyle\int dt \\[3ex] -\Big(\Gamma_{e1} + \dfrac{M}{L_2}\Gamma_{e1}\Big) \displaystyle\int dt & G_2 + \Big(\Gamma_{e1} + \Gamma_{e2} + \dfrac{M}{L_1}\Gamma_{e2} + \dfrac{M}{L_2}\Gamma_{e1}\Big) \displaystyle\int dt \end{bmatrix}
\begin{bmatrix} v_1 \\[2ex] v_2 \end{bmatrix}
\tag{4.57}
$$

Since

$$
\frac{M}{L_1}\Gamma_{e2} = \frac{M}{L_2}\Gamma_{e1} = \frac{M}{L_1 L_2 - M^2}
$$

we will designate it as Γ_M and finally reduce Eq. (4.57) to

$$
\begin{bmatrix} \dfrac{E}{R_1} \\[2ex] 0 \end{bmatrix}
=
\begin{bmatrix} G_1 + \Gamma_{e1} \displaystyle\int dt & -(\Gamma_{e1} + \Gamma_M) \displaystyle\int dt \\[3ex] -(\Gamma_{e1} + \Gamma_M) \, dt & G_2 + (\Gamma_{e1} + \Gamma_{e2} + 2\Gamma_M) \displaystyle\int dt \end{bmatrix}
\begin{bmatrix} v_1 \\[2ex] v_2 \end{bmatrix}
\tag{4.58}
$$

Later we will develop the techniques necessary to obtain the node voltage solutions for networks containing energy-storing elements.

4.4 Loop (Mesh) Analysis

Loop (or mesh) analysis is based upon the application of Kirchhoff's voltage law (KVL) to obtain the solution for the loop currents in a network. Once these variables have been determined, other voltages and currents in the network can be computed. This is easiest to demonstrate by computing the loop currents in the two-loop network shown in Fig. 4.27.

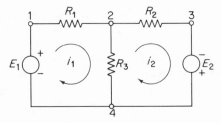

Figure 4.27 Loop analysis introduction.

In applying Kirchhoff's law to this network, we first select (arbitrary) loop currents, such as the loops defined by i_1 and i_2 in Fig. 4.20. Note that both i_1 and i_2 pass through R_3. Then we procede around each loop, summing voltage drops. The drop from node 2 to node 4 is thus $R_3(i_1 - i_2)$. Hence going around loop 1 clockwise (recalling that in general $R = v_R/i_R$), we write

$$R_1 i_1 + R_3(i_1 - i_2) = -E_1 = 0$$

and going around loop 2 clockwise we find that

$$R_2 i_2 - E_2 + R_3(i_2 - i_1) = 0$$

Combining, we obtain the two equations

$$\begin{aligned} (R_1 + R_3)i_1 - R_3 i_2 &= E_1 \\ -R_3 i_1 + (R_2 + R_3)i_2 &= E_2 \end{aligned} \tag{4.59}$$

Equations (4.59) may then be solved using any method desired. The matrix equivalent of Eq. (4.59) is of course the equation

$$\begin{bmatrix} R_1 + R_3 & -R_3 \\ -R_3 & R_2 + R_3 \end{bmatrix} \begin{bmatrix} i_1 \\ i_2 \end{bmatrix} = \begin{bmatrix} E_1 \\ E_2 \end{bmatrix} \tag{4.60}$$

or

$$\mathbf{Ri} = \mathbf{E} \tag{4.61}$$

In writing Eq. (4.60), we selected all loop currents clockwise, just as in the nodal cases we selected all node voltage potentials above the reference node potential. Although this is not required, these selections result in negative signs for the off-diagonal matrix elements in a passive network and give rather simple rules for obtaining the matrix elements, as previously found for the nodal case.

We adopt some definitions analogous to those used in the nodal case. The sum of all resistances around a loop is called the *loop self-resistance*, and the resistance common to two loops we term *mutual resistance*. The loops are numbered in order. Then the diagonal of the matrix **R** in Eq. (4.61) is composed of the loop self-resistances, while the off-diagonal elements are formed from the negative of the mutual resistances. The equivalent voltage source vector on the right-hand side of Eq. (4.61) has as components the independent source voltages. The numerical voltage value is positive if the source voltage *rise* is in the same direction as the loop current chosen (or the source voltage *drop* is in the opposite direction to the loop current). Thus, if we go from a minus to a plus sign through the source as we proceed in the direction of the current, the voltage is positive in the voltage vector.

EXAMPLE 4.11

In Fig. 4.28 the source voltages are specified in volts and the element resistances in ohms. The objective is to determine the currents in each of the elements.

We first define arbitrary loop currents i_1, i_2, and i_3 in a clockwise direction as shown. Then the loop self-resistances are $r_{11} = 7$, $r_{22} = 11$, and $r_{33} = 15$. The mutual resistances are $r_{12} = r_{21} = 3$, $r_{13} = r_{31} = 4$, and $r_{23} = r_{32} = 5$.

Around loop 1 the total voltage source rise in the direction of the current i_1 is

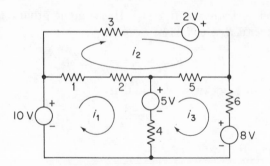

Figure 4.28 Loop analysis example.

10 V − 5 V = 5 V, around loop 2 the total voltage source rise is 2 V, and around loop 3 the total voltage source rise is 5 V − 8 V = −3 V. We now write the loop equations in matrix form as

$$\begin{array}{c} \\ 1 \\ 2 \\ 3 \end{array} \begin{array}{ccc} 1 & 2 & 3 \\ \begin{bmatrix} 7 & -3 & -4 \\ -3 & 11 & -5 \\ -4 & -5 & 15 \end{bmatrix} \end{array} \begin{bmatrix} i_1 \\ i_2 \\ i_3 \end{bmatrix} = \begin{bmatrix} 5 \\ 2 \\ -3 \end{bmatrix} \tag{4.62}$$

The rows and columns of the **R** matrix are numbered to assist in proper placement of self- and mutual-resistance values.

Equation (4.62) can also be written as

$$
\begin{aligned}
7i_1 - 3i_2 - 4i_3 &= 5 \\
-3i_1 + 11i_2 - 5i_3 &= 2 \\
-4i_1 - 5i_2 + 15i_3 &= -3
\end{aligned}
\tag{4.63}
$$

Using Kirchhoff's voltage law in tours around each loop, we find

(loop 1) $3(i_1 - i_2) + 4(i_1 - i_3) + 5 - 10 = 0$ (4.64a)

(loop 2) $3i_2 + 3(i_2 - i_1) + 5(i_2 - i_3) - 2 = 0$ (4.64b)

(loop 3) $6i_3 + 5(i_3 - i_2) + 4(i_3 - i_1) + 8 - 5 = 0$ (4.64c)

Collecting terms in Eq. (4.64) we see that the resulting equations are the same as those in Eq. (4.63), which verifies our matrix expression. The currents i_1, i_2, and i_3 can now be calculated by the Gauss-elimination method or by inverting the **R** matrix and multiplying as shown in Eq. (4.65). Thus,

$$\begin{bmatrix} i_1 \\ i_2 \\ i_3 \end{bmatrix} = \mathbf{R}^{-1} \begin{bmatrix} 5 \\ 2 \\ -3 \end{bmatrix} \tag{4.65}$$

where

$$\mathbf{R}^{-1} \begin{bmatrix} 7 & -3 & -4 \\ -3 & 11 & -5 \\ -4 & -5 & 15 \end{bmatrix} = \mathbf{I}$$

Note that the total current through mutual resistances are the difference of two currents; thus, $i_4 = i_1 - i_3$.

Dependent voltage sources in a loop analysis are handled in a fashion analogous to dependent current sources in the nodal-analysis case; that is, they modify the diagonal and off-diagonal elements of R, causing it to become nonsymmetric. If the dependent source in the network is a current source, we must first translate it into an equivalent dependent voltage source.

In Fig. 4.29 suppose we have a dependent voltage source $E_2 = r_m i_{R_1} = r_m(i_1 - i_2)$ in loop 3. Then the equation around loop 3 becomes

$$R_2(i_3 - i_1) + R_3(i_3 - i_2) + R_4 i_3 + r_m(i_1 - i_2) = 0 \qquad (4.66)$$

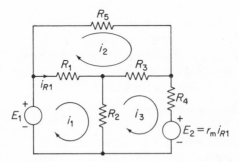

Figure 4.29 Loop analysis example containing dependent sources.

Combining terms in Eq. (4.66) we obtain

$$-(R_2 - r_m)i_1 - (R_3 + r_m)i_2 + (R_2 + R_3 + R_4)i_3 = 0$$

Thus, in the **R** matrix

$$r_{31} = -R_2 + r_m \quad \text{and} \quad r_{32} = -R_3 - r_m$$

Hence, if we have a dependent voltage source in loop j which depends upon the currents in loops i and k such that $E = r_m(i_i - i_k)$, if the dependent source voltage is a drop in the direction of the current in loop j, then we alter the element r_{ji} by $-r_m$ and the element r_{jk} by $+r_m$. As in the nodal case, to prevent errors it may be simpler to first write the dependent sources as a part of the equivalent voltage source vector and then transfer them into the **R** matrix term by term. The procedure for writing loop equations with dependent sources is as follows:

1. Prepare the network for loop analysis by translating all dependent and independent current sources into equivalent voltage sources by use of the Norton–Thévenin transformation.

2. Assign a minimum number of arbitrary loop currents in the network such that at least one loop current passes through each element. The direction of all the loop currents should be either clockwise or counterclockwise. A mixture of directions ruins the recipe.

3. Adopt the unknown currents in the loops of step 2 as the current vector $\mathbf{i} = [i_1, i_2, i_3, \ldots, i_n]^T$, where $i_1, i_2, i_3, \ldots, i_n$ are the loop currents assigned in a common direction.

4. Write the matrix equation for the network as

$$
\begin{array}{c}
\begin{array}{cccc} 1 & 2 & & n \end{array} \\
\begin{array}{c} 1 \\ 2 \\ \cdots \\ n \end{array}
\left[\begin{array}{cccc}
r_{11} & r_{12} & \cdots & r_{1n} \\
r_{21} & r_{22} & \cdots & r_{2n} \\
\cdot & \cdot & \cdots & \cdot \\
r_{n1} & r_{n2} & \cdots & r_{nn}
\end{array}\right]
\left[\begin{array}{c}
i_1 \\ i_2 \\ \cdots \\ i_n
\end{array}\right]
=
\left[\begin{array}{c}
E_1 \\ E_2 \\ \cdots \\ E_n
\end{array}\right]
\end{array}
\tag{4.67}
$$

or

$$\mathbf{Ri} = \mathbf{E} \tag{4.68}$$

where $\mathbf{R}$ is the loop resistance matrix and the elements of the modified equivalent-voltage-source vector $\mathbf{E}$ on the right-hand side of Eq. (4.68) contain both the independent and dependent voltage sources in the respective loops. The rows and columns of the resistance matrix in Eq. (4.67) are numbered to assist in locating the matrix elements.

  a. The diagonal elements $r_{11}, r_{22}, r_{33}, \ldots, r_{nn}$ are the self-resistances of loops 1, 2, 3, .., n, respectively.
  b. The off-diagonal elements $r_{ij} = r_{ji}$ $(i \neq j)$ are equal to the negative of the mutual resistance between loops i and j. That is, each term will contain a negative sign.

The $\mathbf{R}$ matrix for Eq. (4.68) will be symmetric since the matrix does not yet contain any dependent sources or coupled elements such as transformers. The matrix will also be quite sparse for large networks.

  The final two steps are necessary only if the network contains dependent sources or coupled elements.

5. The dependent voltage sources must be written in terms of the loop current variables. That is, if the source is given as $\mathbf{E}_x = \mu v_j$, where $v_j = R_j(i_i - i_k)$, then E_x must be expressed in the source vector as $E_x = \mu R_j(i_i - i_k)$ in the present stage of the development.

6. The dependent source term is then transferred to the loop resistance matrix in the row corresponding to the loop containing the source and into the columns corresponding to the source variables.

The resulting matrix equation is

$$\mathbf{R}_1 \mathbf{i} = \mathbf{E}_1 \tag{4.69}$$

where $\mathbf{R}_1$ is the modified loop resistance matrix, containing the dependent source terms, and $\mathbf{E}_1$ is the voltage source vector, containing only independent source terms. Note that Eq. (4.69) is exactly the same as Eq. (4.68) if there are no dependent terms. In general, the $\mathbf{R}_1$ matrix, containing dependent sources and coupled elements, will not be symmetrical.

  The digital computer can be used to determine the loop currents in the loop method in the same manner as it was used in the nodal-analysis method. A program similar to the analysis program in Fig. 4.21 can be written to obtain the loop currents from a given loop resistance matrix and equivalent voltage source vector. The implementation is left to the reader as an exercise.

Loop Equations of *RLC* Networks

We will use the two-loop network in Fig. 4.30 to demonstrate the method of writing a set of simultaneous differential equations from Kirchhoff's laws.

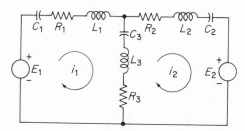

Figure 4.30 *RLC* network loop analysis example.

The voltage across a capacitor, defined by Eq. (2.11), is expressed as

$$v_c(t) = S \int i_c \, dt \tag{4.70}$$

The inductance voltage drop and resistance voltage drop are defined by Eqs. (2.16) and (2.1), respectively.

If we consider only the total voltage drop across R_1, L_1, and C_1 in Fig. 4.30 we see that

$$v = R_1 i + L_1 \frac{di_1}{dt} + S_1 \int i \, dt \tag{4.71}$$

By adopting an equation notation similar to that introduced in Eq. (4.48), we can rewrite Eq. (4.71) as

$$v = \left(R_1 + L_1 \frac{d}{dt} + S_1 \int dt \right) i \tag{4.72}$$

We then proceed with the analysis now much as we did with resistive networks. Applying KVL to loops 1 and 2 of the network in Fig. 4.30, we arrive at the equations

$$\left[R_1 + R_3 + (L_1 + L_3)\frac{d}{dt} + (S_1 + S_3) \int dt \right] i_1$$
$$- \left[R_3 + L_3 \frac{d}{dt} + S_3 \int dt \right] i_2 = E_1 \tag{4.73a}$$

$$- \left[R_3 + L_3 \frac{d}{dt} + S_3 \int dt \right] i_1$$
$$+ \left[R_2 + R_3 + (L_2 + L_3)\frac{d}{dt} + (S_2 + S_3) \int dt \right] i_2 = -E_2 \tag{4.73b}$$

which are of the form

$$
\begin{bmatrix} a_{11} & -a_{12} \\ -a_{21} & a_{22} \end{bmatrix} \begin{bmatrix} i_1 \\ i_2 \end{bmatrix} = \begin{bmatrix} E_1 \\ -E_2 \end{bmatrix} \tag{4.74}
$$

where

$$
a_{11} = R_1 + R_3 + (L_1 + L_3)\frac{d}{dt} + (S_1 + S_3)\int dt
$$

$$
a_{12} = a_{21} = R_3 + L_3\frac{d}{dt} + S_3\int dt
$$

$$
a_{22} = R_2 + R_3 + (L_2 + L_3)\frac{d}{dt} + (S_2 + S_3)\int dt
$$

Matrix Eq. (4.74) defines a set of integrodifferential equations which possess the same qualities as the resistance matrix $\mathbf{R}$ in Eq. (4.69). That is, the matrix is symmetric (assuming no dependent sources), the diagonal elements are all positive, and the off-diagonal elements are negative. Any diagonal element a_{ii} is in an abstract sense the self-impedance in loop i. Similarly, any element a_{ij} is the negative of the mutual impedance between loops i and j.

EXAMPLE 4.12

Write the loop equations for the network in Fig. 4.23.

By inspection, the network equations can be written in matrix form as

$$
\begin{bmatrix} R_1 + L_1\frac{d}{dt} + S_1\int dt & -S_1\int dt \\ -S_1\int dt & R_2 + (S_1 + S_2)\int dt \end{bmatrix} \begin{bmatrix} i_1 \\ i_2 \end{bmatrix} = \begin{bmatrix} E_1 \\ -E_2 \end{bmatrix} \tag{4.75}
$$

The voltage source E_2 is dependent upon the current i_1. That is, $E_2 = kR_1i_1$. We must translate the dependent source into the loop impedance matrix before we have the equation in the proper form. Thus,

$$
\begin{bmatrix} R_1 + L_1\frac{d}{dt} + S_1\int dt & -S_1\int dt \\ kR_1 - S_1\int dt & R_2 + (S_1 + S_2)\int dt \end{bmatrix} \begin{bmatrix} i_1 \\ i_2 \end{bmatrix} = \begin{bmatrix} E_1 \\ 0 \end{bmatrix} \tag{4.76}
$$

Again the dependent source in the "impedance" matrix destroys the symmetry property.

EXAMPLE 4.13

Write the loop equations for the network in Fig. 4.25 which contains a pair of coupled inductances.

We begin by transforming the coupled inductances (transformer) into an equivalent network containing only R, L, and/or C elements and dependent voltage sources. The transformer model which satisfies Eqs. (2.24) and the above requirement is shown in Fig. 4.31. The loop equivalent for the network in Fig. 4.25 is shown in Fig. 4.32. The loop equations for the modified form of the network can be written in matrix form

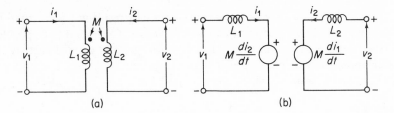

Figure 4.31 Two-winding transformer loop equivalent circuit.

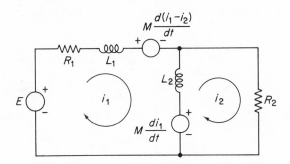

Figure 4.32 Loop analysis form of inductively coupled circuit in Fig. 4.25.

by inspection as

$$
\begin{bmatrix} R + (L_1 + L_2)\dfrac{d}{dt} & -L_2\dfrac{d}{dt} \\ -L_2\dfrac{d}{dt} & R_2 + L_2\dfrac{d}{dt} \end{bmatrix}\begin{bmatrix} i_1 \\ i_2 \end{bmatrix} = \begin{bmatrix} E - 2M\dfrac{di_1}{dt} + M\dfrac{di_2}{dt} \\ M\dfrac{di_1}{dt} \end{bmatrix} \tag{4.77}
$$

Translating the terms due to the dependent from the source vector in Eq. (4.77) to the left side of the equation, we obtain the following proper matrix equation defining the network in Fig. 4.25:

$$
\begin{bmatrix} R_1 + (L_1 + L_2 + 2M)\dfrac{d}{dt} & -(L_2 + M)\dfrac{d}{dt} \\ -(L_2 + M)\dfrac{d}{dt} & R_2 + L_2\dfrac{d}{dt} \end{bmatrix}\begin{bmatrix} i_1 \\ i_2 \end{bmatrix} = \begin{bmatrix} E \\ 0 \end{bmatrix} \tag{4.78}
$$

We will discuss methods in the following chapters for obtaining the frequency- and time-domain responses of networks containing energy-storing elements.

4.5 Loop and/or Nodal Analysis

You might well inquire, "Should I use loop analysis, or nodal analysis, or both in a problem? Theoretically, both loop- and nodal-analysis methods may be applied in any given situation, but practically it usually becomes simpler to use one or the other. In some instances, one of these techniques may result in fewer equations than the other and from this standpoint may be preferable. Thus, an n node problem may be an $n - 1$ loop problem, or vice versa. Sometimes given conditions may result in a simpler application of one analysis than the other; for example, sources may be specified as current sources, which suggests nodal analysis.

EXAMPLE 4.14

As a final example of Kirchhoff analysis we will calculate the gain of a single-stage common-emitter transistor amplifier using the *hybrid-π* transistor model.

The gain can be calculated using dc analysis, assuming that the circuit operates within the linear region and that capacitance and inductance effects are negligible.

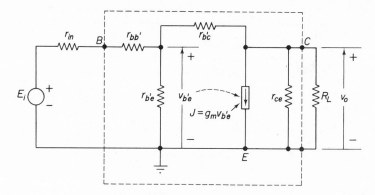

Figure 4.33 Transistor amplifier equivalent circuit.

The dc amplifier equivalent circuit is shown in Fig. 4.33 with the transistor hybrid-π equivalent circuit enclosed in dotted lines. The circuit values are given as follows:

$$r_{\text{in}} = 600\ \Omega$$
$$r_{\text{bb}'} = 20\ \Omega$$
$$r_{\text{b}'\text{e}} = 1000\ \Omega$$
$$r_{\text{b}'\text{c}} = 5 \times 10^6\ \Omega$$
$$r_L = 5000\ \Omega$$
$$r_{\text{b}'\text{c}} = 50 \times 10^3\ \Omega$$
$$g_m = 0.01\ \text{A/V}$$

First, we combine $r_{\text{bb}'}$ and r_{in} and change the voltage source E_i to a current source $E_i/(r_{\text{bb}'} + R_{\text{in}})$. Then we obtain the circuit shown in Fig. 4.34 (using given values in ohms). The self-admittance at node 1 is

$$\frac{1}{620} + \frac{1}{1000} + \frac{10^{-6}}{5} = 2.6131 \times 10^{-3}\ \mho$$

The self-admittance at node 2 is

$$\frac{10^{-3}}{50} + \frac{10^{-3}}{5} + \frac{10^{-6}}{5} = 0.2202 \times 10^{-3}\ \mho$$

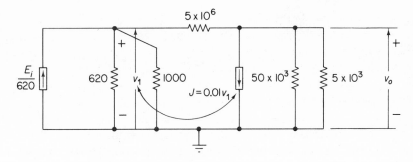

Figure 4.34 Nodal analysis form of amplifier equivalent circuit.

The mutual admittance between nodes 1 and 2 is 0.2×10^{-6} ℧. The resulting nodal conductance matrix and equivalent current vector (without the dependent source) is

$$\begin{bmatrix} 2.6131 \times 10^{-3} & -0.2 \times 10^{-6} \\ -0.2 \times 10^{-6} & 0.2202 \times 10^{-3} \end{bmatrix} \begin{bmatrix} v_1 \\ v_0 \end{bmatrix} = \begin{bmatrix} \dfrac{E_i}{620} \\ 0 \end{bmatrix}$$

The dependent current source defined by $J = g_m v_{b'c} = g_m v_1 = 0.01 v_1$ connected to node 2 is entered into the source vector, resulting in

$$\begin{bmatrix} 2.6131 \times 10^{-3} & -0.2 \times 10^{-6} \\ -0.2 \times 10^{-6} & 0.2202 \times 10^{-3} \end{bmatrix} \begin{bmatrix} v_1 \\ v_0 \end{bmatrix} = \begin{bmatrix} \dfrac{E_i}{620} \\ -0.01 v_1 \end{bmatrix}$$

or finally,

$$\begin{bmatrix} 2.6131 \times 10^{-3} & -0.2 \times 10^{-6} \\ 0.01 & 0.2202 \times 10^{-3} \end{bmatrix} \begin{bmatrix} v_1 \\ v_0 \end{bmatrix} = \begin{bmatrix} \dfrac{E}{620} \\ 0 \end{bmatrix}$$

where we neglect -0.2×10^{-6} compared to $10,000 \times 10^{-6}$. The node voltages of the network in this example are obtained by using the matrix-inversion technique described in Eq. (3.29). Thus,

$$\begin{bmatrix} v_1 \\ v_0 \end{bmatrix} = \frac{10^6}{0.5774} \begin{bmatrix} 0.2202 \times 10^{-3} & 0.2 \times 10^{-6} \\ -0.01 & 2.6131 \times 10^{-3} \end{bmatrix} \begin{bmatrix} \dfrac{E_i}{620} \\ 0 \end{bmatrix}$$

and

$$v_0 = -\frac{0.01 \times 10^{-6}}{(0.5774)(620)} E_i$$

The voltage gain for the amplifier is then

$$\frac{v_0}{E_i} = -27.934$$

If we neglect the 5×10^{-6}-Ω resistor between nodes 1 and 2, then from Fig. 4.34,

$$v_1 = \frac{E_i (10^3)}{620(2.61)} = 0.619 E_i$$

and

$$v_0 = \frac{-(0.619)(0.01) E_i (10^3)}{0.220}$$

or

$$\frac{v_0}{E_i} = -28.1$$

to slide-rule accuracy. The error in omitting the resistance $r_{b'c}$ is very small in this case. A great deal of an engineer's ability is involved in his approximating skill.

It is also important to note that in forming the nodal conductance matrix that some of the terms used in forming the elements are numerically insignificant. Particularly, when the dependent source term is transferred from the source vector to the **G** matrix, the original g_{21} term due to $r_{b'c}$ is omitted from the matrix. This lack of precision in the representation of numbers can cause problems in obtaining meaningful numerical solutions in loop or nodal analysis. This includes solutions obtained using digital-computer techniques.

Additional Reading

DESOER, C. A., and E. S. KUH, *Basic Circuit Theory*, McGraw-Hill Book Company, New York, 1969.

HAYT, W. H., JR., and G. W. HUGHES, *Introduction to Electrical Engineering*, McGraw-Hill Book Company, New York, 1968.

KUO, F. F., *Network Analysis and Synthesis*, John Wiley & Sons, Inc., New York, 1966.

LEON, B. J., and WINTZ, *Basic Linear Networks for Electrical and Electronics Engineers*, Holt, Rinehart and Winston, Inc., New York, 1970.

SESHU, S., and N. BALABANIAN, *Linear Network Analysis*, John Wiley & Sons, Inc., New York, 1959.

VAN VALKENBURG, M. E., *Network Analysis*, Prentice-Hall, Inc., Englewood Cliffs, N.J., 1964.

Problems

4.1. Find the Thévenin and Norton equivalent circuits with respect to terminals *a-b* for the networks shown in Fig. P4.1.

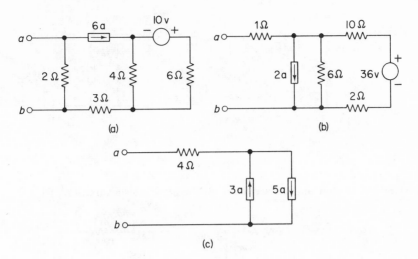

Figure P4.1

4.2. Find the Thévenin and Norton equivalent circuits with respect to terminals *a-b* (if they exist) for the networks shown in Fig. P4.2.

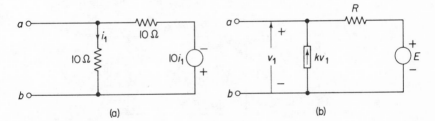

Figure P4.2

4.3. Derive the Thévenin equivalent circuit for the network shown in Fig. P4.3 using the terminal *v-i* characteristics.

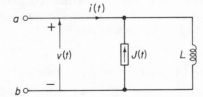

Figure P4.3

4.4. Find the Thévenin equivalent circuit for the networks shown in Fig. P4.4. Note that in determining the driving-point resistance, dependent sources are not set to zero.

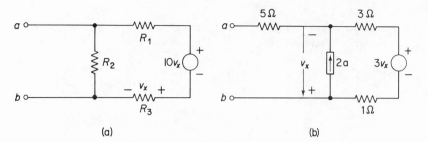

Figure P4.4

4.5.   Find the Norton equivalent circuits for the networks shown in Fig. P4.5.

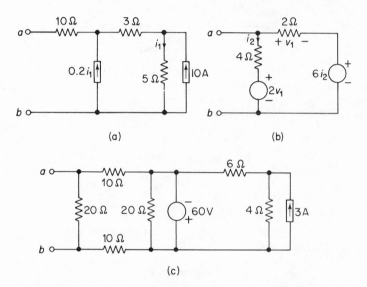

Figure P4.5

4.6.   Given the networks in Fig. P4.6, derive equations which relate R_1, R_2, and R_3 to R_a, R_b, and R_c and vice versa. Use the results to obtain the equivalent resistance at terminals *a-b* of the network in Fig. P4.7.

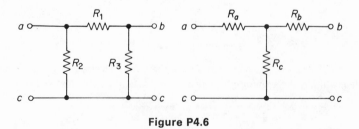

Figure P4.6

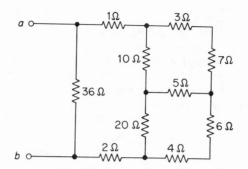

Figure P4.7

4.7. Find the equivalent resistance of the network shown in Fig. P4.8 using KVL and KCL. The network corresponds to a transistor amplifier with emitter degeneration. The equivalent resistance is the input resistance of the amplifier, or v_{in}/i_b.

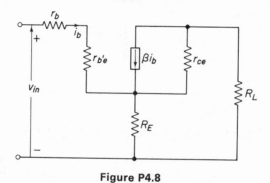

Figure P4.8

4.8. Find the Thévenin equivalent circuit at terminals a-b for the network shown in Fig. P4.9. R_L is the "load." The network is a dc or low-frequency model of a common-emitter transistor amplifier.

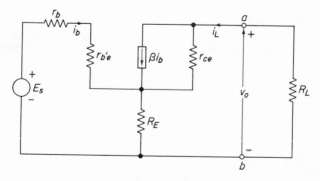

Figure P4.9

4.9. Using the Thévenin circuit derived in Problem 4.8, find the voltage gain v_s/E_s using the simplification $r_{ce} \gg R_E$.

4.10.   (a) Find the Norton equivalent circuit for the network in Fig. P4.9 at terminals a and b. Verify the voltage-gain calculation in Problem 4.9 using the Norton equivalent circuit. (b) Using the Thévenin equivalent voltage of Problem 4.8 and the Norton equivalent current of this problem, find the driving-point resistance by using Eq. (4.23b). (c) Show that if v_0 is plotted as a function of i_L, the Thévenin circuit voltage source is equal to the intersection of the line with the $i_L = 0$ axis, that the Norton current source is equal to the intersection of the line with the $v_0 = 0$ axis, and that the line slope equals the Thévenin driving-point resistance.

4.11.   Given the network of Fig. P4.10.
   (a) Write the node equations for the network in matrix form.
   (b) Solve the matrix equation from part (a) using either the Gauss-elimination or matrix-inversion technique.
   (c) Verify the node voltage solution calculated in part (b) using KCL at each node.

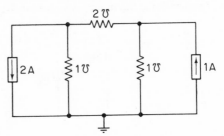

Figure P4.10

4.12.   Repeat Problem 4.11 for the network shown in Fig. P4.11.

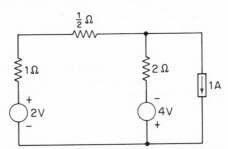

Figure P4.11

4.13.   Repeat Problem 4.11 for the network shown in Fig. P4.12 using any solution method available for part (b).

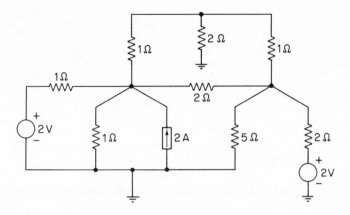

Figure P4.12

4.14. Repeat Problem 4.11 for the network shown in Fig. P4.13.

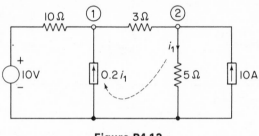

Figure P4.13

4.15. Write the node equations for the network shown in Fig. P4.14.

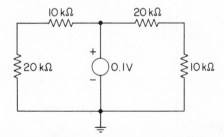

Figure P4.14

4.16. Write the node equations for the network shown in Fig. P4.15. *Hint:* First replace the transformer with a model containing only L elements and transfluences.

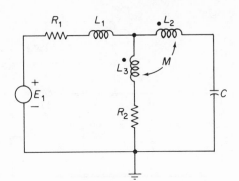

Figure P4.15

4.17. Write the node equation in matrix form for the networks in Fig. P4.16.

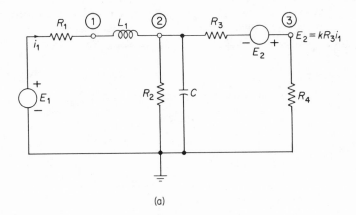

(a)

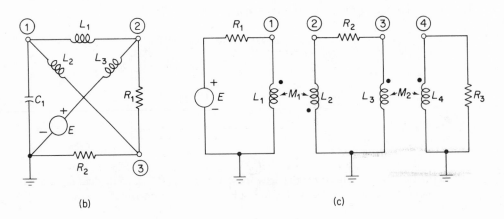

(b)   (c)

Figure P4.16

4.18. Write the node equations in matrix form for the network shown in Fig. P4.17.

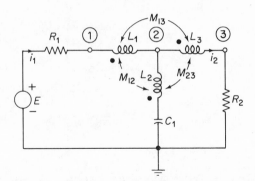

Figure P4.17

4.19. (a) Determine the voltage gain (v_0/E_S) for the network shown in Fig. P4.18 using nodal-analysis techniques. The network is a model of a common-emitter transistor amplifier driven by an input voltage source E_S with a finite input resistance R_S. The transistor is represented by a low-frequency hybrid-π equivalent circuit.

(b) Repeat part (a) for $r_{ce} \longrightarrow \infty$.

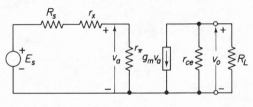

Figure P4.18

4.20. Determine the voltage gain (v_0/E_S) for the network shown in Fig. P4.19 using nodal-analysis techniques.

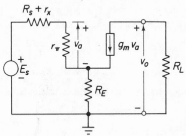

Figure P4.19

4.21. Find the output impedance Z of the cathode follower amplifier shown in Fig. P4.20.

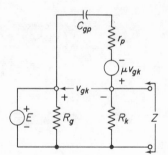

Figure P4.20

4.22.   Write the mesh equations for the network shown in Fig. P4.21 and solve for the voltage v_8.

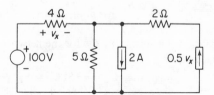

Figure P4.21

4.23.   Write the mesh equations for the network shown in Fig. P4.22 and solve for v_x.

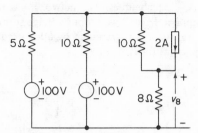

Figure P4.22

4.24.   Write the mesh equations in matrix form for the network shown in Fig. P4.17.

4.25.   Repeat Problem 4.24 for the network shown in Fig. P4.16.

4.26.   Repeat Problem 4.19 using the mesh analysis.

4.27.   Given the electronic circuit and the transistor equivalent circuit shown in Fig. P4.23.

   (a) Write the node equations in matrix form.
   (b) Compute the voltage gain for the circuit using the values given using nodal-analysis techniques.
   (c) Write the mesh equations in matrix form.
   (d) Repeat part (b) using the results of part (c).
   (e) Compute the input resistance (E_S/i_a) of the amplifier using any method.

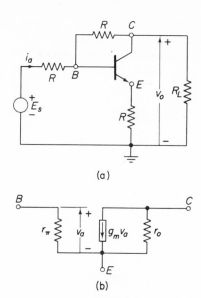

Figure P4.23

(a)

(b)

4.28. Write a Fortran program that will read in the numerical values of R_a, R_b, and R_c of the T-section in Fig. P4.6b and compute the corresponding equivalent π-section values R_1, R_2, and R_3 in Fig. P4.6a. This program performs the wye–delta transformation. Test the program with the following sets of input values:

	R_a	R_b	R_c
(a)	$10\,\Omega$	$10\,\Omega$	$10\,\Omega$
(b)	$1\,k\Omega$	$100\,\Omega$	$500\,\Omega$
(c)	$1\,\Omega$	$2\,\Omega$	$2\,\Omega$

4.29. The dual of Problem 4.28 is the conversion of resistive π-sections Fig. P4.6(a) to equivalent T-sections (delta–wye conversion). Repeat Problem 4.28 to convert values of R_1, R_2, and R_3 to equivalent values R_a, R_b, and R_c. Test the program using the following sets of values:

	R_a	R_b	R_c
(a)	$10\,\Omega$	$10\,\Omega$	$10\,\Omega$
(b)	$1\,k\Omega$	$100\,\Omega$	$500\,\Omega$
(c)	$1\,\Omega$	$90\,\Omega$	$5\,\Omega$

4.30. Problems 4.28 and 4.29 may be modified to convert complex impedance T-sections to complex impedance π-sections and vice versa. The design equations are identical to those used in the resistive case except that each term is complex. Write a Fortran program which will perform both delta–wye and wye–delta conversions, depending upon a data flag read in by the program to specify the type of conversion. Write the program assuming that the computer that will be used will not be capable of doing complex arithmetic. Test the program using the following sets of input data.

(a) $Z_a = 1.0 + j1.0$, $Z_b = 1.0 + j1.0$, $Z_c = 1.0 + j1.0$
(b) $Z_a = 100.0 + j10.0$, $Z_b = -j100.0$, $Z_c = 100.0$
(c) $Z_1 = 1.0 + j1.0$, $Z_2 = 1.0 + j1.0$, $Z_3 = 1.0 + j1.0$
(d) $Z_1 = +j10.0$, $Z_2 = 100.0 - j100.0$, $Z_3 = -j100.0$

4.31.   Write a Fortran program similar to Fig. 4.21 to evaluate node voltages for a network containing complex elements. The program will require either a complex Gauss-elimination subroutine or a complex matrix-inversion routine to obtain a solution.

Chapter 5

DIFFERENTIAL EQUATIONS

There is no substitute for hard work.

THOMAS A. EDISON

5.1 First-Order Differential Equations—*RL* Circuits

Assume that we have an *RL* circuit as shown in Fig. 5.1 with sw_1 closed and sw_2 open. If this condition has existed for a long time, the current in the inductance *L* is

$$i_L = \frac{E}{R_1} \qquad \text{amps} \tag{5.1}$$

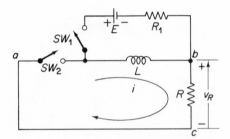

Figure 5.1 *RL* network.

We now close sw_2 and open sw_1 simultaneously at a time *t* arbitrarily designated as zero ($t = 0$). Then from Kirchhoff's voltage law around the loop *abc* we obtain the equation

$$L\frac{di}{dt} + Ri = 0 \qquad t \geq 0 \tag{5.2}$$

This is a linear homogeneous differential equation, since the right-hand side (here the source voltage) is always zero.

Rearranging terms in Eq. (5.2), we obtain

$$\frac{di}{i} = -\frac{R}{L}dt \qquad t \geq 0 \tag{5.3}$$

and if this is integrated, we get

$$\ln i = -\frac{R}{L}t + k \qquad t \geq 0 \tag{5.4}$$

where ln means logarithm to the base $e = 2.718 \cdots$ and *k* is a constant. Let $k = \ln K$ so that

$$\ln i = \ln e^{-(R/L)t} + \ln K$$

or

$$\ln i = \ln(Ke^{-(R/L)t})$$

Therefore, the general solution to Eq. (5.2) is

$$i = Ke^{-(R/L)t} \qquad t \geq 0 \tag{5.5}$$

To evaluate the constant K we must know i at a particular t. For some situations where changes or transients occur, variables are not always continuous at the time of circuit alteration. In this case, i may not be continuous at $t = 0$ and hence cannot be defined at that time. To take care of this situation we can talk about $t = 0_+$, which means $t = \lim_{\epsilon \to 0} (0 + \epsilon)$, and $t = 0_-$, which means $t = \lim_{\epsilon \to 0} (0 - \epsilon)$, where ϵ is a positive number. In the first case, we approach, and finally reach, a function $f(t)$ at $t = 0$ from the right (t decreasing) in Fig. 5.2, and in the second case we approach the

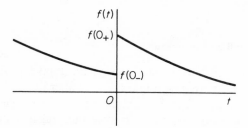

Figure 5.2 Discontinuity in $f(t)$ at $t = 0$.

function at $t = 0$ from the left (t increasing). As shown in Fig. 5.2, $f(0_+)$ may not be the same as $f(0_-)$, and $f(0)$ may be undefined. The equations written above, such as Eqs. (5.2), (5.3), and (5.5), indicate that the equations are valid for $t \geq 0$. More precisely, we mean by this notation that $t \geq 0_+$, and we imply this in future cases. We are usually given the function at $t = 0_-$, and must find the function at $t = 0_+$ in order to obtain K in Eq. (5.5). The value of the function $t = 0_+$ is called the *initial condition*, which must be known before we can obtain the solution to the homogeneous equation.

One way to approach the initial-condition problem is to integrate Eq. (5.2) over the interval from $t = 0_-$ to $t = 0_+$. Then in Eq. (5.2),

$$L \int_{0_-}^{0_+} di + R \int_{0_-}^{0_+} i \, dt = 0 \tag{5.6}$$

or

$$L(i(0_+) - i(0_-)) + R \int_{0_-}^{0_+} i \, dt = 0$$

The term containing R is zero since the integration is over an infinitesimal interval. Thus

$$L(i(0_+) - i(0_-)) = 0$$

or

$$i(0_+) = i(0_-) \tag{5.7}$$

Equation (5.7) states that i is indeed continuous at $t = 0$ in this case.

Engineers usually prefer a more physical approach to the situation. For electrical engineers two continuity relations exist, the first being the continuity of charge, or

$$\sum q_1 = \sum q_2 \qquad (5.8)$$

This charge equation states that the total charge on capacitances before the change (subscript 1) equals the total charge after the change (subscript 2). The summation $\sum$ is over the number of individual charges.

The second relation is the continuity of flux linkages, or

$$\sum L_1 i_1 = \sum L_2 i_2 \qquad (5.9)$$

where again the subscripts refer to before and after conditions. The summation in each case includes all L's (inductances) with their associated currents. One flux linkage occurs when one line of magnetic flux links one turn of a coil, as in Fig. 5.3. Equation

Figure 5.3 Flux linkage.

(5.9) implies that we cannot change the total flux linkages in an inductive circuit instantaneously; that is, it takes time to alter flux linkages. The fundamental basis of this rule lies in the fact that flux linkages represent energy, and changing flux linkages requires an energy change. To alter flux linkages instantaneously requires infinite power, which is unavailable. Flux linkages are related to the inductance L and current i by definition as shown in Chapter 2. Mutual inductance effects, if they exist, must be included in Eq. (5.9).

If only one inductance is involved in Eq. (5.9), or $L_1 = L_2$, then $i_1 = i_2$. Thus, the common expression "current cannot change instantaneously in an inductance." This expression is true for one inductance but may be untrue if other inductances enter the network after switching. Similarly, since $q = Cv$ in Eq. (5.8), then we have the common expression "voltage cannot change instantaneously across a capacitance," again true for one capacitance.

Going back to Fig. 5.1 and applying Eq. (5.9), since $L_1 = L_2$, $i_L(0_+) = i_L(0_-)$. Then in Eq. (5.5), at $t = 0_+$,

$$i = i_L = \frac{E}{R_1} = K$$

or finally,

$$i = \frac{E}{R_1} e^{-(R/L)t} \qquad t \geq 0 \qquad (5.10)$$

The current i through L is depicted in Fig. 5.4 for two L/R values. Note that if the voltage source of Fig. 5.1 has a polarity the reverse of E, the polarity of the current in Eq. (5.10) would be reversed since the initial current $i(0_+) = E/R_1$ would be opposite to the reference current i.

If we analyze the voltage v_R across R, we see that since $i_R = 0$ before $t = 0$, $v_R = 0$ for $t < 0$. Immediately after $t = 0$, $v_R = iR$, or $v_R = (ER/R_1)e^{-(R/L)t}$, $t > 0$.

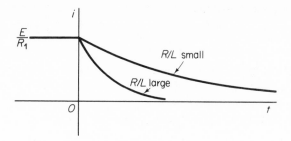

Figure 5.4 Transient response of *RL* network.

The curve in Fig. 5.5 shows this response, and the fact that the voltage v_R is discontinuous at $t = 0$.

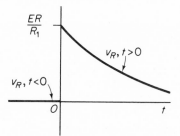

Figure 5.5 Transient response
of v_R in network of Fig. 5.1.

If the circuit of Fig. 5.1 is altered to that of Fig. 5.6 by adding voltage source E_1, and with the switches operating as before, loop analysis gives

$$L\frac{di}{dt} + Ri = E_1 \qquad t \geq 0 \tag{5.11}$$

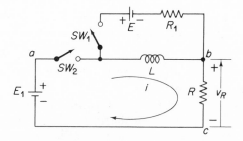

Figure 5.6 *RL* network to
demonstrate steady-state term.

The source or forcing function, E_1, appears to the circuit as a step function of amplitude E_1 in this case. The particular integral (steady-state) solution, using the method of undetermined coefficients, is

$$i = K_2 \tag{5.12}$$

where K_2 is a constant. Then in Eq. (5.11),

$$\frac{di}{dt} = \frac{dK_2}{dt} = 0$$

and

$$RK_2 = E_1$$

or

$$K_2 = \frac{E_1}{R}$$

The complete solution is the sum of the previous homogeneous (transient) solution, and the particular integral (steady-state) solution, or

$$i = Ke^{-(R/L)t} + \frac{E_1}{R} \tag{5.13}$$

Again at $t = 0_-$, $i_L = E/R_1$; hence, $i(0_+) = E/R_1$. Then at $t = 0_+$,

$$\frac{E}{R_1} = K + \frac{E_1}{R} \tag{5.14}$$

or

$$K = \frac{E}{R_1} - \frac{E_1}{R}$$

Finally,

$$i = \frac{E}{R_1}e^{-(R/L)t} + \frac{E_1}{R}[1 - e^{-(R/L)t}] \qquad t \geq 0 \tag{5.15}$$

If E in Fig. 5.6 is reversed, we obtain

$$i = -\frac{E}{R_1} + \frac{E_1}{R}e^{-(R/L)t} + \frac{E_1}{R} \qquad t \geq 0 \tag{5.16}$$

If $E = 0$, then

$$i = \frac{E_1}{R}(1 - e^{-(R/L)t}) \qquad t \geq 0 \tag{5.17}$$

This is called the *zero-state response* (response with zero initial conditions) to a step function of magnitude E_1.

The response of Eq. (5.17) is sketched in Fig. 5.7. The time t necessary to give the exponent of the second term on the right-hand side of Eq. (5.17) [or the exponent of the right-hand term in Eq. (5.10)], a value of unity is called the *circuit time constant, τ*. In this case

$$\tau = \frac{L}{R} \qquad \text{seconds} \tag{5.18}$$

The current i in Fig. 5.7 reaches 63.2 per cent of its final value in one time constant and 99.3 per cent of its final value in five time constants.

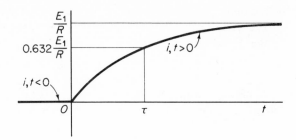

Figure 5.7 Zero-state response of *RL* network
in Fig. 5.6.

5.2 First-Order Differential Equations—*RC* Circuits

The circuit shown in Fig. 5.8 is similar to that of Fig. 5.1 except that now we have a capacitance C in place of the inductance. Switch sw_2 has been open and sw_1 closed for a long time, and they reverse states at $t = 0$. From Kirchhoff's voltage law around

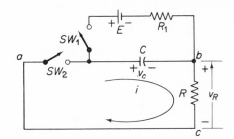

Figure 5.8 *RC* network.

loop *abc* we get

$$\frac{1}{C} \int i \, dt + Ri = 0 \qquad t \geq 0 \tag{5.19}$$

Differentiating Eq. (5.19) we obtain

$$\frac{i}{C} + R\frac{di}{dt} = 0 \qquad t \geq 0 \tag{5.20}$$

Here again $t \geq 0$ implies that $t \geq 0_+$. From the previous discussion, the solution is obviously

$$i = Ke^{-(1/RC)t} \qquad t \geq 0 \tag{5.21}$$

To find $i(0_+)$ we must first find $v_c(0_+)$. We see from Fig. 5.8 that

$$v_c(0_-) = E \tag{5.22}$$

if we proceed around the loop *abc* clockwise, or in the same direction as the given current i. From the continuity-of-charge principle mentioned in the previous section, $v_{c1}C = v_{c2}C$, or since C is common, the capacitance voltage is the same just before as just after the change. Hence

$$v_c(0_+) = v_c(0_-) = E \tag{5.23}$$

Now if we examine the network in Fig. 5.8 after sw_2 closes and sw_1 opens, a Kirchhoff tour around the loop indicates that $v_R(0_+) = -v_c(0_+)$, and therefore

$$i(0_+) = -\frac{E}{R} \tag{5.24}$$

Then in Eq. (5.21) at $t = 0_+$,

$$-\frac{E}{R} = K \tag{5.25}$$

or finally,

$$i = -\frac{E}{R}e^{-(1/RC)t} \qquad t \geq 0 \tag{5.26}$$

The current i is plotted in Fig. 5.9. Note that i has two values at $t = 0$, or is discontinuous at this point. The time constant τ [time necessary to make the exponent of Eq. (5.26) unity] is $\tau = RC$. This is analogous to the time constant $\tau = L/R$ in the previous RL circuit. By adjusting RC or L/R it is possible to obtain time constants in the order of seconds, and which can be observed by ordinary ammeters. Circuits with short time constants can only be observed by an oscilloscope.

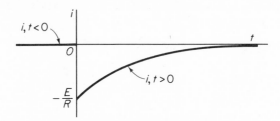

Figure 5.9 Transient response of RC network.

If the circuit of Fig. 5.8 is altered to that of Fig. 5.10 by adding the voltage source or forcing function, E_1, and we have the previous switch operation, we obtain the equation

$$\frac{1}{C}\int i\, dt + Ri = E_1 \tag{5.27}$$

from which the current i is

$$i = Ke^{-(1/RC)t} \tag{5.28}$$

as before. Again $v_c(0_+) = E$. However, this time, inspection of Fig. 5.10 shows that

$$i(0_+) = \frac{E_1 - E}{R} \tag{5.29}$$

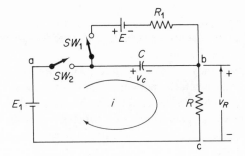

Figure 5.10 *RC* network to demonstrate steady-state solution.

and hence

$$K = \frac{E_1 - E}{R}$$

or

$$i = \frac{E_1 - E}{R} e^{-(1/RC)t} \qquad t \geq 0 \qquad (5.30)$$

The sign of i depends on the relative values of E and E_1, and again i will be discontinuous at $t = 0$.

With *RC* circuits, it is often more convenient to solve for voltages than for currents. Since the current through a capacitance is

$$i = C\frac{dv_c}{dt} \qquad (5.31)$$

the voltage across a resistance in series with C may be expressed as

$$v_R = RC\frac{dv_c}{dt} \qquad (5.32)$$

Now making a Kirchhoff tour clockwise in Fig. 5.10, we obtain

$$RC\frac{dv_c}{dt} + v_c = E_1 \qquad (5.33)$$

Solving this as usual we get a transient solution conditioned on $v_c(0_+)$ as follows:

$$v_c = Ke^{-(1/RC)t} \qquad t \geq 0 \qquad (5.34)$$

The final (steady-state) value of v_c from Fig. 5.10 is obviously E_1. The total solution then becomes

$$v_c = Ke^{-(1/RC)t} + E_1 \qquad t \geq 0 \qquad (5.35)$$

But $v_c(0_+) = v_c(0_-) = E$. Hence at $t = 0_+$,

$$E = K + E_1$$

or

$$K = E - E_1$$

or

$$v_c = (E - E_1)e^{-(1/RC)t} + E_1 \qquad t \geq 0 \tag{5.36}$$

Knowing v_c, i may be readily found by first finding $v_R(t)$.

Note that when the network contains independent sources, the solution must be put in terms of the transient solution plus the steady-state solution *before* the constants are evaluated. In other words, the constants depend on the sources as well as on the initial currents or voltages. Evaluating the constants in the transient solution prior to adding the steady-state solution results in error (this is a common mistake).

5.3 Second-Order Differential Equations—
RLC Circuits

The circuit of Fig. 5.11 represents a series connection of R, L, and C with a source $E(t)$. The inductance L may have an initial current i_0 at $t = 0_-$ and the capacitance C may have an initial charge on it resulting in a voltage v_{c0} at $t = 0_-$. The mechanisms

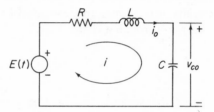

Figure 5.11 Series *RLC* circuit
with initial conditions.

for introducing these initial conditions have already been discussed in the previous sections; we assume that such devices are operating here. The direction of i_0 and the polarity of v_{c0} must be known, and one set of conditions are shown in Fig. 5.11.

Selecting a current i, we make a Kirchhoff tour clockwise, resulting in the equation

$$Ri + L\frac{di}{dt} + S \int i \, dt = E(t) \tag{5.37}$$

where

$$S = \frac{1}{C}$$

Differentiating both sides, we get

$$L\frac{d^2i}{dt^2} + R\frac{di}{dt} + Si = \frac{dE(t)}{dt} \tag{5.38}$$

Using the network shown in Fig. 5.12, and using nodal analysis, we find that

$$Cv + \Gamma \int v \, dt + C\frac{dv}{dt} = J(t) \tag{5.39}$$

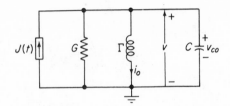

Figure 5.12 Parallel *RLC* network with initial conditions.

Differentiating both sides, we get

$$C\frac{d^2v}{dt^2} + G\frac{dv}{dt} + \Gamma v = \frac{dJ(t)}{dt} \tag{5.40}$$

Equation (5.40) represents the same type of equation (with slightly different symbolism) as Eq. (5.38). The two circuits are called *duals*. If we can solve one of the equations, the solution to the other is obvious. Let us therefore look at Eq. (5.38).

Assume that $E(t)$ in Eq. (5.38) is a constant E. Then we obtain the homogeneous equation

$$L\frac{d^2i}{dt^2} + R\frac{di}{dt} + Si = 0 \tag{5.41}$$

From our previous work, we suspect solutions of the form

$$i = Ke^{\lambda t} \tag{5.42}$$

From Eq. (5.42),

$$\frac{di}{dt} = \lambda Ke^{\lambda t}$$

$$\frac{d^2i}{dt^2} = \lambda^2 Ke^{\lambda t}$$

Substituting these terms into Eq. (5.41), we obtain

$$L\lambda^2 Ke^{\lambda t} + R\lambda Ke^{\lambda t} + SKe^{\lambda t} = 0 \tag{5.43}$$

or

$$Ke^{\lambda t}(L\lambda^2 + R\lambda + S) = 0 \tag{5.44}$$

Unless $K = 0$ in Eq. (5.44), which is a trivial solution, then

$$L\lambda^2 + R\lambda + S = 0 \tag{5.45}$$

Equation (5.45) is termed the characteristic equation of the network. It may be obtained directly by substituting λ for di/dt, λ^2 for d^2i/dt^2, and 1 for i in (5.41). Solving the characteristic equation, we find that

$$\lambda = \frac{-R \pm \sqrt{R^2 - 4LS}}{2L} \tag{5.46}$$

or in terms of C,

$$\lambda = \frac{-R}{2L} \pm \sqrt{\frac{R^2}{4L^2} - \frac{1}{LC}} \tag{5.47}$$

where $C = 1/S$.

There are three solutions to Eq. (5.47); (1) roots real and unequal, (2) roots real and equal, and (3) roots complex. The solution is determined by the corresponding value of the term inside the radical (discriminant): positive, zero, or negative.

EXAMPLE 5.1

Assume that the values of the elements in the RLC network of Fig. 5.11 are (a) $R = 10\ \Omega$, $L = 1$ H, $C = 0.2$ F; then from Eq. (5.47),

$$\lambda = -\frac{10}{2} \pm \sqrt{25 - \frac{10}{2}}$$

$$= -5 \pm \sqrt{25 - 5}$$

$$\lambda_1 = -5 + \sqrt{20} = -0.54$$

$$\lambda_2 = -5 - \sqrt{20} = -9.46$$

(b) $R = 10$, $L = 1$, $C = 4 \times 10^{-2}$. The roots (or eigenvalues) of the equation are

$$\lambda = -\frac{10}{2} \pm \sqrt{25 - \frac{100}{4}}$$

$$\lambda_1 = -5$$

$$\lambda_2 = -5$$

(c) $R = 10$, $L = 1$, $C = 4 \times 10^{-3}$. The roots in this case are

$$\lambda = -\frac{10}{2} \pm \sqrt{25 - \frac{10^3}{4}}$$

$$\lambda_1 = -5 + j15$$

$$\lambda_2 = -5 - j15$$

where $j = \sqrt{-1}$.

If the roots λ_1 and λ_2 are real and unequal, as in part (a) of this example, then the homogeneous solution of Eq. (5.41) becomes

$$i = K_1 e^{\lambda_1 t} + K_2 e^{\lambda_2 t} \tag{5.48}$$

If the roots are equal, $\lambda_1 = \lambda_2 = \lambda$, we have a degenerate solution, which is

$$i = K_1 e^{\lambda t} + K_2 t e^{\lambda t} \tag{5.49}$$

Finally, if the roots are complex,

$$i = K_1 e^{(\sigma_1 + j\omega_d)t} + K_2 e^{(\sigma_1 - j\omega_d)t} \tag{5.50}$$

where $\lambda_1 = \sigma_1 + j\omega_d$ and $\lambda_2 = \sigma_1 - j\omega_d$. [In part c, $\sigma_1 = -5$, $\omega_d = 15$.]

For case 1 (roots real and unequal),

$$i = K_1 e^{\lambda_1 t} + K_2 e^{\lambda_2 t} \tag{5.51a}$$

and

$$\frac{di}{dt} = \lambda_1 K_1 e^{\lambda_1 t} + \lambda_2 K_2 e^{\lambda_2 t} \tag{5.51b}$$

We now need to find K_1 and K_2 from the initial conditions.

From our previous experience, $i(0_+) = i(0_-) = i_0$ at $t = 0$. Now we return to Eq. (5.37) with the right-hand side a constant E. The voltage v_{c0} across the capacitance is continuous at $t = 0$ and represents the term $S \int i\, dt$ at time zero. Hence in Eq. (5.37) at $t = 0_+$ we obtain

$$Ri_0 + L\frac{di}{dt}(0) + v_{c0} = E \tag{5.52}$$

or

$$\frac{di}{dt}(0) = \frac{E - v_{c0} - Ri_0}{L} \tag{5.53}$$

Substituting $t = 0$ into Eqs. (5.51), we find that

$$i_0 = K_1 + K_2 \tag{5.54a}$$

and

$$\frac{di}{dt}(0) = \lambda_1 K_1 + \lambda_2 K_2 \tag{5.54b}$$

EXAMPLE 5.2

Assume that the network of Fig. 5.11 has the element values specified in Example 5.1(a) and the initial conditions $i_0 = 6$ A and $v_{c0} = 20$ V. Also let the source voltage $E(t) = 0$ V. Determine the network loop current. From Eq. (5.53) we have

$$\frac{di}{dt}(0) = \frac{-20 - (10)(6)}{1} = -80$$

Then using Eqs. (5.54), we obtain

$$-80 = -0.54K_1 - 9.46K_2$$
$$6 = K_1 + K_2$$

or

$$K_1 = -2.61$$

and

$$K_2 = 8.61$$

The loop current is

$$i = -2.61e^{-0.54t} + 8.61e^{-9.46t}$$

Figure 5.13 sketches the solution, which is the sum of two decaying exponentials. Note that the value of i becomes negative at slightly over 0.1 second.

For case 2 (roots equal), the constants are determined in a similar way. This network is said to be *critically damped*, for reasons to be explained.

For case 3 (roots complex) we have a more interesting situation. For greater generality assume that in Eq. (5.50) K_1 and K_2 are complex quantities or let

$$K_1 = \alpha + j\beta$$

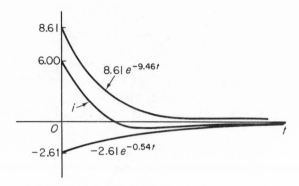

Figure 5.13 Example 5.2 loop current waveform.

We plot the complex quantity $\alpha + j\beta$ in Fig. 5.14, and in this figure we let $r = \sqrt{\alpha^2 + \beta^2}$ and $\tan \theta = \beta/\alpha$. Then from the identity $e^{j\theta} = \cos \theta + j \sin \theta$,

$$\alpha + j\beta = r(\cos \theta + j \sin \theta)$$
$$= re^{j\theta}$$

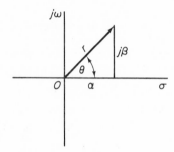

Figure 5.14 Complex quantity
vector representation.

Furthermore, if $K_1 = (\alpha + j\beta)$, then

$$K_2 = (\alpha - j\beta)$$
$$= re^{-j\theta}$$
$$= K_1^*\dagger$$

since in Eq. (5.50) i is a real value at all times, including $t = 0$. Using these ideas, Eq. (5.50) becomes

$$i = \sqrt{\alpha^2 + \beta^2}[e^{\sigma_1 t + j(\omega_d t + \theta)} + e^{\sigma_1 t - j(\omega_d t + \theta)}] \tag{5.55}$$

$\dagger K_1^*$ is the complex conjugate of K_1.

Thus

$$i = re^{\sigma_1 t}[e^{j(\omega_d t + \theta)} + e^{-j(\omega_d t + \theta)}]$$
$$= 2re^{\sigma_1 t}[\cos(\omega_d t + \theta)] \tag{5.56}$$

since $\cos \phi = (e^{j\phi} + e^{-j\phi})/2$, or

$$i = K_3 e^{\sigma_1 t} \cos(\omega_d t + \theta) \tag{5.57}$$

Equation (5.57) has two constants, K_3 and θ, just as Eq. (5.50) has two constants, but Eq. (5.57) has perhaps more physical meaning.

EXAMPLE 5.3

Assume that the network in Fig. 5.11 has the element values specified in Example 5.1(c) and the initial conditions $i_0 = 6$ A and $v_{c0} = 20$ V. Also let $E(t) = 0$ for simplicity. Then

$$i_0 = 6 \quad \text{and} \quad \frac{di}{dt}(0) = -80$$

as in Example 5.2. Using the roots calculated in Example 5.1(c) and the general form of the solution given in Eq. (5.50),

$$i = K_1 e^{(\sigma_1 + j\omega_d)t} + K_2 e^{(\sigma_1 - j\omega_d)t}$$
$$= (\alpha + j\beta)e^{(-5+j15)t} + (\alpha - j\beta)e^{(-5-j15)t}$$

or at $t = 0$,

$$6 = (\alpha + j\beta) + (\alpha - j\beta) = 2\alpha$$

or

$$\alpha = 3$$

Differentiating the equation for i with respect to t,

$$\frac{di}{dt} = (\sigma_1 + j\omega_d)(\alpha + j\beta)e^{(-5+j15)t} + (\sigma_1 - j\omega_d)(\alpha - j\beta)e^{(-5-j15)t}$$

or at $t = 0$,

$$-80 = (-5 + j15)(3 + j\beta) + (-5 - j15)(3 - j\beta)$$
$$= [-15 - 15\beta + j(45 - j5\beta)] + [-15 - 15\beta - j(45 - j5\beta)]$$
$$= -30 - 30\beta$$
$$30\beta = 80 - 30 = 50$$

or

$$\beta = \tfrac{5}{3}$$

Then

$$\sqrt{\alpha^2 + \beta^2} = 3.44$$

and

$$K_3 = 2\sqrt{\alpha^2 + \beta^2} = 6.88$$

From Fig. 5.14,

$$\theta = \tan^{-1} \frac{5}{3(3)}$$

or

$$\theta = 29°$$

The loop current for this underdamped case is†

$$i = 6.88e^{-5t}\cos(15t + 29°)$$

[We could have evaluated the constants of Eq. (5.57) directly by trigonometric manipulation.] The solution is sketched in Fig. 5.15. We see that in the case of complex roots we get a damped oscillatory solution.

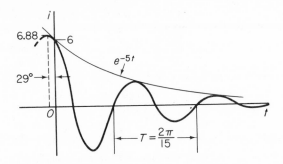

Figure 5.15 Example 5.3 loop current waveform.

Let us now look at the condition of zero state response to a step function input $E(t) = Eu(t)$ for the circuit of Fig. 5.11, but make the capacitance voltage the response of interest. Since $i_c = C(dv_c/dt)$ and $i = i_c$, then substituting in Eq. (5.37) we obtain

$$LC\frac{d^2v_c}{dt^2} + RC\frac{dv_c}{dt} + v_c = E \qquad t \geq 0 \tag{5.58}$$

The characteristic equation is

$$L\lambda^2 + R\lambda + \frac{1}{C} = 0 \tag{5.59}$$

which is the same as Eq. (5.45). This illustrates the important point that the eigenvalues of the network depend only on the network components (in this case, L, R, and C) and not on the sources or the responses desired. From our previous discussion, the transient solution with complex roots is

$$v_c = (\alpha + j\beta)e^{\sigma_1 + j\omega_d} + (\alpha - j\beta)e^{\sigma_1 - j\omega_d} \tag{5.60}$$

or alternatively,

$$v_c = K_3 e^{\sigma_1 t}\cos(\omega_d t + \theta) \tag{5.61}$$

†To evaluate the current i the two terms making up the angle must be in the same units, either degrees or radians. Computer programs usually use radians, hence the 29° should be converted to radians.

where

$$K_3 = 2\sqrt{\alpha^2 + \beta^2}, \qquad \theta = \tan\frac{\beta}{\alpha}$$

$$\sigma_1 = \frac{-R}{2L} \qquad \omega_d = \sqrt{\frac{1}{LC} - \frac{R^2}{4L^2}}$$

The steady-state solution from Eq. (5.58) is

$$v_c = E$$

so the complete solution is

$$v_c = K_3 e^{\sigma_1 t} \cos(\omega_d t + \theta) + E \qquad (5.62)$$

Since v_c is continuous at $t = 0$ and $v_c(0_-) = 0$ (zero state at $t = 0$),

$$v_c(0_+) = 0$$

or

$$0 = K_3 \cos\theta + E$$

$$K_3 \cos\theta = -E$$

From Eq. (5.62),

$$\frac{dv_c}{dt} = \sigma_1 K_3 e^{\sigma_1 t}\cos(\omega_d t + \theta) - \omega_d K_3 e^{\sigma_1 t}\sin(\omega_d t + \theta) \qquad (5.63)$$

Now i_L is continuous at $t = 0$ and $i_L = i_c = C(dv_c/dt)$, and since $i_L(0_-) = 0$, then $i(0_+) = 0$, or

$$C\frac{dv_c}{dt}(0) = 0$$

Hence, in Eq. (5.63) at $t = 0_+$

$$0 = \sigma_1 K_3 \cos\theta - \omega_d K_3 \sin\theta$$

or

$$\omega_d \sin\theta = \sigma_1 \cos\theta$$

and

$$\tan\theta = \frac{\sigma_1}{\omega_d}$$

From a right triangle with sides σ_1 and ω_d (see Fig. 5.16),

$$\cos\theta = \frac{\omega_d}{(\sigma_1^2 + \omega_d^2)^{1/2}}$$

Figure 5.16 Right triangle relationship between θ, ω_d, and σ.

Thus,

$$K_3 = -E\frac{(\sigma_1^2 + \omega_d^2)^{1/2}}{\omega_d}$$

The final solution is then

$$v_c = -E\frac{(\sigma_1^2 + \omega_d^2)^{1/2}}{\omega_d}e^{\sigma_1 t}\cos(\omega_d t + \theta) + E \qquad (5.64)$$

where $\tan\theta = \sigma_1/\omega_d$.

EXAMPLE 5.4

Let $R = 10\,\Omega$, $L = 1$ H, $C = 4 \times 10^{-3}$ F as in Example 5.1(c). Let $E = 100$ V. Applying the results of Example 5.1(c), $\sigma_1 = -5$, $\omega_d = 15$. Hence,

$$v_c = -100\left(\frac{15.8}{15}\right)e^{-5t}\cos(15t - 18.4°) + 100$$

$$= -105.3e^{-5t}\cos(15t - 18.4°) + 100$$

The response of the voltage v_c is sketched in Fig. 5.17.

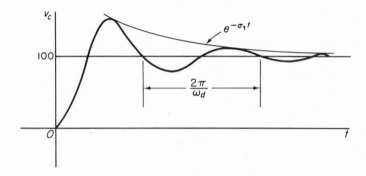

Figure 5.17 Response of v_c.

Figure 5.17 for Example 5.4 shows that the response v_c overshoots the final value and oscillates about the final value at frequency $\omega_d/2\pi$ Hz, with the amplitude decaying at rate e^{-5t}. This overshoot occurs for all underdamped networks with a step input. The response of a second-order network may be further analyzed by defining some new variables in the characteristic equation. The characteristic equation given in Eq. (5.45) may be written

$$\lambda^2 + \frac{R}{L}\lambda + \frac{1}{LC} = 0 \qquad (5.65)$$

If the roots are complex, we may also write Eq. (5.65) as

$$(\lambda - \sigma_1 + j\omega_d)(\lambda - \sigma_1 - j\omega_d) = 0 \qquad (5.66)$$

where $(\sigma_1 + j\omega_d)$ and $(\sigma_1 - j\omega_d)$ are the roots of Eq. (5.65). Expanding Eq. (5.66), we obtain

$$\lambda^2 - 2\sigma_1\lambda + \sigma_1^2 + \omega_d^2 = 0 \tag{5.67}$$

Now we define $\omega_n^2 = (\sigma_1^2 + \omega_d^2)$, and $\zeta = -\sigma_1/\omega_n$, so Eq. (5.67) becomes

$$\lambda^2 + 2\zeta\omega_n\lambda + \omega_n^2 = 0 \tag{5.68}$$

where ζ is called the *damping factor* and ω_n the *natural frequency*. The roots of Eq. (5.68) (if complex) are

$$\lambda = -\zeta\omega_n \pm \sqrt{\zeta^2\omega_n^2 - \omega_n^2}$$

or

$$\lambda_1 = -\zeta\omega_n + j\omega_n\sqrt{1 - \zeta^2} \tag{5.69a}$$

$$\lambda_2 = -\zeta\omega_n - j\omega_n\sqrt{1 - \zeta^2} \tag{5.69b}$$

Thus, the oscillation angular velocity and the damping exponent are

$$\omega_d = \omega_n\sqrt{1 - \zeta^2}$$

$$\sigma_1 = -\zeta\omega_n$$

The two complex roots (eigenvalues of the network) are plotted in Fig. 5.18. We see

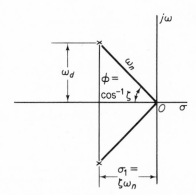

Figure 5.18 Complex roots of a second-order network.

that the complex root $\lambda_1 = \sigma_1 + j\omega_d = -\zeta\omega_n + j\omega_n\sqrt{1 - \zeta^2}$ has a distance ω_n from the origin and an angle with the negative real axis of $\phi = \cos^{-1}\zeta$. If we hold ω_n constant and vary ζ, then the roots form a semicircle in the left half of the complex plane as in Fig. 5.19. In terms of the new variables, Eq. (5.64) becomes

$$v_c = -E\frac{1}{\sqrt{1 - \zeta^2}}e^{-\zeta\omega_n t}\cos(\sqrt{1 - \zeta^2}\,\omega_n t + \theta) + E \tag{5.70}$$

where

$$\tan\theta = \frac{\zeta}{\sqrt{1 - \zeta^2}}$$

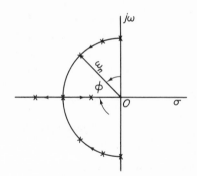

Figure 5.19   Complex   pole
positions in plane for different
values of R in second-order
network.

If $\zeta = 0$, then the eigenvalues lie on the $j\omega$-axis in Fig. 5.19 (no damping) and Eq. (5.70) shows that we have a perfect oscillator with a cosine function centered around E (see Fig. 5.20). As ζ becomes larger, ϕ becomes smaller (following the arrows in

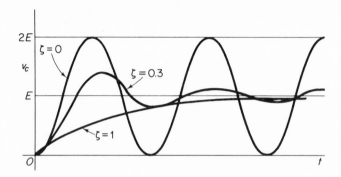

Figure 5.20 Effect of damping factor on
transient response of second-order network.

Fig. 5.19), and with $0 < \zeta < 1$ we get a response similar to that in Example 5.4. When $\zeta = 1$, $\sqrt{1 - \zeta^2} = 0$ and Eq. (5.70) does not apply. If we reexamine Eq. (5.68), with $\zeta = 1$, we have

$$\lambda^2 + 2\omega_n\lambda + \omega_n^2 = 0$$

and

$$(\lambda + \omega_n)(\lambda + \omega_n) = 0$$

or we have equal roots. In this case the network is critically damped. For $\zeta > 1$ the roots in Eq. (5.68) become real and unequal, as shown in Fig. 5.19, and we obtain two real exponentials for the transient solutions. Thus, the critical damping case ($\zeta = 1$) is on the borderline between oscillation and no oscillation, the network oscillating for $\zeta < 1$ and not oscillating for $\zeta > 1$. This is the reason why ζ is called the *damping factor*. Differentiation of Eq. (5.70) gives a formula for the peak (first) overshoot[†]

†See Chapter 9.

which shows that the overshoot depends only on ζ:

$$\% \text{ overshoot} = 100 \exp\left(\frac{-\zeta\pi}{\sqrt{1-\zeta^2}}\right) \tag{5.71}$$

If ζ becomes negative ($\phi > 90°$ in Fig. 5.19), the solution grows exponentially and the network is said to be unstable. For this to occur, R must be negative, that is, must supply power rather than absorb power. Such networks may be constructed by arranging energy feedback paths in the network.

EXAMPLE 5.5

Given the differential equation

$$4\frac{d^2y}{dt^2} + 16\frac{dy}{dt} + 100y = 125u(t) \qquad t \geq 0$$

Find the damping factor ζ, the oscillation frequency, the peak overshoot, and the final value.

Solution. The characteristic equation is

$$\lambda^2 + 4\lambda + 25 = 0$$

Compared to Eq. (5.68),

$$\omega_n^2 = 25 \quad \text{or} \quad \omega_n = 5$$

$$2\zeta\omega_n = 2\zeta(5) = 4 \quad \text{or} \quad \zeta = 0.4$$

$$\omega_d = \omega_n\sqrt{1-\zeta^2} = 5\sqrt{1-0.16} = 4.59$$

The oscillation frequency is

$$f_c = \frac{\omega_d}{2\pi} = \frac{4.59}{2\pi} = 0.733 \text{ Hz}$$

The peak overshoot from Eq. (5.71) is

$$\% \text{ overshoot} = 100 \exp\left(\frac{-0.4\pi}{\sqrt{1-0.6}}\right)$$

$$= 100e^{-1.372}$$

$$= 25.4\%$$

The final value ($t = \infty$) is

$$y = \frac{125}{100} = 1.25$$

5.4 Networks Driven by External Sources

A nonhomogeneous differential equation consists of an equation in which the right-hand side is a nonzero function of time $f(t)$ and defines a forcing function or some derivative of the forcing function. We have already solved some equations of this type in this chapter [for example, Eq. (5.11)]. In studying these equations we note that there are two parts to the solution: (1) the solution i_c to the corresponding homogeneous equation (right side set equal to zero), called the *complementary function*, and

(2) the solution i_p necessary to satisfy the forcing function, called the *particular integral*. Thus, we write the total solution i as the sum of two parts, or

$$i = i_p + i_c$$

We can find the complementary part i_c using methods already described. We now concentrate on the particular integral solution i_p.

In electrical network analysis, the forcing function or derivative of the forcing function that we encounter usually occurs as one of a restricted set of mathematical forms, such as K, $\cos \omega t$, kt, e^{-at}, or linear combinations of these terms. (Driving sources which generate functions such as $\tan \omega t$ are extremely rare.) If we investigate forcing functions or sources of the types mentioned, the method of undetermined coefficients admirably suits our purpose. In the method of undetermined coefficients, we select trial functions of all the possible forms that might satisfy the differential equation. Each term in the trial solution is assigned an arbitrary undetermined coefficient. We then substitute the trial solutions into the differential equation and form a set of linear algebraic equations by equating coefficients of like functions resulting from the substitution. The terms not part of the solution will have zero coefficients.

The required forms of the trial solutions for our restricted set of functions $f(t)$ are tabulated in Table 5.1. Using Table 5.1 the following procedure is suggested to evaluate a solution to nonhomogeneous differential equations:

TABLE 5.1 Particular Integral Trial Solutions for Differential Equations with $f(t)$ on the Right Side

Term in $f(t)$[a]	Trial solution form[b]
1. a_0 (constant)	K
2. $a_1 t^n$	$K_0 t^n + K_1 t^{n-1} + \cdots + K_{n-1} t + K_n$
3. $a_2 e^{bt}$	$K e^{bt}$
4. $a_3 \cos \omega t$	$K_1 \cos \omega t + K_2 \sin \omega t$
5. $a_4 \sin \omega t$	
6. $a_5 t^n e^{bt} \cos \omega t$	$(G_1 t^n + \cdots + G_{n-1} t + G_n) e^{bt} \cos \omega t$
7. $a_6 t^n e^{bt} \sin \omega t$	$\quad + (H_1 t^n + \cdots + H_{n-1} t + H_n) e^{bt} \sin \omega t$

[a]When $f(t)$ consists of a sum of several terms, the appropriate particular integral is the sum of the particular integrals corresponding to the individual terms.

[b]Whenever a term in any of the trial integrals is already a part of the complementary solution of the given equation, it is necessary to modify the required choice by multiplying it by t before using it. If the term appears r times in the complementary solution, the choice must be multiplied by t^r.

1. Compute the complementary solution i_c with arbitrary coefficients as described in Sections 5.1, 5.2, and 5.3 (The coefficients cannot be evaluated until the particular integral or steady-state solution is known.) Compare each term of the solution with the forcing function.

2. Obtain the trial form of the particular integral from Table 5.1. Each term of the trial solution should be given a unique unknown coefficient. We observe footnote b, and modify the trial form accordingly if it is already a part of the complementary solution.

3. Substitute the trial solution into the differential equation. By equating coefficients of like terms, form a set of algebraic equations in the undetermined coefficients.

4. Determine the coefficients for the trial solution of the particular integral. Combine the result with the complementary solution to obtain the total solution. Now we can evaluate the arbitrary coefficients of the complementary solution using the knowledge of the initial conditions and the steady-state solution.

EXAMPLE 5.6

Determine the response $i(t)$ of the series RL network shown in Fig. 5.21 driven by a time-varying voltage source $E(t) = Ee^{-at}$, where E and a are constants. The switch sw closes at $t = 0$, and $i(0) = 0$.

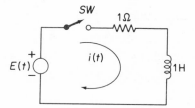

Figure 5.21 Example 5.6 series RL network.

By Kirchhoff's voltage law, the differential equation is

$$L\frac{di}{dt} + Ri = Ee^{-at}$$

or

$$\frac{di}{dt} + \frac{R}{L}i = \frac{E}{L}e^{-at}$$

Using the element values given,

$$\frac{di}{dt} + i = Ee^{-at} \tag{5.72}$$

The complementary solution for this equation becomes

$$i_c = K_1 e^{-(R/L)t}$$
$$= K_1 e^{-t}$$

From Table 5.1, the trial particular integral is

$$i_p = K_2 e^{-at}$$

Assuming that $a \neq 1$, we substitute this i_p into Eq. (5.72), to obtain

$$-aK_2 e^{-at} + K_2 e^{-at} = Ee^{-at}$$

or

$$K_2 = \frac{E}{1-a} \qquad a \neq 1$$

The total solution $i(t)$ is the sum of i_c and i_p, which is

$$i(t) = K_1 e^{-t} + \frac{E}{1-a} e^{-at} \qquad a \neq 1$$

Since $i(0) = 0$, then

$$K_1 = -\frac{E}{1-a}$$

and the final solution is

$$i(t) = \frac{E}{1-a}(e^{-at} - e^{-t}) \qquad a \neq 1 \quad t \geq 0$$

If $a = 1$, the form of the trial solution for i_p should be

$$i_p = K_2 t e^{-t}$$

Substituting this trial solution into Eq. (5.72) gives

$$K_2(-t e^{-t} + e^{-t}) + K_2 t e^{-t} = E e^{-t}$$

or

$$K_2 = E$$

The solution in this case is thus

$$i(t) = E t e^{-t} + K_1 e^{-t}$$

Inserting the initial condition $i(0) = 0$ gives

$$i(t) = E t e^{-t} \qquad t \geq 0$$

5.5 Simultaneous Differential Equations

If more than one loop or node pair exists, we obtain a system of differential equations.

Assume that the switch in Fig. 5.22 is open at $t = 0$, the capacitance has an initial

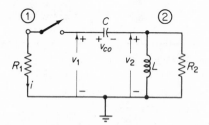

Figure 5.22 Two-node *RLC* network.

voltage v_{c0}, and the initial inductance current is zero. If sw closes at $t = 0$, we may write the two equations

$$G_1 v_1 + C\left(\frac{dv_1}{dt} - \frac{dv_2}{dt}\right) = 0 \qquad (5.73a)$$

$$C\left(\frac{dv_2}{dt} - \frac{dv_1}{dt}\right) + \Gamma \int v_2 \, dt + G_2 v_2 = 0 \qquad (5.73b)$$

These must be solved simultaneously. We assume that v_1 and v_2 have an exponential form, or that

$$v_1 = K_1 e^{\lambda t}$$

$$v_2 = K_2 e^{\lambda t}$$

Substituting these assumed solutions into Eq. (5.73a) we get the equation

$$G_1 K_1 e^{\lambda t} + C\lambda K_1 e^{\lambda t} - C\lambda K_2 e^{\lambda t} = 0 \qquad (5.74a)$$

Taking the time derivative of Eq. (5.73b) and again substituting, we get

$$C\lambda^2 K_2 e^{\lambda t} - C\lambda^2 K_1 e^{\lambda t} + \Gamma K_2 e^{\lambda t} + G_2 \lambda K_2 e^{\lambda t} = 0 \qquad (5.74b)$$

From Eq. (5.74a) the relationship between K_1 and K_2 is

$$K_1(G_1 + C\lambda) = C\lambda K_2$$

or

$$K_1 = \frac{C\lambda K_2}{G_1 + C\lambda}$$

Substituting this relationship into Eq. (5.74b) we find that

$$C\lambda^2 K_2 - \frac{C^2 \lambda^3 K_2}{G_1 + C\lambda} + \Gamma K_2 + G_2 \lambda K_2 = 0$$

or

$$(CG_1 - C^2 + CG_2)\lambda^2 + (G_1 G_2 + C\Gamma)\lambda + G_1 \Gamma = 0$$

This last equation is the characteristic equation. If the roots are λ_1 and λ_2, then

$$v_1 = K_{11} e^{\lambda_1 t} + K_{12} e^{\lambda_2 t}$$

and

$$v_2 = K_{21} e^{\lambda_1 t} + K_{22} e^{\lambda_2 t}$$

are the solutions. If the roots are complex conjugate, then the associated constants will be complex conjugate. The constants are determined by the initial conditions. The inductance in the network of Fig. 5.22 is an open circuit at $t = 0$, so $i(0_+) = v_{c0}/(R_1 + R_2)$. Then $v_1(0_+) = v_{c0} R_1/(R_1 + R_2)$ and $v_2(0_+) = -v_{c0} R_2/(R_1 + R_2)$. Since we discuss the subject of initial conditions more fully in the next section, we need not continue it here.

Simultaneous differential equations may be handled more easily on the digital computer in state-space form, which we introduce in Section 5.7 and discuss more fully in Chapters 10 through 12.

5.6 Initial Conditions

In the previous discussion of differential equations, we pointed out that the evaluation of constants in transient terms depends on the initial conditions. In electri-

cal networks these initial conditions consist of the initial currents through the inductances, the initial voltages across the capacitances, and the initial source currents and
voltages (both independent and dependent). By "initial" we mean the values of these
currents and voltages at $t = 0_+$, or just after time changes are initiated, which we take
arbitrarily as $t = 0$. We use the $t = 0_+$ symbolism since in many cases some quantities
may not be continuous at $t = 0$ and hence are undefined at this time. We must know,
or be given, the conditions at $t = 0_-$, or just before making circuit changes, and from
these determine the conditions at $t = 0_+$. In previous sections we gave two principles
for making the determinations: (1) the time continuity of charge and (2) the time
continuity of flux linkages. For other engineering systems, other physical principles
apply; for example, in translational mechanical systems the momentum is continuous,
and in hydraulic systems the quantity of fluid is continuous. In this section we outline
some techniques and examples for finding initial conditions in complex networks.

A capacitance with an initial charge may be represented by a capacitance with no
initial charge in series with a voltage source $v_c(0_+)$, where $v_c(0_+)$ is the initial voltage on
the capacitance as shown in Fig. 2.8. Similarly, an inductance with initial current
$i_L(0_+)$ may be shown, as in Fig. 2.13, as a zero current inductance in parallel with a
constant current source $i_L(0_+)$. In addition to the initial values, the polarity of the
voltage $v_c(0_+)$ and the direction of the current $i_L(0_+)$ must be clearly specified. Thus, a
capacitance at $t = 0_+$ (and only then) may be shown as in Fig. 5.23 and an inductance

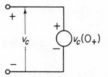

Figure 5.23 A capacitance
model valid at $t = 0_+$.

as in Fig. 5.24. The polarity of $v_c(0_+)$ may be the opposite of that shown in Fig. 5.23,
and the constant current may be the opposite of that shown in Fig. 5.24 for the given
conditions. If $v_c(0_+) = 0$, the capacitance representation is a short circuit at $t = 0_+$,
and if $i_L(0_+) = 0$, the inductance representation is an open circuit at $t = 0_+$ (the
internal resistance of a constant current source is infinite). We may use these ideas to
assist in determining initial conditions, which we illustrate in the following examples.

Figure 5.24 An inductance
model valid at $t = 0_+$.

EXAMPLE 5.7

The capacitance C in Fig. 5.25 has a voltage of $v_c(0_-)$ at $t = 0_-$ and the inductance
has a current of $i_L(0_-)$ at $t = 0_-$ with polarity and directions as shown, and the switch
sw is closed at $t = 0$. The capacitance voltage remains at its value at $t = 0$ due to the
continuity of charge and the inductance current remains at its value due to the continuity
of flux linkages. The situation at $t = 0_+$ may therefore be as shown in Fig. 5.26.

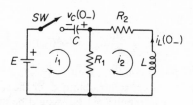

Figure 5.25 Two-loop *RLC* network for Example 5.7.

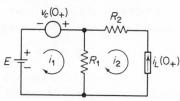

Figure 5.26 Example 5.7 equivalent circuit valid at $t = 0_+$.

The differential equations describing the network of Fig. 5.25 after the switch sw closes are obtained from Kirchhoff's voltage law. We proceed around each loop clockwise using reference currents i_1 and i_2 to obtain

$$\frac{1}{C}\int i_1\,dt + R_1(i_1 - i_2) = E \qquad t \geq 0 \tag{5.75a}$$

$$R_1(i_2 - i_1) + R_2 i_2 + L\frac{di_2}{dt} = 0 \qquad t \geq 0 \tag{5.75b}$$

These equations apply at any time t after the switch closes, including $t = 0_+$. At $t = 0_+$ (and at no other time), we see by inspection of Fig. 5.26 that

$$i_2 = -i_L(0_+) \qquad t = 0_+ \tag{5.76a}$$

$$R_1(i_1 - i_2) = E + v_c(0_+) \qquad t = 0_+ \tag{5.76b}$$

Substituting Eq. (5.76a) in Eq. (5.76b), we obtain

$$R_1 i_1 = E + v_c(0^+) + R_1(-i_L(0_+))$$

or finally

$$i_1 = \frac{1}{R_1}[E + v_c(0_+) - R_1 i_L(0_+)] \qquad t = 0_+ \tag{5.77a}$$

$$i_2 = -i_L(0_+) \qquad t = 0_+ \tag{5.77b}$$

For the sake of brevity, let us define i_1 at $t = 0_+$ as $i_1(0_+)$ and i_2 at $t = 0_+$ as $i_2(0_+)$. Then in Eq. (5.75b) (which applies at $t = 0_+$),

$$R_1(i_2(0_+) - i_1(0_+)) + R_2 i_2(0_+) + L\frac{di_2}{dt}(0_+) = 0 \qquad t = 0_+$$

or

$$\frac{di_2}{dt}(0_+) = \frac{R_1(i_1(0_+) - (R_1 + R_2)(i_2(0_+))}{L} \qquad t = 0_+ \tag{5.78}$$

This evaluates $di_2/dt(0_+)$. To find $di_1/dt(0_+)$, differentiate Eq. (5.75a) to obtain

$$\frac{1}{C}i_1 + R_1\left(\frac{di_1}{dt} - \frac{di_2}{dt}\right) = 0 \qquad t \geq 0 \tag{5.79}$$

Now substituting in the previously found values, we get

$$R_1\frac{di_1}{dt}(0_+) = R_1\frac{di_2}{dt}(0_+) - \frac{1}{C}i_1(0_+)$$

or

$$\frac{di_1}{dt}(0_+) = \frac{di_2}{dt}(0_+) - \frac{1}{R_1 C}i_1(0_+) \qquad (5.80)$$

We now have the initial conditions on i_1, i_2, di_1/dt, and di_2/dt, all of which will be necessary to evaluate the constants obtained for the solution of the original Eqs. (5.75). The solutions here look complicated but would be much simpler in appearance if we used numeric values rather than algebraic symbols. The initial conditions on higher derivatives may be found by continued differentiation.

EXAMPLE 5.8

For our next example, take the circuit shown in Fig. 5.27. Let the current source have a constant value J, which has been flowing for a long time. The switch sw closes $t = 0$. Find the voltages $v_1(t)$ and $v_2(t)$.

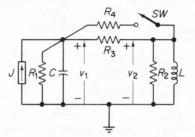

Figure 5.27 Example 5.8
RLC network.

Solution. At $t = 0_-$, the capacitance C has become fully charged and is an open circuit, while in the inductance L the flux linkages have ceased changing and L is a short circuit. The equivalent circuit valid at $t = 0_-$ is as shown in Fig. 5.28. Then $v_2 = 0$ and $J = v_1(G_1 - G_3)$. The initial inductance current is

$$i_L = v_1 G_3$$

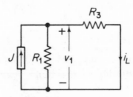

Figure 5.28 Example 5.8 equivalent network at $t = 0_-$ used to compute initial conditions.

or

$$i_L = \frac{JG_3}{G_1 + G_3}$$

where $G_1 = 1/R_1$ and $G_3 = 1/R_3$. The voltage across the capacitance is

$$v_C = v_1 = \frac{J}{G_1 + G_3}$$

Since the voltage across the capacitance and the inductance current will be continuous at $t = 0$, then

$$i_L(0_+) = i_L(0_-) = \frac{JG_3}{G_1 + G_3}$$

$$v_C(0_+) = v_1(0_-) = \frac{J}{G_1 + G_3}$$

The circuit at $t = 0_+$ will be as shown in Fig. 5.29. The equations for the network of Fig. 5.27 after sw closes are

$$v_1 G_1 + C\frac{dv_1}{dt} + G_5(v_1 - v_2) = J \tag{5.81a}$$

$$v_2 G_2 + G_5(v_2 - v_1) + \frac{1}{L}\int v_2\, dt = 0 \tag{5.81b}$$

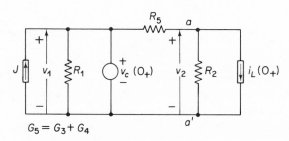

$$G_5 = G_3 + G_4$$

Figure 5.29 Example 5.8 equivalent circuit valid at $t = 0_+$.

where

$$G_5 = G_3 + G_4$$

From Fig. 5.29, $v_1(0_+) = v_c(0_+)$. Using Thévenin's theorem on the network in Fig. 5.29 at terminals aa', we get the simplified network in Fig. 5.30a and finally the network in Fig. 5.30b. From Fig. 5.30,

$$v_2(0_+) = [v_C(0_+)G_5 - i_L(0_+)](G_2 + G_5)$$

Substituting in Eq. (5.81a), we get

$$G_1 v_1(0_+) = C\frac{dv_1}{dt}(0_+) + G_5(v_1(0_+) - v_2(0_+)) = J$$

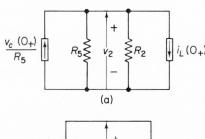

(a)

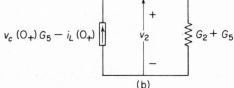

(b)

Figure 5.30 Simplified equivalent circuits for Fig. 5.29.

or

$$\frac{dv_1}{dt}(0_+) = \frac{J - v_1(0_+)(G_1 + G_5) + G_5 v_2(0_+)}{C}$$

Differentiating Eq. (5.81b), we get

$$G_2 \frac{dv_2}{dt} + G_5 \left(\frac{dv_2}{dt} - \frac{dv_1}{dt}\right) + \frac{1}{L} v_2 = 0$$

or

$$\frac{dv_2}{dt}(0_+) = \frac{G_5}{G_2 + G_5} \frac{dv_1}{dt}(0_+) - \frac{1}{L(G_2 + G_5)} v_2(0_+)$$

This example shows that for transients the final condition of a capacitance is an open circuit and that of an inductance is a short circuit, or somewhat the opposite of the initial conditions (we study steady-state conditions separately). This is often useful in making a quick check of final conditions.

If a source voltage is varying with time, its value at $t = 0$ enters into the picture.

EXAMPLE 5.9

For the next example, we take an RC network driven by a sinusoidal voltage source as shown in Fig. 5.31. The network at $t = 0_+$ becomes that of Fig. 5.32. Then

$$i(0_+) = \frac{A \sin \theta - v_C(0_+)}{R}$$

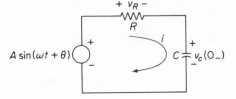

Figure 5.31 Example 5.9 RC network.

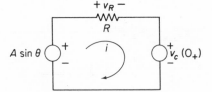

Figure 5.32 Example 5.9 initial condition network valid at $t = 0_+$.

and hence

$$v_R(0_+) = RC\frac{dv_C}{dt}(0_+) = A \sin \theta - v_C(0_+) \tag{5.82a}$$

or

$$\frac{dv_C}{dt}(0_+) = \frac{A \sin \theta - v_C(0_+)}{RC} \tag{5.82b}$$

"Pathological" Cases

Let C_1 in Fig. 5.33 be charged to q_1 and C_2 be charged to q_2 to give the initial voltages shown. When switch sw is closed at $t = 0$, the differential equation becomes

$$RC_T \frac{dv_{CT}}{dt} + v_{C_1} - v_{C_2} = 0$$

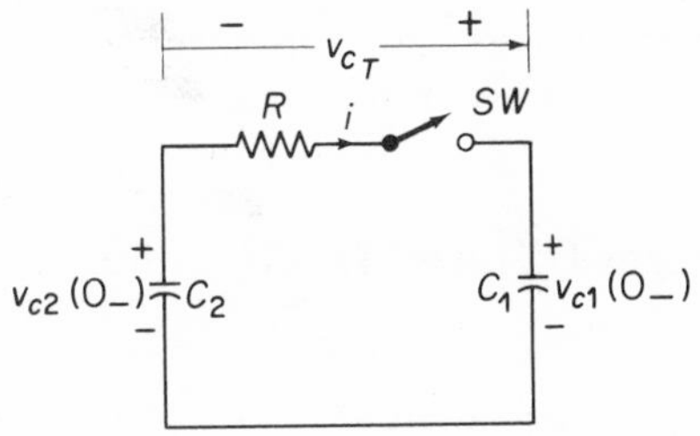

Figure 5.33 Pathological RC network.

where

$$C_T = \frac{C_1 C_2}{C_1 + C_2}$$

The charge $q_1 = C_1 v_{C_1}(0_-)$ and charge $q_2 = C_2 v_{C_2}(0_-)$. When $t = \infty$, $i = C_T(dv_{CT}/dt) = 0$, or $v_{C_1} = v_{C_2}$. Also since charge is conserved, $q_1 + q_2 = C_2 v_{C_1}(0_-) + C_2 v_{C_2}(0_-)$ is constant. Thus,

$$C_1 v_{C_1}(0_-) + C_2 v_{C_2}(0_-) = (C_1 + C_2) v_{C_1}(\infty)$$

or

$$v_{C_1}(\infty) = v_{C_2}(\infty) = \frac{C_1 v_{C_1}(0_-) + C_2 v_{C_2}(0_-)}{C_1 + C_2}$$

The equation for v_{C_1} is of the form

$$v_{C_1} = K e^{-t/RC_T} + v_{C_1}(\infty) \qquad t \geq 0$$

Now as $R \rightarrow 0$, then $K e^{-t/RC_T}$ is of very short duration, or $v_{C_1}(0_+) \rightarrow v_{C_1}(\infty)$. Thus,

$$v_{C_1}(0_+) = v_{C_2}(0_+) = \frac{C_1 v_{C_1}(0_-) + C_2 v_{C_2}(0_-)}{C_1 + C_2}$$

and the voltage on each capacitance is not a continuous function of time at $t = 0$. Under this condition, i becomes very large, and in the limit is an impulse at $t = 0$ with an area sufficient to account for the transfer of the charge from one capacitance to the other.

If we close the switch sw at $t = 0$ in Fig. 5.34, the current in L_1 and L_2 must change from zero to some value instantaneously if $R \rightarrow \infty$ and if J is a constant current source. The voltage v will be an impulse in the limit.

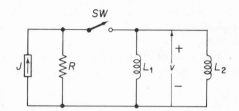

Figure 5.34 Pathological *RL*
network.

Such pathological cases arise due to taking expressions to a limit. Limits become very tricky, but since impulses are convenient concepts, they are sometimes necessary operations. In such cases, usual rules may seem to be violated.

5.7 Second-Order Differential Equations in State Space

Given the differential equation

$$\frac{d^2q(t)}{dt^2} + P\frac{dq(t)}{dt} + Qq(t) = Rf(t) \tag{5.83}$$

which is a general form for the second-order linear differential equation, let $x_1(t) = q(t)$, $x_2(t) = dq/dt$. Then dropping the functional notation,

$$\frac{dx_1}{dt} = x_2 \tag{5.84a}$$

$$\frac{dx_2}{dt} = -Qx_1 - Px_2 + Rf(t) \tag{5.84b}$$

$$q = x_1 \tag{5.84c}$$

In Equations (5.84) if we let $\dot{x}_1 = dx_1/dt$, $\dot{x}_2 = dx_2/dt$, these equations may be written in matrix form as

$$\begin{bmatrix} \dot{x}_1 \\ \dot{x}_2 \end{bmatrix} = \begin{bmatrix} 0 & 1 \\ -Q & -P \end{bmatrix} \begin{bmatrix} x_1 \\ x_2 \end{bmatrix} + \begin{bmatrix} 0 & 0 \\ 0 & R \end{bmatrix} \begin{bmatrix} 0 \\ f(t) \end{bmatrix} \tag{5.85a}$$

$$\begin{bmatrix} q_1 \\ q_2 \end{bmatrix} = \begin{bmatrix} 1 & 0 \\ 0 & 0 \end{bmatrix} \begin{bmatrix} x_1 \\ x_2 \end{bmatrix} + \begin{bmatrix} 0 & 0 \\ 0 & 0 \end{bmatrix} \begin{bmatrix} 0 \\ f(t) \end{bmatrix} \tag{5.85b}$$

or

$$\dot{\mathbf{x}} = \mathbf{A}\mathbf{x} + \mathbf{B}\mathbf{m} \tag{5.86a}$$

$$\mathbf{q} = \mathbf{C}\mathbf{x} + \mathbf{D}\mathbf{m} \tag{5.86b}$$

where

$$\dot{\mathbf{x}} = \begin{bmatrix} \dot{x}_1 \\ \dot{x}_2 \end{bmatrix}, \qquad \mathbf{x} = \begin{bmatrix} x_1 \\ x_2 \end{bmatrix}, \qquad \mathbf{m} = \begin{bmatrix} 0 \\ f(t) \end{bmatrix}$$

and the matrices **A**, **B**, **C**, and **D** may be identified from Eqs. (5.85a) and (5.85b). Equations (5.85) [or in shorthand form, Eqs. (5.86)] present the system of Eq. (5.83) in state-space form. That is, x_1 and x_2 are considered to be vectors defining a space (in this case of two dimensions). This is precisely the form we need for digital-computer solution (see Chapter 12). The elements of **A**, **B**, **C**, and **D** may be time varying in general, but matrices cannot be used to describe nonlinear networks. The vector $\mathbf{x}(t)$, which describes the excursions of the network with some source function $f(t)$ and initial condition $\mathbf{x}(0)$, may be shown on the state-space plane. For example, with the second-order equation of Eq. (5.83) and $f(t) = 0$, we show an underdamped path and an overdamped path from a given initial condition $x_1(0)$, $x_2(0)$ in Fig. 5.35. Time

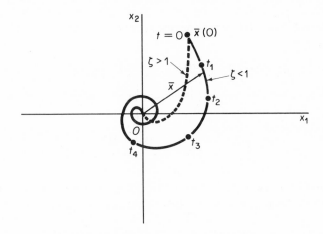

Figure 5.35 State space representation of network response.

unfolds along the curve (trajectory) in Fig. 5.35, increasing in a clockwise direction. The vector $\mathbf{x}(t) = [x_1(t)\ x_2(t)]^\mathsf{T}$ gives the condition or state of the network at any time $t = \tau$. Since the form of the state-space equations is not unique, there could be other equally valid descriptions of the second-order network than Eqs. (5.85). However, in this particular case, $x_2 = \dot{x}_1$, and the time between any two points on the trajectory can be found by the equation

$$t = \int_{\mathbf{x}_a}^{\mathbf{x}_b} \frac{dx_1}{x_2} \tag{5.87}$$

where $\mathbf{x}_a$ and $\mathbf{x}_b$ are the points in question. The form of state-space representation where $x_2 = \dot{x}_1$ is widely used for analyzing nonlinear second-order differential equations. State-space description is useful for networks of higher than second order, but graphical representation can be used only by analogy. We discuss general methods for finding state-space forms subsequently.

Although we do not discuss computer solutions of first-order ordinary differential equations in detail until Chapter 12, we provide a superficial introduction here to demonstrate the importance of the state-space form. Look at the first-order differential equation

$$\frac{dx(t)}{dt} + Ax(t) = f(t) \qquad x(t_1) = x_1 \tag{5.88}$$

where A and x_1 are constants.

The derivative dx/dt is defined as

$$\frac{dx}{dt} = \lim_{\Delta t \to 0} \frac{x(t + \Delta t) - x(t)}{\Delta t}$$

Letting Δt be a finite (but small) increment in t, we can then write an approximation of Eq. (5.88) as

$$x(t + \Delta t) = x(t) - Ax(t)\,\Delta t + f(t)\,\Delta t \tag{5.89}$$

Some value, called the *initial* or *boundary value* of $x(t)$ at $t = t_1$, must be given. Since $x(t_1) = x_1$ is known, Eq. (5.89) becomes

$$x(t_1 + \Delta t) = x(t_1) - Ax(t_1)\,\Delta t + f(t_1)\,\Delta t \tag{5.90}$$

from which $x(t_1 + \Delta t)$ can be approximated.

We may now proceed to a new iteration and find $x(t_1 + 2\Delta t)$ by reapplying Eq. (5.90). If we let $x(t_1 + \Delta t) = x(t_2)$ and $x(t_1 + 2\Delta t) = x(t_3)$, we compute

$$x(t_3) = x(t_2) - Ax(t_2)\,\Delta t + f(t_2)\,\Delta t$$

or, in general,

$$x(t_{n+1}) = x(t_n) - Ax(t_n)\,\Delta t + f(t_n)\,\Delta t \tag{5.91}$$

Equation (5.91) provides an algorithm for continued iteration on n to find $x(t_n)$. We can usually write an ordinary differential equation of any order (or any set of such equations) in state-space form, which means that we have a set of simultaneous first-order equations. We can now solve this set of equations, since we can solve each first-order differential equation in the set. Computer solution techniques for ordinary differential equations of second order or higher are rare; hence, the state-space form becomes necessary for computer application. The algorithm shown by Eq. (5.91) is crude but demonstrates the concept. Better techniques are described in Chapter 12.

Additional Reading

DESOER, C. A., and E. S. KUH, *Basic Circuit Theory*, McGraw-Hill Book Company, New York, 1969.

SMITH, R. J., *Circuits, Devices, and Systems*, John Wiley & Sons, Inc., New York, 1971.

SOKOLNIKOFF, I. S., and E. S. SOKOLNIKOFF, *Higher Mathematics for Engineers and Physicists*, McGraw-Hill Book Company, New York, 1941.

WYLIE, C. R., Jr., *Advanced Engineering Mathematics*, McGraw-Hill Book Company, New York, 1960.

VAN VALKENBURG, M. E., *Network Analysis*, Prentice-Hall, Inc., Englewood Cliffs, N.J., 1964.

Problems

5.1. Find the general solution of each of the following equations:

(a) $\frac{d^2i}{dt^2} + 3\frac{di}{dt} + 2i = 0$

(b) $\frac{d^2i}{dt^2} + 5\frac{di}{dt} + 6i = 0$

(c) $\frac{d^2i}{dt^2} + 7\frac{di}{dt} + 12i = 0$

(d) $\frac{d^2i}{dt^2} + 2\frac{di}{dt} + 26i = 0$

(e) $\frac{d^2i}{dt^2} + 4\frac{di}{dt} + 13i = 0$

5.2. Find the specific homogeneous solution for $t \geq 0$ of each equation in Problem 5.1 if

(a) $i(0_+) = 1$ and $\frac{di}{dt}(0_+) = 0$

(b) $i(0_+) = 0$ and $\frac{di}{dt}(0_+) = 1$

5.3. Solve the following nonhomogeneous differential equations for $t \geq 0$ if $i(0_+) = di/dt(0_+) = 0$:

(a) $\frac{d^2i}{dt^2} + 2\frac{di}{dt} + i = 1u(t)$

(b) $\frac{d^2i}{dt^2} + 3\frac{di}{dt} + 2i = 5tu(t)$

(c) $\frac{d^2i}{dt^2} + 5\frac{di}{dt} + 6i = e^{-2t} + 5e^{-3t}$

(d) $\frac{di}{dt} + 3i = 6\cos 5t + 4e^{-3t} + 3e^{-2t} + 5u(t)$

5.4. We wish to multiply the differential equation

$$\frac{di}{dt} + P(t)i = Q(t)$$

by an integrating factor R such that the left-hand side of the equation equals the derivative $d(Ri)/dt$.

(a) Show that the required integrating factor is $R = \exp[\int P(t)\, dt]$.

(b) Using this integrating factor, find the solution to the differential equation that corresponds to

$$i = e^{-Pt} \int Q e^{Pt}\, dt + K e^{-Pt}$$

where P is constant.

5.5.   Consider the differential equation

$$\frac{di}{dt} + ai = f_k(t)$$

where a is real and positive. Find the general solution of this equation if $f_k = 0$ for $t < 0$ and for $t \geq 0$ has the following values:

(a) $f_1 = k_1 t$
(b) $f_2 = \cos w_0 t$
Hint: Find the integrating factor.

5.6.   With the switch K in position 1 in the network of Fig. P5.1, a steady current $i(t)$ flows. Switch K is then moved from position 1 to position 2 at $t = 0$. Find $i(t) = t \geq 0$.

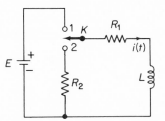

Figure P5.1

5.7.   The switch K in Fig. P5.2 has been in position 1 for a long time. At $t = 0$, the switch is moved from position 1 to position 2 using a make-before-break mechanism. Find $v_2(t)$ using the numerical values given in the network. Assume that the current in the 2-H inductance is zero at $t = 0$.

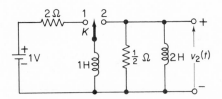

Figure P5.2

5.8.   The switch K in Fig. P5.3 is moved from position 1 to position 2 at $t = 0$, having been in position 1 for a long time before $t = 0$. Capacitance C_2 is uncharged at $t = 0$.
(a) Find the solution for $i(t)$ for $t > 0$.
(b) Find the solution for $v_2(t)$ for $t > 0$.

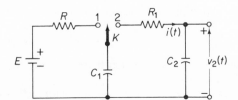

Figure P5.3

5.9. In a network it has been determined that the output voltage is given by the expression

$$v_0(t) = k_1 e^{-at} - k_2 e^{-bt} \qquad t > 0 \quad a > b$$

Show that $v_0(t)$ reaches its maximum value at time t given by the equation

$$t = \frac{1}{a - b} \ln\!\left(\frac{ak_1}{bk_2}\right)$$

5.10. Consider a series RLC network which is excited by a voltage source.
   (a) Determine the characteristic equation for the series current $i(t)$.
   (b) Suppose that L and C are fixed in value but that R varies from 0 to ∞. What will be the locus of the roots of the characteristic equation?
   (c) Plot the roots of the characteristic equation in the complex plane if $L = 1$ H, $C = 1$ μF, and R has the following values: 500 Ω, 1000 Ω, 3000 Ω, and 5000 Ω.

5.11. Again considering the circuit of Problem 5.10, plot the locus of the roots of the characteristic equation if L, C, and R all vary but
   (a) the ratio R/L remains constant.
   (b) $R\sqrt{C/L}$ remains constant.

5.12. The switch K in Fig. P5.4 has been in position 1 for a long time. At $t = 0$ the switch is moved (make-before-break) from position 1 to position 2. Find the initial conditions at $t = 0_+$ for i, di/dt, and d^2i/dt^2.

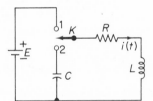

Figure P5.4

5.13. The switch K in Fig. P5.5 is closed at $t = 0$, connecting the battery E to an unenergized network.
   (a) Determine i, di/dt, and d^2i/dt^2 at $t = 0_+$.
   (b) Determine v_1, dv_1/dt, and d^2v_1/dt^2 at $t = 0_+$.
   (c) After the network arrives at a steady state under the new conditions, open the switch at a new $t = 0$. (Shift the time origin.) Solve for the quantities listed in parts (a) and (b) at the new $t = 0_+$.

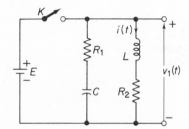

Figure P5.5

5.14. The capacitance in the network of Fig. P5.6 has an initial voltage $v_c(0_-) = v_1$, and at the same time $i_L(0_-) = 0$. At $t = 0$ switch K is closed. Determine an expression for the voltage $v_2(t)$, $t \geq 0$.

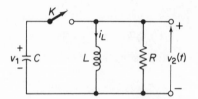

Figure P5.6

5.15. In the network shown in Fig. P5.7, the switch K is closed and a steady state is established. At $t = 0$, the switch is opened with no sparking. Find an expression for $i_2(t)$.

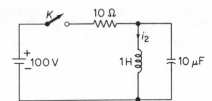

Figure P5.7

5.16. For the RLC network in Fig. P5.8, find the resistance value to cause the network to be critically clamped. What is the natural frequency ω_n of the network?

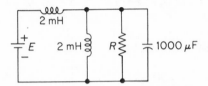

Figure P5.8

5.17. The initial conditions shown in the network of Fig. P5.9 have been established at $t = 0_-$ by circuits not shown. Determine the element current $i_c(t)$ for $t > 0$. The circuit is driven by $E(t) = \sin t\, u(t)$.

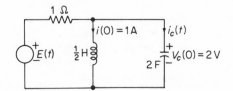

Figure P5.9

5.18. Consider a parallel RLC network driven by a time-varying current source $J(t) = \cos t\, u(t)$ connected across it. Let the element values be $L = 1$ H, $C = 1$ F, and $R = 10\,\Omega$. Determine the complete response for the initial conditions $v_c(0) = 2$ V and $i_L(0) = 5$ A. Indicate clearly the transient part and the steady-state part of the solution.

5.19. The voltage source $E(t)$ in the network shown in Fig. P5.10 is described by $E(t) = 2\cos 2t\, u(t)$. Determine $v_1(t)$. Repeat for $E(t) = k_1 t\, u(t)$. All initial conditions zero for both cases.

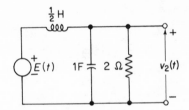

Figure P5.10

5.20. Put the equations of Problem 5.3 in state-space form. Use the variables $x_1 = i$ and $x_2 = di/dt$.

5.21. Put the equations for the circuit of Problem 5.16 (Fig. P5.8) in state-space form.

5.22. Using Eq. (5.91), develop a computer program to solve the equation

$$\frac{di(t)}{dt} + 3i(t) = 0$$

where $i(0) = 2$. Take t out to about 1 second, and let $\Delta t = 0.01$. Use various values of Δt and compare results with the exact solution.

5.23. Use Eq. (5.91) to write a computer program to solve the equation

$$\frac{di(t)}{dt} + 3i(t) = 5 \sin (10t + 45°) \qquad i(0) = 2$$

5.24. Use the state-space form of the equations found in Problem 5.20 to solve Problem 5.3.

Chapter 6

SINUSOIDAL
STEADY-STATE ANALYSIS

*Nature is an endless combination
and repetition of very few laws.*

RALPH WALDO EMERSON

6.1 Steady-State Solutions

In this section we examine the steady-state response of linear networks described by differential equations with sinusoidal time-varying sources. Periodic functions, as pointed out in Chapter 1, may be expressed as the sum of sine (or cosine) terms, making steady-state solutions very useful in engineering. We again look at Eq. (5.37) with $E(t)$ a sinusoidal function of time. Let us investigate the equation

$$Ri + L\frac{di}{dt} + S \int i \, dt = E \cos \omega t \tag{6.1}$$

and find the steady-state solution $i(t)$ (particular integral). If we differentiate Eq. (6.1), we obtain

$$L\frac{d^2i}{dt^2} + R\frac{di}{dt} + Si = -\omega E \sin \omega t \tag{6.2}$$

At this point we introduce a new concept which proves to be very useful in sinusoidal steady-state analysis. This concept involves a rotating line segment in the complex plane which we call a *phasor*. The utility of this concept will become more clear as we proceed. To describe a phasor we draw lines from the origin of the complex plane to the points

$$e^{j\omega t} = \cos \omega t + j \sin \omega t \tag{6.3a}$$

and

$$e^{-j\omega t} = \cos \omega t - j \sin \omega t \tag{6.3b}$$

The right-hand side of Eqs. (6.3) gives two directed lines in the complex plane, one rotating at angular velocity ω counterclockwise and the other rotating at the same velocity ω clockwise, as in Fig. 6.1. The angle ϕ between each line and the real axis at any time t is given by $\phi = \omega t$. These rotating lines are called *phasors*, and the angle ϕ is called a *phase angle*. Figure 6.1 shows the two phasors as though a picture were taken at a specific time $t = t_1$. Since

$$\cos \omega t = \frac{1}{2}(e^{j\omega t} + e^{-j\omega t}) \tag{6.4a}$$

$$\sin \omega t = \frac{1}{2j}(e^{j\omega t} - e^{-j\omega t}) \tag{6.4b}$$

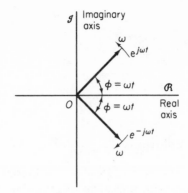

Figure 6.1   Rotating phasors in
   the complex plane.

cosine ωt_1 is apparently half the sum of the projections of the two phasors on the real axis. Also from Eq. (6.3),

$$\cos \omega t = \Re e^{j\omega t} \tag{6.5a}$$

$$\sin \omega t = \mathcal{g} e^{j\omega t} \tag{6.5b}$$

where $\Re$ means "the real part of" and $\mathcal{g}$ means "the imaginary part of." Thus, if we can find a steady-state solution for a source $Ee^{j\omega t}$ rather than $E \cos \omega t$, then by applying Eq. (6.4a) or (6.5a) we can translate this into a solution for the latter. Furthermore, by using Eq. (6.4b) or (6.5b), we can establish solutions for sources of the form $E \sin \omega t$. Let us therefore put $Ee^{j\omega t}$ on the right-hand side of Eq. (6.1), again differentiate, and examine the equation

$$L\frac{d^2 i}{dt^2} + R\frac{di}{dt} + Si = j\omega E e^{j\omega t} \tag{6.6}$$

The method of undetermined coefficients assumes that the particular integral is the sum of an exponential similar to that on the right-hand side of Eq. (6.6) and its derivative. Then let

$$i(t) = k_1 e^{j\omega t} + j\omega k_1 e^{j\omega t}$$

where k_1 is a constant. Continuing, define new constants $A = k_1$ and $B = -\omega k_1$. Then

$$i(t) = (A - jB)e^{j\omega t}$$
$$= Ke^{-j\theta}e^{j\omega t}$$

where

$$K = \sqrt{A^2 + B^2}$$

and

$$\tan \theta = \frac{B}{A}$$

Then

$$\frac{di}{dt} = j\omega K e^{-j\theta} e^{j\omega t}$$

$$\frac{d^2i}{dt^2} = (j\omega)^2 Ke^{-j\theta}e^{j\omega t}$$

Substituting the assumed solution and its derivatives in Eq. (6.6), we obtain

$$-\omega^2 LKe^{-j\theta}e^{j\omega t} + j\omega RKe^{-j\theta}e^{j\omega t} + SKe^{-j\theta}e^{j\omega t} = j\omega Ee^{j\omega t}$$

whence

$$K[\omega^2 L - S - j\omega R]e^{j\omega t - j\theta} = -j\omega Ee^{j\omega t}$$

or

$$K[(\omega^2 L - S) - j\omega R] = \omega Ee^{j\theta}e^{-j\pi/2} \qquad (6.7)$$

Changing the quantity in the brackets to polar form, we find that

$$K[(\omega^2 L - S)^2 + (\omega R)^2]^{1/2}e^{-j\gamma} = \omega Ee^{j(\theta - \pi/2)} \qquad (6.8)$$

where

$$\tan \gamma = \frac{\omega R}{\omega^2 L - S} = \frac{R}{\omega L - 1/\omega C}$$

From Eq. (6.8),

$$-\gamma = \theta - \frac{\pi}{2}$$

or

$$\tan \theta = \frac{\omega L - 1/\omega C}{R}$$

and

$$K = \frac{\omega E}{[(S - L\omega^2)^2 + (\omega R)^2]^{1/2}}$$

$$= \frac{E}{[R^2 + (L\omega - 1/\omega C)^2]^{1/2}}$$

Then

$$i(t) = \frac{E}{[R^2 + (\omega L - 1/\omega C)^2]^{1/2}} e^{-j\theta}e^{j\omega t} \qquad (6.9)$$

where

$$\tan \theta = \frac{\omega L - 1/\omega C}{R}$$

If we repeat the process for a source $Ee^{-j\omega t}$ in Eq. (6.1) we obtain the same result as Eq. (6.9), except that the exponential terms on the right are

$$e^{j\theta}e^{-j\omega t}$$

The solution for the current i when driven by the function

$$E \cos \omega t = \tfrac{1}{2}(e^{j\omega t} + e^{-j\omega t})E$$

is obtained by adding Eq. (6.9) to the modified solution for $Ee^{-j\omega t}$ and dividing the result by 2. Thus,

$$i = \frac{1}{2} \frac{E}{[R^2 + (\omega L - 1/\omega C)^2]^{1/2}} [e^{j(\omega t - \theta)} + e^{-j(\omega t - \theta)}]$$

or

$$i = \frac{E}{[R^2 + (\omega L - 1/\omega C)^2]^{1/2}} \cos(\omega t - \theta) \tag{6.10}$$

where

$$\theta = \tan^{-1} \frac{\omega L - 1/\omega C}{R}$$

We may now define a complex impedance Z valid for sinusoidal sources as

$$Z = R + jX \qquad \text{ohms} \tag{6.11a}$$
$$= \sqrt{R^2 + X^2} \, e^{j\theta} \qquad \text{ohms} \tag{6.11b}$$

where

$$X = \omega L - \frac{1}{\omega C} \tag{6.12a}$$

and

$$\tan \theta = \frac{X}{R} \tag{6·12b}$$

where X is known as the reactance.
Then

$$Z = [R^2 + X^2]^{1/2} e^{j\theta} \tag{6.13a}$$

$$= \left[R^2 + \left(\omega L - \frac{1}{\omega C} \right)^2 \right]^{1/2} e^{j\theta} \tag{6.13b}$$

where

$$\theta = \tan^{-1} \frac{X}{R} \tag{6.14a}$$

$$= \tan^{-1} \frac{\omega L - 1/\omega C}{R} \tag{6.14b}$$

Equation (6.9) becomes

$$i = \frac{E}{Z} e^{j\omega t} \tag{6.15}$$

or

$$i = \hat{I} e^{j\omega t} \tag{6.16}$$

where $\hat{I} = E/Z$ is a complex quantity. We may work with voltages, currents, and impedances as though they were complex numbers using Eq. (6.15). Once we find the relationships between the complex numbers, the actual sinusoidal variations can be written by inspection, following the procedure used to obtain Eq. (6.10). Clearly the voltage can be represented by a complex number in general since if the source voltage $E(t)$ is given by

$$E(t) = E \cos(\omega t + \phi)$$

then we use $Ee^{j\phi} e^{j\omega t}$ on the right-hand side of Eq. (6.1) and proceed with the development as before. The response i is shifted in phase by the angle ϕ from the previous result, whence Eq. (6.15) becomes

$$\hat{I} = \frac{Ee^{j\phi}}{Z}$$

or

$$\hat{I} = \frac{\hat{E}}{Z} \tag{6.17}$$

where $\hat{E}$ is the complex number $Ee^{j\phi} = E(\cos \phi + j \sin \phi)$.

In dealing with complex number representation, the important aspects of the currents and voltages in steady-state linear sinusoidal analysis consist essentially of two factors: magnitude and phase. In Fig. 6.2 we represent the voltage by a phasor

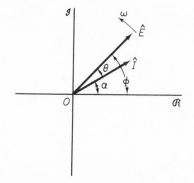

Figure 6.2 The rotating phasors $\hat{E}$ and $\hat{I}$ in the complex plane at $t = 0$.

$\hat{E}e^{j\omega t}$ rotating counterclockwise at angular velocity ω. When $t = 0$ the angle of the voltage phasor with the real axis is ϕ. The corresponding current phasor $\hat{I}e^{j\omega t}$ is related to $\hat{E}$ by the complex impedance Z, such that the magnitude of $\hat{I}$ (designated $|\hat{I}|$) is given by

$$|\hat{I}| = \frac{|\hat{E}|}{|Z|}$$

and the phase angle of $\hat{I}$ is given by

$$\alpha = \phi - \theta$$

where

$$Z = |Z| e^{j\theta}$$

The angle θ is the angle between $\hat{E}$ and $\hat{I}$ and remains constant for a fixed ω. The complete time expression for the voltage (and similarly for the current), is obtained by taking the real part ($\mathcal{R}\hat{E}e^{j\omega t}$) for the cosine, or the imaginary part ($\mathcal{I}\hat{E}e^{j\omega t}$) for the sine, according to Eqs. (6.5) and writing this as we did in Eq. (6.10).

EXAMPLE 6.1

Given the network in Fig. 5.11, let $E(t) = 100 \cos(\omega t + 30°)$, $R = 1000\,\Omega$, $L = 2\,\text{H}$, $C = 1\,\mu\text{F}$ and the frequency f be 39.7 Hz. Find $i(t)$. Increase f to 318 Hz and repeat.

Solution

$$\omega = 2\pi(39.7) = 250\,\text{rad/s}$$

$$\omega L = 2(250) = 500\,\Omega$$

$$\frac{1}{\omega C} = \frac{10^6}{250} = 4000\,\Omega$$

$$X = 500 - 4000 = -3500\,\Omega$$

$$Z = 100 - j3500 = 3640e^{-j74.1°}$$

Since

$$\hat{E} = 100e^{j30°} \qquad \text{(given)}$$

then

$$\hat{I} = \frac{100e^{j30°}}{3640e^{-j74.1°}} = 0.0282e^{j114.1°}$$

Thus,

$$i = 0.0282 \cos(250t + 114.1°)$$

The instantaneous voltage at $t = 0$ is $100 \cos 30° = 86.6$ V. The instantaneous current at $t = 0$ is $0.0282 \cos 114.1° = -0.0155$ A. The voltage and current waves are shown in Fig. 6.3. The phasor relation is shown in Fig. 6.4 at time $t = 0$. The current scale

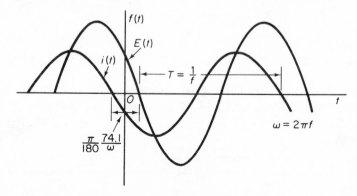

Figure 6.3   Current and voltage as a function of time for Example 6.1.

and the voltage scale are not the same for the waveforms in the figure. The instantaneous voltage and current values at $t = 0$ in Fig. 6.4 are the projections of the phasors on the real axis. In this case, the reactance X is predominantly capacitive, and the current phasor leads the voltage in phase angle.

If $f = 318$ Hz, $\omega = 2\pi(318) = 2000$ rad/s.

$$\omega L = 2000(2) = 4000\,\Omega$$

$$\frac{1}{\omega C} = \frac{10^6}{2000} = 500\,\Omega$$

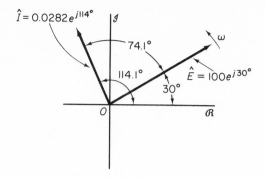

Figure 6.4 Current and voltage phasors for Example 6.1 at $f = 39.7$ Hz.

$$X = 4000 - 500 = 3500 \ \Omega$$

$$Z = 3640e^{j74.1°}$$

Hence,

$$\hat{I} = \frac{100e^{j30°}}{3640e^{j74.1°}} = 0.0282e^{-j44.1°}$$

Thus,

$$i = 0.0282 \cos(2000t - 44.1°)$$

The phasor diagram at $t = 0$ for $f = 318$ Hz is shown in Fig. 6.5. The voltage waveform is the same as that shown in Fig. 6.4; the current waveform of Fig. 6.4 is shifted to the right 180°. The reactance X is predominately inductive in this case, and the current phasor lags the voltage phasor in phase angle.

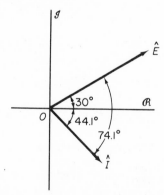

Figure 6.5 Current and voltage phasors, Example 6.1, 318 Hz.

Note that the impedance Z in Example 6.1 varies in magnitude and phase angle with the angular velocity ω. The minimum value of the impedance Z occurs when $X = 0$ [see Eqs. (6.11)], and at this condition the phase angle θ of the impedance is zero, from Eq. (6.12a). Using Eq. (6.12a), when $X = 0$,

$$\omega L - \frac{1}{\omega C} = 0$$

or

$$\omega^2 = \frac{1}{LC}$$

$$\omega = \omega_R = \sqrt{\frac{1}{LC}} \tag{6.18}$$

If ω has the value given by Eq. (6.18),

$$Z = R$$

and

$$\hat{I} = \frac{\hat{E}}{R}$$

or $\hat{I}$ and $\hat{E}$ are in phase. This is called the *condition of resonance*. If we look at the characteristic equation of the network, which is

$$\lambda^2 + \frac{R}{L}\lambda + \frac{1}{LC} = 0$$

the natural angular velocity ω_n of this equation has the same value as ω_R,

$$\omega_n = \sqrt{\frac{1}{LC}}$$

Resonance therefore occurs when the angular velocity of the source is the same as the natural frequency ω_n. The transient oscillation angular velocity $\omega_d = \sqrt{1 - \zeta^2}\,\omega_n$, if it exists, does not equal the resonance angular velocity ω_R unless $\zeta = 0$ ($R = 0$ here).

EXAMPLE 6.2

Find the resonance frequency for the circuit of Example 6.1.

Solution

$$\omega_R = \frac{1}{\sqrt{10^{-6}(2)}} = \frac{10^3}{\sqrt{2}} = 706 \text{ rad/s}$$

$$f_R = \frac{10^3}{2\pi\sqrt{2}} = 113 \text{ Hz}$$

$$\omega_n = 706 \text{ rad/s} \qquad \zeta = 0.354 \qquad \omega_d = 658 \text{ rad/s}$$

$$\max |\hat{I}| = \frac{100}{1000} = 0.1 \text{ A when } \omega = \omega_R$$

Nodal steady-state analysis proceeds similarly to loop analysis. The nodal differential equation in Eq. (5.39) describes the voltage response of the network. If the driving current $J(t)$ is a cosine function, we may represent it by the phasor $\hat{J} = Je^{j\phi}$. The voltage phasor $\hat{E}$ is then given by the formula

$$\hat{E} = \frac{\hat{J}}{Y} \tag{6.19a}$$

where

$$Y = G + jB \qquad \text{mhos} \tag{6.19b}$$

Y is called the *complex admittance* and B is the *susceptance.* By following the previous line of attack, the admittance $Y(\omega)$ for the *RLC* network shown in Fig. 6.6 is

$$Y(\omega) = \frac{\hat{I}}{\hat{E}} \tag{6.20a}$$

$$= G + jB \tag{6.20b}$$

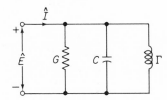

Figure 6.6 Parallel *RLC* network.

where

$$B = \omega C - \frac{1}{\omega L}$$

We find the solution for the node voltage in the same way in which the current was computed in Example 6.1.

We may invert an impedance Z to find the equivalent admittance Y in the same way a resistance can be inverted to find an equivalent conductance. In general,

$$Y = |Y|e^{j\phi} = \frac{1}{Z} = \frac{1}{|Z|e^{j\theta}} = \frac{1}{|Z|}e^{-j\theta}$$

or

$$|Y| = \frac{1}{|Z|} \tag{6.21}$$

and

$$\phi = -\theta$$

In rectangular form,

$$Y = G + jB = \frac{1}{Z} \tag{6.22}$$

$$= \frac{1}{R + jX} \tag{6.23}$$

$$= \frac{R - jX}{(R + jX)(R - jX)}$$

$$= \frac{R - jX}{R^2 + X^2} \tag{6.24}$$

Hence,

$$G = \frac{R}{R^2 + X^2} \tag{6.25a}$$

and

$$B = \frac{-X}{R^2 + X^2} \tag{6.25b}$$

In these complex expressions we see that $Y = 1/Z$, but Eq. (6.25) shows that $G \neq 1/R$ unless $X = 0$ and that $B \neq 1/X$ unless $R = 0$. Similarly,

$$Z = R + jX = \frac{1}{Y}$$

$$= \frac{1}{G + jB}$$

or

$$Z = \frac{G - jB}{G^2 + B^2}$$

Thus

$$R = \frac{G}{G^2 + B^2} \tag{6.26a}$$

$$X = \frac{-B}{G^2 + B^2} \tag{6.26b}$$

Analogous to the case of resistive networks, impedance applies most naturally to KVL and elements in series. Thus, if Z_1 and Z_2 are in series, as in Fig. 6.7, the total impedance Z_T is given by

$$\begin{aligned} Z_T &= Z_1 + Z_2 \\ &= R_1 + jX_1 + R_2 + jX_2 \end{aligned} \tag{6.27a}$$

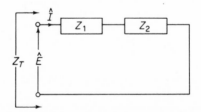

Figure 6.7 Series impedance.

or

$$Z_T = (R_1 + R_2) + j(X_1 + X_2) \tag{6.27b}$$

where

$$Z_T = \frac{\hat{E}}{\hat{I}}$$

On the other hand, admittance has more use in KCL and to elements in parallel. If Y_1 and Y_2 are in parallel, as in Fig. 6.8, the total admittance is given by

$$\begin{aligned} Y_T &= Y_1 + Y_2 \\ &= G_1 + jB_1 + G_2 + jB_2 \end{aligned} \tag{6.28a}$$

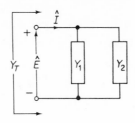

Figure 6.8 Parallel admittances.

or

$$Y_T = (G_1 + G_2) + j(B_1 + B_2) \qquad (6.28b)$$

where

$$Y_T = \frac{\hat{I}}{\hat{E}}$$

6.2 Networks with Sinusoidal (AC) Sources

The concept of complex impedance and admittance immediately allows us to obtain steady-state solutions for the response of networks with sinusoidal sources in a manner similar to that used with resistive sources. We use Kirchhoff's voltage and/or current laws to establish a system of simultaneous equations. Using phasor and impedance or admittance concepts, the equations are algebraic, but with complex coefficients. We now have Y and Z matrices with complex elements which are respectively analogous to the previous G and R matrices. The solution of the system of equations is conceptually no more difficult but much more arduous (at least by hand), which provides more motivation to use the computer. We can best illustrate the difficulty in obtaining a hand solution by showing the following example.

EXAMPLE 6.3

The resistance and reactance values of the elements at a particular frequency ω_1 have been calculated for the network in Fig. 6.9 and are shown in ohms. The two driving voltage sources are given by $E_1(t) = 50 \cos(\omega_1 t + 20°)$ and $E_2(t) = 60 \cos(\omega_1 t - 30°)$. (The symbols given in each source show that the source is sinusoidal.) Find the steady-state values of i_1 and i_2. [*Note:* The polarity marks on $E_1(t)$ and $E_2(t)$ indicate a reference polarity at the particular time t which we use in writing the loop equations. Here $t = 0$.]

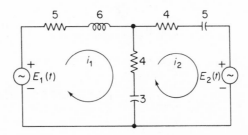

Figure 6.9 Network with sinusoidal voltage sources.

Solution

$$E_1(t) = \Re[50e^{j20°}e^{j\omega_1 t}] \quad \text{or} \quad \hat{E}_1 = 50e^{j20°}$$

$$E_2(t) = \Re[60e^{-j30°}e^{j\omega_1 t}] \quad \text{or} \quad \hat{E}_2 = 60e^{-j30°}$$

Proceeding clockwise around each loop in turn, we get

$$(5 + j6)\hat{I}_1 + (4 - j3)(\hat{I}_1 - \hat{I}_2) = 50e^{j20°}$$

$$(4 - j3)(\hat{I}_2 - \hat{I}_1) + (4 - j5)(\hat{I}_2) = -60e^{-j30°}$$

or

$$(9 + j3)\hat{I}_1 - (4 - j3)\hat{I}_2 = 50e^{j20°} \qquad (6.29a)$$

$$-(4 - j3)\hat{I}_1 + (8 - j8)\hat{I}_2 = -60e^{-j30°} \qquad (6.29b)$$

We may solve Eqs. (6.29a) and (6.29b) simultaneously in any desired manner. Choosing the matrix method, we get

$$\begin{bmatrix} 9 + j3 & -(4 - j3) \\ -(4 - j3) & 8 - j8 \end{bmatrix} \begin{bmatrix} \hat{I}_1 \\ \hat{I}_2 \end{bmatrix} = \begin{bmatrix} 50e^{j20°} \\ -60e^{-j30°} \end{bmatrix} \qquad (6.30)$$

or

$$\mathbf{Z}\hat{\mathbf{I}} = \hat{\mathbf{E}} \qquad (6.31)$$

The **Z** matrix is similar to the previous **R** matrix, except that the elements are complex quantities. From Eq. (6.30),

$$\begin{bmatrix} \hat{I}_1 \\ \hat{I}_2 \end{bmatrix} = \frac{1}{\det \mathbf{Z}} \begin{bmatrix} 8 - j8 & 4 - j3 \\ 4 - j3 & 9 + j3 \end{bmatrix} \begin{bmatrix} 50e^{j20°} \\ -60e^{-j30°} \end{bmatrix}$$

where $\det \mathbf{Z} = (8 - j8)(9 + j3) - (4 - j3)^2$. To complete the solution, we can change the matrix elements to polar form or the voltage elements to rectangular form. Multiplication may be performed using either form, but the polar form is more commonly used when a large number of multiplications occur. Addition, on the other hand, favors the rectangular form, while division again favors the polar form. For illustration, we find $\det \mathbf{Z}$ using the rectangular form only. Thus,

$$
\begin{array}{cc}
8 - j8 & 4 - j3 \\
9 + j3 & 4 - j3 \\
\hline
72 - j72 & 16 - j12 \\
+ j24 + 24 & - j12 - 9 \\
\hline
96 - j48 & 7 - j24
\end{array}
$$

or

$$\det \mathbf{Z} = (96 - j48) - (7 - j24) = 89 - j24 = 92.2e^{-j15.1°}$$

or

$$\begin{bmatrix} \hat{I}_1 \\ \hat{I}_2 \end{bmatrix} = \frac{e^{j15.1°}}{92.2} \begin{bmatrix} 11.3e^{-j45°} & 5e^{-j36.9°} \\ 5^{-j36.9°} & 9.49e^{j18.5°} \end{bmatrix} \begin{bmatrix} 50e^{j20°} \\ -60e^{-j30°} \end{bmatrix}$$

Then

$$\hat{I}_1 = \frac{e^{j15.1°}}{92.2}[(11.3e^{-j45°})(50e^{j20°}) - (5e^{-j36.9°})(60e^{-j30°})]$$

$$= \frac{e^{j15.1°}}{92.2}(565e^{-j25°} - 300e^{-j66.9°})$$

$$= \frac{e^{j15.1°}}{92.2}[512 - j239 - (117.5 - j2750)]$$

$$= \frac{e^{j15.1°}}{92.2}(395e^{j5.2°})$$

$$= 4.28e^{j20.3°}$$

Thus $i_1(t) = 4.28 \cos(\omega_1 t + 20.3°)$. We leave the computation of $\hat{I}_2$ to the reader. We suggest leaving the elements in rectangular form during the matrix multiplication to compare the methods.

Example 6.3 illustrates the impracticality of extensive complex quantity calculations. Digital-computer programs deal easily with complex numbers, making the hand calculation of voltages and currents for large networks unnecessary. The procedure for constructing the system matrices for computing the response of networks with sinusoidal driving sources is similar to the rules discussed in Chapter 4. The solutions to the complex network equations give the steady-state values (particular integral) of voltage and current only. For nodal analysis the procedure is as follows:

1. Select one node as ground, or reference node.

2. Assign sequential numbers to the remaining nodes in any arbitrary order. In some cases minor nodes may be eliminated (see step 3b).

3. Calculate the admittance $Y = G + jB$ of each branch at the frequency f desired.
   (a) For a capacitance, $B = \omega C = 2\pi f C$, and for an inductance, $B = -1/\omega L = -\Gamma/\omega = -\Gamma/2\pi f$. The conductance $G = 1/R$.
   (b) Minor nodes may be eliminated by calculating the impedance $Z = R + jX$ of a series combination and then finding $Y = 1/Z$. If a minor node is removed, the remaining nodes should be renumbered to correct the node number sequence.

4. Adopt the unknown voltages of the nodes of step 3 as a voltage vector

$$\hat{\mathbf{V}} = [\hat{V}_1 \hat{V}_2 \cdots \hat{V}_n]^\mathsf{T} \qquad \text{where } \hat{V}_1, \hat{V}_2, \ldots, \hat{V}_n$$

are above the reference node in potential. These voltages have complex number values.

5. Write the matrix $\mathbf{Y}$ of the network as

$$\mathbf{Y} = \begin{matrix} & \begin{matrix} 1 & 2 & \cdots & n \end{matrix} \\ \begin{matrix} 1 \\ 2 \\ \cdot \\ \cdot \\ \cdot \\ n \end{matrix} & \begin{bmatrix} Y_{11} & Y_{12} & \cdots & Y_{1n} \\ Y_{21} & Y_{22} & \cdots & Y_{2n} \\ & & \cdots & \\ & & & \\ Y_{n1} & Y_{n2} & \cdots & Y_{nn} \end{bmatrix} \end{matrix} \qquad (6.32)$$

The diagonal elements $Y_{11}, Y_{22}, Y_{33}, \ldots, Y_{nn}$ are the sum of the admittances connected to nodes $1, 2, 3, \ldots, n$, respectively. Each of the elements $Y_{ij} = Y_{ji}, i \neq j$, consists of the mutual admittance between node i and node j, but with a negative sign.

6. Change all voltage sources to current sources by using the Norton–Thévenin transformation.

7. Write the source vector in proper order; that is, $\hat{\mathbf{J}} = [\hat{J}_1, \hat{J}_2, \hat{J}_3, \ldots, \hat{J}_n]^\mathsf{T}$,

where $\hat{J}_1, \hat{J}_2, \hat{J}_3, \ldots, \hat{J}_n$ are the net source currents into nodes 1, 2, 3, ..., n, respectively, and includes both independent and dependent sources. The source currents are expressed as complex numbers.

8. Put the dependent sources of step 7 in terms of the node voltages $\hat{V}_1, \hat{V}_2, \ldots,$ $\hat{V}_n$.

9. Keeping each dependent source term found in step 8 in the row corresponding to their row position in 8, move each term into the proper column of the admittance matrix **Y**; that is, terms involving $\hat{V}_1$ go into column 1, those involving $\hat{V}_2$ go into column 2, and so on. In general the resultant admittance matrix $\mathbf{Y}_1$ will be nonsymmetric.

10. Solve the matrix equation

$$\mathbf{Y}_1\hat{\mathbf{V}} = \hat{\mathbf{J}}_1 \tag{6.33}$$

where $\hat{\mathbf{J}}_1$ contains independent source currents only as elements. Since $\mathbf{Y}_1$ and $\hat{\mathbf{J}}_1$ will contain complex elements, a computer algorithm must provide for complex arithmetic operations. The implementation of a complex arithmetic version of subroutine LINQ (Fig. 3.4) is straightforward and left to the reader as an exercise.

EXAMPLE 6.4

Given the *RLC* active network in Figure 6.10, develop Eq. (6.33). The resistance values in the figure are specified in ohms, capacitance values in farads, and inductance values in henries. The frequency of both independent sources is $5/\pi$ Hz. (The rather unrealistic numbers are for purposes of illustration.)

Solution. We number the nodes as in Fig. 6.10 and select node 6 as the ground or reference node. We eliminate nodes 4 and 5 by finding the equivalent impedance of the

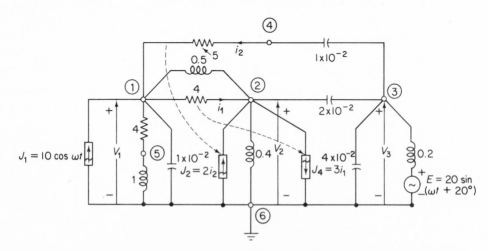

Figure 6.10 Example of steady-state nodal analysis with sinusoidal sources.

two branches Z_1 and Z_2 with series elements connected to node 1. These are

$$Z_1 = 4 + j2\pi \frac{5}{\pi}(1) = 4 + j10$$

$$Z_2 = 5 - j\frac{\pi}{2\pi(5)(1 \times 10^{-2})} = 5 - j10$$

Then

$$Y_1 = \frac{1}{4 + j10} = \frac{4 - j10}{16 + 100} = \frac{1}{116}(4 - j10)$$

$$Y_2 = \frac{1}{5 - j10} = \frac{5 + j10}{25 + 100} = \frac{1}{125}(5 + j10)$$

The admittance calculations of the other branches are straightforward. To put the voltage source at node 3 on the same basis as the current source at node 1, we respecify $E = 20 \sin(\omega t + 20°) = 20 \cos(\omega t - 70°)$. Using the real part of rotating phasors, these sources are represented by $J_1 = 10e^{j(\omega t + 0°)}$ and $E = 20e^{j(\omega t - 70°)}$, respectively. The source phasors at $t = 0$ are shown in the complex plane in Fig. 6.11 as $\hat{J}_1 = 10e^{j0°}$

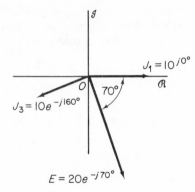

Figure 6.11 Phasors for Example 6.4 at $t = 0$.

and $\hat{E} = 20e^{-j70°}$. Changing the voltage source $\hat{E}$ to a current source, we obtain $\hat{J}_3 = 20(-j0.5)e^{-j70°} = 10e^{-j90°}e^{-j70°} = 10e^{-j160°} = -10e^{j20°}$.

The phasors $\hat{E}$, $\hat{J}_1$, and $\hat{J}_3$ shown at $t = 0$ in Fig. 6.11 are not to scale. Figure 6.12 (in which all admittances are shown in mhos) is equivalent to Fig. 6.10.

From Fig. 6.12 and step 4 we write the following admittance matrix $\mathbf{Y}$ to three-digit accuracy:

$$\begin{array}{c} \\ 1 \\ 2 \\ 3 \end{array} \begin{array}{ccc} 1 & 2 & 3 \end{array}$$
$$\begin{bmatrix} 0.325 - j0.106 & -0.25 + j0.2 & -0.04 - j0.08 \\ -0.25 + j0.2 & 0.25 - j0.25 & -j0.2 \\ -0.04 - j0.08 & -j0.2 & 0.04 + j0.18 \end{bmatrix} = \mathbf{Y}$$

The current vector $\hat{\mathbf{J}}$ from step 6 is

$$\hat{\mathbf{J}} = \begin{bmatrix} 10 + j0 \\ 2\hat{I}_2 - 3\hat{I}_1 \\ -9.40 - j3.42 \end{bmatrix}$$

From Fig. 6.12, $\hat{I}_1 = (\hat{V}_1 - \hat{V}_2)(0.25)$ and $\hat{I}_2 = (\hat{V}_1 - \hat{V}_3)(0.04 + j0.08)$. The complex

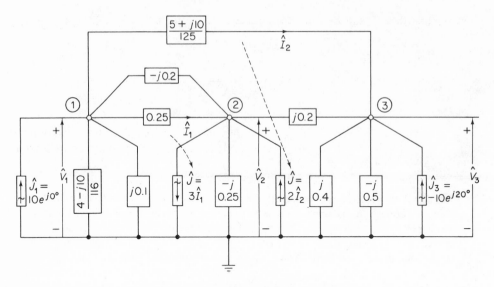

Figure 6.12 Network for Example 6.4 with
calculated admittances.

equivalent current vector $\hat{\mathbf{J}}$ then becomes

$$\hat{\mathbf{J}} = \begin{bmatrix} 10 + j0 \\ \hat{V}_1(-0.67 + j0.16) + \hat{V}_2(0.75 + j0) - \hat{V}_3(0.08 - j0.16) \\ -9.40 - j3.42 \end{bmatrix}$$

Since

$$\mathbf{Y}\hat{\mathbf{V}} = \hat{\mathbf{J}}$$

the coefficients of the voltages change sign as they are transferred from $\hat{\mathbf{J}}$ to $\mathbf{Y}$. Hence we obtain the altered admittance matrix $\mathbf{Y}_1$, where

$$\mathbf{Y} = \begin{bmatrix} 0.325 - j0.106 & -0.25 + j0.2 & -0.04 - j0.08 \\ 0.42 + j0.04 & -0.05 - j0.25 & +0.08 - j0.04 \\ -0.04 - j0.08 & -j0.2 & 0.04 + j0.18 \end{bmatrix}$$

Finally,

$$\mathbf{Y}_1\hat{\mathbf{V}} = \hat{\mathbf{J}}_1$$

where $\mathbf{Y}_1$ is shown above and

$$\hat{\mathbf{J}} = \begin{bmatrix} 10 + j0 \\ 0 \\ -9.40 - j3.42 \end{bmatrix}$$

Solving for $\hat{\mathbf{V}}$ by hand clearly would be exceedingly fatiguing.

For loop analysis, we may develop similar rules as follows:

1. Assign a minimum number of loop currents to the network such that at least one current passes through each element. Make each loop current direction

the same, say clockwise. Number the currents in any arbitrary order. Eliminate minor loops wherever possible (see step 3b).

2. Adopt the currents in step 1 as the current vector $\hat{\mathbf{I}} = [\hat{I}_1 \ \hat{I}_2 \ \hat{I}_3 \ \cdots \ \hat{I}_n]^\mathsf{T}$. These current elements will be expressed as complex quantities.

3. Calculate the impedance of each branch at the frequency f desired.
   (a) For an inductance, $X = \omega L = 2\pi f L$, and for a capacitance, $X = -1/\omega C = -1/2\pi f C = -S/2\pi f$. The resistance $R = 1/G$. A mutual inductance introduces in series with one inductance a voltage of $j\omega M$ times the current in the coupled inductance. (See Example 6.5.)
   (b) Minor loops can be eliminated by calculating the admittance $Y = G + jB$ of a parallel combination and then finding $Z = 1/Y$. Elimination of a minor loop requires a renumbering of the remaining loop currents.

4. Write the matrix $\mathbf{Z}$ of the network as

$$
\mathbf{Z} = \begin{array}{c} \\ 1 \\ 2 \\ \cdot \\ \cdot \\ \cdot \\ n \end{array}
\begin{array}{cccc}
1 & 2 & \cdots & n \\
\left[\begin{array}{cccc}
Z_{11} & Z_{12} & \cdots & Z_{1n} \\
Z_{21} & Z_{22} & \cdots & Z_{2n} \\
\multicolumn{4}{c}{\cdot \quad \cdot \quad \cdot \quad \cdot \quad \cdot \quad \cdot} \\
Z_{n1} & Z_{n2} & \cdots & Z_{nn}
\end{array}\right]
\end{array}
\qquad (6.34)
$$

The elements $Z_{11}, Z_{22}, Z_{33}, \ldots, Z_{nn}$ are the sum of the impedances around loops $1, 2, 3, \ldots, n$, respectively. Each of the elements $Z_{ij} = Z_{ji}$, $i \neq j$, is the mutual impedance common to loops i and j but with a negative sign.

5. Change all current sources to voltage sources by using the Norton–Thévenin transformation. Thus, $\hat{J} = \hat{E}/Y$ in parallel with $Y = 1/Z$ becomes a voltage source $\hat{E}$ in series with an impedance Z.

6. Write the source vector in proper order, that is, $\hat{\mathbf{E}} = [\hat{E}_1, \hat{E}_2, \hat{E}_3, \ldots, \hat{E}_n]^\mathsf{T}$, where $\hat{E}_1, \hat{E}_2, \hat{E}_3, \ldots, \hat{E}_n$ are the net source voltages around loops $1, 2, 3, \ldots, n$, respectively. A voltage source in this vector is considered positive if the loop current direction proceeds from $-$ to $+$ through the source. The vector includes both independent and dependent sources. The voltages are expressed as complex numbers.

7. Put the dependent sources of step 6 in terms of the loop currents $\hat{I}_1, \hat{I}_2, \ldots, \hat{I}_n$.

8. Keeping each dependent source term found in step 7 in the row position found, move each term into the proper column of the impedance matrix $\mathbf{Z}$; that is, terms involving $\hat{I}_1$ go into column 1, terms involving $\hat{I}_2$ go into column 2, and so on. In general the resultant impedance matrix $\mathbf{Z}_1$ will be nonsymmetric.

9. Solve the matrix equation

$$\mathbf{Z}_1 \hat{\mathbf{I}} = \hat{\mathbf{E}}_1 \qquad (6.35)$$

where $\hat{\mathbf{E}}_1$ contains independent source voltages only as elements.

EXAMPLE 6.5

Given the network of Fig. 6.13, in which $R_1 = 30\,\Omega$, $L_1 = 0.1$ H, $L_2 = 0.4$ H, $M = 0.15$ H, $R_2 = 20\,\Omega$, $C_1 = 40 \times 10^{-6}$ F, $R_3 = 40\,\Omega$, and $E = 10\cos(200t + 30°)V$, find $\hat{I}_1$, $\hat{I}_2$, and $\hat{I}_3$.

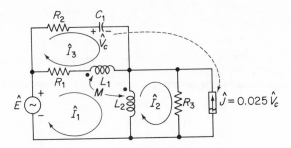

Figure 6.13 Steady-state loop analysis example with sinusoidal sources.

Solution. Two network transformations must be made before the loop equation of the network can be written. First, the parallel combination of J and R_3 must be converted to an equivalent network with no current sources. Second, the transformer must be replaced by an equivalent circuit containing only passive elements and dependent voltage sources as described in Chapter 2. The network shown in Fig. 6.14 contains the two required transformations and is in the proper form for loop analysis.

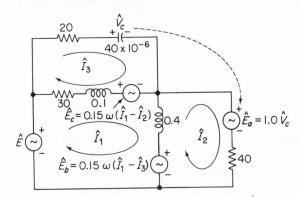

Figure 6.14 Mesh analysis equivalent circuit for Example 6.5.

The matrix **Z** for the network may be written by inspection as

$$\mathbf{Z} = \begin{bmatrix} 30 + j100 & -j80 & -(30 + j20) \\ -j80 & 40 + j80 & 0 \\ -(30 + j20) & 0 & 50 - j105 \end{bmatrix} \begin{matrix} 1 \\ 2 \\ 3 \end{matrix}$$

with column headers 1, 2, 3 above.

The voltage source vector $\hat{\mathbf{E}}$ is given by computing

$$\hat{E}_1 = 8.66 + j5 - [j30(\hat{I}_1 - \hat{I}_2)] - [j30(\hat{I}_1 - \hat{I}_3)]$$
$$\hat{E}_2 = j30(\hat{I}_1 - \hat{I}_3) - 1.0\hat{V}_c$$
$$\hat{E}_3 = j30(\hat{I}_1 - \hat{I}_2)$$

Since $\hat{V}_c = -j125(\hat{I}_3)$, we get

$$\hat{\mathbf{E}} = \begin{bmatrix} (8.66 + j5) - j60\hat{I}_1 + j30\hat{I}_2 + j30\hat{I}_3 \\ j30\hat{I}_1 + j95\hat{I}_3 \\ j30\hat{I}_1 - j30\hat{I}_2 \end{bmatrix}$$

Transferring the dependent terms to the $\mathbf{Z}$ matrix, we obtain

$$\mathbf{Z}_1 = \begin{array}{c} \\ \begin{bmatrix} 30 + j160 & -j110 & -30 - j50 \\ -j110 & 40 + j80 & -j95 \\ -30 - j50 & +j30 & 50 - j105 \end{bmatrix} \end{array} \begin{array}{c} 1 \\ 2 \\ 3 \end{array}$$

$$\begin{array}{ccc} 1 & 2 & 3 \end{array}$$

and the final equation is

$$\mathbf{Z}_1\hat{\mathbf{I}} = \hat{\mathbf{E}}_1$$

where

$$\hat{\mathbf{E}}_1 = \begin{bmatrix} 8.66 + j5 \\ 0 \\ 0 \end{bmatrix}$$

Complex Arithmetic on the Digital Computer

In Fortran IV, a complex quantity is defined as an ordered pair of real quantities, the first being the real part and the second the imaginary part. For example, the expression $a = 1 + 2i$ is written as $A = (1.0, 2.0)$, or complex data in input and output appear as a pair of real values. The input and output of complex numbers therefore require special formats. We may specify that a variable is a complex number by using a COMPLEX statement such as

```
COMPLEX A, B, C, R
```

The compiler will then take A, B, C, and R to be complex numbers. The operations of addition, subtraction, multiplication, and division are then performed by writing the normal signs. Thus,

```
C = A + B
```

finds the sum of A and B and stores it in C. The operation of exponentiation is allowed but may be used only with real integer exponents. The normal functions supplied are

$$\text{CSQRT}(a + bi) = r^{1/2}\left(\cos\frac{\theta}{2} + \sin\frac{\theta}{2}i\right)$$

$$CABS(a + bi) = \sqrt{a^2 + b^2}$$

$$CEXP(a + bi) = e^a(\cos b + i \sin b)$$

$$CSIN(a + bi) = \sin a \frac{e^b + e^{-b}}{2} - \cos a \frac{e^b - e^{-b}}{2} i$$

$$CCOS(a + bi) = \cos a \frac{e^b - e^{-b}}{2} - \sin a \frac{e^b - e^{-b}}{2} i$$

but some installations may use other implementations. A complex number may be defined in the program by using the function CMPLX. Thus,

$$C = CMPLX(X, Y)$$

where X and Y are real variables previously defined, would store $X + Yi$ in C. $B = CONJG(A)$ would store $a - bi$ in B if A were $a + bi$.

The programmer must take great care in input and output statements. Since a complex number has two parts, the FORMAT statement must provide for two numbers to be read or to be written for each variable specified. The program segment

```
      COMPLEX A, B, C, R
      READ (5,100) A,B
      C = A + B
      R = A * B + C
      WRITE (6,100) C,R
  100 FORMAT (4F10.5)
```

illustrates the formatting required for complex input and output. The program of course expects two numbers on the data card for each complex variable specified, and these must be punched in proper order. The DIMENSION statement usually automatically provides proper storage space if the variables have previously been properly defined.

6.3 Sinusoidal Sources: Effective Current, Power, and Power Factor

Let R in Fig. 6.15 represent a linear resistance of constant value and let

$$i(t) = I_p \cos \omega t \tag{6.36}$$

where I_p is the maximum or peak value of the curent. The instantaneous electrical power $p(t)$ dissipated in R (which as time passes becomes converted to heat energy)

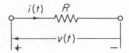

Figure 6.15 Linear resistance.

is given by Eq. (2.3) as

$$p(t) = i^2(t)R \tag{6.37}$$

Thus, the average power P dissipated over one cycle of the current $i(t)$ becomes

$$P = \frac{1}{T} \int_0^T i^2(t)R \, dt \tag{6.38}$$

where T is the period. Inserting Eq. (6.36) into Eq. (6.38), we get

$$
\begin{aligned}
P &= \frac{1}{T} \int_0^T I_p^2 R \cos^2(\omega t) \, dt \\
&= \frac{I_p^2}{T\omega} [\tfrac{1}{2}(\omega t) + \tfrac{1}{4} \sin 2(\omega t)] \Big|_0^T \\
&= \frac{I_p^2 R}{T\omega} \tfrac{1}{2}\omega T \\
&= \frac{I_p^2 R}{2}
\end{aligned}
\tag{6.39}
$$

We now define an effective or root-mean-square (rms) value of current I_e such that the average power in a constant resistance R is

$$P = I_e^2 R \tag{6.40}$$

We declare this definition valid regardless of the waveform of the current. For the sinusoidal case, the effective current is obviously

$$I_e = \frac{I_p}{\sqrt{2}} \tag{6.41}$$

Since the voltage $v(t)$ across the resistance R is given by

$$v(t) = i(t)R$$

then for the sinusoidal case,

$$
\begin{aligned}
P &= \frac{I_p I_p R}{2} \\
&= \frac{V_p^2}{2R}
\end{aligned}
\tag{6.42}
$$

We may also define an effective or rms value of the voltage V_e such that

$$P = \frac{V_e^2}{R} \tag{6.43}$$

regardless of the wave shape of $e(t)$. For the sinusoidal case, the effective voltage V_e is

$$V_e = \frac{V_p}{\sqrt{2}} \qquad (6.44)$$

We may perform calculations and describe network response with rms values of current and voltage in a manner similar to the way we used peak values in the previous section. Since a constant ratio exists between the rms value V_e and V_p, we may shrink the latter by a scale factor of $\sqrt{2}$, and work all sinusoidal steady-state problems with rms values.

We may consider $I_e e^{j\omega t}$ as being a rotating phasor which is always in phase with $I_p e^{j\omega t}$ as shown in Fig. 6.16 and where

$$I_e = \frac{I_p}{\sqrt{2}}$$

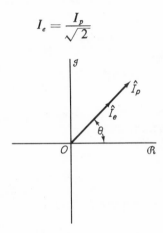

Figure 6.16 Phasor relationship between I_e and I_p.

This last step seems rather artificial but is valid since we obtain the same ratio for I_e to I_p regardless of the phase of the original $i(t)$.

Because of the early predominance of the power industry, rms values of sinusoidal voltages and currents have become accepted as standard rather than peak values. Thus, if a lighting circuit is said to be "120 volts," this means that the rms value $V_e = 120$ V, whereas the peak value is

$$V_p = 120\sqrt{2} \text{ V}$$
$$= 169.8 \text{ V}$$

as shown in Fig. 6.17.

For network response solutions as outlined in Section 6.2, we simply divide all voltages V and currents I by $\sqrt{2}$, which is equivalent to a $\sqrt{2}$ scale change. In writing the current as a function of time, however, a serious mistake would result unless we multiply the rms result by $\sqrt{2}$ to return the value to the peak value.

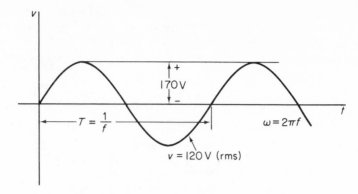

Figure 6.17 Peak and effective values of sinusoidal voltage.

EXAMPLE 6.6

If the rms value of a sinusoidal voltage across an impedance $Z = 10e^{j30°} \, \Omega$ is $50e^{j10°}$ V, the rms current $\hat{I}_e$ is

$$\hat{I}_e = \frac{50e^{j10°}}{10e^{j30°}} = 5e^{-j20°} \text{ A}$$

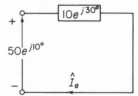

Figure 6.18 Example 6.6 network.

then the instantaneous current i becomes

$$i = \sqrt{2}(5) \cos(\omega t - 20°) \text{ A}$$

if we are using the real part of the phasors.

The value used (rms, peak, or instantaneous) in a particular situation must be specified or inferred. If no data are given, then the rms value may be inferred in problems where the sources are of the form $e^{j\omega t}$.

Extending the results obtained from Eqs. (6.38) and (6.40), the effective or rms current I_e for any periodic waveform $i(t)$ may be written

$$I_e = \left[\frac{1}{T} \int_0^T i^2(t) \, dt \right]^{1/2} \tag{6.45}$$

Similarly, the effective or rms voltages V_e for any periodic waveform $v(t)$ may be written

$$V_e = \left[\frac{1}{T} \int_0^T v^2(t) \, dt \right]^{1/2} \tag{6.46}$$

By a periodic current we mean one in which $i(t + T) = i(t)$ for any t. Effective values may be found by integration if $i(t)$ or $v(t)$ are known, and either hand integration or a computer integration routine may be used.

The energy $W(t)$ absorbed by the network of Fig. 6.19 from some arbitrary time zero time to t may be written

$$W = \int_0^t v(\tau)i(\tau)\, d\tau \tag{6.47}$$

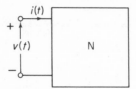

Figure 6.19 Energy absorbing
                network.

while the instantaneous power p at any time t is given by

$$p(t) = v(t)i(t) \tag{6.48}$$

Thus, in a resistance, where $v = Ri$,

$$W_R = \int_0^t Ri^2\, d\tau$$

and

$$p_R = Ri^2(t) = \frac{v^2(t)}{R} \tag{6.49}$$

as previously shown.

If $i = I_p \sin \omega t$, then

$$
\begin{aligned}
W_R &= \int_0^t RI_p^2 \sin^2 \omega\tau\, d\tau \\
&= \frac{RI_p^2}{2} \int_0^t (1 - \cos 2\omega\tau)\, d\tau \\
&= \frac{RI_p^2}{2}\left(t - \frac{\sin 2\omega t}{2\omega}\right) \\
&= RI_e^2\left(t - \sin \frac{2\omega t}{2\omega}\right) \qquad \text{joules} \\
p_R &= RI_p^2 \sin^2 \omega t \\
&= \frac{RI_p^2}{2}(1 - \cos 2\omega t) \\
&= RI_e^2(1 - \cos 2\omega t) \qquad \text{watts}
\end{aligned}
\tag{6.50}
$$

$$\tag{6.51}$$

Sketching these results in Fig. 6.20 we see that the power and energy vary at twice the frequency of the current and voltage and that p and W are ≥ 0. We have already

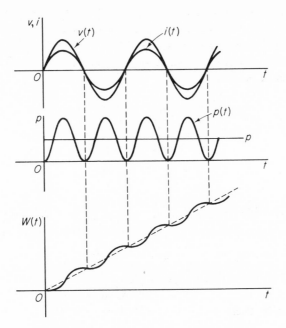

Figure 6.20 Relationship between voltage, current, power, and energy in a resistance.

found the energy in an inductance in Chapter 2 at any current i to be

$$W_L = \tfrac{1}{2}Li^2$$

and the energy in a capacitance at any voltage v to be

$$W_C = \tfrac{1}{2}Cv^2$$

If we let $i = I_P \sin \omega t$, such that v at $t = 0$, $i = 0$, then

$$
\begin{aligned}
W_L &= \tfrac{1}{2}LI_p^2 \sin^2 \omega t \\
&= \tfrac{1}{4}LI_p^2(1 - \cos 2\omega t) \\
&= \tfrac{1}{2}LI_e^2(1 - \cos 2\omega t)
\end{aligned}
\tag{6.52}
$$

and

$$
\begin{aligned}
p_L &= \frac{dW_L}{dt} \\
&= \tfrac{1}{2}LI_p^2\omega \sin 2\omega t \\
&= LI_e^2\omega \sin 2\omega t
\end{aligned}
\tag{6.53}
$$

Similarly, if we let $v = V_p \sin \omega t$, such that at $t = 0$, $v = 0$, then

$$
\begin{aligned}
W_c &= \tfrac{1}{2} C V_p^2 \sin^2 \omega t \\
&= \tfrac{1}{2} C V_e^2 (1 - \cos \omega t)
\end{aligned}
\tag{6.54}
$$

and

$$
\begin{aligned}
p_c &= \frac{dW_C}{dt} \\
&= \tfrac{1}{2} C V_p^2 \omega \sin 2\omega t \\
&= C V_e^2 \omega \sin 2\omega t
\end{aligned}
\tag{6.55}
$$

The power and energy are periodic at twice the period of v and i. From Eqs. (6.53) and (6.55) the average power over a complete cycle absorbed by either L or C is zero. The energy varies from zero to a maximum twice during each cycle of v or i. The capacitance or inductance thus absorbs energy part of the time, but delivers it back again, while the average power is zero. The resistance absorbs electrical energy continuously and converts this energy to heat.

Assume that we have a voltage $\hat{V}$ across an impedance Z in Fig. 6.21 and a current $\hat{I}$ into it. We take $\hat{V}$ and $\hat{I}$ to be effective or rms values in the sinusoidal case and omit the subscript.

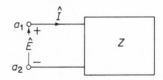

Figure 6.21 Figure for defining driving-point impedance and vector power.

The impedance $Z = \hat{V}/\hat{I}$ is termed the *driving-point impedance*, or the impedance looking into the terminals $a_1 a_2$ in Fig. 6.21. We now define complex power $\hat{S}$ by the equation

$$
\hat{S} = \hat{V} \hat{I}^*
\tag{6.56}
$$

where $\hat{I}^*$ means conjugate $\hat{I}$.

Let $\hat{I} = I e^{j\alpha}$, $\hat{V} = V e^{j\phi}$, and $Z = |Z| e^{j\theta}$. Then since by definition

$$
Z = \frac{\hat{V}}{\hat{I}}
$$

then

$$
|Z| = \frac{V}{I}
$$

and

$$
\theta = \phi - \alpha
$$

From Eq. (6.56) $\hat{S}$ now becomes

$$\begin{aligned}
\hat{S} &= (Ve^{j\phi})(Ie^{-j\alpha}) \\
&= VIe^{j(\phi-\alpha)} \\
&= VIe^{j\theta} \\
&= VI(\cos\theta + j\sin\theta)
\end{aligned} \tag{6.57}$$

However, the average power P is given by

$$\begin{aligned}
P &= I^2 R \\
&= I^2 \Re(|Z|(\cos\phi + j\sin\theta)) \\
&= I^2 |Z| \cos\theta
\end{aligned}$$

But $|Z| = V/I$; hence,

$$P = VI\cos\theta \qquad \text{watts} \tag{6.58}$$

In Eq. (6.58) $\cos\theta$ is defined as the power factor (Pf). The power factor is also defined to be *leading* if the current leads the voltage ($\alpha > \phi$), and *lagging* if the current lags the voltage ($\alpha < \phi$). If we now define a quantity Q as

$$Q = EI\sin\theta \qquad \text{reactive volt-amps} \tag{6.59}$$

then in Eq. (6.57)

$$\hat{S} = P + jQ \tag{6.60}$$

or

$$P = \Re(\hat{S}) \tag{6.61}$$

and

$$Q = \mathcal{I}(\hat{S}) \tag{6.62}$$

The magnitude of $\hat{S}$ is

$$\begin{aligned}
|\hat{S}| &= \sqrt{P^2 + Q^2} \\
&= VI \qquad \text{volt-amps}
\end{aligned} \tag{6.63}$$

where

$$VI = |\hat{V}||\hat{I}|$$

The reactive volt-amps, Q, is positive for an inductive Z and negative for a capacitive Z. Some texts define $\hat{S}$ by the relation $\hat{S} = \hat{V}^*\hat{I}$, which reverses the sign of Q.

Since $\hat{S} = \hat{V}\hat{I}^* = \hat{I}\hat{I}^* Z = I^2 Z$ and $Z = R + jX$,

$$P = I^2 R \qquad \text{watts} \tag{6.64a}$$

$$Q = I^2 X \qquad \text{vars} \tag{6.64b}$$

Similarly, $\hat{S} = \hat{V}\hat{I}^* = \hat{V}\hat{V}^* Y^* = V^2 Y^*$ and $Y = G + jB$, or $Y^* = G - jB$. So

$$P = V^2 G \qquad \text{watts} \tag{6.65a}$$

$$Q = -V^2 B \qquad \text{vars} \tag{6.65b}$$

EXAMPLE 6.7

Let $Z_1 = 1 + j5$, $Z_2 = 4 + j2$, $Z_3 = 6 - j1$, and $\hat{I} = 5e^{j0°}$ in the network of Fig. 6.22. Find the total power and power factor, branch powers, and voltage V.

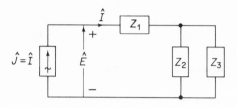

Figure 6.22 Example 6.7 network.

Solution

$$Z_T = Z_1 + \frac{Z_2 Z_3}{Z_2 + Z_3}$$

$$= 1 + j5 + \frac{(4 + j2)(6 - j1)}{(4 + j2) + (6 - j1)}$$

$$= 1 + j5 + \frac{24 + 2 + j(12 - 4)}{10 + j1}$$

$$\simeq 1 + j5 + \frac{(26 + j8)(10 - j1)}{100}$$

$$= 1 + j5 + 2.68 + j0.54$$

or

$$Z_T \simeq 3.68 + j5.54$$

Then

$$P = I^2 \Re(Z) = 25(3.68) = 92.0 \text{ W}$$

$$Q = I^2 \Im(Z) = 25(5.54) = 138.5 \text{ vars}$$

$$\hat{S} = 92.0 + j138.5$$

or

$$|\hat{S}| = 166 \text{ V-amps}$$

$$V = \frac{166}{5} = 33.2 \text{ V}$$

Note that the voltage V is actually specified in rms volts, although the designation rms is omitted.

$$\text{Pf} = \cos\theta = \cos\left(\tan^{-1}\frac{138.5}{92.0}\right) = \cos 56.3°$$

$$\simeq 0.556 \text{ lagging}$$

The power absorbed by Z_1 is $I^2 R = 25$ W.
The current in Z_2 is $5[|Z_3|/(|Z_2 + Z_3|)] = 5[(|6 - j1|)/(|10 + j1|)] \simeq 5(6.08/10) = 3.04$ A.
The current in Z_3 is $5[|Z_2|/(|Z_2 + Z_3|)] = 5[(|4 + j2|)/(|10 + j1|)] \simeq 5(4.48/10) = 2.24$ A.
The power in Z_2 is $P_2 \simeq (3.04)^2(4) = 37$ W.
The power in Z_3 is $P_3 \simeq (2.24)^2(6) = 30$ W.
The total power $= 37 + 30 + 25 = 92$ W.

Power-Factor Correction

The network in Fig. 6.23 represents a source in series with a line of resistance R, reactance X, and a load Z. The power lost in R is I^2R, where I is the effective current. For a given effective voltage V at Z, $I = |\hat{S}_z|/V$, and since $|\hat{S}|_z = \sqrt{P^2 + Q^2}$, it is clear that if P, the power absorbed by Z, is fixed (that is, we assume a certain rate of work to be done by Z), then to reduce I^2R to a minimum, we need to make $Q = 0$, or to have a unity power factor Z. In most electrical devices, Z is inductive (motors have around 0.80 Pf lagging). The power factor can be improved by putting a capacitance in parallel with Z.

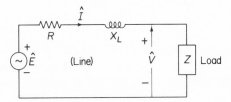

Figure 6.23 Load impedance driven through a lossy line.

Figure 6.24a shows an inductive load Z with $\hat{S} = P + jQ$ in parallel with a capacitance C, while Fig. 6.24b shows a power diagram. The line segments P, Q, and S in Fig. 6.24b are sometimes called *vectors* which do not rotate with time. The power factor $\cos\theta$ of Z alone is given by

$$\text{Pf} = \cos\left(\tan^{-1}\frac{P}{Q}\right) \tag{6.66}$$

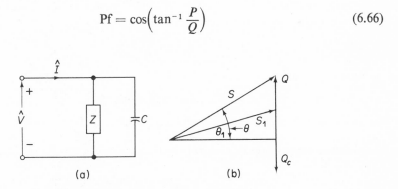

Figure 6.24 Vector power diagram.

The power factor $\cos\theta_1$ of Z in parallel with C is given by

$$\text{Pf}_1 = \cos\left(\tan^{-1}\frac{P}{Q - Q_c}\right) \tag{6.67}$$

since Q_c has the opposite sign to the original Q. With P constant (fixed work), then $|\hat{S}_1| < |\hat{S}|$, and I is reduced (constant V). The power supplier usually charges a power-factor penalty, or levies a charge for Q as well as for P. Hence, the customer with load Z may find it economically advantageous to install a power-factor correction capacitance C to reduce his Q charges. (The power P is considered useful and not reduceable.) To correct the original power factor Pf to the new Pf_1, we calculate the required Q_c from Eq. (6.67). Since $|Q_c| = V^2 B_c = V^2 \omega C$, then the required capacitance is

$$C = \frac{|Q_c|}{V^2 \omega} \tag{6.68}$$

Normally it is not economical to make $Pf_1 = 1.0$ since the required C becomes so large that the capacitance cost overshadows the penalty charge savings. To find the most economical size of C, we need to make a number of calculations comparing costs, and here the computer becomes attractive.

EXAMPLE 6.8

Take a given load as 1 kW (1000 W) at a Pf $= 0.707$ lagging. What correction capacitance C is required to correct the Pf to 0.9 lagging? What C is required to obtain a Pf $= 1$?

Solution

$$\text{original } P + jQ = 1000[1 + \tan(\cos^{-1} 0.707)]$$

$$= 1000 + j1000$$

If Pf $= 0.9$ (P constant),

$$P + jQ = 1000[1 + \tan(\cos^{-1} 0.9)]$$

$$= 1000 + j486$$

Hence, $Q_C = 1000 - 486 = 514$

$$C = \frac{514}{V^2 \omega}$$

If Pf $= 1$ (P constant),

$$Q_C = 1000 \text{ vars}$$

$$C = \frac{1000}{V^2 \omega}$$

or almost twice as much.

Since the line in Fig. 6.23 also contains inductive reactance, from the power supplier's viewpoint a second reason for altering load power factor stems from the fact that the voltage difference between $\hat{E}$ and $\hat{V}$ depends on the phase of $\hat{I}$. Thus, if we let $\hat{V}$ be the reference voltage or $V = Ve^{j0°}$, then in Fig. 6.23

$$\hat{E} = Ve^{j0°} + I(\cos \theta - j \sin \theta)(R + jX)$$

$$= V + j0 + I(R \cos \theta + X \sin \theta) + j(X \cos \theta - R \sin \theta))$$

where θ is the angle of $\hat{I}$ lagging $\hat{V}$. Then

$$\hat{E} = [V + I(R \cos \theta + X \sin \theta)] + jI(X \cos \theta - R \sin \theta) \tag{6.69}$$

The imaginary part of Eq. (6.69) is normally small compared to V, hence as a first approximation

$$|\hat{E}| = V + I(R \cos \theta + X \sin \theta)$$

In many cases, $X > R$; hence $IX \sin \theta$ may be an important part of the voltage difference. (If $\cos \theta = 0.9$, $\sin \theta = 0.346$, for example.) Thus the power supplier may install either fixed capacitance, or synchronous machines which supply reactive volt-amperes, on his lines to maintain good voltage conditions in addition to reducing line power losses.

Maximum Power Transfer

It is frequently desirable in electronic equipment that the generator deliver as much power to the load as possible. For example, the load may represent an electro-mechanical speaker, and we want to obtain the most efficient transfer of electrical source energy to audio energy. In the network of Fig. 6.25, let E represent the Thévenin equivalent of the source and Z_1 the Thévenin impedance. The load Z_2 has a fixed impedance $Z_2 = R_2 + jX_2$. How should Z_2 be adjusted such that the electrical power P_2 absorbed by R_2 is a maximum?

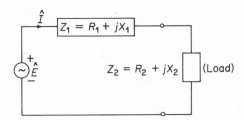

Figure 6.25 Network with complex source impedance and a complex load.

First of all, for the adjustment of X_1, since $P_2 = I^2 R_2$, it is clear that we need to make I as large as possible. From our previous discussion, this occurs at unity Pf, or we make

$$X_1 = -X_2 \tag{6.70}$$

Under these conditions, $I = E/(R_1 + R_2)$ and

$$P_2 = \frac{E^2 R_2}{(R_1 + R_2)^2}$$

Then setting dP_2/dR_2 to zero, we get

$$\frac{dP_2}{dR_1} = \frac{(R_1 + R_2)^2 E^2 - 2(R_1 + R_2)E^2 R_2}{(R_1 + R_2)^4} = 0$$

or

$$R_1^2 - R_2^2 = 0$$

and

$$R_1 = R_2 \tag{7.71}$$

Combining Eqs. for R_1, $R_2 > 0$, we get the requirement

$$Z_1 = Z_2^* \tag{6.72}$$

where Z_2^* is the complex conjugate of Z_2 for maximum power transfer from source to load. Note in this derivation that we are not concerned with power efficiency since the power loss in R_1 equals the power absorbed in R_2. This arrangement might be disastrous in large power applications.

Suppose that in the network of Fig. 6.25 Z_1 is fixed and we wish to obtain maximum power in R_2, but only $|Z_2|$ is adjustable. In this case,

$$\begin{aligned} P_2 &= I^2 |Z_2| \cos \theta \\ &= \frac{E^2 |Z_2| \cos \theta}{(R_1 + |Z_2| \cos \theta)^2 + (X_1 + |Z^2| \sin \theta)^2} \end{aligned} \tag{6.73}$$

Set the derivitive $dP_2/d|Z_2| = 0$ in Eq. (6.73). With some algebra and patience, we get

$$\begin{aligned} |Z_2| &= \sqrt{R_1^2 + X_1^2} \\ &= |Z_1| \end{aligned} \tag{6.74}$$

Now suppose that $|Z_1|$ is fixed but cannot be made equal to $|Z_1|$, owing to other requirements. We can still satisfy Eq. (6.74) from the standpoint of the source (and the power in the fixed R_2) by inserting an ideal transformer as in Fig. 6.26. An ideal transformer is defined by

$$\frac{V_1}{V_2} = \frac{N_1}{N_2} = n \tag{6.75}$$

and

$$\frac{I_1}{I_2} = \frac{N_2}{N_1} = \frac{1}{n} \tag{6.76}$$

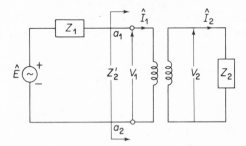

Figure 6.26 Ideal transformer matching network.

where N_1 and N_2 are the number of turns on coils 1 and 2, respectively. Since

$$Z_2 = \frac{V_2}{I_2}$$

and

$$Z_2' = \frac{V_1}{I_1}$$

then

$$Z'_2 = \frac{nV_2}{I_2/n} \tag{6.77}$$

$$= n^2 Z_2$$

where Z'_2 is the driving-point impedance at terminals $a_1 a_2$. Z'_2 is the impedance seen by the source E. Hence we insert a transformer between Z_1 and Z_2 such that

$$|Z'_2| = |Z_1|$$

or

$$|Z_2| = \frac{|Z_1|}{n^2}$$

Thus, if

$$n = \left(\frac{|Z_1|}{|Z_2|}\right)^{1/2} \tag{6.78}$$

we will have the required "match" to provide maximum power in R_2. This will not be as large as the maximum power in the previous case, since in that case we could manipulate X_2. In this development we have been somewhat loose with voltage and current polarities, but these need to be observed if V_2 and I_2 are required to have certain phase relationships with V_1 and I_1. Although we cannot obtain an "ideal" transformer, we may use this approximation, and then calculate the effects on the departures from the ideal case separately, if required.

In devices operating at low power levels, as in communications, "matching" sources and loads by means of transformers or other devices becomes a vital necessity. For large power situations, power efficiency becomes the dominating consideration.

We complete this section with the observation that we could find steady-state solutions to Eq. (6.1) or (6.6) with a more general exponential source function $E(t) = Ee^{st}$, where $s = \sigma + j\omega$, a general complex quantity. If we substitute this in Eq. (6.6) and assume that $i(t) = Ae^{st}$, then proceeding in the same manner, we can define a general impedance

$$Z(s) = Ls + R + \frac{1}{Cs}$$

which forms a relation between input and response quantities. We study this in more rigorous detail in Chapter 8.

6.4 Complete Solution, Sinusoidal Sources

The complete solution to a differential equation consists of the sum of the transient solution and the steady-state (particular integral) solution. To illustrate, let $E(t) = 25\sqrt{2} \sin(10t + 30°)$ V in Fig. 6.27. The initial conditions are shown at time $t = 0$.

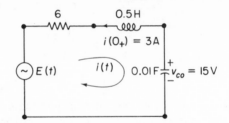

Figure 6.27 *RLC* series network.

First we calculate the steady-state current i using effective values.

$$Z = 6 + j5 - j10$$
$$= 6 - j5$$

Then

$$\hat{I} = \frac{25e^{j30}}{7.81e^{-j39.8°}} = 3.20e^{j69.8}$$

or

$$i = 3.20\sqrt{2} \ (\sin 10t + 69.8°) \qquad \text{(slide-rule accuracy)}$$

Next, we find the transient current i. The differential equation is

$$0.5\frac{di}{dt} + 6i + 100 \int i \, dt = 25\sqrt{2} \ \sin(10t + 30°) \qquad (6.79)$$

Differentiating and setting the right-hand side of Eq. (6.79) equal to zero, we obtain

$$0.5\frac{d^2i}{dt^2} + 6\frac{di}{dt} + 100i = 0$$

The characteristic equation is

$$\lambda^2 + 12\lambda + 200 = 0$$

or

$$\omega_n = \sqrt{200}$$
$$\zeta = 0.424$$

The transient solution is of the form

$$i = Ke^{-6t}\sin(12.8t + \phi)$$

The complete solution, $i(t)$, then becomes

$$i(t) = 3.20\sqrt{2} \ \sin(10t + 69.8°) + Ke^{-6t}\sin(12.8t + \phi) \qquad (6.80)$$

The constant K is found by evaluating the initial conditions at $t = 0_+$: $i = -3$, or

$$-3 = 3.20\sqrt{2} \ \sin(69.8) + K\sin\phi$$

Thus,

$$K\sin\phi = -7.24 \qquad (6.81)$$

This gives us one equation with unknowns K and ϕ. We formulate a second equation by taking the derivative of Eq. (6.80), obtaining

$$\frac{di}{dt} = 45.2 \cos(10t + 69.8°) + K(-6)e^{-6t} \sin(12.8t + \phi)$$
$$+ Ke^{-6t}(12.8) \cos(12.8t + \phi) \tag{6.82}$$

and writing a loop equation based on KVL at $t = 0_+$. The loop equation is

$$0.5 \frac{di}{dt}(0_+) - 6(3) + 15 = 25\sqrt{2} \sin 30°$$

which reduces to

$$\frac{di}{dt}(0_+) = 41.4$$

Substituting the derivation at $t = 0_+$ into Eq. (6.82) we form the second equation in K and ϕ

$$41.4 = 45.2 \cos 69.8° - 6K \sin \phi + 12.8K \cos \phi$$
$$12.8K \cos \phi - 6K \sin \phi = 25.8 \tag{6.83}$$

Substituting Eq. (6.81) into Eq. (6.83), we find that

$$12.8K \cos \phi + 43.4 = 25.8$$

or

$$K \cos \phi = -1.372$$

Then

$$\tan \phi = \frac{K \sin \phi}{K \cos \phi} = \frac{-7.24}{-1.372}$$

or

$$\phi = 180° + 79.2°$$
$$= -100.8°$$

(observe that we must keep the signs on the tangent components to correctly identify the angle)
and

$$K \sin \phi = -7.24$$

or

$$K = \frac{-7.24}{-0.983} = 7.34$$

The complete solution, assuming that we have made no errors, is

$$i_t = 3.20\sqrt{2} \sin(10t + 69.8°) + 7.34e^{-6t} \sin(12.8t - 100.8°) \tag{6.84}$$

The transient solution constants thus involve the steady-state solution. This leads us to believe that it might be possible to have no transient if the conditions were properly adjusted. We illustrate this in the next example.

EXAMPLE 6.9

Let $E_1 = 10\sqrt{2}\,\sin(100t)$ in the network of Fig. 6.28. At what time t should the switch sw be closed such that no transient current will occur?

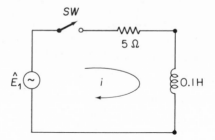

Figure 6.28 A case for no transients.

Solution. First, we find the steady-state current:

$$Z = 5 + j10$$

Thus,

$$\hat{I} = \frac{10}{5 + j10}$$
$$= \frac{50 - j100}{125}$$

or

$$i = 8.93\sqrt{2}\,\sin(\omega t - 63.4°)$$

The transient current is

$$i = Ke^{-50t}$$

The total current therefore becomes

$$i_t = 8.93\sqrt{2}\,\sin(100t - 63.4°) + Ke^{-50t}$$

It is desired to make $K = 0$ by making t something other than zero at switch closing. Whenever the switch is closed, $i = 0$ (no current in the inductance). Thus, if the switch is thrown at time $t \neq 0$, and $K = 0$,

$$0 = 8.93\sqrt{2}\,\sin(100t - 63.4°) + 0$$

Hence,

$$(100t - 63.4°) = K\pi \qquad K = 0, 1, 2, 3, \ldots$$

or

$$100t = K\pi + 1.106$$
$$t = \frac{K\pi + 1.106}{100}$$

In any situation involving sinusoidal steady-state currents we may similarly find a switch closing time which will cause at least one transient to not exist.

Additional Reading

CLOSE, C. M., *The Analysis of Linear Circuits*, Harcourt Brace Jovanovich, Inc., 1966.

LEON, B. J., and P. A. WINTZ, *Basic Linear Networks for Electrical and Electronics Engineers*, Holt, Rinehart and Winston, Inc., New York, 1970.

McCRACKEN, D. D., *Guide to FORTRAN IV Programming*, John Wiley & Sons, Inc., New York, 1965.

SMITH, R. J., *Circuits, Devices, and Systems*, John Wiley & Sons, Inc., New York, 1971.

VAN VALKENBURG, M. E., *Network Analysis*, Prentice-Hall, Inc., Englewood Cliffs, N.J., 1964.

Problems

6.1. Find the steady-state solution (particular integral) for the following equations:

(a) $\dfrac{di}{dt} + 3i = \cos t$

(b) $\dfrac{di}{dt} + 2i = \cos(2t + 30°)$

(c) $\dfrac{d^2 i}{dt^2} + i = 4 \sin t$

(d) $\dfrac{d^2 i}{dt^2} + 2\dfrac{di}{dt} + 2i = 3 \cos(t + 30°)$

(e) $\dfrac{di}{dt} + 2i = \sin 2t + \cos t$

6.2. Find the impedance at terminals *a-b* of the circuits in Fig. P6.1.

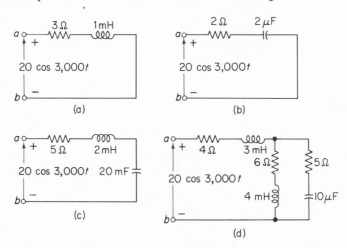

Figure P6.1

6.3.   In Problem 6.2 take the angular velocity ω as one half the given value and recalculate the impedances.

6.4.   In Problem 6.2 double the angular velocity ω and recalculate the impedances. Make a table comparing impedances at the three different angular velocities specified in Problems 6.2, 6.3, and 6.4.

6.5.   In Fig. P6.2 calculate the power loss in the 10-Ω resistance at the angular velocities ω of 4500, 5000, and 5500 rad/s.

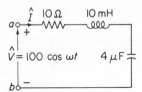

Figure P6.2

6.6.   Repeat Problem 6.5 if the 10-Ω resistance is changed to a 1-Ω resistance. Use angular velocities of 4950, 5000, and 5050 rad/s.

6.7.   Find the frequency of resonance to sinusoidal inputs of the network in Fig. P6.3. The condition of resonance is defined as the condition of unity power factor or when the current $\hat{I}$ is in phase with the voltage $\hat{V}$. Hence at resonance the impedance $Z = \hat{V}/\hat{I}$ has no imaginary component. Compare this frequency with the resonant frequency if the two parallel branches in Fig. P6.3 are connected in series.

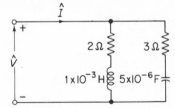

Figure P6.3

6.8.   In Problem 6.5 calculate the voltages across the inductance and the capacitance at the three given angular velocities. Draw a phasor diagram showing the source voltage, each element voltage, and the current for all three cases.

6.9.   In Fig. P6.4 V_1 is a sinusoidal voltage. $R = 1\ \Omega$, $L = 1$ mH, $C = 1 \times 10^{-9}$ F. Find the ratio $|V_1/V_2|$ and the phase angle between V_1 and V_2 at $\omega_2 = 10^6$ rad/s and $\omega_2 = 1.1 \times 10^6$ rad/s. Discuss.

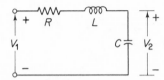

Figure P6.4

6.10. Using the network in Fig. P6.5, repeat Problem 6.9 using the same element values.

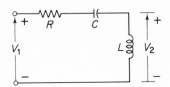

Figure P6.5

6.11. The resonance curve for a parallel RLC circuit is shown in Fig. P6.6.
   (a) Find the values of R, L, and C.
   (b) The same resonance behavior is desired around a center frequency of 50 kHz. The maximum value of $|Z(j\omega)|$ is to be 10 kΩ. Find the proper values of R, L, and C.

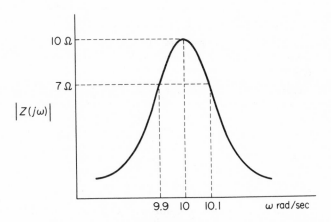

Figure P6.6

6.12. Assume we are given Z_A, Z_B, and Z_C in the delta circuit of Fig. P6.7(a) as complex impedances to sinusoidal inputs. We wish to form a wye circuit as in Fig. P6.7(b), which is equivalent to (a) as far as terminals a, b, and c are concerned (delta–wye conversion). Find Z_1, Z_2, and Z_3 in terms of Z_A, Z_B, and Z_C. If $Z_A = 1 + j3$, $Z_B = 2 - j2$, and $Z_C = 3 - j1$, calculate Z_1, Z_2, and Z_3.

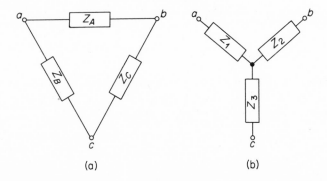

Figure P6.7

6.13.   Repeat Problem 6.12 except find Z_A, Z_B, and Z_C in terms of Z_1, Z_2, and Z_3 (wye–delta conversion). If $Z_1 = 1 + j3$, $Z_2 = 2 - j2$, and $Z_3 = 3 - j1$, calculate Z_A, Z_B, and Z_C.

6.14.   Find the equivalent total impedance at terminals *a-b* of the network shown in Fig. P6.8.

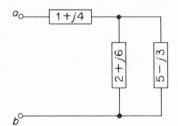

Figure P6.8

6.15.   (a) Find the current $\hat{I}$ in the network of Fig. P6.9.
        (b) If the angular velocity of the source voltage doubles, find the new current $\hat{I}$.

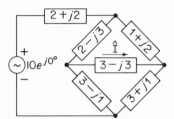

Figure P6.9

6.16.   Find the Thévenin equivalent circuit for the network in Fig. P6.10 with respect to the terminals *a-b*. Find the current $\hat{I}_L$ through the impedance Z_L if $Z_L = 1 - j7$. Repeat if $Z_L = 7 - j1$. Compare this technique with the methods of finding $\hat{I}_L$ by either loop analysis or by reducing the impedance as seen by the generator to a single element.

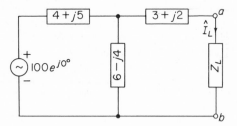

Figure P6.10

6.17.   Find either the Thévenin or the Norton equivalent circuit as viewed from terminals *a-b* in the network of Fig. P6.11. Find $\hat{I}_z$ if $Z = 2 + j4$. What value should Z have for maximum power in Z? What is this maximum power?

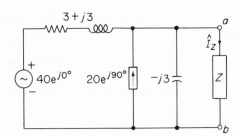

Figure P6.11

6.18. In the network shown in Fig. P6.12, $L_1 = 1 \times 10^{-3}$ H, $L_2 = 20 \times 10^{-3}$ H and the coefficient of coupling $k = 0.90$. The impedances of the other elements at $\omega = 1000$ are as shown. Find the impedance of the circuit as viewed from the terminals a-b.

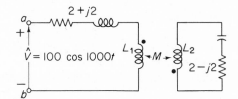

Figure P6.12

6.19. Write the node equations in matrix form for the network shown in Fig. P6.13.

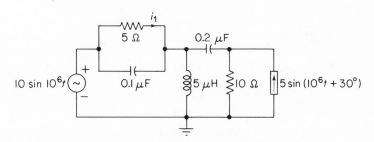

Figure P6.13

6.20. The network of Fig. P6.14 is operating in the steady state with $\hat{E}_1 = 2 \sin 2t$ and $k = -0.333$. Under these conditions
   (a) Write the network loop equations in matrix form.
   (b) Determine $i_2(t)$.

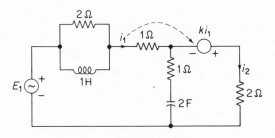

Figure P6.14

6.21. The inductively coupled network of Fig. P6.15 is operating in the sinusoidal steady state with $E(t) = 2 \cos t$.
(a) Write the loop equations for the network in matrix form.
(b) Write the node equations for the network in matrix form.
(c) Determine $V_a(t)$ using the equations of part (a).
(d) Verify the value obtained in part (c) by solving for $V_a(t)$ using the equations of part (b).

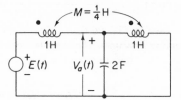

Figure P6.15

6.22. The load Z in Fig. P6.16 requires 50 kW at 80 percent Pf lagging. Find the capacitance C required to give a 90 per cent Pf at terminals a-b if the frequency is 60 Hz. There are two possible answers. Which of the two solutions would you choose and why?

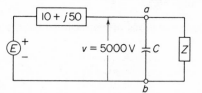

Figure P6.16

6.23. (a) Calculate the voltage E and the power lost in the 10-Ω line resistance for Problem 6.22 without the capacitance C.
(b) Repeat part (a) for both possible C values in Problem 6.22. If the per cent voltage regulation is $(|E| - |V|)100/|V|$, find the voltage regulation for all three cases.

6.24. In Fig. P6.17 the load L is 5000 W at a power factor of 80 per cent inductive. R and C vary simultaneously so that the power factor of the R-C branch is always 0.3 leading. Find the values of R and X_C to produce an overall power factor of 90 per cent lagging at the terminals a-b. Draw the power diagram.

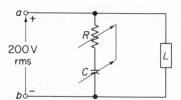

Figure P6.17

6.25. Find the transformer turns ratio n in Fig. P6.18 which produces maximum power in Z_L.

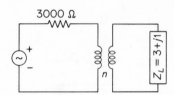

Figure P6.18

6.26. In the network of Fig. P6.19 let $C_p = 1 \times 10^{-9}$ F, $L_p = 1/(40\pi^2)$ H, $Q_p = \omega_R L_p / R_p$ = 100, where $\omega_R = 2\pi(100,000)$. Let $C_s = 1 \times 10^{-12}$ F, $L_s = 10^2/(4\pi^2)$ H, $Q_s = \omega_R L_s / R_s = 100$, and $M = k\sqrt{L_p L_s}$. Write a computer program to find $|V_2/V_1|$ as a function of angular velocity ω. Plot this ratio for $k = 0.01$ in the region of ω_R. Repeat for $k = 0.015, 0.005$, and 0.03. Discuss. (See F. E. Terman, *Radio Engineering*, McGraw-Hill Book Company, New York, 1947, p. 66.)

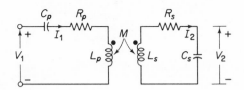

Figure P6.19

6.27. Complex arithmetic is not available in all versions of Fortran. Assuming that your computer installation is one of the facilities without complex arithmetic capabilities, write a set of subroutine subprograms which will give you a basic set of complex arithmetic capabilities using only real arithmetic. The basic set of subroutines should include

(a) addition (CADD) to compute $C = A + B$
(b) subtraction (CSUB) to compute $C = A - B$
(c) multiplication (CMPY) to compute $C = A \cdot B$
(d) division (CDIV) to compute $C = A/B$
(e) exponentiation (CEXPR) to compute $C = A^D$
(f) exponentiation (CEXPI) to compute $C = A^I$
(g) rectangular–polar conversion (RECPØL)
(h) polar–rectangular conversion (PØLREC)
(i) square root (CSQRT) to compute $\sqrt{A}$
(j) absolute value (CABS) to compute $|A|$
(k) exponent (CEXP) to compute e^A
(l) sine (CSIN) to compute $\sin A$
(m) cosine (CCØS) to compute $\cos A$

The variables A, B, and C are complex. The variable D is real. The variable I is imaginary.

6.28. Using the set of subroutines written in Problem 6.27, write a Fortran program to compute the loop currents of a network from a user-supplied impedance matrix **Z** and equivalent voltage source vector **E**. Remember, the computer is assumed to lack a complex arithmetic capability so the input and output must use only real variables. *Hint:* The program should be similar to the program in Fig. 4.21.

Chapter 7

NETWORK TOPOLOGY

I think that I shall never see
A billboard lovely as a tree.
Perhaps, unless the billboards fall,
I'll never see a tree at all.

OGDEN NASH, *Song of the Open Road*

7.1 Introduction

The first six chapters of the text presented network-analysis procedures adapted to hand analysis. We developed systematic methods of writing network equations for which numerical solutions of the node voltages and loop currents by computer techniques became advantageous when evaluating more than three unknowns.

It would be even more useful to develop systematic procedures in which the computer would accept as input the description of the network a branch at a time, generate the network equations, and then determine the unknown voltages and currents. In a few words, we seek to automate the generation and solution of the network equations.

As usual, when we try to systematize an engineering process the first step involves abstraction. Kirchhoff's laws lay the foundation of network analysis, and since these laws do not make assumptions concerning the nature of the network elements, it seems completely natural to overlook element types and reduce the network to the simpler form of a graph, where the elements reduce to line segments. We devote the first section of this chapter to the development of graph theory, which we apply to network analysis in this and the following chapters. We do not present a rigorous treatise on graph theory but only present the material necessary to accomplish our objectives. (A more comprehensive treatment on graph theory can be found in the references listed at the end of the chapter.) We use the graph theory, or topological techniques, to formulate equations describing arbitrary networks.

7.2 Definitions

The topology of a network is important in the systematic formulation of the network equations. We develop the topological properties of a network by replacing all its elements by simple line segments to obtain the *graph* of the network. This graph contains a set of branches and nodes defined as follows:

Branch. A branch is a line segment which represents a network element or a combination of elements connected between two nodes. A branch also can be sometimes referred to as an *edge* of the graph.

Node. A node is located at each end of a branch. Thus, a *node*, *vertex*, or *junction* exists at the junction between two branches or at the end of an isolated branch.

In general, a branch represents the location of any single network element such as a resistance R or a voltage source E or it may represent any combination of these elements. Typical branches are shown in Fig. 7.1.

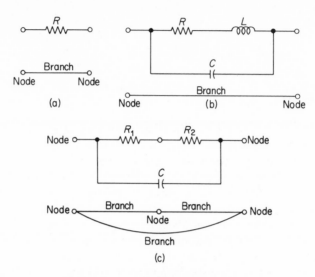

Figure 7.1 Typical branches and nodes.

Oriented Graph. In network analysis applications we number the branches of the graph, and each branch carries an arrow to indicate its orientation. We term such a graph *oriented*. In electrical networks the branch orientation indicates the arbitrary assumed positive reference polarity of the branch current or voltage. An example of a network and its oriented graph is shown in Fig. 7.2. The relationship between the orientation of a branch and its assumed positive voltage and current is shown in Fig. 7.3.

Subgraph. A subgraph of a network is a subset of the branches and nodes of a graph. The subgraph is considered *proper* if it consists of less than all the branches and nodes of the graph.

Connected Graph. A graph is connected if there exists at least one path between any two nodes. The graph in Fig. 7.2 is connected but the graph in Fig. 7.4 is *disconnected*.

Loop. A loop is a collection of branches in a graph which forms a closed path. For example, nodes 1, 2, and 3 in Fig. 7.2 together with branches 2, 3, and 4 form a loop. When specifying a loop, either the set of all the branches in the loop or the set of all nodes can be specified (when no two branches are in parallel). Each of these sets uniquely specify the loop; thus, nodes {1, 2, 3} or branches {2, 3, 4} specify a loop in Fig. 7.2.

Tree. A tree is a connected subgraph of a graph which contains all the nodes of the graph but contains no loops. The branches of the tree are called *tree* branches or

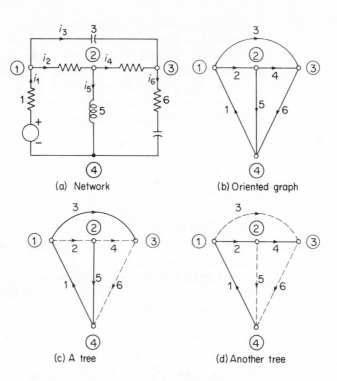

(a) Network

(b) Oriented graph

(c) A tree

(d) Another tree

Figure 7.2 A network and its oriented graph.

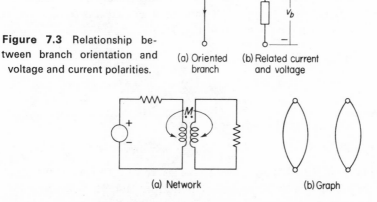

Figure 7.3 Relationship between branch orientation and voltage and current polarities.

(a) Oriented branch

(b) Related current and voltage

(a) Network

(b) Graph

Figure 7.4 Example of a disconnected graph.

twigs. The branches of the graph not part of the tree being considered are *link* branches. Since a graph may contain several trees, the link branches depend upon the selection of a particular tree. For example, if we select branches 1, 3, and 4 of the graph as the tree, as in Fig. 7.2c, branches 2, 4 and 6 become links, or link branches. If we select branches 1, 2, and 4 as a tree, as in Fig. 7.2d, then branches 3, 5, and 6 are links.

The number of tree branches of a graph containing n_t nodes always equals $n_t - 1$, as may be easily proved by induction. There must always be a total of $b - (n_t - 1)$ links in a graph, where b is the total number of network branches. Since trees are formed with the maximum number of branches in a graph without forming any loops, we note that the tree plus any one of its associated link branches always forms a loop. In Fig. 7.2b if we arbitrarily select the tree consisting of branches 1, 3, and 5 we see that the addition of any one of the links (branches 2, 4, and 6) will form a loop in the graph [see Fig. 7.2c.]

7.3 Incidence Matrix and Formulation
of Node Equations

Given an oriented graph, such as the one in Fig. 7.2b, we may describe the graph completely in a compact and convenient matrix form. In this matrix we specify the orientation of each branch in the graph and the nodes at which this branch is incident.

For a graph having n_t nodes and b branches, the complete node-branch incidence matrix (or complete incidence matrix) $\mathbf{A}_a = [a_{ij}]$ is an $n_t \times b$ rectangular matrix whose (i, j)th element is defined by the following:

$$a_{ij} = \begin{cases} 1 & \text{if branch } j \text{ is incident at node } i \text{ and is oriented away from node} \\ & i \text{ (branch } j \text{ "leaves" node } i) \\ -1 & \text{if branch } j \text{ is incident at node } i \text{ and is oriented toward node } i \\ 0 & \text{if branch } j \text{ is not incident at node } i \end{cases}$$

Since each branch leaves a single node and enters a single node, each column of matrix $\mathbf{A}_a$ contains a single 1 and a single -1, with all other elements equal to zero. The complete incidence matrix for the graph in Fig. 7.2 is

$$\mathbf{A}_a = \begin{array}{c} \text{node} \\ \downarrow \\ \begin{array}{c} 1 \\ 2 \\ 3 \\ 4 \end{array} \end{array} \overset{\begin{array}{cccccc} 1 & 2 & 3 & 4 & 5 & 6 \leftarrow \text{branch} \end{array}}{\begin{bmatrix} -1 & 1 & 1 & 0 & 0 & 0 \\ 0 & -1 & 0 & 1 & 1 & 0 \\ 0 & 0 & -1 & -1 & 0 & 1 \\ 1 & 0 & 0 & 0 & -1 & -1 \end{bmatrix}} \tag{7.1}$$

If all rows of the matrix in Eq. (7.1) are added to the last row of the matrix, the result is a row of zeros, indicating that the rows are not all independent. Thus, we can eliminate one row from the matrix, since it is the negative sum of all the other rows. The rank n of the resulting matrix then becomes $n = n_t - 1$.

The matrix obtained from $\mathbf{A}_a$ by eliminating one row is called the *incidence matrix* $\mathbf{A}$. Let us define the row eliminated from the incidence matrix as the row which corresponds to the datum (ground) node of the network.

What does network analysis have to do with this $\mathbf{A}$ matrix? Kirchhoff's current law for a network states that the sum of the currents leaving a node is zero. We now assert that in a network this may be expressed by the equation

$$\mathbf{A}\mathbf{i}_b = 0 \tag{7.2}$$

where $\mathbf{i}_b$ is the vector of branch currents. The positive reference direction of the branch currents corresponds to the orientation of the graph branches. To demonstrate this, let node 4 be the datum node in the network of Fig. 7.2a. Then Eq. (7.2) states that

$$\mathbf{A}\mathbf{i}_b = 0$$

or

$$\begin{bmatrix} -1 & 1 & 1 & 0 & 0 & 0 \\ 0 & -1 & 0 & 1 & 1 & 0 \\ 0 & 0 & -1 & -1 & 0 & 1 \end{bmatrix} \cdot \begin{bmatrix} i_1 \\ i_2 \\ i_3 \\ i_4 \\ i_5 \\ i_6 \end{bmatrix} = \begin{bmatrix} 0 \\ 0 \\ 0 \\ 0 \\ 0 \\ 0 \end{bmatrix} \tag{7.3}$$

which corresponds to the KCL equations written by inspection as follows:

$$\text{At node 1: } -i_1 + i_2 + i_3 = 0 \tag{7.4a}$$

$$\text{At node 2: } -i_2 + i_4 + i_5 = 0 \tag{7.4b}$$

$$\text{At node 3: } -i_3 - i_4 + i_6 = 0 \tag{7.4c}$$

In this case we see that the equations in Eq. (7.4) are linearly independent, since each equation contains a variable not contained in any of the other equations.

We adopt the convention that node voltages $v_{n1}, v_{n2}, \ldots, v_{nn}$ of network nodes $1, 2, \ldots, n$ are measured with respect to the datum node. Kirchhoff's voltage law guarantees that these node voltages are defined unambiguously. Hence, if we calculate the voltage of any node with respect to that of the datum node by computing the algebraic sum of the branch voltages along a path from the datum node to the node in question, the sum will be independent of the path chosen.

If we let $\mathbf{v}_n$ be the vector of node voltages $v_{n1}, v_{n2}, \ldots, v_{nn}$, the branch voltages $v_{b1}, v_{b2}, \ldots, v_{bb}$, designated $\mathbf{v}_b$, are related to $\mathbf{v}_n$ by the relationship

$$\mathbf{v}_b = \mathbf{A}^\mathsf{T}\mathbf{v}_n \tag{7.5}$$

where the $b \times n$ matrix $\mathbf{A}^\mathsf{T}$ is the transpose of the incidence matrix $\mathbf{A}$.

To show this we refer again to the network of Fig. 7.2a and the corresponding incidence matrix $\mathbf{A}$ of Fig. 7.2b, which is given by

$$\mathbf{A} = \begin{bmatrix} -1 & 1 & 1 & 0 & 0 & 0 \\ 0 & -1 & 0 & 1 & 1 & 0 \\ 0 & 0 & -1 & -1 & 0 & 1 \end{bmatrix} \begin{matrix} 1 \\ 2 \\ 3 \end{matrix} \tag{7.6}$$

$$\text{branch} \rightarrow \quad 1 \quad 2 \quad 3 \quad 4 \quad 5 \quad 6 \underset{\text{node}}{\uparrow}$$

By Eq. (3.7) we have

$$\mathbf{A}^{\mathsf{T}} = \begin{bmatrix} -1 & 0 & 0 \\ 1 & -1 & 0 \\ 1 & 0 & -1 \\ 0 & 1 & -1 \\ 0 & 1 & 0 \\ 0 & 0 & 1 \end{bmatrix} \begin{matrix} 1 \\ 2 \\ 3 \\ 4 \\ 5 \\ 6 \end{matrix} \qquad (7.7)$$

$$\text{node} \rightarrow 1 \quad 2 \quad 3 \underset{\uparrow}{\ } \text{branch}$$

Using Eq. (7.7), Eq. (7.5) states that

$$\mathbf{v}_b = \mathbf{A}^{\mathsf{T}}\mathbf{v}_n$$

$$= \begin{bmatrix} -1 & 0 & 0 \\ 1 & -1 & 0 \\ 1 & 0 & -1 \\ 0 & 1 & -1 \\ 0 & 1 & 0 \\ 0 & 0 & 1 \end{bmatrix} \begin{bmatrix} v_{n1} \\ v_{n2} \\ v_{n3} \end{bmatrix} \qquad (7.8)$$

or

$$v_{b1} = -v_{n1} \qquad (7.9a)$$

$$v_{b2} = v_{n1} - v_{n2} \qquad (7.9b)$$

$$v_{b3} = v_{n1} - v_{n3} \qquad (7.9c)$$

$$v_{b4} = v_{n2} - v_{n3} \qquad (7.9d)$$

$$v_{b5} = v_{n2} \qquad (7.9e)$$

$$v_{b6} = v_{n3} \qquad (7.9f)$$

We recognize the six scalar equations in Eq. (7.9) as expressions of KVL for the network.

Thus, Eqs. (7.2) and (7.5) give

$$\mathbf{A}\mathbf{i}_b = 0 \qquad \text{(KCL)}$$

and

$$\mathbf{v}_b = \mathbf{A}^{\mathsf{T}}\mathbf{v}_n \qquad \text{(KVL)}$$

which are the two basic equations of node analysis. These are obtained directly from Kirchhoff's laws and the oriented network graph, and are independent of the elements of the network. Now we must characterize the network branches, that is, write the b branch equations which relate the branch voltages $\mathbf{v}_b$ to the branch currents $\mathbf{i}_b$. The network elements enter into the analysis only through these equations.

The sole restriction that we have placed upon the composition of a branch so far is that it must have two terminals; hence, we can allow the branch to contain a collection of passive elements and dependent and independent sources. At the start, let us assume that the branch contains only resistances, conductances, and independent voltage and current sources. The general form of such a branch is shown in Fig. 7.5.

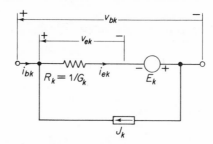

Figure 7.5 Simplified general branch.

The currents and voltages for the simplified general branch of Fig. 7.5 are defined as

i_{bk} = branch current through branch k

J_k = independent current source in branch k

$i_{ek} = i_{bk} + J_k$ = element current through the passive element in branch k

v_{bk} = branch voltage across branch k

E_k = independent voltage source in branch k

$v_{ek} = v_{bk} + E_k$ = element voltage across the passive element in branch k

Two important points should be emphasized here concerning the general branch shown in Fig. 7.5. First, the orientation of the elements in the branch exists for definition purposes. The polarities of the sources could have been reversed without introducing difficulties in the network equation formulation. We will see as we proceed that alterations in the general branch will only change some of the signs in the matrix equations. Also the positive branch and element voltage and current polarity assignments could have been made differently without creating formulation problems. The proof of these statements is left to the reader as an exercise.

The second important point is that a branch may contain more than a single element, and in fact may contain any combination of passive elements and sources. The more elements a branch contains, the more complex become the relationships between the branch current i_b, the branch voltage v_b, and the independent sources. Thus, in referring to sources in branch k, we mean just that.

Using the orientations shown in Fig. 7.5, the branch current i_{bk} through the simplified general branch k is

$$
\begin{aligned}
i_{bk} &= i_{ek} - J_k \\
&= G_k v_{ek} - J_k \\
&= G_k(v_{bk} + E_k) - J_k \\
&= G_k v_{bk} - J_k + G_k E_k
\end{aligned}
\tag{7.10}
$$

Equation (7.10) defines i_{bk} in terms of the branch voltage v_{bk} and the source voltages and currents.

In matrix notation the network branch current may be expressed as

$$\mathbf{i}_b = \mathbf{G}\mathbf{v}_b - \mathbf{J} + \mathbf{G}\mathbf{E} \tag{7.11}$$

where $\mathbf{G}$ is the branch conductance matrix. The matrix $\mathbf{G}$ is a diagonal matrix of order b given by

$$\mathbf{G} = \begin{bmatrix} G_1 & 0 & \cdots\cdots & 0 \\ 0 & G_2 & & \vdots \\ \vdots & & \ddots & 0 \\ 0 & \cdots\cdots & 0 & G_b \end{bmatrix} \tag{7.12}$$

where $G_1, G_2, \ldots, G_b$ are the branch conductances in the order indicated by the previously chosen oriented graph.

The branch source vectors $\mathbf{J}$ and $\mathbf{E}$ are of dimension b; that is,

$$\mathbf{J} = \begin{bmatrix} J_1 \\ \cdots \\ J_b \end{bmatrix} \qquad \mathbf{E} = \begin{bmatrix} E_1 \\ \cdots \\ E_b \end{bmatrix} \tag{7.13}$$

Again the order of the elements in the vector is determined by the oriented graph. Positive entries in the source vectors $\mathbf{E}$ and $\mathbf{J}$ correspond to sources in the general branch oriented in the directions shown in Fig. 7.5. Sources with polarities opposite the assumed directions enter the source vectors with negative signs. Zeros are entered for branches containing no sources.

We now combine Eqs. (7.2), (7.5), and (7.11) to obtain a matrix equation describing the network in terms of the network variable $\mathbf{v}_n$. First premultiplying Eq. (7.11) by $\mathbf{A}$, we obtain

$$\mathbf{A}\mathbf{i}_b = \mathbf{A}\mathbf{G}\mathbf{v}_b - \mathbf{A}\mathbf{J} + \mathbf{A}\mathbf{G}\mathbf{E}$$
$$= \mathbf{A}\mathbf{G}\mathbf{v}_b - \mathbf{A}(\mathbf{J} - \mathbf{G}\mathbf{E})$$

Using Eq. (7.2) we obtain the equation

$$\mathbf{A}\mathbf{G}\mathbf{v}_b = \mathbf{A}(\mathbf{J} - \mathbf{G}\mathbf{E}) \tag{7.14}$$

Now substituting Eq. (7.5) for $\mathbf{v}_b$ into Eq. (7.14), we get

$$\mathbf{A}\mathbf{G}\mathbf{A}^\mathsf{T}\mathbf{v}_n = \mathbf{A}(\mathbf{J} - \mathbf{G}\mathbf{E}) \tag{7.15}$$

The node voltages $\mathbf{v}_n$ can thus be determined from the expression

$$\mathbf{v}_n = (\mathbf{A}\mathbf{G}\mathbf{A}^\mathsf{T})^{-1}\mathbf{A}(\mathbf{J} - \mathbf{G}\mathbf{E}) \tag{7.16}$$

In Eq. (7.15) $\mathbf{AGA^T}$ is an $n \times n$ square matrix, whereas $\mathbf{A(J - GE)}$ is an nth-order vector. If we introduce the notation

$$\mathbf{G}_n \triangleq \mathbf{AGA^T} \qquad \text{(nodal conductance matrix)} \tag{7.17a}$$

$$\mathbf{J}_n \triangleq \mathbf{A(J - GE)} \qquad \text{(equivalent current vector)} \tag{7.17b}$$

then Eq. (7.15) becomes

$$\mathbf{G}_n \mathbf{v}_n = \mathbf{J}_n \tag{7.18}$$

which is identical to Eq. (4.38).

EXAMPLE 7.1

Write the node equations for the resistive network shown in Fig. 7.6 using Eq. (7.15).

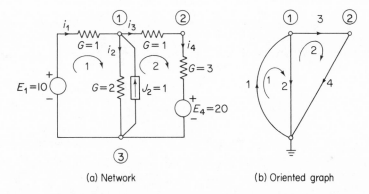

(a) Network   (b) Oriented graph

Figure 7.6 Passive resistive network examples.

The incidence matrix for the network is

$$\mathbf{A} = \begin{bmatrix} -1 & 1 & 1 & 0 \\ 0 & 0 & -1 & 1 \end{bmatrix}$$

and the branch conductance matrix is

$$\mathbf{G} = \begin{bmatrix} 1 & 0 & 0 & 0 \\ 0 & 2 & 0 & 0 \\ 0 & 0 & 1 & 0 \\ 0 & 0 & 0 & 3 \end{bmatrix}$$

Note that $\mathbf{G}$ here is not the same as the conductance matrix of Chapter 4. The branch source vectors are given by

$$\mathbf{J} = \begin{bmatrix} 0 \\ 1 \\ 0 \\ 0 \end{bmatrix} \qquad \mathbf{E} = \begin{bmatrix} 10 \\ 0 \\ 0 \\ -20 \end{bmatrix}$$

where again these vectors differ from those defined in Chapter 4. Also notice that the polarity of the component E_4 is negative because of the positive reference voltage source polarity defined by the general branch of Fig. 7.5.

Combining the matrices $\mathbf{A}$ and $\mathbf{G}$ to obtain the nodal conductance matrix in Eq. (7.17a), we have

$$\mathbf{G}_n = \mathbf{AGA}^\mathsf{T}$$

$$= \begin{bmatrix} -1 & 1 & 1 & 0 \\ 0 & 0 & -1 & 1 \end{bmatrix} \begin{bmatrix} 1 & 0 & 0 & 0 \\ 0 & 2 & 0 & 0 \\ 0 & 0 & 1 & 0 \\ 0 & 0 & 0 & 3 \end{bmatrix} \begin{bmatrix} -1 & 0 \\ 1 & 0 \\ 1 & -1 \\ 0 & 1 \end{bmatrix}$$

$$= \begin{bmatrix} -1 & 2 & 1 & 0 \\ 0 & 0 & -1 & 3 \end{bmatrix} \begin{bmatrix} -1 & 0 \\ 1 & 0 \\ 1 & -1 \\ 0 & 1 \end{bmatrix}$$

$$= \begin{bmatrix} 4 & -1 \\ -1 & 4 \end{bmatrix}$$

The equivalent current vector $\mathbf{J}_n$ is

$$\mathbf{J}_n = \mathbf{A}(\mathbf{J} - \mathbf{GE})$$

$$= \begin{bmatrix} -1 & 1 & 1 & 0 \\ 0 & 0 & -1 & 1 \end{bmatrix} \left\{ \begin{bmatrix} 0 \\ 1 \\ 0 \\ 0 \end{bmatrix} - \begin{bmatrix} 1 & 0 & 0 & 0 \\ 0 & 2 & 0 & 0 \\ 0 & 0 & 1 & 0 \\ 0 & 0 & 0 & 3 \end{bmatrix} \begin{bmatrix} 10 \\ 0 \\ 0 \\ -20 \end{bmatrix} \right\}$$

$$= \begin{bmatrix} -1 & 1 & 1 & 0 \\ 0 & 0 & -1 & 1 \end{bmatrix} \begin{bmatrix} -10 \\ 1 \\ 0 \\ 60 \end{bmatrix}$$

$$= \begin{bmatrix} 11 \\ 60 \end{bmatrix}$$

The resulting set of node equations in matrix form thus becomes

$$\begin{bmatrix} 4 & -1 \\ -1 & 4 \end{bmatrix} \begin{bmatrix} v_{n1} \\ v_{n2} \end{bmatrix} = \begin{bmatrix} 11 \\ 60 \end{bmatrix}$$

which is identical to the set of equations obtained by the classical methods described in Chapter 4.

The purpose of our labors consists in writing a simple computer program which reads the network description a branch at a time, constructs the nodal conductance matrix $\mathbf{G}_n$, the equivalent current vector $\mathbf{J}_n$, and then solves either the system of equations described by Eq. (7.18) using a technique such as Gauss-elimination or the the system described by Eq. (7.16) using inversion techniques.

The input data required by the methods developed in this section include the branch type, the branch orientation and connections, and the branch parameter value.

If the element is a controlled source, the controlling branch number should also be included. A typical branch data card for a dc analysis is

$$\text{R} \quad 7 \quad 1 \quad 3 \quad 450.5$$

which specifies a resistor (R) in branch 7 between nodes 1 and 3 (oriented from node 1 to node 3) of value 450.5 Ω. There are several other ways these data could be presented. For example, the branch could be specified as

$$\text{B7} \quad \text{N(1, 3)}, \quad \text{R} = 450.5$$

as in ECAP† or

$$\text{R7, } 1 - 3 = 450.5$$

as in SCEPTRE.‡ The input format depends primarily upon the degree of complexity designed into the *language processor*. The particular form of language processor used (or designed by the reader) is left to the reader. We will use an elementary processor, which can be easily replaced or improved if desired.

A flowchart for a computer program which analyzes networks of up to 50 branches and 20 independent nodes is shown in Fig. 7.7. The program forms the nodal conductance matrix and equivalent current vector from the information supplied by the rudimentary language processor, then the node voltage solution is obtained using the LINQ subroutine (Gauss-elimination) described in Chapter 3.

The listing of the basic program is shown in Fig. 7.8.

Notice that the basic program does not allow dependent voltage or current sources. This capability will be included as the development progresses. Also the program allows only node voltage calculations. Since the branch voltages are defined by

$$\mathbf{v}_b = \mathbf{A}^\mathsf{T} \mathbf{v}_n \tag{7.19}$$

We may find the element voltages by this equation and the equation

$$\mathbf{v}_e = \mathbf{v}_b + \mathbf{E} \tag{7.20}$$

Also the element currents are given by

$$\mathbf{i}_e = \mathbf{G} \mathbf{v}_e \tag{7.21}$$

and the branch currents are then given by the equation

$$\mathbf{i}_b = \mathbf{i}_e - \mathbf{J} \tag{7.22}$$

---

†R. W. Jensen and M. D. Lieberman, *IBM Electronic Circuit Analysis Program: Techniques and Applications*, Prentice-Hall, Inc., Englewood Cliffs, N.J., 1968.

‡H. W. Mathers, S. R. Sedore, and J. R. Sents, *Automated Digital Computer Program for Determining Responses of Electronic Circuits to Transient Nuclear Radiation* (*SCEPTRE*), AFWL–TR–66–126, Electronic Systems Center, International Business Machines Corporation, Owego, N.Y., Feb. 1967.

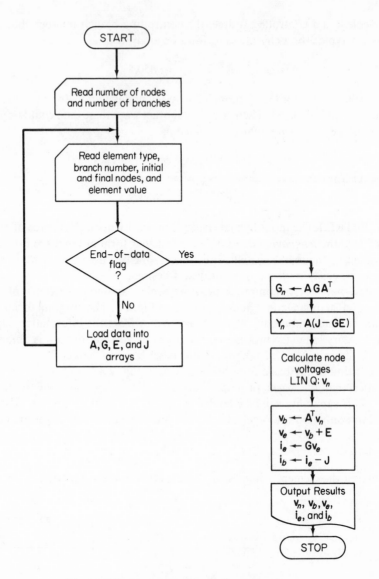

Figure 7.7   Flowchart of basic dc node voltage
analysis program.

The implementation of these branch and element parameter calculations in the program of Fig. 7.8 is left to the reader as an exercise.

The branch illustrated in Fig. 7.5 can be extended as shown in Fig. 7.9 so that it is completely general. The voltages and currents in the general branch are the same as defined for the simplified branch in Fig. 7.5, with the following exceptions:

E_{dk} = dependent voltage source in branch k; this voltage source can be either a transpotential or a transresistance, that is, either voltage or current dependent

```
C       BASIC DC ANALYSIS PROGRAM                                        DC02    1
C       IMPLEMENTATION OF EQUATION 7.15 WHICH IS BASED UPON              DC02    2
C       THE SIMPLIFIED GENERAL BRANCH IN FIGURE 7.59                     DC02    3
C                                                                        DC02    4
C       DEFINITION OF VARIABLES                                          DC02    5
C           NODES    - NUMBER OF NETWORK NODES                           DC02    6
C           BRNCHS   - NUMBER OF NETWORK BRANCHES                        DC02    7
C           MISC     - OUTPUT FLAG                                       DC02    8
C                    = 0    NODE VOLTAGES ONLY                           DC02    9
C                    = 1    SAME AS 0 PLUS GN PLUS JN                    DC02   10
C                    = 2    DEBUG - SAME AS 1 PLUS A,G,ES AND JS         DC02   11
C           ETYPE    - BRANCH TYPE      ( R G C L E J TC TF TP TR EX )   DC02   12
C           NTYPE    - BRANCH TYPE CODE ( 1 2 3 4 5 6  7  8  9 10 11 )   DC02   13
C           BRANCH   - BRANCH NUMBER                                     DC02   14
C           NODE1    - INITIAL NODE                                      DC02   15
C           NODE2    - FINAL NODE                                        DC02   16
C           VALUE    - BRANCH ELEMENT VALUE                              DC02   17
C           DEPBR    - "FROM" BRANCH NUMBER FOR DEPENDENT SOURCE         DC02   18
C           SHORT    - DC INDUCTANCE VALUE                               DC02   19
C           OPEN     - DC CAPACITANCE VALUE                              DC02   20
C           A        - INCIDENCE MATRIX                                  DC02   21
C           G        - BRANCH CONDUCTANCE MATRIX                         DC02   22
C           ES       - INDEPENDENT VOLTAGE SOURCE VECTOR                 DC02   23
C           JS       - INDEPENDENT CURRENT SOURCE VECTOR                 DC02   24
C           G1       - TEMPORARY WORKING MATRIX TO FORM GN              DC02   25
C           GN       - NODAL CONDUCTANCE MATRIX                          DC02   26
C           J1       - TEMPORARY WORKING VECTOR TO FORM JN              DC02   27
C           JN       - EQUIVALENT CURRENT VECTOR                         DC02   28
C           VLTGS    - NODE VOLTAGE VECTOR                               DC02   29
C           DET      - DETERMINANT OF GN                                 DC02   30
C       INPUT DATA FORMAT                                                DC02   31
C           ELEMENT TYPE   COLUMNS 1-3                                   DC02   32
C           BRANCH NUMBER COLUMNS 5-6                                    DC02   33
C           INITIAL NODE   COLUMNS 9-10                                  DC02   34
C           FINAL NODE     COLUMNS 12-13                                 DC02   35
C           ELEMENT VALUE COLUMNS 16-30                                  DC02   36
C           DEP. FROM BR. COLUMNS 34-35                                  DC02   37
C                                                                        DC02   38
C                                                                        DC02   39
        REAL JS,JN,J1                                                    DC02   40
        INTEGER ETYPE,BRANCH,DEPBR,BRNCHS                                DC02   41
        DIMENSION NTYPE(11),A(20,50),ES(50),JS(50),J1(20)               DC02   42
        DIMENSION G(50,50),GN(20,20),JN(20),G1(20,50),VLTGS(20)         DC02   43
        DATA NTYPE/'R ','G ','C ','L ','E ','J ','TC','TF','TP','TR','EX'/DC02  44
        DATA A/1000*0.0/,ES/50*0.0/,JS/50*0.0/,G/2500*0.0/              DC02   45
C       INITIALIZE SHORT AND OPEN VALUES                                 DC02   46
        DATA SHORT/0.01/,OPEN/1.0E7/                                    DC02   47
C                                                                        DC02   48
C       READ NUMBER OF INDEPENDENT NODES AND NO. OF NETWORK BRANCHES     DC02   49
        READ (5,10) NODES,BRNCHS,MISC                                   DC02   50
   10   FORMAT (3I5)                                                     DC02   51
        WRITE (6,20) NODES,BRNCHS                                        DC02   52
   20   FORMAT (1H1,5X,'DC ANALYSIS PROBLEM'///5X,'NUMBER OF NODES =',I3/5DC02  53
       1X,'NUMBER OF BRANCHES =',I3//)                                   DC02   54
C       READ BRANCH DATA STATEMENT                                       DC02   55
   30   READ (5,40) ETYPE,BRANCH,NODE1,NODE2,VALUE,DEPBR                 DC02   56
   40   FORMAT (A2,2X,I2,2X,I2,I3,2X,F15.0,3X,I2)                       DC02   57
C       TEST BRANCH FOR ELEMENT TYPE                                     DC02   58
        DO 50 I=1,11                                                     DC02   59
        IF (ETYPE.EQ.NTYPE(I)) GO TO 70                                  DC02   60
   50   CONTINUE                                                         DC02   61
C       BRANCH TYPE ERROR                                                DC02   62
        WRITE (6,60) ETYPE                                               DC02   63
   60   FORMAT (5X,A2,' IS NOT A VALID BRANCH TYPE.'/)                   DC02   64
        GO TO 30                                                         DC02   65
   70   IF (I.EQ.11) WRITE (6,80)                                        DC02   66
   80   FORMAT (5X,'ENTERING EXECUTION. ')                              DC02   67
        IF (I.EQ.11) GO TO 240                                           DC02   68
        GO TO (90,110,120,140,260,180,190,190,190,190),I                DC02   69
```

Figure 7.8 Listing of basic dc node voltage
analysis program.

```
C     LOAD INCIDENCE MATRIX A, CONDUCTANCE MATRIX G, AND SOURCE              DC02  70
C     VECTORS ES AND JS.                                                     DC02  71
C     OUTPUT BRANCH DATA STATEMENT                                           DC02  72
C     RESISTANCE ELEMENT                                                     DC02  73
   90 G(BRANCH,BRANCH)=1.0/VALUE                                             DC02  74
      WRITE (6,100) ETYPE,BRANCH,NODE1,NODE2,VALUE                           DC02  75
  100 FORMAT (5X,A2,' BRANCH',I3,'     NODE',I3,' TO',I3,'    VALUE = ',1DC02  76
     1PE15.5)                                                                DC02  77
      GO TO 230                                                              DC02  78
C     CONDUCTANCE ELEMENT                                                    DC02  79
  110 G(BRANCH,BRANCH)=VALUE                                                 DC02  80
      WRITE (6,100) ETYPE,BRANCH,NODE1,NODE2,VALUE                           DC02  81
      GO TO 230                                                              DC02  82
C     CAPACITANCE ELEMENT                                                    DC02  83
  120 G(BRANCH,BRANCH)=1.0/OPEN                                              DC02  84
      WRITE (6,100) ETYPE,BRANCH,NODE1,NODE2,VALUE                           DC02  85
      WRITE (6,130) OPEN                                                     DC02  86
  130 FORMAT (5X,'** CAPACITANCE VALUE REPLACED BY',1PE10.3,' OHMS RESISDC02  87
     1TANCE')                                                                DC02  88
      GO TO 230                                                              DC02  89
C     INDUCTANCE ELEMENT                                                     DC02  90
  140 G(BRANCH,BRANCH)=1.0/SHORT                                             DC02  91
      WRITE (6,100) ETYPE,BRANCH,NODE1,NODE2,VALUE                           DC02  92
      WRITE (6,150) SHORT                                                    DC02  93
  150 FORMAT (5X,'** INDUCTANCE VALUE REPLACED BY',1PE10.3,' OHMS RESISTDC02  94
     1ANCE')                                                                 DC02  95
      GO TO 230                                                              DC02  96
C     INDEPENDENT VOLTAGE SOURCE (E ELEMENT)                                 DC02  97
  260 ES(BRANCH)=VALUE                                                       DC02  98
      WRITE (6,170) ETYPE,BRANCH,VALUE                                       DC02  99
  170 FORMAT (5X,A2,' BRANCH',I3,20X,' VALUE = ',1PE15.5)                    DC02 100
      GO TO 30                                                               DC02 101
C     INDEPENDENT CURRENT SOURCE (J ELEMENT)                                 DC02 102
  180 JS(BRANCH)=VALUE                                                       DC02 103
      WRITE (6,170) ETYPE,BRANCH,VALUE                                       DC02 104
      GO TO 30                                                               DC02 105
C     DEPENDENT SOURCES (ELEMENT TYPES TC,TF,TP, AND TR)                     DC02 106
C     ARE NOT IMPLEMENTED IN THIS VERSION OF THE DC ANALYSIS PROGRAM.        DC02 107
  190 GO TO 30                                                               DC02 108
C     LOAD ELEMENTS OF INCIDENCE MATRIX A                                    DC02 109
  230 IF (NODE1.GT.0)A(NODE1,BRANCH)=1.0                                     DC02 110
      IF (NODE2.GT.0)A(NODE2,BRANCH)=-1.0                                    DC02 111
      GO TO 30                                                               DC02 112
  240 IF (MISC.LE.1) GO TO 350                                               DC02 113
C     OUTPUT A, G, ES, JS, MATRICES                                          DC02 114
      WRITE (6,250)                                                          DC02 115
  250 FORMAT (1H1,5X,'NETWORK INCIDENCE MATRIX'/)                            DC02 116
  260 C(IP,J)=C(IP,J)/CL                                                     DC02 117
C     ZERO PIVOT ELEMENT COLUMNM ADJUST MATRIX                               DC02 118
      DO 180 I=IST,N                                                         DC02 119
      CL=C(I,IP)                                                             DC02 120
      DO 270 I=1,NODES                                                       DC02 121
  270 WRITE (6,280) (A(I,J),J=1,BRNCHS)                                      DC02 122
  280 FORMAT (1H0,2X,25F5.0)                                                 DC02 123
      WRITE (6,290)                                                          DC02 124
  290 FORMAT (///5X,'E SOURCE VECTOR',10X,'J SOURCE VECTOR'/)                DC02 125
      DO 300 I=1,BRNCHS                                                      DC02 126
  300 WRITE (6,310) I,ES(I),JS(I)                                            DC02 127
  310 FORMAT (3X,I2,3X,1PE12.5,13X,1PE12.5)                                  DC02 128
      WRITE (6,320)                                                          DC02 129
  320 FORMAT (///5X,'BRANCH CONDUCTANCE MATRIX'/)                            DC02 130
      DO 330 I=1,BRNCHS                                                      DC02 131
  330 WRITE (6,340) (G(I,J),J=1,BRNCHS)                                      DC02 132
  340 FORMAT (5X,1P10E12.5)                                                  DC02 133
C     COMPUTE NODAL CONDUCTANCE MATRIX GN=A*G*AT                             DC02 134
C        FORM PARTIAL PRODUCT  G1=A*G                                        DC02 135
```

Figure 7.8—Cont.

```
  350 DO 360 I=1,NODES                                        DC02 136
      DO 360 J=1,BRNCHS                                       DC02 137
      G1(I,J)=0.0                                             DC02 138
      DO 360 K=1,BRNCHS                                       DC02 139
  360 G1(I,J)=G1(I,J)+A(I,K)*G(K,J)                           DC02 140
C         COMPLETE GN=G1*AT                                   DC02 141
      DO 370 I=1,NODES                                        DC02 142
      DO 370 J=1,NODES                                        DC02 143
      GN(I,J)=0.0                                             DC02 144
      DO 370 K=1,BRNCHS                                       DC02 145
  370 GN(I,J)=GN(I,J)+G1(I,K)*A(J,K)                          DC02 146
      IF (MISC.LT.1) GO TO 410                                DC02 147
C     OUTPUT NODAL CONDUCTANCE MATRIX GN                      DC02 148
      WRITE (6,380)                                           DC02 149
  380 FORMAT (1H1,4X,'NODAL CONDUCTANCE MATRIX'/)             DC02 150
      DO 390 I=1,NODES                                        DC02 151
  390 WRITE (6,400) (GN(I,J),J=1,NODES)                       DC02 152
  400 FORMAT (1P8E15.5)                                       DC02 153
C     COMPUTE EQUIVALENT CURRENT VECTOR   JN=A*(JS-G*ES)      DC02 154
C         FORM PARTIAL PRODUCT   J1=(J-G*E)                   DC02 155
  410 DO 420 I=1,BRNCHS                                       DC02 156
      J1(I)=JS(I)                                             DC02 157
      DO 420 J=1,BRNCHS                                       DC02 158
  420 J1(I)=J1(I)-G(I,J)*ES(J)                                DC02 159
C         COMPLETE JN=A*J1                                    DC02 160
      DO 430 I=1,NODES                                        DC02 161
      JN(I)=0.0                                               DC02 162
      DO 430 J=1,BRNCHS                                       DC02 163
  430 JN(I)=JN(I)+A(I,J)*J1(J)                                DC02 164
      IF (MISC.LT.1) GO TO 470                                DC02 165
      WRITE (6,440)                                           DC02 166
  440 FORMAT (//////5X,'EQUIVALENT CURRENT VECTOR'/)          DC02 167
      DO 450 I=1,NODES                                        DC02 168
C     OUTPUT EQUIVALENT CURRENT VECTOR                        DC02 169
  450 WRITE (6,460) JN(I)                                     DC02 170
  460 FORMAT (1PE20.5)                                        DC02 171
C     CALCULATE NODE VOLTAGES BY GAUSS ELIMINATION            DC02 172
  470 CALL LINQ(GN,JN,VLTGS,NODES,DET)                        DC02 173
C     TEST FOR SINGULAR SYSTEM OF EQUATIONS                   DC02 174
      IF (DET.NE.0.0) GO TO 490                                DC02 175
C     ERROR OUTPUT                                            DC02 176
      WRITE (6,480)                                           DC02 177
  480 FORMAT (5X,'INVALID SYSTEM OF EQUATIONS'///)            DC02 178
      STOP                                                    DC02 179
C     OUTPUT NODE VOLTAGES                                    DC02 180
  490 WRITE (6,500)                                           DC02 181
  500 FORMAT (1H1///5X,'NODE VOLTAGES'//)                     DC02 182
      DO 510 I=1,NODES                                        DC02 183
  510 WRITE (6,520) I,VLTGS(I)                                DC02 184
  520 FORMAT (5X,'NODE',I2,'  VOLTAGE = ',1PE12.5)            DC02 185
      WRITE (6,530)                                           DC02 186
  530 FORMAT (///5X,'END OF ANALYSIS')                        DC02 187
      STOP                                                    DC02 188
      END                                                     DC02 189
```

Figure 7.8—Cont.

$v_{ek} = v_{bk} + E_k + E_{dk}$ = element voltage across the passive element in branch k

J_{dk} = dependent current source in branch k. The source can be either a transfluence or a transconductance, that is, either current or voltage dependent

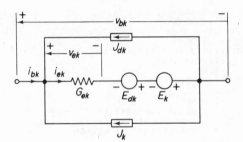

Figure 7.9 General branch.

The branch current i_{bk} through the general branch k in Fig. 7.9 is

$$
\begin{aligned}
i_{bk} &= i_{ek} - J_k - J_{dk} \\
&= G_k v_{ek} - J_k - J_{dk} \\
&= G_k(v_{bk} + E_k) - J_k - J_{dk} \\
&= G_k v_{bk} - J_k - J_{dk} + G_k E_k
\end{aligned}
\tag{7.23}
$$

(We assume that dependent voltage sources are zero for simplicity in this stage of the discussion.) First consider that only transfluences and transconductances are present. Then the sources are defined by

$$
J_{dk} = \alpha_{kj} i_{ej} \qquad \text{(transfluence)} \tag{7.24}
$$

where the multiplication constant α_{kj} represents the strength of the source in branch k due to the element current i_{ej} in branch j, and

$$
J_{dk} = g_{mkj} v_{ej} \qquad \text{(transconductance)} \tag{7.25}
$$

where g_{mkj} represents the strength of the source in branch k due to the element voltage v_{ej} in branch j.

In matrix notation the branch currents may be expressed as

$$
\mathbf{i}_b = \mathbf{G}\mathbf{v}_b - \mathbf{J} + \mathbf{G}\mathbf{E} - [g_m]\mathbf{v}_e - [\alpha]\mathbf{i}_e \tag{7.26}
$$

where $[g_m]$ is a square matrix of order b in which an element g_{mkj} corresponds to the transconductance gain constant from branch j to branch k. The transfluence matrix $[\alpha]$ is also a square matrix of order b in which an element α_{kj} corresponds to the gain constant from branch j to branch k.

However, $\mathbf{i}_e = \mathbf{G}\mathbf{v}_e$ and $\mathbf{v}_e = \mathbf{v}_b + \mathbf{E}$, which allows us to rewrite Eq. (7.26) as

$$
\mathbf{i}_b = (\mathbf{G} - [g_m] - [\alpha]\mathbf{G})(v_b + \mathbf{E}) - \mathbf{J} \tag{7.27}
$$

$$
= \mathbf{G}_1 \mathbf{v}_b + \mathbf{G}_1 \mathbf{E} - \mathbf{J} \tag{7.28}
$$

where

$$
\mathbf{G}_1 = \mathbf{G} - [g_m] - [\alpha]\mathbf{G}
$$

Branch k will frequently be referred to as the "to" branch in our discussion. Branch j will likewise be referred to as the "from" branch.

Since it is not normal to find an element α_{kk} or $g_{m_{kk}}$ in a network (this corresponds to a dependent source in a branch depending upon its own element current or voltage) the entries in $\mathbf{G}_1$ due to the dependent sources are all off-diagonal. Thus, one can include the dependent current sources directly in the branch conductance matrix rather than spending time performing the matrix operations indicated in Eqs. (7.27) and (7.28) to form $\mathbf{G}_1$. This is possible only if the branch conductance is known before introduction of the associated transfluence (see Example 7.2).

Equation (7.28) can now be combined with Eqs. (7.2) and (7.5) to form

$$
\begin{aligned}
\mathbf{A}\mathbf{i}_b &= \mathbf{A}\mathbf{G}_1\mathbf{v}_b - \mathbf{A}\mathbf{J} + \mathbf{A}\mathbf{G}_1\mathbf{E} \\
&= \mathbf{A}\mathbf{G}_1\mathbf{v}_b - \mathbf{A}(\mathbf{J} - \mathbf{G}_1\mathbf{E}) \\
&= \mathbf{A}\mathbf{G}_1\mathbf{A}^\mathsf{T}\mathbf{v}_n - \mathbf{A}(\mathbf{J} - \mathbf{G}_1\mathbf{E})
\end{aligned}
$$

or

$$
\mathbf{A}\mathbf{G}_1\mathbf{A}^\mathsf{T}\mathbf{v}_n = \mathbf{A}(\mathbf{J} - \mathbf{G}_1\mathbf{E}) \tag{7.29}
$$

The resulting node voltage solution

$$
\mathbf{v}_n = (\mathbf{A}\mathbf{G}_1\mathbf{A}^\mathsf{T})^{-1}\mathbf{A}(\mathbf{J} - \mathbf{G}_1\mathbf{E}) \tag{7.30}
$$

is very similar to the solution obtained for the simplified general branch in Eq. (7.16). Again, dependent current sources introduce off-diagonal elements in the branch conductance matrix which can be incorporated directly without any matrix operation.

EXAMPLE 7.2

Given the network in Fig. 7.10, write the node equations using the methods used in deriving Eq. (7.29). The incidence matrix for the network is

$$
\mathbf{A} = \begin{bmatrix} -1 & 1 & 0 & 0 & 0 \\ 0 & -1 & 1 & 1 & 0 \\ 0 & 0 & 0 & -1 & 1 \end{bmatrix}
$$

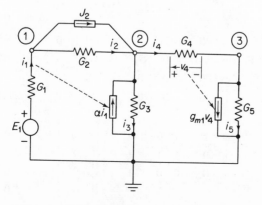

Figure 7.10 Network example containing dependent current sources.

The branch conductance matrix $\mathbf{G}_1$ is

$$\mathbf{G}_1 = \begin{bmatrix} G_1 & 0 & 0 & 0 & 0 \\ 0 & G_2 & 0 & 0 & 0 \\ 0 & 0 & G_3 & 0 & 0 \\ 0 & 0 & 0 & G_4 & 0 \\ 0 & 0 & 0 & 0 & G_5 \end{bmatrix} - \begin{bmatrix} 0 & 0 & 0 & 0 & 0 \\ 0 & 0 & 0 & 0 & 0 \\ 0 & 0 & 0 & 0 & 0 \\ 0 & 0 & 0 & 0 & 0 \\ 0 & 0 & 0 & -g_{m_1} & 0 \end{bmatrix} - \begin{bmatrix} 0 & 0 & 0 & 0 & 0 \\ 0 & 0 & 0 & 0 & 0 \\ \alpha_1 G_1 & 0 & 0 & 0 & 0 \\ 0 & 0 & 0 & 0 & 0 \\ 0 & 0 & 0 & 0 & 0 \end{bmatrix}$$

$$= \begin{bmatrix} G_1 & 0 & 0 & 0 & 0 \\ 0 & G_2 & 0 & 0 & 0 \\ -\alpha_1 G_1 & 0 & G_3 & 0 & 0 \\ 0 & 0 & 0 & G_4 & 0 \\ 0 & 0 & 0 & g_{m_1} & G_5 \end{bmatrix}$$

$$\mathbf{A}\mathbf{G}_1\mathbf{A}^{\mathsf{T}} = \begin{bmatrix} -1 & 1 & 0 & 0 & 0 \\ 0 & -1 & 1 & 1 & 0 \\ 0 & 0 & 0 & -1 & 0 \end{bmatrix} \begin{bmatrix} G_1 & 0 & 0 & 0 & 0 \\ 0 & G_2 & 0 & 0 & 0 \\ -\alpha_1 G_1 & 0 & G_3 & 0 & 0 \\ 0 & 0 & 0 & G_4 & 0 \\ 0 & 0 & 0 & g_{m_1} & G_5 \end{bmatrix} \begin{bmatrix} -1 & 0 & 0 \\ 1 & -1 & 0 \\ 0 & 1 & 0 \\ 0 & 1 & -1 \\ 0 & 0 & 1 \end{bmatrix}$$

$$= \begin{bmatrix} -G_1 & G_2 & 0 & 0 & 0 \\ -\alpha_1 G_1 & -G_2 & G_3 & G_4 & 0 \\ 0 & 0 & 0 & -G_4 + g_{m_1} & G_5 \end{bmatrix} \begin{bmatrix} -1 & 0 & 0 \\ 1 & -1 & 0 \\ 0 & 1 & 0 \\ 0 & 1 & -1 \\ 0 & 0 & 1 \end{bmatrix}$$

$$= \begin{bmatrix} G_1 + G_2 & -G_2 & 0 \\ \alpha_1 G_1 - G_2 & G_2 + G_3 + G_4 & -G_4 \\ 0 & -G_4 + g_{m_1} & G_4 + G_5 - g_{m_1} \end{bmatrix}$$

this result corresponds to the nodal conductance matrix obtained using the classical methods described in Chapter 4. The equivalent current vector is defined by

$$\mathbf{J}_n = \mathbf{A}(\mathbf{J} - \mathbf{G}_1\mathbf{E})$$

$$\mathbf{G}_1\mathbf{E} = \begin{bmatrix} G_1 & 0 & 0 & 0 & 0 \\ 0 & G_2 & 0 & 0 & 0 \\ -\alpha_1 G_1 & 0 & G_3 & 0 & 0 \\ 0 & 0 & 0 & G_4 & 0 \\ 0 & 0 & 0 & g_{m_1} & G_5 \end{bmatrix} \begin{bmatrix} E_1 \\ 0 \\ 0 \\ 0 \\ 0 \end{bmatrix} = \begin{bmatrix} E_1 G_1 \\ 0 \\ -\alpha G_1 E_1 \\ 0 \\ 0 \end{bmatrix}$$

$$\mathbf{A}(\mathbf{J} - \mathbf{G}_1\mathbf{E}) = \begin{bmatrix} -1 & 1 & 0 & 0 & 0 \\ 0 & -1 & 1 & 1 & 0 \\ 0 & 0 & 0 & -1 & 1 \end{bmatrix} \left\{ \begin{bmatrix} 0 \\ -J_2 \\ 0 \\ 0 \\ 0 \end{bmatrix} - \begin{bmatrix} E_1 G_1 \\ 0 \\ -\alpha_1 G_1 E_1 \\ 0 \\ 0 \end{bmatrix} \right\}$$

$$= \begin{bmatrix} E_1 G_1 - J_2 \\ J_2 + \alpha_1 G_1 E_1 \\ 0 \end{bmatrix}$$

which also corresponds to the classical results obtained by the methods of Chapter 4. The node equations in matrix form are finally

$$\begin{bmatrix} G_1 + G_2 & -G_2 & 0 \\ \alpha_1 G_1 - G_2 & G_2 + G_3 + G_4 & -G_4 \\ 0 & -G_4 + g_{m_1} & G_4 + G_5 - g_{m_1} \end{bmatrix} \begin{bmatrix} v_1 \\ v_2 \\ v_3 \end{bmatrix} = \begin{bmatrix} E_1 G_1 - J_2 \\ J_2 + \alpha_1 G_1 E_1 \\ 0 \end{bmatrix}$$

The dependent voltage source, which was omitted from the development of Eq. (7.30), can now be inserted in the node voltage formulation. (The reason for the earlier omission will quickly become apparent.)

The branch voltage v_{bk} for branch k is

$$v_{bk} = v_{ek} - E_{dk} - E_k$$

or

$$v_{bk} + E_k = v_{ek} - E_{dk}$$

In matrix form

$$\mathbf{v}_b + \mathbf{E} = (\mathbf{I} - \mathbf{D})\mathbf{v}_e \tag{7.31}$$

where $\mathbf{D} = [\mu] + [r_m]\mathbf{G}$ is the dependent voltage-source-strength matrix and $[\mu]$ is a square matrix of order b in which the element μ_{ij} corresponds to the transpotential gain constant from branch j to branch i. The transresistance matrix $[r_m]$ is also a square matrix of order b in which the element $r_{m_{ij}}$ corresponds to the gain constant from branch j to branch i.

The branch current $\mathbf{i}_b$ is given by

$$\begin{aligned} \mathbf{i}_b &= \mathbf{i}_e - \mathbf{J} - \mathbf{J}_d \\ &= \mathbf{G}\mathbf{v}_e - \mathbf{J} - [g_m]\mathbf{v}_e - [\alpha]\mathbf{i}_e \\ &= \mathbf{G}\mathbf{v}_e - \mathbf{J} - ([g_m] + [\alpha]\mathbf{G})\mathbf{v}_e \\ &= (\mathbf{G} - [g_m] - [\alpha]\mathbf{G})\mathbf{v}_e - \mathbf{J} \end{aligned} \tag{7.32}$$

Solving Eq. (7.31) for $\mathbf{v}_e$, we have

$$\mathbf{v}_e = (\mathbf{I} - \mathbf{D})^{-1}(\mathbf{v}_b + \mathbf{E}) \tag{7.33}$$

so that

$$\begin{aligned} \mathbf{i}_b &= (\mathbf{G} - [g_m] - [\alpha]\mathbf{G})(\mathbf{I} - \mathbf{D})^{-1}(\mathbf{v}_b + \mathbf{E}) - \mathbf{J} \\ &= \mathbf{G}_1(\mathbf{I} - \mathbf{D})^{-1}(\mathbf{v}_b + \mathbf{E}) - \mathbf{J} \\ &= \mathbf{G}_1(\mathbf{I} - \mathbf{D})^{-1}(\mathbf{A}^T\mathbf{v}_n + \mathbf{E}) - \mathbf{J} \end{aligned} \tag{7.34}$$

where $\mathbf{G}_1$ is defined in Eq. (7.28). Kirchhoff's current law then yields

$$\begin{aligned} \mathbf{A}\mathbf{i}_b &= \mathbf{A}\mathbf{G}_1(\mathbf{I} - \mathbf{D})^{-1}(\mathbf{A}^T\mathbf{v}_n + \mathbf{E}) - \mathbf{A}\mathbf{J} \\ &= 0 \end{aligned} \tag{7.35}$$

from which

$$\mathbf{v}_n = \{\mathbf{A}\mathbf{G}_1(\mathbf{I} - \mathbf{D})^{-1}\mathbf{A}^T\}^{-1}\mathbf{A}\{\mathbf{J} - \mathbf{G}_1(\mathbf{I} - \mathbf{D})^{-1}\mathbf{E}\} \tag{7.36}$$

By defining an equivalent conductance matrix $\mathbf{G}_2$ such that

$$\mathbf{G}_2 = (\mathbf{G} - [g_m] - [\alpha]\mathbf{G})(\mathbf{I} - \mathbf{D})^{-1} \tag{7.37}$$

the resulting matrix equation for the node voltages is

$$\mathbf{v}_n = (\mathbf{A}\mathbf{G}_2\mathbf{A}^\mathsf{T})^{-1}\mathbf{A}(\mathbf{J} - \mathbf{G}_2\mathbf{E}) \tag{7.38}$$

which is equivalent to the results in Eqs. (7.16) and (7.30).

EXAMPLE 7.3

Consider the circuit shown in Fig. 7.11. Find the node voltage equations in matrix form.

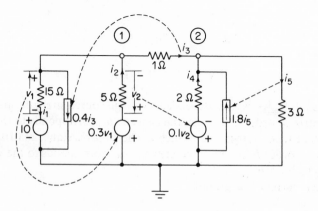

Figure 7.11 Network example containing both
dependent voltage and current sources.

Solution. The incidence matrix $\mathbf{A}$ for the network is

$$\mathbf{A} = \begin{bmatrix} 1 & -1 & 1 & 0 & 0 \\ 0 & 0 & -1 & -1 & 1 \end{bmatrix}$$

and the branch conductance matrix $\mathbf{G}_1$ is

$$\mathbf{G}_1 = \begin{bmatrix} \frac{1}{15} & 0 & 0.4 & 0 & 0 \\ 0 & \frac{1}{5} & 0 & 0 & 0 \\ 0 & 0 & 1 & 0 & 0 \\ 0 & 0 & 0 & \frac{1}{2} & 0.6 \\ 0 & 0 & 0 & 0 & \frac{1}{3} \end{bmatrix}$$

The dependent voltage source strength matrix $\mathbf{D}$ combines with the unit matrix to form

$$(\mathbf{I} - \mathbf{D}) = \begin{bmatrix} 1 & 0 & 0 & 0 & 0 \\ 0.3 & 1 & 0 & 0 & 0 \\ 0 & 0 & 1 & 0 & 0 \\ 0 & 0.1 & 0 & 1 & 0 \\ 0 & 0 & 0 & 0 & 1 \end{bmatrix}$$

Notice that the entries for the dependent voltage sources are negative in the **D** matrix if the polarity of the source is opposite that assumed in the general branch of Fig. 7.9. For example, $d_{21} = -0.3$ in the dependent source matrix. The inverse of $(I - D)$ is

$$(I - D)^{-1} = \begin{bmatrix} 1 & 0 & 0 & 0 & 0 \\ -0.3 & 1 & 0 & 0 & 0 \\ 0 & 0 & 1 & 0 & 0 \\ 0.03 & -0.1 & 0 & 1 & 0 \\ 0 & 0 & 0 & 0 & 1 \end{bmatrix}$$

The equivalent conductance matrix G_2 is next calculated and is given by

$$G_2 = G_1(I - D)^{-1}$$

$$= \begin{bmatrix} \frac{1}{15} & 0 & 0.4 & 0 & 0 \\ 0 & \frac{1}{5} & 0 & 0 & 0 \\ 0 & 0 & 1 & 0 & 0 \\ 0 & 0 & 0 & \frac{1}{2} & 0.6 \\ 0 & 0 & 0 & 0 & \frac{1}{3} \end{bmatrix} \begin{bmatrix} 1 & 0 & 0 & 0 & 0 \\ -0.3 & 1 & 0 & 0 & 0 \\ 0 & 0 & 1 & 0 & 0 \\ 0.03 & -0.1 & 0 & 1 & 0 \\ 0 & 0 & 0 & 0 & 1 \end{bmatrix}$$

$$= \begin{bmatrix} 0.067 & 0 & 0.4 & 0 & 0 \\ -0.06 & 0.2 & 0 & 0 & 0 \\ 0 & 0 & 1 & 0 & 0 \\ 0.015 & -0.05 & 0 & 0.5 & 0.6 \\ 0 & 0 & 0 & 0 & 0.333 \end{bmatrix}$$

The nodal conductance matrix G_n is then formed from AG_2A^T, so

$$G_n = \begin{bmatrix} 1 & -1 & 1 & 0 & 1 \\ 0 & 0 & -1 & -1 & 1 \end{bmatrix} \begin{bmatrix} 0.067 & 0 & 0.4 & 0 & 0 \\ -0.06 & 0.2 & 0 & 0 & 0 \\ 0 & 0 & 1 & 0 & 0 \\ 0.015 & -0.05 & 0 & 0.5 & 0.6 \\ 0 & 0 & 0 & 0 & 0.333 \end{bmatrix} \begin{bmatrix} 1 & 0 \\ -1 & 0 \\ 1 & -1 \\ 0 & -1 \\ 0 & 1 \end{bmatrix}$$

$$= \begin{bmatrix} 1.727 & -1.4 \\ -1.065 & 1.233 \end{bmatrix}$$

The equivalent current vector J_n is calculated from

$$J_n = -AG_2E$$

$$= \begin{bmatrix} 0.127 & -0.2 & 1.4 & 0 & 0 \\ -0.015 & 0.05 & -1 & -0.5 & -0.267 \end{bmatrix} \begin{bmatrix} -10 \\ 0 \\ 0 \\ 0 \\ 0 \end{bmatrix}$$

$$= \begin{bmatrix} 1.27 \\ -0.16 \end{bmatrix}$$

and the node equations are

$$\begin{bmatrix} 1.7267 & -1.4 \\ -1.065 & 1.233 \end{bmatrix} \begin{bmatrix} v_{n1} \\ v_{n2} \end{bmatrix} = \begin{bmatrix} 1.267 \\ -0.15 \end{bmatrix}$$

from which $\mathbf{v}_n = [2.1176 \quad 1.707]^T$.

The utility of a good computer program is apparent even in simple problems like this one.

7.4   Network Sensitivity Calculations

One of the most important steps in the design of networks or circuits is the analysis of component tolerance effects on circuit operation. This type of analysis, which we refer to as *sensitivity analysis*, makes it possible to produce reliable circuits at minimum cost. For example, if we are designing an amplifier which should have a constant dc output voltage with a ± 3 per cent tolerance over a wide range of operating temperatures, we will undoubtedly select components with very small temperature coefficients and tolerances of less than ± 1 per cent for some portions of the circuit (for example, biasing network) and components with larger tolerances, say ± 10 per cent, in portions of the circuit where component tolerances are not critical. In order to determine the relative output voltage sensitivity to each of the individual component tolerances and thus select component tolerances which will satisfy our design requirement without excessive costs, we must perform a sensitivity analysis. This analysis involves calculating the sensitivity of important circuit variables, such as output voltages, to changes in the component tolerances. To do this we compute a *sensitivity coefficient*

$$S = \frac{\partial v}{\partial P_j}$$

where S is the sensitivity coefficient, v is the variable, and P_j is the circuit parameter. The calculation of the output voltage variation Δv_0 of an amplifier due to a change in the input signal ΔE_s is also important. This could be represented by

$$\frac{\Delta v_0}{\Delta E_s} = \frac{\partial v_0}{\partial E_s}$$
$$= A_v$$

where A_v is the dc voltage gain of the amplifier.

We will consider two of the many methods of computing circuit sensitivity. The first method is as follows: The nominal voltages and currents in the circuit are computed using the techniques already discussed. Then we change a component value a small amount ΔP_j and recompute the voltages and currents. The ratio of the change in the variable of interest Δv and the component value deviation ΔP_j is

$$\frac{\Delta v}{\Delta P_j} \doteq \frac{\partial v}{\partial P_j}$$

which approximates the sensitivity coefficient. This procedure, known as the *perturbation method*, becomes tedious if the circuit contains more than three or four components and has several parameters of interest in the analysis.

The second method for obtaining sensitivity coefficients allows us to apply the topological methods developed earlier in this chapter to directly obtain network sensitivity coefficients. This direct method is more efficient and accurate than the perturbation method and much easier to implement on the computer.

We will base the development of the direct method on the general branch in Fig. 7.6 and the network formulation described in Eq. (7.30), which is repeated here for reference:

$$\mathbf{v}_n = (\mathbf{A}\mathbf{G}_1\mathbf{A}^\mathsf{T})^{-1}\mathbf{A}(\mathbf{J} - \mathbf{G}_1\mathbf{E}) \tag{7.30}$$

Our development considers only the node voltages as variables of interest since the development for other variables follows nearly identical lines. The partial derivative of the node voltage vector $\mathbf{v}_n$ with respect to the branch resistance R_i defines the node voltage sensitivity to the resistance R_i, and is given by

$$\frac{\partial \mathbf{v}_n}{\partial R_i} = \frac{\partial (\mathbf{A}\mathbf{G}_1\mathbf{A}^\mathsf{T})^{-1}}{\partial R_i}\mathbf{A}(\mathbf{J} - \mathbf{G}_1\mathbf{E}) + (\mathbf{A}\mathbf{G}_1\mathbf{A}^\mathsf{T})^{-1}\mathbf{A}\frac{\partial (\mathbf{J} - \mathbf{G}_1\mathbf{E})}{\partial R_i} \tag{7.39}$$

The derivative of the inverse of a matrix is defined as[†]

$$\frac{d}{dt}\mathbf{F}^{-1}(t) = -\mathbf{F}^{-1}(t)\frac{d}{dt}[\mathbf{F}(t)]\mathbf{F}^{-1}(t) \tag{7.40}$$

Applying this to Eq. (7.39), we have

$$\begin{aligned}
\frac{\partial \mathbf{v}_n}{\partial R_i} = &-(\mathbf{A}\mathbf{G}_1\mathbf{A}^\mathsf{T})^{-1}\mathbf{A}\frac{\partial \mathbf{G}_1}{\partial R_i}\mathbf{A}^\mathsf{T}(\mathbf{A}\mathbf{G}_1\mathbf{A}^\mathsf{T})^{-1}\mathbf{A}(\mathbf{J} - \mathbf{G}_1\mathbf{E}) \\
&+ (\mathbf{A}\mathbf{G}_1\mathbf{A}^\mathsf{T})^{-1}\mathbf{A}\left[\frac{\partial \mathbf{J}}{\partial R_i} - \frac{\partial \mathbf{G}_1}{\partial R_i}\mathbf{E} - \mathbf{G}_1\frac{\partial \mathbf{E}}{\partial R_i}\right]
\end{aligned} \tag{7.41}$$

Inserting Eq. (7.30) into Eq. (7.41) and observing that $\partial \mathbf{J}/\partial R_i = \partial \mathbf{E}/\partial R_i = 0$, the partial derivative reduces to

$$\begin{aligned}
\frac{\partial \mathbf{v}_n}{\partial R_i} &= -(\mathbf{A}\mathbf{G}_1\mathbf{A}^\mathsf{T})^{-1}\mathbf{A}\frac{\partial \mathbf{G}_1}{\partial R_i}\mathbf{A}^\mathsf{T}\mathbf{v}_n - (\mathbf{A}\mathbf{G}_1\mathbf{A}^\mathsf{T})^{-1}\mathbf{A}\frac{\partial \mathbf{G}_1}{\partial R_i}\mathbf{E} \\
&= -(\mathbf{A}\mathbf{G}_1\mathbf{A}^\mathsf{T})^{-1}\mathbf{A}\frac{\partial \mathbf{G}_1}{\partial R_i}(\mathbf{A}^\mathsf{T}\mathbf{v}_n + \mathbf{E})
\end{aligned} \tag{7.42}$$

Similarly, the node voltage sensitivities with respect to the other network parameters are

$$\frac{\partial \mathbf{v}_n}{\partial \alpha_{ji}} = -(\mathbf{A}\mathbf{G}_1\mathbf{A}^\mathsf{T})^{-1}\mathbf{A}\frac{\partial \mathbf{G}_1}{\partial \alpha_{ji}}(\mathbf{A}^\mathsf{T}\mathbf{v}_n + \mathbf{E}) \tag{7.43}$$

[†]L. A. Pipes, *Matrix Methods for Engineering*, Prentice-Hall, Inc., Englewood Cliffs, N.J., 1963, p. 89.

$$\frac{\partial \mathbf{v}_n}{\partial E_i} = -(\mathbf{AG}_1\mathbf{A}^\mathsf{T})^{-1}\mathbf{AG}_1 \frac{\partial \mathbf{E}}{\partial E_i} \tag{7.44}$$

$$\frac{\partial \mathbf{v}_n}{\partial J_i} = -(\mathbf{AG}_1\mathbf{A}^\mathsf{T})^{-1}\mathbf{A} \frac{\partial \mathbf{J}}{\partial J_i} \tag{7.45}$$

We now define a column vector $\boldsymbol{\epsilon}_i$ of dimension b as

$$\boldsymbol{\epsilon}_i = \begin{bmatrix} 0 \\ \cdots \\ 0 \\ 1 \\ 0 \\ \cdots \\ 0 \end{bmatrix} \tag{7.46}$$

The vector $\boldsymbol{\epsilon}_i$ has a 1 in the ith row of the vector and zeros elsewhere. Then the product $\boldsymbol{\epsilon}_i\boldsymbol{\epsilon}_j^\mathsf{T}$ is a matrix with a 1 in the ijth location and zeros elsewhere.

With this notation

$$\begin{aligned}
\frac{\partial \mathbf{G}_1}{\partial R_i} &= -\frac{1}{R_i^2}\boldsymbol{\epsilon}_i\boldsymbol{\epsilon}_i^\mathsf{T} - \sum_{j \neq i}\frac{\alpha_{ji}}{R_i^2}\boldsymbol{\epsilon}_j\boldsymbol{\epsilon}_i^\mathsf{T} \\
&= -\frac{1}{R_i^2}\sum_{j=1}^{m}\alpha_{ji}\boldsymbol{\epsilon}_j\boldsymbol{\epsilon}_i^\mathsf{T} \qquad \alpha_{ii} = 1
\end{aligned} \tag{7.47}$$

where i is the column in the $\mathbf{G}_1$ matrix corresponding to the ith branch, and j is the row corresponding to the jth branch. The term

$$\frac{\alpha_{ji}}{R_i^2}\boldsymbol{\epsilon}_j\boldsymbol{\epsilon}_i^\mathsf{T} \qquad j \neq i$$

in Eq. (7.47) exists only if the ith branch is the controlling (or from-) branch for a transfluence assigned to the jth branch. If the ith branch is the from-branch for dependent current sources assigned to other branches as well, then additional similar terms must appear in the above expression. Using a similar approach

$$\frac{\partial \mathbf{G}_1}{\partial \alpha_{ji}} = \frac{1}{R_i}\boldsymbol{\epsilon}_j\boldsymbol{\epsilon}_i^\mathsf{T} \tag{7.48}$$

$$\frac{\partial \mathbf{E}}{\partial E_i} = \boldsymbol{\epsilon}_i \tag{7.49}$$

$$\frac{\partial \mathbf{J}}{\partial J_i} = \boldsymbol{\epsilon}_i \tag{7.50}$$

The formulas for the partial derivatives of the node voltages with respect to the network elements now become

$$\frac{\partial \mathbf{v}_n}{\partial R_i} = (\mathbf{AG}_1\mathbf{A}^\mathsf{T})^{-1}\mathbf{A}(\boldsymbol{\epsilon}_i + \sum_{j \neq i} \boldsymbol{\epsilon}_j \alpha_{ji}) \frac{\boldsymbol{\epsilon}_i^\mathsf{T}}{R_i^2}(\mathbf{A}^\mathsf{T}\mathbf{v}_n + \mathbf{E})$$

$$= (\mathbf{AG}_1\mathbf{A}^\mathsf{T})^{-1}\mathbf{A}(\boldsymbol{\epsilon}_i + \sum_{j \neq i} \boldsymbol{\epsilon}_j \alpha_{ji}) \frac{1}{R_i^2}[(\mathbf{A}^\mathsf{T}\mathbf{v}_n)_i + E_i]$$

(7.51)

and

$$\frac{\partial \mathbf{v}_n}{\partial \alpha_{ji}} = -(\mathbf{AG}_1\mathbf{A}^\mathsf{T})^{-1}\mathbf{A}\boldsymbol{\epsilon}_j \frac{1}{R_i}[E_i + (\mathbf{A}^\mathsf{T}\mathbf{v}_n)_i]$$

(7.52)

$$\frac{\partial \mathbf{v}_n}{\partial E_i} = -(\mathbf{AG}_1\mathbf{A}^\mathsf{T})^{-1}\mathbf{AG}_1\boldsymbol{\epsilon}_i$$

(7.53)

$$\frac{\partial \mathbf{v}_n}{\partial J_i} = (\mathbf{AG}_1\mathbf{A}^\mathsf{T})^{-1}\mathbf{A}\boldsymbol{\epsilon}_i$$

(7.54)

The α_{ji} quantities stored in the $\mathbf{G}_1$ matrix can be converted to effective $g_{m_{ji}}$ quantities; that is, $g_{m_{ji}} = \alpha_{ji}G_i$. The conductance G_i is obtained from the from-branch of the dependent current source assigned to the jth branch. Hence, the partial derivative with respect to the $g_{m_{ji}}$ is related to the partial derivative with respect to α_{ji} as

$$\frac{\partial \mathbf{v}_n}{\partial g_{m_{ji}}} = \frac{\partial \mathbf{v}_n}{\partial \alpha_{ji}}\frac{\partial \alpha_{ji}}{\partial g_{m_{ji}}} + \frac{\partial \mathbf{v}_n}{\partial R_i}\frac{\partial R_i}{\partial g_{m_{ji}}}$$

$$= \frac{\partial \mathbf{v}_n}{\partial \alpha_{ji}}R_i$$

(7.55)

since $\partial R_i/\partial g_{m_{ji}}$ is zero.

EXAMPLE 7.4

Calculate the sensitivity $\partial v_2/\partial R_3$ of the voltage at node 2 to variations in the value of R_3 and the sensitivity $\partial v_2/\partial E$ of the voltage source E for the network shown in Fig. 7.12 using the topological methods developed in this section.

Solution. The sensitivity of the node voltage to variations of resistance is defined in Eq. (7.51) for the general case. For this example

$$\frac{\partial v_2}{\partial R_3} = (\mathbf{AGA}^\mathsf{T})^{-1}\mathbf{A}\boldsymbol{\epsilon}_3 \frac{(\mathbf{A}^\mathsf{T}\mathbf{v}_n)_3}{R_3^2} \qquad \text{volts/ohm}$$

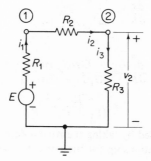

Figure 7.12 Resistive network for sensitivity analysis example.

The incidence and conductance matrices for the network in Fig. 7.12 are

$$\mathbf{A} = \begin{bmatrix} -1 & 1 & 0 \\ 0 & -1 & 1 \end{bmatrix} \quad \mathbf{G} = \begin{bmatrix} G_1 & 0 & 0 \\ 0 & G_2 & 0 \\ 0 & 0 & G_3 \end{bmatrix}$$

The nodal conductance matrix $\mathbf{G}_n = (\mathbf{AGA}^\mathsf{T})$ for the network is then

$$\mathbf{G}_n = \begin{bmatrix} G_1 + G_2 & -G_2 \\ -G_2 & G_2 + G_3 \end{bmatrix}$$

The nodal impedance matrix $(\mathbf{AGA}^\mathsf{T})^{-1}$ is

$$(\mathbf{AGA}^\mathsf{T})^{-1} = \frac{1}{\Delta} \begin{bmatrix} G_2 + G_3 & G_2 \\ G_2 & G_1 + G_2 \end{bmatrix}$$

where $\Delta = G_1 G_2 + G_1 G_3 + G_2 G_3$. Then

$$(\mathbf{AGA}^\mathsf{T})^{-1}\mathbf{A\epsilon}_3 = \frac{1}{\Delta} \begin{bmatrix} -(G_2 + G_3) & G_3 & G_2 \\ -G_2 & -G_1 & G_1 + G_2 \end{bmatrix} \begin{bmatrix} 0 \\ 0 \\ 1 \end{bmatrix}$$

$$= \frac{1}{\Delta} \begin{bmatrix} G_2 \\ G_1 + G_2 \end{bmatrix}$$

$$\mathbf{v}_n = \frac{1}{\Delta} \begin{bmatrix} G_2 + G_3 & G_2 \\ G_2 & G_1 + G_2 \end{bmatrix} \begin{bmatrix} -1 & 1 & 0 \\ 0 & -1 & 1 \end{bmatrix} \begin{bmatrix} -G_1 E \\ 0 \\ 0 \end{bmatrix}$$

$$= \frac{1}{\Delta} \begin{bmatrix} EG_1(G_2 + G_3) \\ EG_1 G_2 \end{bmatrix}$$

Then

$$\mathbf{A}^\mathsf{T}\mathbf{v}_n = \frac{1}{\Delta} \begin{bmatrix} -EG_1(G_2 + G_3) \\ EG_1 G_3 \\ EG_1 G_2 \end{bmatrix}$$

and

$$(\mathbf{A}^\mathsf{T}\mathbf{v}_n)_3 = \frac{EG_1 G_2}{\Delta}$$

Finally,

$$\frac{\partial \mathbf{v}_n}{\partial R_3} = \frac{1}{\Delta^2} \begin{bmatrix} G_2 \\ G_1 + G_2 \end{bmatrix} EG_1 G_2 G_3^2$$

and

$$\frac{\partial v_2}{\partial R_3} = \frac{EG_1 G_2 G_3^2 (G_1 + G_2)}{(G_1 G_2 + G_1 G_3 + G_2 G_3)^2}$$

$$= \frac{E(R_1 + R_2)}{(R_1 + R_2 + R_3)}$$

This result can be verified by taking the partial derivative of voltage v_2 with respect to R_3 using the direct relationship for the voltage $v_2 = ER_3/(R_1 + R_2 + R_3)$.

The second sensitivity coefficient is defined by Eq. (7.53). In this example

$$\frac{\partial \mathbf{v}_n}{\partial E} = [-(\mathbf{AGA^T})^{-1}\mathbf{AG}]_1$$

$$= \frac{-1}{\Delta}\begin{bmatrix} G_2 + G_3 & G_2 \\ G_2 & G_1 + G_2 \end{bmatrix}\begin{bmatrix} -1 & 1 & 0 \\ 0 & -1 & 1 \end{bmatrix}\begin{bmatrix} G_1 & 0 & 0 \\ 0 & G_2 & 2 \\ 0 & 0 & G_3 \end{bmatrix}\begin{bmatrix} 1 \\ 0 \\ 0 \end{bmatrix}$$

$$= \frac{1}{\Delta}\begin{bmatrix} G_1(G_2 + G_3) \\ G_1 G_2 \end{bmatrix}$$

Thus,

$$\frac{\partial v_2}{\partial E} = \frac{G_1 G_2}{G_1 G_2 + G_1 G_3 + G_2 G_3}$$

$$= \frac{R_3}{R_1 + R_2 + R_3}$$

which can also be easily verified by direct differentiation of v_2 with respect to E.

The topological method may appear to be much more cumbersome than direct differentiation for obtaining sensitivity information. However, the systematic approach shown is easily implemented with a circuit analysis program based on the topological incidence matrix techniques derived in this chapter. The actual implementation of this technique outlined in Eqs. (7.52) through (7.55) is left to the reader as an exercise (see Problem 7.23).

7.5 *AC* Node Analysis of Linear Networks

The node-voltage formulation of the network equations given in Section 7.3 also applies in ac steady-state analysis. The basic differences between the dc and ac analysis methods reside in the elements of the source vectors and in the admittance matrix. In the dc analysis of Section 7.3 the admittance matrix was simplified to the conductance matrix $\mathbf{G}$. The formal circuit equations for ac analysis are

$$\mathbf{AY}_b(\omega)\mathbf{A^T v}_n = \mathbf{A}(\mathbf{J} - \mathbf{Y}_b(\omega)\mathbf{E}) \tag{7.56}$$

where $\mathbf{Y}_b(\omega)$ is the frequency-dependent branch admittance matrix, which includes the effect of dependent current and voltage sources as well as passive conductance, capacitance, inductance, and mutual inductance. The source vectors $\mathbf{J}$ and $\mathbf{E}$ are also frequency dependent in the sense that phase differences often exist between sources in a network.

The elements of the branch admittance matrix $\mathbf{Y}_b(\omega)$ in the ac analysis are complex, which causes all the elements of the nodal admittance matrix $\mathbf{Y}_n(\omega)$, given by

$$\mathbf{Y}_n(\omega) = \mathbf{AY}_b(\omega)\mathbf{A^T} \tag{7.57}$$

and the equivalent current vector $\mathbf{J}_n(\omega)$, given by

$$\mathbf{J}_n(\omega) = \mathbf{A}(\mathbf{J} - \mathbf{Y}_b(\omega)\mathbf{E}) \tag{7.58}$$

to be complex quantities. Thus, most of the operations in ac analysis utilize complex quantities. Moreover, an ac analysis does not normally consist of the calculation of the network voltages and currents at a single fixed frequency but usually requires the calculation of the voltages and currents over a wide range of frequencies. Thus, at each frequency step in the range the $\mathbf{Y}_b(\omega)$ matrix must be recalculated to conform to the new frequency.

We determine the incidence matrix $\mathbf{A}$ just as we did in the dc analysis. The source vectors $\mathbf{J}$ and $\mathbf{E}$ are found as in the dc analysis except that we must include data to specify the phase-angle differences between sources. For example, two sources in the network might be specified by

$$\hat{J}_1 = 5e^{j0°}$$
$$\hat{J}_2 = 10e^{j30°}$$

The source $\hat{J}_1$ is a 5-A source at the reference phase angle of zero degrees. The second source J_2 leads J_1 by 30 degrees and has amplitude 10 A. The source currents in a network program might be computed as

```
J(I) = CMPLX (AMP*COS(THETA),AMP*SIN(THETA))
```

where AMP is the amplitude of the source and THETA is the phase angle specified for the source.

The procedure for constructing the $\mathbf{Y}_b(\omega)$ matrix is considerably more complicated in ac analysis than in dc analysis. The matrix consists of a set of complex numbers containing admittances calculated by using the circuit element values and the complex angular velocity $j\omega = j2\pi f$. The frequency f must be selected before $\mathbf{Y}_b$ can be constructed, and each time that the frequency changes in a frequency-response calculation, a new $\mathbf{Y}_b$ must be constructed. Thus, $\mathbf{Y}_b(\omega)$ can be represented as

$$\mathbf{Y}_b(\omega) = \mathbf{G} + j\left(\omega\mathbf{C} - \frac{\mathbf{\Gamma}}{\omega}\right) \tag{7.59}$$

where $\mathbf{G}$ is a square matrix of order b containing the branch conductances in the order indicated by the oriented graph. Similarly, the matrices $\mathbf{C}$ and $\mathbf{\Gamma}$ of order b contain the capacitance and reciprocal inductance values.

In the simple case where the inductance elements are not coupled, the matrix $\mathbf{\Gamma}$ is a matrix in which the elements $\Gamma_{ii} = L_{ii}^{-1}$; that is, the diagonal element Γ_{ii} is the reciprocal of the inductance value in branch i. When mutual coupling exists between inductances in a network, the elements of $\mathbf{\Gamma}$ are not reciprocals of the corresponding elements of the inductance matrix $\mathbf{L}$. The network in Fig. 7.13 illustrates the problem.

The voltage–current relationship between the three magnetically coupled inductance is

$$\mathbf{v}_b = j\omega\mathbf{L}\mathbf{i}_L \tag{7.60}$$

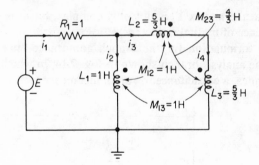

Figure 7.13 Network containing coupled inductances.

where

$$\mathbf{L} = \begin{bmatrix} 1 & -1 & 1 \\ -1 & \frac{5}{3} & -\frac{4}{3} \\ 1 & -\frac{4}{3} & \frac{5}{3} \end{bmatrix}$$

Since we need the inductance current in terms of the inductance voltage, we must invert the **L** matrix to obtain

$$\mathbf{i}_L = \frac{1}{j\omega}\mathbf{L}^{-1}\mathbf{v}_b$$

or

$$\mathbf{i}_L = \frac{-j}{\omega}\mathbf{\Gamma}\mathbf{v}_b \qquad (7.61)$$

where $\mathbf{\Gamma} = \mathbf{L}^{-1}$. The inductance currents are

$$\mathbf{i}_L = \frac{-j}{\omega}\begin{bmatrix} 3 & 1 & -1 \\ 1 & 2 & 1 \\ -1 & 1 & 2 \end{bmatrix}\mathbf{v}_b$$

in terms of the branch voltages.

The inverse of the inductance matrix exists if and only if the mutual inductance M_{ij} between each pair of coupled inductances L_i and L_j is such that

$$M_{ij} < \sqrt{L_iL_j}$$

In other words, the coupling coefficient between each pair of magnetically coupled elements must be less than unity. The complete $\mathbf{Y}_b$ matrix for the network of Fig. 7.13 is then

$$\mathbf{Y}_b(\omega) = \begin{bmatrix} 1 & 0 & 0 & 0 \\ 0 & -\dfrac{j3}{\omega} & -\dfrac{j}{\omega} & +\dfrac{j}{\omega} \\ 0 & -\dfrac{j}{\omega} & -\dfrac{j2}{\omega} & -\dfrac{i}{\omega} \\ 0 & +\dfrac{j}{\omega} & -\dfrac{j}{\omega} & -\dfrac{j2}{\omega} \end{bmatrix}$$

Note that the matrix $\mathbf{Y}_b$ is not only complex, but it is no longer diagonal, owing to the presence of mutual inductances.

The formulation for the branch admittance matrix allows us to establish the general ac analysis branch shown in Fig. 7.14, in which the passive element can be a conductance, a capacitance, or an inductance.

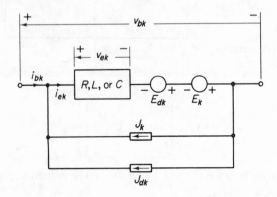

Figure 7.14 General ac analysis branch.

EXAMPLE 7.5

Consider the network shown in Fig. 7.15. Kirchhoff's current law can be written as follows:

$$\begin{bmatrix} 1 & 0 & 1 & 0 & 0 \\ 0 & 1 & -1 & 1 & 0 \\ 0 & 0 & 0 & -1 & 1 \end{bmatrix} \begin{bmatrix} i_1 \\ i_2 \\ i_3 \\ i_4 \\ i_5 \end{bmatrix} = \begin{bmatrix} 0 \\ 0 \\ 0 \end{bmatrix}$$

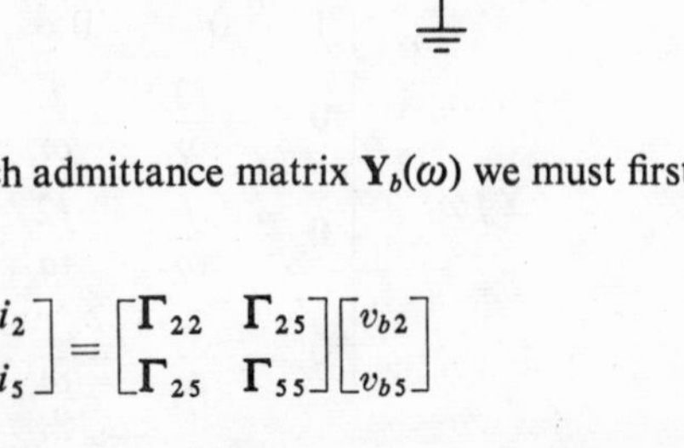

Figure 7.15 Coupled RLC network for Example 7.5.

Before we can form the branch admittance matrix $\mathbf{Y}_b(\omega)$ we must first invert the inductance matrix to obtain

$$\begin{bmatrix} i_2 \\ i_5 \end{bmatrix} = \begin{bmatrix} \boldsymbol{\Gamma}_{22} & \boldsymbol{\Gamma}_{25} \\ \boldsymbol{\Gamma}_{25} & \boldsymbol{\Gamma}_{55} \end{bmatrix} \begin{bmatrix} v_{b2} \\ v_{b5} \end{bmatrix}$$

Then the branch admittance matrix $\mathbf{Y}_b(\omega)$, including the dependent source, becomes

$$\mathbf{Y}_b(\omega) = \begin{bmatrix} G_1 & 0 & 0 & 0 & 0 \\ 0 & \dfrac{\Gamma_{22}}{j\omega} & -g_m & 0 & \dfrac{\Gamma_{25}}{j\omega} \\ 0 & 0 & j\omega C_3 & 0 & 0 \\ 0 & 0 & 0 & G_4 & 0 \\ 0 & \dfrac{\Gamma_{25}}{j\omega} & 0 & 0 & \dfrac{\Gamma_{55}}{j\omega} \end{bmatrix}$$

The nodal admittance matrix $\mathbf{Y}_n(\omega) = \mathbf{A}\mathbf{Y}_b(\omega)\mathbf{A}^\mathsf{T}$ is then

$$\mathbf{Y}_n(\omega) = \begin{bmatrix} G_1 + j\omega C_3 & -j\omega C_3 & 0 \\ -g_m - j\omega C_3 & \dfrac{\Gamma_{22}}{j\omega} + j\omega C_3 + g_m + G_4 & -G_4 + \dfrac{\Gamma_{25}}{j\omega} \\ 0 & \dfrac{\Gamma_{25}}{j\omega} - G_4 & G_4 + \dfrac{\Gamma_{55}}{j\omega} \end{bmatrix}$$

and the equivalent source vector $\mathbf{J}_n(\omega)$ is

$$\mathbf{J}_n(\omega) = \begin{bmatrix} -J_1 \\ 0 \\ 0 \end{bmatrix}$$

Notice that before the branch admittance matrix can be formed an inductance matrix containing the network inductances and the couplings between them must first be inverted and the results from the inverted $\mathbf{L}$ matrix multiplied by $-j/\omega$ and inserted in the appropriate rows and columns of the $\mathbf{Y}_b(\omega)$ matrix.

Basic AC Analysis Program

The construction of the basic ac analysis program is similar to the basic dc analysis program described in Fig. 7.7. The first important difference between the two programs is the ac analysis program requires COMPLEX arithmetic calculations rather than the REAL calculations used in the dc analysis program.

Second, the solution (for example, node voltages) is a function of the analysis frequency. If we perform the analysis at a single frequency, then we can form the branch admittance matrix directly when the input data are read as in the dc analysis program in Fig. 7.8. However, an ac analysis normally determines the response of a network over a range of frequencies, which forces us to construct the branch admittance matrix $\mathbf{Y}_b(\omega)$ at each analysis frequency. Because of this requirement, we store the input R, G, L, and C values in vectors from which we form the $\mathbf{Y}_b(\omega)$ matrix at each analysis step.

We present a simple method for storing and manipulating the mutual inductance values. First, we assume that the self-inductance and mutual-inductance values are stored as L and MUT vectors, respectively, as shown in Fig. 7.16. The inductance vector L contains the inductance values as read from branch data cards. When a

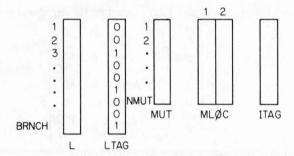

Figure 7.16 Layout of inductance vectors.

mutual inductance data card is read from the input, we store the value in the vector MUT, the coupled branch numbers are stored in columns 1 and 2 of the matrix MLØC and each coupled inductance branch is tagged with a "1" in the LTAG vector. The ITAG vector is used to set up the **L** matrix before inversion and to expand the $\mathbf{L}^{-1}$ matrix after inversion. After the input branch data are completed, we load the ITAG vector with the branch numbers of the coupled branches, and the LTAG vector flags are changed from 1's to the corresponding row numbers of the ITAG vector. Next, we construct the inductance matrix LMAT using the data from the MUT, L, and LTAG vectors, and the MLØC matrix. Finally, the LMAT matrix is inverted as described in the ac analysis program illustrated in Fig. 7.17. The LMAT matrix is used in each frequency step of the analysis to supply reciprocal inductance values to the $\mathbf{Y}_b(\omega)$ matrix. The best way to fully understand the implementation of the preceding algorithm is to follow the program steps related to setting up the inductance matrix in Fig. 7.17.

```
C      BASIC AC ANALYSIS PROGRAM                                    ACO2   1
C      IMPLEMENTATION OF EQUATION 7.56 WHICH IS BASED UPON          ACO2   2
C      THE SIMPLIFIED GENERAL BRANCH IN FIGURE 7.14                 ACO2   3
C                                                                   ACO2   4
C      DEFINITION OF VARIABLES                                      ACO2   5
C        A       - INCIDENCE MATRIX                                 ACO2   6
C        BRANCH  - BRANCH NUMBER                                    ACO2   7
C        BRNCHS  - NUMBER OF NETWORK BRANCHES                       ACO2   8
C        C       - BRANCH CAPACITANCE VECTOR                        ACO2   9
C        DET     - DETERMINANT OF YN                                ACO2  10
C        ES      - INDEPENDENT VOLTAGE SOURCE VECTOR                ACO2  11
C        ETYPE   - BRANCH TYPE      ( R G C L E J TC TF TP TR M  EX )  ACO2  12
C        FH      - HIGHEST ANALYSIS FREQUENCY                       ACO2  13
C        FL      - LOWEST ANALYSIS FREQUENCY                        ACO2  14
C        FM      - FREQUENCY MULTIPLE                               ACO2  15
C        G       - BRANCH CONDUCTANCE VECTOR                        ACO2  16
C        ITAG    - INDUCTANCE MATRIX EXPANSION VECTOR               ACO2  17
C        JS      - INDEPENDENT CURRENT SOURCE VECTOR                ACO2  18
C        JN      - EQUIVALENT CURRENT VECTOR                        ACO2  19
C        J1      - TEMPORARY WORKING VECTOR TO FORM JN              ACO2  20
C        L       - BRANCH INDUCTANCE VECTOR                         ACO2  21
C        LMAT    - INDUCTANCE MATRIX, ALSO STORES INVERSE OF LMAT   ACO2  22
C        LTAG    - COUPLED INDUCTANCE TAG VECTOR                    ACO2  23
C        MISC    - OUTPUT FLAG                                      ACO2  24
C                  = 0   NODE VOLTAGES ONLY                         ACO2  25
C                  = 1   SAME AS 0 PLUS YN PLUS JN                  ACO2  26
C                  = 2   DEBUG - SAME AS 1 PLUS A,Y,ES AND JS       ACO2  27
```

Figure 7.17 Basic ac analysis program listing.

```
C        MLOC    - COUPLED INDUCTANCE LOCATION MATRIX                AC02  28
C        MUT     - MUTUAL INDUCTANCE VALUE VECTOR                    AC02  29
C        NODES   - NUMBER OF NETWORK NODES                           AC02  30
C        NODE1   - INITIAL NODE                                      AC02  31
C        NODE2   - FINAL NODE                                        AC02  32
C        NTAG    - NUMBER OF COUPLED INDUCTANCES                     AC02  33
C        NTYPE   - BRANCH TYPE CODE ( 1 2 3 4 5 6  7  8  9 10 11 12 ) AC02  34
C        VALUE   - BRANCH ELEMENT VALUE                              AC02  35
C        VALUE2  - PHASE VALUE FOR IND. SOURCES (DEGREES)            AC02  36
C        VLTGS   - NODE VOLTAGE VECTOR                               AC02  37
C        Y       - BRANCH ADMITTANCE MATRIX                          AC02  38
C        YN      - NODAL ADMITTANCE MATRIX                           AC02  39
C        Y1      - TEMPORARY WORKING MATRIX TO FORM YN               AC02  40
C     INPUT DATA FORMAT                                              AC02  41
C     DATA CARD ONE CONTAINS (RIGHT JUSTIFIED)                       AC02  42
C        NO. OF NODES      COLUMNS 1-5                               AC02  43
C        NO. OF BRANCHES COLUMNS 6-10                                AC02  44
C        OUTPUT FLAG       COLUMNS 11-15                             AC02  45
C     DATA CARD TWO CONTAINS                                         AC02  46
C        LOWEST FREQUENCY (HZ)      COLUMNS 1-20                     AC02  47
C        HIGHEST FREQUENCY (HZ)     COLUMNS 21-40                    AC02  48
C        FREQUENCY MULTIPLE         COLUMNS 41-60                    AC02  49
C     THE REMAINING DATA CARDS USE THE FOLLOWING FORMAT              ACC2  50
C        ELEMENT TYPE   COLUMNS 1-2 (LEFT JUSTIFIED)                 AC02  51
C        BRANCH NUMBER COLUMNS 5-6 (RIGHT JUSTIFIED)                 AC02  52
C        INITIAL NODE   COLUMNS 9-10 (RIGHT JUSTIFIED)               AC02  53
C        FINAL NODE     COLUMNS 12-13 (RIGHT JUSTIFIED)              AC02  54
C        ELEMENT VALUE COLUMNS 16-30,31-45                          AC02  55
C     DEPENDENT SOURCE CARDS USE NODE1 FOR 'FROM'-BRANCH VALUE AND   AC02  56
C     NODE2 FOR 'TO'-BRANCH VALUE.                                   AC02  57
C                                                                    AC02  58
C                                                                    AC02  59
      COMPLEX JS,ES,Y,Y1,YN,JN,J1,VLTGS,DET                          AC02  60
      INTEGER BRNCHS,ETYPE,BRANCH                                    AC02  61
      REAL L,MUT,LMAT                                                AC02  62
      DIMENSION G(50),C(50),L(50),MUT(10),NTYPE(12),LTAG(50),MLOC(10,2) AC02 63
      DIMENSION ITAG(5),LMAT(5,5),A(20,50),ES(50),JS(50),J1(20)     AC02  64
      DIMENSION Y(50,50),Y1(20,50),YN(20,20),VLTGS(20),JN(20)       AC02  65
      DATA JS/50*(0.0,0.0)/,ES/50*(0.0,0.0)/                         AC02  66
      DATA A/1000*0.0/,Y/2500*(C.C,0.0)/                             AC02  67
      DATA G/50*0.0/,C/50*0.0/,L/50*0.0/,MUT/10*0.0/                AC02  68
      DATA LTAG/50*0/,MLOC/20*0/,ITAG/5*0/                           AC02  69
      DATA LMAT/25*0.0/,NTAG/0/,NMUT/0/                              AC02  70
      DATA NTYPE/'R ','G ','C ','L ','E ','J ','TC','TF','TP','TR','M ', AC02 71
     1'EX'/                                                          AC02  72
C                                                                    AC02  73
C     READ NUMBER OF INDEPENDENT NODES AND NO. OF NETWORK BRANCHES   AC02  74
      READ (5,10) NODES,BRNCHS,MISC                                  AC02  75
   10 FORMAT (3I5)                                                   AC02  76
      WRITE (6,20) NODES,BRNCHS                                      AC02  77
   20 FORMAT (1H1,5X,'AC ANALYSIS PROBLEM'///5X,'NUMBER OF NODES =',I3/5A C02 78
     1X,'NUMBER OF BRANCHES =',I3//)                                 AC02  79
C     READ FREQUENCY DATA                                            AC02  80
      READ (5,30) FL,FH,FM                                           AC02  81
   30 FORMAT (3F20.0)                                                AC02  82
      WRITE (6,40) FL,FH,FM                                          AC02  83
   40 FORMAT (5X,'LOWEST FREQUENCY= ',1PE15.5/5X,'HIGHEST FREQUENCY= ',1A C02 84
     1PE15.5/5X,'FREQUENCY MULTIPLE= ',1PE15.5//)                    AC02  85
C     READ BRANCH DATA STATEMENT                                     AC02  86
   50 READ (5,60) ETYPE,BRANCH,NODE1,NODE2,VALUE,VALUE2              AC02  87
   60 FORMAT (A2,2X,I2,2X,I2,I3,2X,2F15.0)                           AC02  88
C     TEST BRANCH FOR ELEMENT TYPE                                   AC02  89
      DO 70 I=1,12                                                   AC02  90
      IF (ETYPE.EQ.NTYPE(I)) GO TO 90                                AC02  91
   70 CONTINUE                                                       AC02  92
C     BRANCH TYPE ERROR                                              AC02  93
      WRITE (6,80) ETYPE                                             AC02  94
   80 FORMAT (5X,A2,' IS NOT A VALID BRANCH TYPE.'/)                 AC02  95
      GO TO 50                                                       AC02  96
   90 IF (I.EQ.12) WRITE (6,100)                                     AC02  97
```

Figure 7.17—*Cont.*

```
      100 FORMAT (5X,'ENTERING EXECUTION. ')                              ACO2  98
          IF (I.EQ.12) GO TO 260                                         ACO2  99
          GO TO (110,130,140,150,160,180,190,190,190,190,230),I         ACO2 100
C     LOAD INCIDENCE MATRIX A, BRANCH VALUE VECTORS, AND  SOURCE         ACO2 101
C     VECTORS ES AND JS.                                                ACO2 102
C     OUTPUT BRANCH DATA STATEMENT                                      ACO2 103
C     RESISTANCE ELEMENT                                                ACO2 104
      110 G(BRANCH)=1.0/VALUE                                           ACO2 105
          WRITE (6,120) ETYPE,BRANCH,NODE1,NODE2,VALUE                  ACO2 106
      120 FORMAT (5X,A2,' BRANCH',I3,' NODE',I3,' TO',I3,'    VALUE = ',1PE15ACO2 107
         1.5)                                                           ACO2 108
          GO TO 250                                                     ACO2 109
C     CONDUCTANCE ELEMENT                                               ACO2 110
      130 G(BRANCH)=VALUE                                               ACO2 111
          WRITE (6,120) ETYPE,BRANCH,NODE1,NODE2,VALUE                  ACO2 112
          GO TO 250                                                     ACO2 113
C     CAPACITANCE ELEMENT                                               ACO2 114
          GO TO 250                                                     ACO2 117
C     INDUCTANCE ELEMENT                                                ACO2 118
      150 L(BRANCH)=VALUE                                               ACO2 119
          WRITE (6,120) ETYPE,BRANCH,NODE1,NODE2,VALUE                  ACO2 120
      140 C(BRANCH)=VALUE                                               ACO2. 115
          WRITE (6,120) ETYPE,BRANCH,NODE1,NODE2,VALUE                  ACO2 116
          GO TO 250                                                     ACO2 121
C     INDEPENDENT VOLTAGE SOURCE (E ELEMENT)                            ACO2 122
      160 WRITE (6,170) ETYPE,BRANCH,VALUE,VALUE2                       ACO2 123
      170 FORMAT (5X,A2,' BRANCH',I3,14X,' VALUE = ',1PE15.5,10X,'PHASE = ',ACO2 124
         11PE15.5)                                                      ACO2 125
          VALUE2=VALUE2/57.29578                                        ACO2 126
          ES(BRANCH)=CMPLX(VALUE*COS(VALUE2),VALUE*SIN(VALUE2))         ACO2 127
          GO TO 50                                                      ACO2 128
C     INDEPENDENT CURRENT SOURCE (J ELEMENT)                            ACO2 129
      180 WRITE (6,170) ETYPE,BRANCH,VALUE,VALUE2                       ACO2 130
          VALUE2=VALUE2/57.29578                                        ACO2 131
          JS(BRANCH)=CMPLX(VALUE*COS(VALUE2),VALUE*SIN(VALUE2))         ACO2 132
          GO TO 50                                                      ACO2 133
C     DEPENDENT SOURCES (ELEMENT TYPES TC,TF,TP, AND TR)               ACO2 134
C     ARE NOT IMPLEMENTED IN THIS VERSION OF THE AC ANALYSIS PROGRAM.   ACO2 135
      190 GO TO 50                                                      ACO2 136
C     MUTUAL INDUCTANCE AND COUPLED BRANCH LOCATIONS                    ACO2 137
      230 NMUT=NMUT+1                                                   ACO2 138
          MUT(BRANCH)=VALUE                                             ACO2 139
          MLOC(BRANCH,1)=NODE1                                          ACO2 140
          MLOC(BRANCH,2)=NODE2                                          ACO2 141
          LTAG(NODE1)=1                                                 ACO2 142
          LTAG(NODE2)=1                                                 ACO2 143
          WRITE (6,240) ETYPE,BRANCH,NODE1,NODE2,VALUE                  ACO2 144
      240 FORMAT (5X,A2,' NUMBER',I3,' BRANCH',I3,' TO',I3,' VALUE = ',1PE15ACO2 145
         1.5)                                                           ACO2 146
          GO TO 50                                                      ACO2 147
C     LOAD ELEMENTS OF INCIDENCE MATRIX A                               AC/2 148
      250 IF (NODE1.GT.0)A(NODE1,BRANCH)=1.0                            ACO2 149
          IF (NODE2.GT.0)A(NODE2,BRANCH)=-1.0                           ACO2 150
          GO TO 50                                                      ACO2 151
      260 IF (MISC.LE.1) GO TO 360                                      ACO2 152
C     OUTPUT A,G,C,L,ES,JS MATRICES                                     ACO2 153
          WRITE (6,270)                                                 ACO2 154
      270 FORMAT (1H1,5X,'NETWORK INCIDENCE MATRIX'/)                   ACO2 155
          DO 280 I=1,NODES                                             ACO2 156
      280 WRITE (6,290) (A(I,J),J=1,BRNCHS)                            ACO2 157
      290 FORMAT (1H0,2X,25F5.0)                                        ACO2 158
          WRITE (6,300)                                                 ACO2 159
      300 FORMAT (///10X,'E SOURCE VECTOR',22X,'J SOURCE VECTOR'/)      ACO2 160
          DO 310 I=1,BRNCHS                                            ACO2 161
      310 WRITE (6,320) I,ES(I),JS(I)                                  ACO2 162
      320 FORMAT (3X,I2,3X,1P2E12.5,13X,1P2E12.5)                       ACO2 163
          WRITE (6,330)                                                 ACO2 164
      330 FORMAT (///7X,'G VECTOR',5X,'C VECTOR',5X,'L VECTOR',5X,'COUPLED IACO2 165
         1NDUCTORS'/)                                                   ACO2 166
          DO 340 I=1,BRNCHS                                            ACO2 167
```

Figure 7.17—Cont.

```
      340 WRITE (6,350) I,G(I),C(I),L(I),LTAG(I)              ACO2 168
      350 FORMAT (I2,1P3E13.5,12X,I1)                         ACO2 169
C         SET UP INDUCTANCE MATRIX WITH MUTUAL TERMS          ACO2 170
C         DETERMINE NO. OF COUPLED INDUCTORS,                 ACO2 171
C         LOAD TAG VECTOR, AND LMAT DIAGONAL                  ACO2 172
      360 DO 370 I=1,BRNCHS                                   ACO2 173
          NTAG=NTAG+LTAG(I)                                   ACO2 174
          IF (LTAG(I).LE.0) GO TO 370                         ACO2 175
          ITAG(NTAG)=I                                        ACO2 176
          LTAG(I)=NTAG                                        ACO2 177
          LMAT(NTAG,NTAG)=L(I)                                ACO2 178
      370 CONTINUE                                            ACO2 179
          IF (NTAG.LE.0) GO TO 450                            ACO2 180
C         LOAD OFF-DIAGONAL TERMS IN LMAT                     ACO2 181
          DO 380 I=1,NMUT                                     ACO2 182
          J=MLOC(I,1)                                         ACO2 183
          K=MLOC(I,2)                                         ACO2 184
          IMF=LTAG(J)                                         ACO2 185
          IMT=LTAG(K)                                         ACO2 186
          LMAT(IMF,IMT)=MUT(I)                                ACO2 187
      380 LMAT(IMT,IMF)=MUT(I)                                ACO2 188
          IF (MISC.LE.1) GO TO 420                            ACO2 189
          WRITE (6,390)                                       ACO2 190
      390 FORMAT (///5X,'INDUCTANCE MATRIX'/)                 ACO2 191
          DO 400 I=1,NTAG                                     ACO2 192
      400 WRITE (6,410) (LMAT(I,J),J=1,NTAG)                  ACO2 193
      410 FORMAT (1P5E15.5)                                   ACO2 194
C         INVERT INDUCTANCE MATRIX                            ACO2 195
      420 CALL INVERT(LMAT,NTAG,5)                            ACO2 196
          IF (MISC.LE.1) GO TO 450                            ACO2 197
          WRITE (6,430)                                       ACO2 198
      430 FORMAT (///5X,'INVERSE INDUCTANCE MATRIX'/)         ACO2 199
          DO 440 I=1,NTAG                                     ACO2 200
      440 WRITE (6,410) (LMAT(I,J),J=1,NTAG)                  ACO2 201
C                                                             ACO2 202
C         INPUT READY FOR ANALYSIS PHASE                      ACO2 203
C         INITIALIZE FREQUENCY                                ACO2 204
      450 FREQ=FL                                             ACO2 205
C         SET UP Y MATRIX FOR PRESENT FREQUENCY               ACO2 206
      460 W=6.283185*FREQ                                     ACO2 207
          DO 480 I=1,BRNCHS                                   ACO2 208
          DO 470 J=1,BRNCHS                                   ACO2 209
      470 Y(I,J)=(0.0,0.0)                                    ACO2 210
          IF (L(I).EQ.0.0)REAC=W*C(I)                         ACO2 211
          IF (L(I).NE.0.0)REAC=W*C(I)-1.0/W/L(I)              ACO2 212
      480 Y(I,I)=CMPLX(G(I),REAC)                             ACO2 213
C         LOAD COUPLED INDUCTANCE TERMS INTO Y MATRIX         ACO2 214
          IF (NTAG.LE.0) GO TO 500                            ACO2 215
          DO 490 I=1,NTAG                                     ACO2 216
          DO 490 J=1,NTAG                                     ACO2 217
          II=ITAG(I)                                          ACO2 218
          JJ=ITAG(J)                                          ACO2 219
      490 Y(II,JJ)=CMPLX(0.0,-LMAT(I,J)/W)                    ACO2 220
      500 IF (MISC.LE.1) GO TO 540                            ACO2 221
          WRITE (6,510)                                       ACO2 222
      510 FORMAT (///5X,'BRANCH ADMITTANCE MATRIX'/)          ACO2 223
          DO 520 I=1,BRNCHS                                   ACO2 224
      520 WRITE (6,530) (Y(I,J),J=1,BRNCHS)                   ACO2 225
      530 FORMAT (5X,1P10E12.5)                               ACO2 226
C         COMPUTE NODAL ADMITTANCE MATRIX YN=A*Y*AT           ACO2 227
C            FORM PARTIAL PRODUCT   Y1=A*Y                    ACO2 228
      540 DO 550 I=1,NODES                                    ACC2 229
          DO 550 J=1,BRNCHS                                   ACO2 230
          Y1(I,J)=(0.0,0.0)                                   ACO2 231
          DO 550 K=1,BRNCHS                                   ACO2 232
      550 Y1(I,J)=Y1(I,J)+A(I,K)*Y(K,J)                       ACO2 233
C            COMPLETE YN=Y1*AT                                ACO2 234
          DO 560 I=1,NODES                                    ACO2 235
          DO 560 J=1,NODES                                    ACO2 236
          YN(I,J)=(0.0,0.0)                                   ACO2 237
```

Figure 7.17—*Cont.*

```
         DO 560 K=1,BRNCHS                                      ACO2 238
   560 YN(I,J)=YN(I,J)+Y1(I,K)*A(J,K)                           ACO2 239
         IF (MISC.LT.1) GO TO 600                               ACO2 240
C        OUTPUT NODAL ADMITTANCE MATRIX YN                      ACO2 241
         WRITE (6,570)                                          ACO2 242
   570 FORMAT (1H1,4X,'NODAL ADMITTANCE MATRIX'/)               ACO2 243
         DO 580 I=1,NODES                                       ACO2 244
   580 WRITE (6,590) (YN(I,J),J=1,NODES)                        ACO2 245
   590 FORMAT (1P8E15.5)                                        ACO2 246
C        COMPUTE EQUIVALENT CURRENT VECTOR  JN=A*(JS-Y*ES)      ACO2 247
C        FORM PARTIAL PRODUCT   J1=(J-Y*E)                      ACO2 248
   600 DO 610 I=1,BRNCHS                                        ACO2 249
         J1(I)=JS(I)                                            ACO2 250
         DO 610 J=1,BRNCHS                                      ACO2 251
   610 J1(I)=J1(I)-Y(I,J)*ES(J)                                 ACO2 252
C         COMPLETE JN=A*J1                                      ACO2 253
         DO 620 I=1,NODES                                       ACO2 254
         JN(I)=(0.0,0.0)                                        ACO2 255
         DO 620 J=1,BRNCHS                                      ACO2 256
   620 JN(I)=JN(I)+A(I,J)*J1(J)                                 ACO2 257
         IF (MISC.LT.1) GO TO 660                               ACO2 258
         WRITE (6,630)                                          ACO2 259
   630 FORMAT (//////5X,'EQUIVALENT CURRENT VECTOR'/)           ACO2 260
         DO 640 I=1,NODES                                       ACO2 261
C        OUTPUT EQUIVALENT CURRENT VECTOR                       ACO2 262
   640 WRITE (6,650) JN(I)                                      ACO2 263
   650 FORMAT (1P2E20.5)                                        ACO2 264
C        CALCULATE NODE VOLTAGES BY GAUSS ELIMINATION           ACO2 265
   660 CALL CLINQ(YN,JN,VLTGS,NODES,DET)                        ACO2 266
C        TEST FOR SINGULAR SYSTEM OF EQUATIONS                  ACO2 267
         IF (CABS(DET).GT.1.E-60) GO TO 680                     ACO2 268
C        ERROR OUTPUT                                           ACO2 269
         WRITE (6,670)                                          ACO2 270
   670 FORMAT (5X,'INVALID SYSTEM OF EQUATIONS'//)              ACO2 271
         STOP                                                   ACO2 272
C        OUTPUT NODE VOLTAGES                                   ACO2 273
   680 WRITE (6,690) FREQ                                       ACO2 274
   690 FORMAT (///5X,'FREQUENCY = ',1PE13.5//5X,'NODE VOLTAGES'//26X,'MAGACO2 275
        1',10X,'PHA'/)                                          ACO2 276
         DO 700 I=1,NODES                                       ACO2 277
         VREAL=REAL(VLTGS(I))                                   ACO2 278
         VIMAG=AIMAG(VLTGS(I))                                  ACO2 279
         VMAG=SQRT(VREAL**2+VIMAG**2)                           ACO2 280
         VPHASE=57.29578*ATAN2(VIMAG,VREAL)                     ACO2 281
   700 WRITE (6,710) I,VMAG,VPHASE                              ACO2 282
   710 FORMAT (5X,'NODE',I2,'  VCLTAGE = ',1P2E12.5)            ACO2 283
C        INCREMENT FREQUENCY                                    ACO2 284
         IF (FREQ.GE.FH) GO TO 720                              ACO2 285
         FREQ=FM*FREQ                                           ACO2 286
         GO TO 460                                              ACO2 287
C        END OF ANALYSIS                                        ACO2 288
   720 WRITE (6,730)                                            ACO2 289
   730 FORMAT (///5X,'END OF ANALYSIS')                         ACO2 290
         STOP                                                   ACO2 291
         END                                                    ACO2 292
         SUBROUTINE INVERT(A,N,NR)                              INV2   1
C                                                               INV2   2
C     PURPOSE                                                   INV2   3
C        COMPUTE THE INVERSE OF A MATRIX OF ORDER N (N )= 20)   INV2   4
C     DEFINITION OF VARIABLES                                   INV2   5
C        A - INPUT NXN MATRIX                                   INV2   6
C        N - ORDER OF MATRIX A                                  INV2   7
C        NR - DIMENSION OF A IN CALLING PROGRAM                 INV2   8
C        C - NX(2N) WORKING MATRIX                              INV2   9
C        EPS - ZERO TOLERANCE                                   INV2  10
C     NOTE ..... VARIABLES A MUST BE DIMENSIONED IN CALLING PROGRAM  INV2  11
C                                                               INV2  12
         DIMENSION A(NR,NR),C(20,40)                            INV2  13
```

Figure 7.17—Cont.

```
C       INITIALIZE SUBROUTINE VARIABLES                      INV2  14
        DATA EPS/1.0E-6/                                     INV2  15
        N1=2*N                                               INV2  16
C       LOAD A MATRIX AND CONSTRUCT IDENTITY MATRIX          INV2  17
        DO 10 I=1,N                                          INV2  18
        DO 10 J=1,N                                          INV2  19
     10 C(I,J)=A(I,J)                                        INV2  20
        DO 30 I=1,N                                          INV2  21
        DO 30 J=1,N                                          INV2  22
        IF (I.EQ.J) GO TO 20                                 INV2  23
        C(I,J+N)=0.0                                         INV2  24
        GO TO 30                                             INV2  25
     20 C(I,J+N)=1.0                                         INV2  26
     30 CONTINUE                                             INV2  27
C                                                            INV2  28
C       INVERT MATRIX                                        INV2  29
C                                                            INV2  30
        DO 130 IP=1,N                                        INV2  31
C       FIND PIVOT ELEMENT IN COLUMN IP                      INV2  32
        IM=IP                                                INV2  33
        IST=IP+1                                             INV2  34
        IF (IP.GE.N) GO TO 50                                INV2  35
        DO 40 I=IST,N                                        INV2  36
        IF (ABS(C(IM,IP)).GE.ABS(C(I,IP))) GO TO 40          INV2  37
        IM=I                                                 INV2  38
     40 CONTINUE                                             INV2  39
     50 IF (ABS(C(IM,IP)).GE.EPS) GO TO 70                   INV2  40
C       NEAR ZERO DIAGONAL ELEMENT FLAG                      INV2  41
        WRITE (6,60) IP,C(IM,IP)                             INV2  42
     60 FORMAT (1H0,17HDIAGONAL ELEMENT ,I2,2H =,1PE13.5)    INV2  43
        IF (C(IM,IP).EQ.0.0) RETURN                          INV2  44
     70 IF (IM.EQ.IP) GO TO 90                               INV2  45
C       INTERCHANGE ROWS TO POSITION PIVOT ELEMENT           INV2  46
        DO 80 J=IP,N1                                        INV2  47
        CL=C(IP,J)                                           INV2  48
        C(IP,J)=C(IM,J)                                      INV2  49
     80 C(IM,J)=CL                                           INV2  50
C       FIND MULTIPLICATION CONSTANT, SET C(I,I)=1           INV2  51
     90 CL=C(IP,IP)                                          INV2  52
        C(IP,IP)=1.0                                         INV2  53
C       DIVIDE ELEMENT IN ROW BY PIVOT ELEMENT               INV2  54
        DO 100 J=IST,N1                                      INV2  55
    100 C(IP,J)=C(IP,J)/CL                                   INV2  56
C       ZERO COLUMN OF PIVOT ELEMENT                         INV2  57
        DO 120 I=1,N                                         INV2  58
        IF (I.EQ.IP) GO TO 120                               INV2  59
        IP1=IP+1                                             INV2  60
        CL=C(I,IP)                                           INV2  61
        C(I,IP)=0.0                                          INV2  62
        DO 110 J=IP1,N1                                      INV2  63
    110 C(I,J)=C(I,J)-CL*C(IP,J)                             INV2  64
    120 CONTINUE                                             INV2  65
    130 CONTINUE                                             INV2  66
C       LOAD INVERSE MATRIX INTO A                           INV2  67
        M=N+1                                                INV2  68
        DO 140 I=1,N                                         INV2  69
        DO 140 J=M,N1                                        INV2  70
        K=J-N                                                INV2  71
    140 A(I,K)=C(I,J)                                        INV2  72
        RETURN                                               INV2  73
        END                                                  INV2  74
        SUBROUTINE CLINQ(A,B,D,N,DET)                        CLNQ   1
C                                                            CLNQ   2
C       PURPOSE                                              CLNQ   3
C       SOLUTION OF AN NTH ORDER LINEAR SYSTEM OF EQUATIONS  (N )= 20 )   CLNQ   4
C       USING GAUSS' UPPER TRIANGULATION AND BACK SUBSTITUTION METHODS    CLNQ   5
C    DEFINITION OF VARIABLES                                 CLNQ   6
C       A - NXN COEFFICIENT MATRIX (DOUBLE SUBSCRIPTED)      CLNQ   7
```

Figure 7.17—*Cont.*

```
C       B - RIGHT HAND VECTOR FROM AX=B                            CLNQ    8
C       C - NXN+1 WORKING MATRIX                                   CLNQ    9
C       D - SOLUTION VECTOR X                                      CLNQ   10
C       N - NUMBER OF EQUATIONS IN SYSTEM                          CLNQ   11
C       DET - DETERMINANT OF MATRIX A                              CLNQ   12
C       EPS - ZERO TOLERANCE                                       CLNQ   13
C    NOTE ..... VARIABLES A,B,D MUST BE DIMENSIONED IN CALLING PROGRAM   CLNQ   14
C                                                                  CLNQ   15
        COMPLEX A,B,C,D,DET,CL                                     CLNQ   16
        DIMENSION A(20,20),B(20),C(20,21),D(20)                    CLNQ   17
C       INITIALIZE SUBROUTINE PARAMETERS                           CLNQ   18
        N1=N+1                                                     CLNQ   19
        N0=N-1                                                     CLNQ   20
        EPS=1.0E-5                                                 CLNQ   21
        DET=(1.0,0.0)                                              CLNQ   22
C       LOAD A AND B INTO WORKING MATRIX C                         CLNQ   23
        DO 10 I=1,N                                                CLNQ   24
        C(I,N1)=B(I)                                               CLNQ   25
        DO 10 J=1,N                                                CLNQ   26
     10 C(I,J)=A(I,J)                                              CLNQ   27
C                                                                  CLNQ   28
C       TRIANGULARIZE SYSTEM OF EQUATIONS                          CLNQ   29
C                                                                  CLNQ   30
        DO 130 IP=1,N                                              CLNQ   31
        IM=IP                                                      CLNQ   32
        IST=IP+1                                                   CLNQ   33
        IF (IST.LE.N) GO TO 20                                     CLNQ   34
C       LAST ROW CALCULATIONS                                      CLNQ   35
        IF (CABS(C(N,N)).LT.EPS) GO TO 40                          CLNQ   36
        C(N,N1)=C(N,N1)/C(N,N)                                     CLNQ   37
        DET=C(N,N)*DET                                             CLNQ   38
        C(N,N)=(1.0,0.0)                                           CLNQ   39
        GO TO 110                                                  CLNQ   40
C       DETERMINE PIVOT ELEMENT AND UPDATE DETERMINANT             CLNQ   41
     20 DO 30 I=IST,N                                              CLNQ   42
        IF (CABS(C(IM,IP)).GE.CABS(C(I,IP))) GO TO 30              CLNQ   43
        IM=I                                                       CLNQ   44
     30 CONTINUE                                                   CLNQ   45
        IF (CABS(C(IM,IP)).GE.EPS) GO TO 70                        CLNQ   46
C       ERROR OUTPUT SECTION                                       CLNQ   47
     40 WRITE (6,50) IP,C(IM,IP)                                   CLNQ   48
     50 FORMAT (1H ,///10X,23HPIVOT ELEMENT IN COLUMN,I3,2H =,1P2E12.5)   CLNQ   49
        IF (CABS(C(IM,IP)).EQ.0.0) WRITE (6,60)                    CLNQ   50
     60 FORMAT (1H0,9X,46HSOLUTION IMPOSSIBLE, DIAGONAL ELEMENT IS ZERO.)   CLNQ   51
        IF (CABS(C(IM,IP)).EQ.0.0) RETURN                         CLNQ   52
     70 DET=C(IM,IP)*DET                                           CLNQ   53
C       TEST FOR DIAGONAL ELEMENT AND SWAP ROWS                    CLNQ   54
        IF (IM.EQ.IP) GO TO 90                                     CLNQ   55
C       SWAP ROWS TO LOCATE PIVOT ELEMENT                          CLNQ   56
        DET=-DET                                                   CLNQ   57
        DO 80 J=IP,N1                                              CLNQ   58
        CL=C(IP,J)                                                 CLNQ   59
        C(IP,J)=C(IM,J)                                            CLNQ   60
     80 C(IM,J)=CL                                                 CLNQ   61
C       FIND MULTIPLICATION CONSTANT                               CLNQ   62
     90 CL=C(IP,IP)                                                CLNQ   63
        C(IP,IP)=(1.0,0.0)                                         CLNQ   64
C       DIVIDE ROW BY PIVOT ELEMENT                                CLNQ   65
        DO 100 J=IST,N1                                            CLNQ   66
    100 C(IP,J)=C(IP,J)/CL                                         CLNQ   67
C       ZERO PIVOT ELEMENT COLUMNM ADJUST MATRIX                   CLNQ   68
    110 DO 130 I=1,N                                               CLNQ   69
        IF (I.EQ.IP) GO TO 130                                     CLNQ   70
        CL=C(I,IP)                                                 CLNQ   71
        C(I,IP)=(0.0,0.0)                                          CLNQ   72
        DO 120 J=IST,N1                                            CLNQ   73
    120 C(I,J)=C(I,J)-CL*C(IP,J)                                   CLNQ   74
```

Figure 7.17—*Cont.*

```
    130 CONTINUE                                              CLNQ  75
C       LOAD D VECTOR = SOLUTION VECTOR X                     CLNQ  76
        DO 140 I=1,N                                          CLNQ  77
    140 D(I)=C(I,N1)                                          CLNQ  78
        RETURN                                                CLNQ  79
        END                                                   CLNQ  80

        6    8    2
        6.28318                    10.0                2.0
C       1    0    3     0.16667
R       2    0    4     4.0
C       3    4    1     0.125
R       4    1    5     3.0
L       5    5    2     2.0
L       6    2    6     5.0
R       7    6    0     5.0
R       8    3    2     4.0
E       1              92.1954       1.35119
E       2              119.269      -0.576395
EX
```

Figure 7.17—Cont.

After all input data are supplied to the program, a complete node voltage analysis is performed (similar to the dc analysis) at a series of fixed frequencies from f_L to f_H in multiplicative increments f_m. The program prints the results at the end of each step, which can produce a rather voluminous amount of output unless we use care in calculating the constant f_m. The value of f_m is obtained from the relationship

$$f_m = \left(\frac{f_H}{f_L}\right)^{1/(n-1)} \tag{7.62}$$

where f_L is the low frequency, f_H is the upper frequency for the analysis, and n is the total number of frequency steps to be used.

The program of Fig. 7.17 does not provide for dependent sources, this implementation being left as an exercise (see Problem 7.20). The initialization of data in Fig. 7.17 does provide for the dependent sources with the following identifications: TC, transconductance (g_m); TF, transfluence (α); TP, transpotential (μ); TR, transresistance (r_m).

7.6 Mesh and Loop Analysis

Classical network-analysis textbooks usually place greater emphasis on loop (or mesh)† analysis than on nodal analysis. In practice, both loop and nodal methods

---

†As demonstrated in this section, there are important differences between loop- and mesh-analysis methods. However, they are similar enough in concept that we combine them in this introductory section.

are useful in the analysis of networks. The choice of the analysis method is governed in almost all cases by a comparison of the number of nodes and the number of loops in the network. For example, if the number of loops is less than the number of nodes, obviously, loop analysis is easier to apply.

There is very little difference between the implementation of loop- and node-analysis methods on the digital computer, and the topological development of the network loop equations follows closely the derivations made earlier in this chapter for nodal analysis. One of the primary differences is in the relationship between the input data and the network equations. In the last section, we supplied branch data for the nodal analysis in a natural form as follows:

> branch type—branch number—initial node—final node—branch parameter value

The element type, the topological connections, and the element value are supplied to provide a complete picture of the element and its relationship to the network. The data supplied also directly provide the information required by the network incidence matrix.

However, we note that loop equations have no obvious relationship to the network nodes. We are forced to sketch the loop currents on the network drawing to maintain some sanity while writing loop equations by hand. The first problem we must face in the automated formulation of loop equations then becomes: How can we present the branch data to the program so that the data will be meaningful in setting up the network loop equations?

If we agree that the input data format previously used for nodal analysis is the best of all approaches, we must then solve the problem: How can the computer select an independent set of network loops from the topology of the network? The second problem we encounter in the formulation of the network loop equations is illustrated in Fig. 7.18. In nodal analysis, we found n_t nodes of which $n = n_t - 1$ nodes had independent voltages. We can see from Fig. 7.18 that there are seven possible loops, of which only three are independent.

We can ignore both problems for the present since neither problem affects the formulation of the network loop equations. We will then solve both problems in the formulation of a dc network analysis program based upon the loop method of analysis.

7.7 Mesh Analysis of Linear Networks

As an alternative to the previous method of graph description, an oriented network graph can be completely described by specifying the orientation of each branch and the meshes to which the branch is incident. We now define additional terms used to describe the network graph in this way.

Planar. A graph is said to be planar if it can be drawn on a two-dimensional surface in such a way that no two branches intersect at a point that is not a node. An

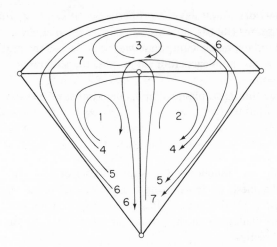

Figure 7.18 Possible network loops in a network of four nodes.

example of a planar graph is a cube in which the edges of the cube are branches. The cube becomes nonplanar if the interior diagonals become branches, as shown in Fig. 7.19.

Mesh. Consider the planar graph of Fig. 7.20. We call any loop of this graph for which there is *no* branch in its interior a mesh. For example, the loop consisting of branches 1, 4, and 3 is not a mesh, whereas loop 2, 4, 3 is a mesh. If the loop has no branches to its exterior, it is an *outer mesh*.

Figure 7.19 Planar and non-planar graphs.

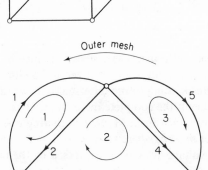

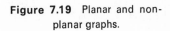

Figure 7.20 Planar oriented graph.

Hinged. If a network graph has the property that it can be partitioned into two nondegenerate subgraphs (that is, not an isolated node) which are connected by only one node, it is said to be hinged. Thus, an *unhinged graph* has the property that where it is partitioned into two nondegenerate subgraphs, the subgraphs will have at least two nodes in common.

For a connected unhinged planar graph, the number of meshes m (not counting the outer mesh) is equal to

$$m = b - n_t + 1 \tag{7.63}$$

where b is the number of branches and n_t is the number of nodes in the network. We arbitrarily select the mesh currents using a clockwise reference direction. This is analogous to selecting all the node voltages positive with respect to ground. The outer mesh becomes analogous to the ground or reference node and is selected in a counterclockwise direction.

For a graph having m_t meshes and b branches, the complete mesh–branch incidence matrix, $\mathbf{M}_a = [M_{ij}]_{m_t,b}$ is a rectangular matrix whose (i, j)th element is defined by the following:

$$M_{ij} = \begin{cases} 1 & \text{if branch } j \text{ is in mesh } i \text{ and their reference directions coincide} \\ -1 & \text{if branch } j \text{ is in mesh } i \text{ and their reference directions do not} \\ & \text{coincide} \\ 0 & \text{if branch } j \text{ is not in mesh } i \end{cases}$$

Since each branch can be common to only two meshes, each column in matrix $\mathbf{M}_a$ contains only a single 1 and a single -1 with all other elements equal to zero. The complete mesh matrix for the graph in Fig. 7.20 is

$$
\begin{array}{c}
\text{mesh} \\
\downarrow
\end{array}
\quad
\begin{array}{ccccc}
1 & 2 & 3 & 4 & 5 \leftarrow \text{branch}
\end{array}
$$

$$\mathbf{M}_a = \begin{array}{c} 1 \\ 2 \\ 3 \\ 4 \end{array} \begin{bmatrix} 1 & 1 & 0 & 0 & 0 \\ 0 & -1 & 1 & 1 & 0 \\ 0 & 0 & 0 & -1 & 1 \\ -1 & 0 & -1 & 0 & -1 \end{bmatrix} \tag{7.64}$$

The complete mesh matrix is degenerate since if all rows are added to the last row of the matrix the result is a row of zeros. To eliminate the degeneracy, we eliminate the row corresponding to the outer mesh as we did with the row corresponding to the ground node in the complete incidence matrix in Section 7.2.

The matrix obtained from $\mathbf{M}_a$ by eliminating one row is called the *mesh matrix* $\mathbf{M}$ and is of order $m \times b$, where $m = m_t - 1 = b - n_t + 1$.

Kirchhoff's voltage law can be expressed by the use of the mesh matrix as

$$\mathbf{M}\mathbf{v}_b = 0 \tag{7.65}$$

or the sum of the branch voltages around a closed path is equal to zero. The positive reference voltage of the branches corresponds to the orientation of the graph branches. For the graph in Fig. 7.20 KVL is expressed as

$$\mathbf{M}\mathbf{v}_b = \mathbf{0}$$

$$\begin{bmatrix} 1 & 1 & 0 & 0 & 0 \\ 0 & -1 & 1 & 1 & 0 \\ 0 & 0 & 0 & -1 & 1 \end{bmatrix} \begin{bmatrix} v_{b1} \\ v_{b2} \\ v_{b3} \\ v_{b4} \\ v_{b5} \end{bmatrix} = \begin{bmatrix} 0 \\ 0 \\ 0 \\ 0 \\ 0 \end{bmatrix}$$

or

$$v_{b1} + v_{b2} = 0 \qquad (7.66a)$$

$$-v_{b2} + v_{b3} + v_{b4} = 0 \qquad (7.66b)$$

$$-v_{b4} + v_{b5} = 0 \qquad (7.66c)$$

In this case we see that Eqs. (7.66) are linearly independent and clearly represent the equations obtained by applying KVL to each mesh of the graph.

Let us define $\mathbf{i}_m$ to be the vector of mesh currents $i_{m1}, i_{m2}, \ldots, i_{m1}$, in which each component is selected with a clockwise reference direction. Kirchhoff's current law states that the current in any branch of the network must be a linear combination of the mesh currents associated with that branch. Thus, we assert that KCL can be represented by the relationship

$$\mathbf{i}_b = \mathbf{M}^{\mathsf{T}}\mathbf{i}_m \qquad (7.67)$$

where the $b \times m$ matrix $\mathbf{M}^{\mathsf{T}}$ is the transpose of the matrix $\mathbf{M}$.

To demonstrate this, we again refer to the network of Fig. 7.20 and the corresponding mesh matrix

$$\begin{matrix} & & & & & & \text{mesh} \\ & & & & & & \downarrow \\ \mathbf{M} = & \begin{bmatrix} 1 & 1 & 0 & 0 & 0 \\ 0 & -1 & 1 & 1 & 0 \\ 0 & 0 & 0 & -1 & 1 \end{bmatrix} & \begin{matrix} 1 \\ 2 \\ 3 \end{matrix} \\ \text{branch} \to & 1 \quad 2 \quad 3 \quad 4 \quad 5 \end{matrix} \qquad (7.68)$$

Applying KCL, we find

$$\mathbf{i}_b = \mathbf{M}^{\mathsf{T}}\mathbf{i}_m$$

$$= \begin{bmatrix} 1 & 0 & 0 \\ 1 & -1 & 0 \\ 0 & 1 & 0 \\ 0 & 1 & -1 \\ 0 & 0 & 1 \end{bmatrix} \begin{bmatrix} i_{m1} \\ i_{m2} \\ i_{m3} \end{bmatrix} \qquad (7.69)$$

or

$$i_{b1} = i_{m1} \tag{7.70a}$$

$$i_{b2} = i_{m1} - i_{m2} \tag{7.70b}$$

$$i_{b3} = i_{m2} \tag{7.70c}$$

$$i_{b4} = i_{m2} - i_{m3} \tag{7.70d}$$

$$i_{b5} = i_{m3} \tag{7.70e}$$

The five scalar equations in Eq. (7.70) are expressions of KCL for the network graph.
We can summarize by saying that

$$\mathbf{M}\mathbf{v}_b = \mathbf{0} \qquad \text{(KVL)}$$

and

$$\mathbf{i}_b = \mathbf{M}^T \mathbf{i}_m \qquad \text{(KCL)}$$

are the two basic equations of mesh analysis. These equations are obtained directly
from Kirchhoff's laws and the oriented network graph and are independent of the
network elements. A complete expression of the network equations requires a knowledge of the elements in the network (definition of the network branch) or a set of
b independent branch equations which relate the branch currents $\mathbf{i}_b$ to the branch
voltages $\mathbf{v}_b$.

As a first step in the mesh equation development, we assume a simplified branch
identical to the branch in Fig. 7.5. Using the orientation of branch elements shown
in Fig. 7.5, the branch voltage across the simplified branch k is

$$\begin{aligned}
v_{bk} &= v_{ek} - E_k \\
&= R_k i_{ek} - E_k \\
&= R_k(i_{bk} + J_k) - E_k \\
&= R_k i_{bk} - R_k J_k - E_k
\end{aligned} \tag{7.71}$$

Equation (7.71) defines branch voltage v_{bk} in terms of the branch current i_{bk} and the
source voltages and currents.

In matrix notation the network branch voltage is expressed as

$$\mathbf{v}_b = \mathbf{R}\mathbf{i}_b - \mathbf{E} + \mathbf{R}\mathbf{J} \tag{7.72}$$

where $\mathbf{R}$ is the branch resistance matrix, $\mathbf{J}$ is the branch independent-current source
vector, and $\mathbf{E}$ is the branch independent-voltage source vector. The matrix $\mathbf{R}$ is a diagonal matrix of order b given by

$$\mathbf{R} = \begin{bmatrix} R_1 & 0 & \cdots & & 0 \\ 0 & R_2 & & & \\ \vdots & & R_3 & & \\ \vdots & & & \ddots & 0 \\ 0 & \cdots & & 0 & R_b \end{bmatrix}$$

where $R_1, R_2, \ldots, R_b$ are the branch resistances in the order indicated by the oriented network graph. Both vectors $\mathbf{J}$ and $\mathbf{E}$ are of order b, in which positive entries correspond to sources in the general branch oriented in the directions shown in Fig. 7.5. Sources with polarities opposite the assumed directions enter the source vectors with negative signs. Zeros are entered for branches containing no sources.

We now combine Eqs. (7.65), (7.67), and (7.72) to obtain a matrix equation describing the network in terms of the mesh current $\mathbf{i}_m$.

First, premultiplying Eq. (7.72) by $\mathbf{M}$, we obtain

$$\mathbf{M}\mathbf{v}_b = \mathbf{M}\mathbf{R}\mathbf{i}_b - \mathbf{M}\mathbf{E} + \mathbf{M}\mathbf{R}\mathbf{J}$$
$$= \mathbf{M}\mathbf{R}\mathbf{i}_b - \mathbf{M}(\mathbf{E} - \mathbf{R}\mathbf{J})$$

Using Eq. (7.65) we obtain the equation

$$\mathbf{M}\mathbf{R}\mathbf{i}_b = \mathbf{M}(\mathbf{E} - \mathbf{R}\mathbf{J}) \tag{7.73}$$

Substituting Eq. (7.67) into Eq. (7.73), we get

$$\mathbf{M}\mathbf{R}\mathbf{M}^\mathsf{T}\mathbf{i}_m = \mathbf{M}(\mathbf{E} - \mathbf{R}\mathbf{J}) \tag{7.74}$$

or

$$\mathbf{i}_m = (\mathbf{M}\mathbf{R}\mathbf{M}^\mathsf{T})^{-1}\mathbf{M}(\mathbf{E} - \mathbf{R}\mathbf{J}) \tag{7.75}$$

In Eq. (7.75) $\mathbf{M}\mathbf{R}\mathbf{M}^\mathsf{T}$ is an $m \times m$ square matrix and $\mathbf{M}(\mathbf{E} - \mathbf{R}\mathbf{J})$ is a vector of order $\mathbf{m}$. Using the notation

$$\mathbf{R}_m = \mathbf{M}\mathbf{R}\mathbf{M}^\mathsf{T} \qquad \text{(mesh resistance matrix)} \tag{7.76a}$$

$$\mathbf{E}_m = \mathbf{M}(\mathbf{E} - \mathbf{R}\mathbf{J}) \qquad \text{(equivalent voltage vector)} \tag{7.76b}$$

then Eq. (7.75) becomes

$$\mathbf{R}_m\mathbf{i}_m = \mathbf{E}_m \tag{7.77}$$

which is identical to the results obtained in Chapter 4 and is an expression of Ohm's law.

EXAMPLE 7.6

Write the mesh equations for the network of Fig. 7.6 using Eq. (7.74).

The mesh matrix for the network is

$$\mathbf{M} = \begin{bmatrix} 1 & 1 & 0 & 0 \\ 1 & -1 & 1 & 1 \end{bmatrix}$$

and the branch resistance matrix is

$$\mathbf{R} = \begin{bmatrix} 1 & 0 & 0 & 0 \\ 0 & 0.5 & 0 & 0 \\ 0 & 0 & 1 & 0 \\ 0 & 0 & 0 & 0.333 \end{bmatrix}$$

The branch source vectors are

$$\mathbf{J} = \begin{bmatrix} 0 \\ 1 \\ 0 \\ 0 \end{bmatrix} \quad \mathbf{E} = \begin{bmatrix} 10 \\ 0 \\ 0 \\ -20 \end{bmatrix}$$

Combining the matrices $\mathbf{M}$ and $\mathbf{R}$ to obtain $\mathbf{MRM^T}$, we have

$$\mathbf{R}_m = \mathbf{MRM^T}$$

$$= \begin{bmatrix} 1 & 1 & 0 & 0 \\ 0 & -1 & 1 & 1 \end{bmatrix} \begin{bmatrix} 1 & 0 & 0 & 0 \\ 0 & 0.5 & 0 & 0 \\ 0 & 0 & 1 & 0 \\ 0 & 0 & 0 & 0.333 \end{bmatrix} \begin{bmatrix} 1 & 0 \\ 1 & -1 \\ 0 & 1 \\ 0 & 1 \end{bmatrix}$$

$$= \begin{bmatrix} 1.5 & -0.5 \\ -0.5 & 1.833 \end{bmatrix}$$

The equivalent voltage source vector $\mathbf{E}_m$ is

$$\mathbf{E}_m = \mathbf{M(E - RJ)}$$

$$= \begin{bmatrix} 1 & 1 & 0 & 0 \\ 0 & -1 & 1 & 1 \end{bmatrix} \left\{ \begin{bmatrix} 10 \\ 0 \\ 0 \\ -20 \end{bmatrix} - \begin{bmatrix} 1 & 0 & 0 & 0 \\ 0 & 0.5 & 0 & 0 \\ 0 & 0 & 1 & 0 \\ 0 & 0 & 0 & 0.333 \end{bmatrix} \begin{bmatrix} 0 \\ 1 \\ 0 \\ 0 \end{bmatrix} \right\}$$

$$= \begin{bmatrix} 9.5 \\ -19.5 \end{bmatrix}$$

The resulting set of mesh equations in matrix form for the network of Fig. 7.6 is

$$\begin{bmatrix} 1.5 & -0.5 \\ -0.5 & 1.833 \end{bmatrix} \begin{bmatrix} i_{m1} \\ i_{m2} \end{bmatrix} = \begin{bmatrix} 9.5 \\ -19.5 \end{bmatrix}$$

which is identical to the equations we would obtain using the classical methods described in Chapter 4. The mesh currents are $\mathbf{i}_m = [3.066 \quad -9.8]^T$.

The branch illustrated in Fig. 7.5 can be extended as shown in Fig. 7.9 so that it is completely general. The branch voltage $\mathbf{v}_{bk}$ across general branch k is

$$v_{bk} = v_{ek} - E_{dk} - E_k \tag{7.78}$$
$$= R_k i_{ek} - E_{dk} - E_k$$

The element current through branch k is

$$i_{ek} = i_{bk} + J_{dk} + J_k$$

where

$$J_{dk} = \sum_{j=1}^{b} \alpha_{kj} i_{ej} + \sum_{j=1}^{b} g_{mkj} v_{ej}$$
$$= \sum_{j=1}^{b} (\alpha_{kj} + g_{mkj} R_j) i_{ej}$$

The dependent current sources J_{dk} can be functions of as many as b element currents and b element voltages or $2b$ quantities. The element current i_{ek} is then

$$i_{ek} = \left[1 + \sum_{j=1}^{b} (\alpha_{kj} + g_{m_{kj}} R_j) \right]^{-1} (i_{bk} + J_k)$$

The dependent voltage source E_{dk} is given by

$$E_{dk} = \sum_{j=1}^{b} (r_{m_{kj}} + \mu_{kj} R_j) i_{ej}$$

so the branch voltage in branch k is

$$\mathbf{v}_{bk} = \left[R_k - \sum_{j=1}^{b} (r_{m_{kj}} + \mu_{kj} R_j) \right] \left[1 + \sum_{j=1}^{b} (\alpha_{kj} + g_{m_{kj}} R_j) \right]^{-1} (i_{bk} + J_k) - E_k \tag{7.79}$$

The relationship between the branch voltage and branch current becomes so complex that many circuit-analysis methods based on the mesh approach limit the dependent source capability to only dependent voltage sources. This limitation eliminates the α and g_m terms from Eq. (7.79), reducing it to

$$\mathbf{v}_{bk} = \left[R_k - \sum_{j=1}^{b} (r_{m_{kj}} + \mu_{kj} R_j) \right] (i_{bk} + J_k) - E_k \tag{7.80}$$

In matrix form the branch voltage–current relationship is somewhat easier to visualize. First, we let

$$\mathbf{v}_b = \mathbf{R} \mathbf{i}_e - \mathbf{E}_d - \mathbf{E} \tag{7.81}$$

and

$$\mathbf{i}_e = \mathbf{i}_b + \mathbf{J}_d + \mathbf{J} \tag{7.82}$$

The dependent current source vector $\mathbf{J}_d$ is given by

$$\begin{aligned} \mathbf{J}_d &= [\alpha] \mathbf{i}_e - [g_m] \mathbf{v}_e \\ &= [\alpha] \mathbf{i}_e + [g_m] \mathbf{R} \mathbf{i}_e \\ &= \mathbf{F} \mathbf{i}_e \end{aligned} \tag{7.83}$$

where $\mathbf{F} = [\alpha] + [g_m] \mathbf{R}$ and matrices $[\alpha]$ and $[g_m]$ are defined in Eq. (7.26). Combining Eqs. (7.82) and (7.83), we find

$$\mathbf{i}_b + \mathbf{J} = (\mathbf{I} - \mathbf{F}) \mathbf{i}_e$$

or

$$\mathbf{i}_e = (\mathbf{I} - \mathbf{F})^{-1} (\mathbf{i}_b + \mathbf{J}) \tag{7.84}$$

The dependent voltage source vector given by

$$\mathbf{E}_d = \{ [r_m] + [\mu] \mathbf{R} \} \mathbf{i}_e \tag{7.85}$$

is then inserted into Eq. (7.81) to give

$$\mathbf{v}_b = \{\mathbf{R} - [r_m] - [\mu]\mathbf{R}\}\mathbf{i}_e - \mathbf{E} \tag{7.86}$$

Substituting Eq. (7.84) into Eq. (7.86) gives the desired relationship between branch voltage and current:

$$\mathbf{v}_b = \{\mathbf{R} - [r_m] - [\mu]\mathbf{R}\}(\mathbf{I} - \mathbf{F})^{-1}(\mathbf{i}_b + \mathbf{J}) - \mathbf{E} \tag{7.87}$$

If we let $\mathbf{R}_2 = \{\mathbf{R} - [r_m] - [\mu]\mathbf{R}\}(\mathbf{I} - \mathbf{F})^{-1}$ the branch voltage expression reduces to

$$\mathbf{v}_b = \mathbf{R}_2\mathbf{i}_b + \mathbf{R}_2\mathbf{J} - \mathbf{E} \tag{7.88}$$

Kirchhoff's voltage law applied to Eq. (7.88) yields

$$\begin{aligned} \mathbf{M}\mathbf{v}_b &= \mathbf{M}\mathbf{R}_2\mathbf{i}_b - \mathbf{M}(\mathbf{E} - \mathbf{R}_2\mathbf{J}) \\ &= \mathbf{0} \end{aligned} \tag{7.89}$$

Using KCL, we finally obtain the equation

$$\mathbf{M}\mathbf{R}_2\mathbf{M}^{\mathsf{T}}\mathbf{i}_m = \mathbf{M}(\mathbf{E} - \mathbf{R}_2\mathbf{J}) \tag{7.90}$$

or

$$\mathbf{i}_m = (\mathbf{M}\mathbf{R}_2\mathbf{M}^{\mathsf{T}})^{-1}\mathbf{M}(\mathbf{E} - \mathbf{R}_2\mathbf{J}) \tag{7.91}$$

Equations (7.90) and (7.91) define the mesh equations for a general resistive network which may contain any or all of the four types of dependent voltage and current sources.

EXAMPLE 7.7

Consider the circuit shown in Fig. 7.11. Write the mesh equations in matrix form using the topological techniques described in this section.

The oriented graph for the circuit is shown in Fig. 7.21. From this graph the mesh matrix is

$$\mathbf{M} = \begin{bmatrix} -1 & -1 & 0 & 0 & 0 \\ 0 & 1 & 1 & -1 & 0 \\ 0 & 0 & 0 & 1 & 1 \end{bmatrix}$$

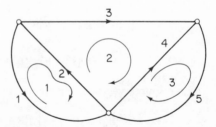

Figure 7.21 Oriented graph for the network of Fig. 7.11.

The branch resistance matrix $\mathbf{R}_1$ is

$$\mathbf{R}_1 = \mathbf{R} - [r_m] - [\mu]\mathbf{R}$$

$$= \begin{bmatrix} 15 & 0 & 0 & 0 & 0 \\ 4.5 & 5 & 0 & 0 & 0 \\ 0 & 0 & 1 & 0 & 0 \\ 0 & 0.5 & 0 & 2 & 0 \\ 0 & 0 & 0 & 0 & 3 \end{bmatrix}$$

The dependent current source strength matrix $\mathbf{F}$ combines with the unit matrix $\mathbf{I}$ to form

$$\mathbf{I} - \mathbf{F} = \begin{bmatrix} 1 & 0 & 0.4 & 0 & 0 \\ 0 & 1 & 0 & 0 & 0 \\ 0 & 0 & 1 & 0 & 0 \\ 0 & 0 & 0 & 1 & 1.8 \\ 0 & 0 & 0 & 0 & 1 \end{bmatrix}$$

Notice that the entries in the $\mathbf{F}$ matrix for the dependent current sources are negative if the direction is opposite that assumed in the general branch of Fig. 7.9. The inverse of $(\mathbf{I} - \mathbf{F})$ is

$$(\mathbf{I} - \mathbf{F})^{-1} = \begin{bmatrix} 1 & 0 & -0.4 & 0 & 0 \\ 0 & 1 & 0 & 0 & 0 \\ 0 & 0 & 1 & 0 & 0 \\ 0 & 0 & 0 & 1 & -1.8 \\ 0 & 0 & 0 & 0 & 1 \end{bmatrix}$$

The composite branch resistance matrix $\mathbf{R}_2$ is next calculated and is given by

$$\mathbf{R}_2 = \mathbf{R}_1(\mathbf{I} - \mathbf{F})^{-1}$$

$$= \begin{bmatrix} 15 & 0 & -6 & 0 & 0 \\ 4.5 & 5 & -1.8 & 0 & 0 \\ 0 & 0 & 1 & 0 & 0 \\ 0 & 0.5 & 0 & 2 & -3.6 \\ 0 & 0 & 0 & 0 & 3 \end{bmatrix}$$

The mesh resistance matrix $\mathbf{R}_m$ is then formed from the matrix product $\mathbf{MR}_2\mathbf{M}^\mathsf{T}$ so that

$$\mathbf{R}_m = \begin{bmatrix} -1 & -1 & 0 & 0 & 0 \\ 0 & 1 & 1 & -1 & 0 \\ 0 & 0 & 0 & 1 & 1 \end{bmatrix} \begin{bmatrix} 15 & 0 & -6 & 0 & 0 \\ 4.5 & 5 & -1.8 & 0 & 0 \\ 0 & 0 & 1 & 0 & 0 \\ 0 & 0.5 & 0 & 2 & -3.6 \\ 0 & 0 & 0 & 0 & 3 \end{bmatrix} \begin{bmatrix} -1 & 0 & 0 \\ -1 & 1 & 0 \\ 0 & 1 & 0 \\ 0 & -1 & 1 \\ 0 & 0 & 1 \end{bmatrix}$$

$$= \begin{bmatrix} 24.5 & 2.8 & 0 \\ -9 & 5.3 & 1.6 \\ -0.5 & -1.5 & 1.4 \end{bmatrix}$$

The equivalent voltage vector $\mathbf{E}_m$ is given by

$$\mathbf{E}_m = \mathbf{ME}$$

$$= \begin{bmatrix} -1 & -1 & 0 & 0 & 0 \\ 0 & 1 & 1 & -1 & 0 \\ 0 & 0 & 0 & 1 & 1 \end{bmatrix} \begin{bmatrix} -10 \\ 0 \\ 0 \\ 0 \\ 0 \end{bmatrix}$$

$$= \begin{bmatrix} 10 \\ 0 \\ 0 \end{bmatrix}$$

The resulting mesh equations are then

$$\begin{bmatrix} 24.5 & 2.8 & 0 \\ -9 & 5.7 & -2 \\ -0.5 & -1.5 & 1.4 \end{bmatrix} \begin{bmatrix} i_{m1} \\ i_{m2} \\ i_{m3} \end{bmatrix} = \begin{bmatrix} 10 \\ 0 \\ 0 \end{bmatrix}$$

Again the need for a computer algorithm is apparent. The mesh currents are $\mathbf{i}_m = [0.361 \quad 0.4106 \quad 0.569]^\mathsf{T}$.

A dc analysis program for linear networks can be written based upon Eq. (7.91), which is analogous to the program shown in Fig. 7.8. The mesh currents are unknown variables rather than node voltages, and Kirchhoff's laws are based upon the mesh matrix $\mathbf{M}$ instead of the incidence matrix $\mathbf{A}$. In nodal-analysis programs we specify a branch by giving the branch number, its topological connections, and the elements contained in it. For example, a branch 7 can be specified in one input terminology as

B7 N(1,2), R = 5, E = −2

This branch is oriented from node 1 to node 2 and contains a 5-unit† resistance and a 2-unit voltage source oriented in the branch opposite the source orientation in the general branch definition. If we limit ourselves to analyzing planar networks, we might then propose a mesh input scheme such as

B7 M(1,2), R = 5, E = −2

in which branch 7, oriented in phase with mesh 1 and opposite mesh 2, contains a 5-unit resistance and a 2-unit voltage source as described above. A *nonplanar* network is invalid with this type of input because a given branch may coincide with more than two meshes. (Consider a network whose graph is shown in Fig. 7.19.)

---

†The user might assume one of several compatible sets of units. For example, in volt-amp-sec units, R is specified in ohms, C in farads, L in henries. In volt-mA-ns units, R is in kΩ, C in pF, and L in μH.

7.8 Loop Analysis of Linear Networks

In the preceding section we indicated that, although mesh-analysis methods are useful, mesh analysis is only valid for a class of networks with planar graphs. Most networks are planar, but the restriction which prohibits the analysis of nonplanar circuits motivates us to search for another analysis approach.

Recalling the fact that every link branch forms an independent loop in a graph (Section 7.1), we return to graph theory to formulate a method to write loop equations utilizing this property of a link branch.

Consider a connected graph with b branches and n_t nodes. Select an arbitrary tree T. The tree will contain $n = n_t - 1$ tree branches and $l = b - n$ link branches. Every link defines a fundamental *loop* of the network, that is, the loop formed by the link and the unique tree path between the nodes of the links. For example, the graph in Fig. 7.22 presents the $b - n = 3$ fundamental loops of the tree (abe). The loops

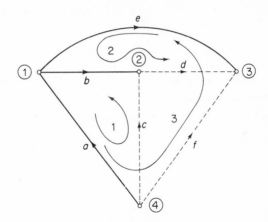

Figure 7.22 Fundamental loops for tree (abe).

do not correspond to the meshes for the graph but do form an independent set of equations. The selection of tree (bcd) gives another set of equations, which corresponds to the mesh currents (except for reference direction). We have allowed the loop currents to cross branches of the graph, a condition we did not allow in the mesh-equation development. Removing this restriction allows us to consider nonplanar graphs.

In order to apply KVL to each fundamental loop, we adopt a reference direction for the loop which coincides with the reference direction of the link defining the loop. For the network in Fig. 7.22,

Loop 1: $v_c - v_b - v_a = 0$
Loop 2: $v_d - v_e + v_b = 0$
Loop 3: $v_f - v_e - v_a = 0$

or

$$
\begin{array}{c}
\text{loop} \\
\downarrow
\end{array}
\begin{array}{c}
1 \\
2 \\
3 \\
\end{array}
\begin{bmatrix}
-1 & -1 & 1 & 0 & 0 & 0 \\
0 & 1 & 0 & 1 & -1 & 0 \\
-1 & 0 & 0 & 0 & 1 & 1
\end{bmatrix}
\begin{array}{c}
\\ \text{branch} \to a \quad b \quad c \quad d \quad e \quad f
\end{array}
\begin{bmatrix}
v_a \\ v_b \\ v_c \\ v_d \\ v_e \\ v_f
\end{bmatrix}
=
\begin{bmatrix}
0 \\ 0 \\ 0 \\ 0 \\ 0 \\ 0
\end{bmatrix}
$$

or

$$\mathbf{B}_f \mathbf{v}_b = \mathbf{0} \qquad \text{(KVL)} \tag{7.92}$$

where $\mathbf{B}_f$ is an $l \times b$ matrix called the *fundamental loop matrix*. The (i, j) element of the matrix is defined to be

$$
b_{ij} =
\begin{cases}
1 & \text{if branch } j \text{ is in loop } i \text{ and their reference directions coincide} \\
-1 & \text{if branch } j \text{ is in loop } i \text{ and their reference directions do not coincide} \\
0 & \text{if branch } j \text{ is not in loop } i
\end{cases}
$$

Since each fundamental loop contains only one link branch and the orientation of the loop and link agree, we can partition the link and tree branches so that

$$\mathbf{B}_f = [\mathbf{B}_t \mid \mathbf{B}_l] \tag{7.93}$$

$$= [\mathbf{H} \mid \mathbf{I}] \tag{7.94}$$

where $\mathbf{B}_t = \mathbf{H}$. For the loop matrix corresponding to Fig. 7.22, the rearranged and partitioned loop matrix is

$$
\mathbf{B}_f =
\begin{array}{c}
\text{loop} \\
\downarrow
\end{array}
\begin{array}{c}
1 \\
2 \\
3
\end{array}
\left[
\begin{array}{ccc|ccc}
-1 & -1 & 0 & 1 & 0 & 0 \\
0 & 1 & -1 & 0 & 1 & 0 \\
-1 & 0 & -1 & 0 & 0 & 1
\end{array}
\right]
$$

$$
\text{branch} \to \underbrace{a \quad b \quad e}_{\substack{n \text{ tree} \\ \text{branches}}} \quad \underbrace{c \quad d \quad f}_{l \text{ links}}
$$

Kirchhoff's current law from the fundamental loop matrix $\mathbf{B}_f$ is

$$\mathbf{i}_b = \mathbf{B}_f^{\mathsf{T}} \mathbf{i}_l \qquad \text{(KCL)} \tag{7.95}$$

which is easily verified and corresponds to the mesh representation given in Eqs. (7.65) and (7.67). We can now rewrite the mesh equations [Eq. (7.91)] in terms of fundamental loop currents as

$$\mathbf{i}_l = (\mathbf{B}_f\mathbf{R}_2\mathbf{B}_f^\mathsf{T})^{-1}\mathbf{B}_f(\mathbf{E} - \mathbf{R}_2\mathbf{J}) \tag{7.96}$$

Note that the elements in matrix $\mathbf{R}_2$ and vectors $\mathbf{E}$ and $\mathbf{J}$ must correspond to the branch arrangement in the columns of matrix $\mathbf{B}_f$. We have not discarded any of the mesh-equation development but have extended the results to nonplanar networks using a "link" approach.

7.9 DC Loop-Analysis Program for Linear Networks

A dc analysis program based upon loop-analysis methods appears to be more versatile than a program based upon mesh methods. As usual, we do not obtain something for nothing. The loop-analysis method forces us to supply information defining the network tree (unnecessary in the mesh approach). Realistically we cannot require that the loops associated with each branch be specified [for example, B6 L$(-1, -2, 3)$, R = 7 to define branch 6 and its relationship to loops 1, 2, and 3 in Fig. 7.22]. However, if we can demonstrate a relationship between the network topology (incidence matrix $\mathbf{A}$) and the fundamental loop matrix $\mathbf{B}_f$, we can use the input data format from the dc analysis program based upon nodal methods. The nodal-analysis input data format is much more natural to use because we do not have to sketch a tree and establish loop orientations, nor do we have to allow for several loop currents common to a single branch.

The existence of a relationship between $\mathbf{A}$ and $\mathbf{B}_f$ can be demonstrated as follows: First, we are given that

$$\mathbf{B}_f\mathbf{v}_b = 0 \qquad \text{(KVL)} \tag{7.97}$$

Assuming that $\mathbf{v}_b$ is partitioned into $[\mathbf{v}_{bt} \mid \mathbf{v}_{bl}]^\mathsf{T}$, where $\mathbf{v}_{bl}$ contains the link branch voltages and $\mathbf{v}_{bt}$ contains the tree branch voltages, we can rearrange $\mathbf{A}$ so that the columns of $\mathbf{A}$ correspond to the same ordering without invalidating the relation

$$\mathbf{v}_b = \mathbf{A}^\mathsf{T}\mathbf{v}_n \tag{7.98}$$

Substituting Eq. (7.98) into Eq. (7.97), we obtain

$$\mathbf{B}_f\mathbf{A}^\mathsf{T}\mathbf{v}_n = 0 \tag{7.99}$$

for all $\mathbf{v}_n$. This implies that

$$\mathbf{B}_f\mathbf{A}^\mathsf{T} = 0 \tag{7.100}$$

We have partitioned $\mathbf{A}$ into the form

$$\mathbf{A} = [\underset{\text{tree}}{\underbrace{\mathbf{A}_t}} \mid \underset{\text{link}}{\underbrace{\mathbf{A}_l}}] \tag{7.101}$$

to conform with the multiplication of Eq. (7.100).

Kirchhoff's current law can be written as

$$[\mathbf{A}_t \mid \mathbf{A}_l]\begin{bmatrix} \mathbf{i}_{bt} \\ \mathbf{i}_{bl} \end{bmatrix} = \mathbf{A}_t\mathbf{i}_{bt} + \mathbf{A}_l\mathbf{i}_{bl} = 0$$

Then

$$\mathbf{A}_t\mathbf{i}_{bt} = -\mathbf{A}_l\mathbf{i}_{bl}$$

or

$$\mathbf{i}_{bt} = -\mathbf{A}_t^{-1}\mathbf{A}_l\mathbf{i}_{bl}$$

In matrix form, we have

$$[\mathbf{I} \mid \mathbf{A}_t^{-1}\mathbf{A}_l]\mathbf{i}_b = 0 \qquad (7.102)$$

[We will learn later that Eq. (7.102) involves the fundamental cut-set matrix $\mathbf{Q}$. The proper name is unimportant at this time.] Equation (7.100) can be rewritten as

$$[\mathbf{H} \mid \mathbf{I}]\begin{bmatrix} \mathbf{I} \\ \hline \mathbf{X}^\mathsf{T} \end{bmatrix} = \begin{bmatrix} \mathbf{0} \\ \hline \mathbf{0} \end{bmatrix}$$

where $\mathbf{X} = \mathbf{A}_t^{-1}\mathbf{A}_l$ or

$$\mathbf{HI} + \mathbf{IX}^\mathsf{T} = 0$$
$$\mathbf{H} + \mathbf{X}^\mathsf{T} = 0$$

Hence,

$$\mathbf{H} = -\mathbf{X}^\mathsf{T}$$

and transposing,

$$-\mathbf{H}^\mathsf{T} = \mathbf{A}_t^{-1}\mathbf{A}_l \qquad (7.103)$$

The fundamental loop matrix $\mathbf{B}_f$ is then given in terms of the incidence matrix by

$$\mathbf{B}_f = [-(\mathbf{A}_t^{-1}\mathbf{A}_l)^\mathsf{T} \mid \mathbf{I}] \qquad (7.104)$$

Using the results of Eq. (7.104) we see that the input could have a form identified to the data used by the nodal-analysis programs; that is, the input data could be used to directly construct the $\mathbf{A}$ matrix.

After the $\mathbf{A}$ matrix is constructed from the input data, it must be partitioned into $[\mathbf{A}_t \mid \mathbf{A}_l]$ by either specifying the tree branches with lower numbers or by swapping columns. The partitioned matrix is then manipulated into the form $[\mathbf{I} \mid \mathbf{A}_t^{-1}\mathbf{A}_l]$ using elementary row operations so that the $\mathbf{B}_f = [-(\mathbf{A}_t^{-1}\mathbf{A}_l)^\mathsf{T} \mid \mathbf{I}]$ matrix can be formed.

The algorithm shown in Fig. 7.23 can be used to partition $\mathbf{A}$ into $[\mathbf{A}_t \mid \mathbf{A}_l]$.

Once the matrix partitioning algorithm is complete, the new partitioned matrix $\mathbf{A}_p = [\mathbf{A}_t \mid \mathbf{A}_l]$ must be diagonalized to form

$$\begin{aligned} \mathbf{A}_f &= [\mathbf{I} \mid \mathbf{A}_t^{-1}\mathbf{A}_l] \\ &= [I \mid -\mathbf{H}^\mathsf{T}] \end{aligned} \qquad (7.105)$$

where $\mathbf{A}_f$ might be called the *fundamental incidence matrix* and $\mathbf{H}^\mathsf{T}$ is an $n \times l$ matrix.

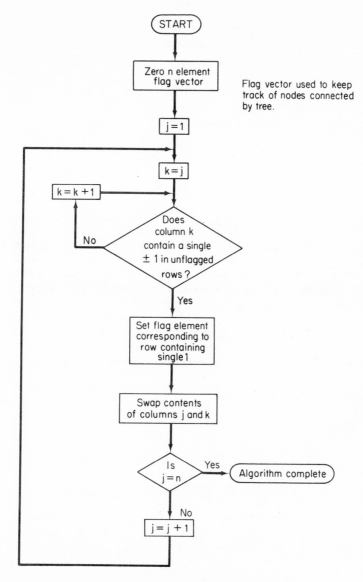

Figure 7.23 Incident matrix tree-link partitioning algorithm.

The $l \times b$ fundamental loop matrix $\mathbf{B}_f$ is then formed as

$$\mathbf{B}_f = [\mathbf{H} \mid \mathbf{I}] \tag{7.106}$$

This function can be performed in a single subroutine of the form

CALL LOOP (A, B, b, n)

where A and B are the incidence and fundamental loop matrices, respectively, and b and n are the number of branches and independent nodes in the network, respectively.

The number of link branches is computed from $l = b - n$. The implementation of the algorithm is left to the reader as an exercise.

EXAMPLE 7.8

Given the network graph shown in Fig. 7.22 with node 4 as a reference. Find a fundamental loop matrix.

Solution. The incidence matrix for the graph is

$$
\begin{array}{c}
\text{node} \\
\downarrow \\
\mathbf{A} = \begin{array}{c} 1 \\ 2 \\ 3 \end{array}
\begin{bmatrix}
-1 & 1 & 0 & 0 & 1 & 0 \\
0 & -1 & -1 & 1 & 0 & 0 \\
0 & 0 & 0 & -1 & -1 & -1
\end{bmatrix} \\
\text{branch} \rightarrow \quad a \quad\ b \quad\ c \quad\ d \quad\ e \quad\ f
\end{array}
$$

First, we find column a contains a single ± 1. We then set a flag vector, FLAG = [1 0 0], since we have located a tree branch which connects a node (in this case, node 1) to the reference. Starting the column search over again at column b, we find that this column also introduces a single -1 in row 2. Row 1 is not scanned since node 1 is already present in the tree. The flag vector is updated to FLAG = [1 1 0], to indicate that nodes 1 and 2 are now in the tree. Starting the search again at column c, we observe that column c introduces no 1's while column d does introduce a -1 in row 3. The flag vector is then set to [1 1 1] and columns c and d are swapped.

Since $n = 3$ the partitioning algorithm is complete and the partitioned matrix $\mathbf{A}_p$ is

$$
\begin{array}{c}
\text{node} \\
\downarrow \\
\mathbf{A}_p = \begin{array}{c} 1 \\ 2 \\ 3 \end{array}
\left[
\begin{array}{ccc|ccc}
-1 & 1 & 0 & 0 & 1 & 0 \\
0 & -1 & 1 & -1 & 0 & 0 \\
0 & 0 & -1 & 0 & -1 & -1
\end{array}
\right] \\
\text{branch} \rightarrow \underbrace{a \quad\ b \quad\ d}_{\mathbf{A}_t} \quad \underbrace{c \quad\ e \quad\ f}_{\mathbf{A}_e}
\end{array}
$$

Next $\mathbf{A}_t$ is diagonalized so that $\mathbf{A}_f$ is given by

$$
\mathbf{A}_f = \left[
\begin{array}{ccc|ccc}
1 & 0 & 0 & 1 & 0 & 1 \\
0 & 1 & 0 & 1 & 1 & 1 \\
0 & 0 & 1 & 0 & 1 & 1
\end{array}
\right]
$$

$$
\underbrace{}_{\mathbf{I}} \quad \underbrace{}_{-\mathbf{H}^\mathsf{T}}
$$

The fundamental loop matrix $\mathbf{B}_f$ is finally

$$
\begin{array}{c}
\text{loop} \\
\downarrow \\
\mathbf{B}_f = \begin{array}{c} 1 \\ 2 \\ 3 \end{array}
\left[
\begin{array}{ccc|ccc}
-1 & -1 & 0 & 1 & 0 & 0 \\
0 & -1 & -1 & 0 & 1 & 0 \\
-1 & -1 & -1 & 0 & 0 & 1
\end{array}
\right] \\
\text{branch} \rightarrow \underbrace{a \quad\ b \quad\ d}_{\mathbf{B}_t} \quad \underbrace{c \quad\ e \quad\ f}_{\mathbf{B}_e}
\end{array}
$$

In order to be useful in a computer algorithm the columns of $\mathbf{B}_f$ must be returned to their original order to coincide with the ordering of the branches in the $\mathbf{R}_2$ matrix and the $\mathbf{E}$ and $\mathbf{J}$ vectors. Therefore, the proper form for $\mathbf{B}_f$ is

$$\mathbf{B}_f = \begin{bmatrix} -1 & -1 & 1 & 0 & 0 & 0 \\ 0 & -1 & 0 & -1 & 1 & 0 \\ -1 & -1 & 0 & -1 & 0 & 1 \end{bmatrix}$$

The fundamental loops and the selected tree are shown in Fig. 7.24.

The dc loop-analysis computer program is almost identical to the dc program shown in Fig. 7.8 except for the call to the LOOP transformation after the completion of the data input. Again the implementation is left to the reader as an exercise.

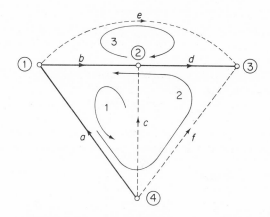

Figure 7.24 Fundamental loops and tree for Example 7.8.

7.10 Cut-Set Analysis

A *cut-set* of a network graph is a collection of branches and nodes such that, if these branches are removed from the graph, the graph becomes divided into two separate subgraphs or parts. Restoring any one of the branches from the cut-set reunites the subgraphs into a single graph. The cut-set must contain at least one tree branch since the tree of a graph connects every node of the graph. Obviously a large number of cut-sets can be drawn for a given graph. However, for a given tree of the network, a unique cut-set, called a *fundamental cut-set*, can be formed by forcing the cut-set to contain only one tree branch. The cut-set is then made up of links and a single tree branch.

We conclude that, given a tree configuration, the total number of cut-sets in the network is equal to the number of tree branches. As a simple example of the cut-set, consider the graph shown in Fig. 7.25. In this example the four cut-sets are drawn as closed contours which cut through a set of branches which includes one tree branch. Cut-set 1 is the set of branches {*1, 5, 6*} while cut-set 3 is {*3, 6, 7, 8*}. Note that cut-set

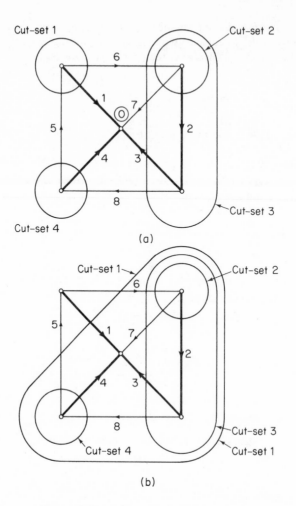

Figure 7.25 Cut-sets for a network graph.

1 in parts (a) and (b) of the figure cut through the same set of the branches but enclose opposite subgraphs. We will resolve this difficulty shortly.

Kirchhoff's current law states that the sum of the currents leaving a node equals zero. We intuitively extend this idea to a gaussian† surface which encloses one or more nodes of the network such as one of the cut-sets in Fig. 7.25. Kirchhoff's current law now implies that the current leaving the gaussian surface (cut-set) equals zero. In other words, the branches that cross the gaussian surface comprise a cut-set.

Cut-Set Matrix

Using the cut-set terminology, we noted in the preceding paragraph that KCL will be valid for a gaussian surface as well as for a single node. We state KCL for a

---

†Gauss's law states that the electric flux out of a closed surface is equal to the charge enclosed.

cut-set as follows: For any lumped network, for *any* of its cut-sets, and at any time, the algebraic sum of all the branch currents traversing the cut-set branches is zero. Before we can apply KCL to the cut-set we must assign a reference direction to the cut-set which in this development is such that current flow from inside the gaussian surface to the outside is positive. This choice is arbitrary, but once accepted it must be followed in later developments.

For consistency, we select the polarity of the cut-set to agree with the orientation of the tree branch defining the cut-set. This results in a unique collection of cut-sets. Under these conditions, if we apply KCL to the four cut-sets in Fig. 7.25a, we obtain

Cut-set 1: $\qquad\qquad\qquad i_1 - i_5 + i_6 = 0$

Cut-set 2: $\qquad\qquad\qquad i_2 - i_6 + i_7 = 0$

Cut-set 3: $\qquad\qquad i_3 - i_6 + i_7 + i_8 = 0$

Cut-set 4: $\qquad\qquad\qquad i_4 + i_5 - i_8 = 0$

In matrix form, the equation is

$$
\begin{matrix} \text{cut-set} \\ \downarrow \end{matrix}
\begin{matrix} 1 \\ 2 \\ 3 \\ 4 \end{matrix}
\begin{bmatrix} 1 & 0 & 0 & 0 & -1 & 1 & 0 & 0 \\ 0 & 1 & 0 & 0 & 0 & -1 & 1 & 0 \\ 0 & 0 & 1 & 0 & 0 & -1 & 1 & 1 \\ 0 & 0 & 0 & 1 & 1 & 0 & 0 & -1 \end{bmatrix}
\begin{bmatrix} i_1 \\ i_2 \\ i_3 \\ i_4 \\ i_5 \\ i_6 \\ i_7 \\ i_8 \end{bmatrix}
=
\begin{bmatrix} 0 \\ 0 \\ 0 \\ 0 \end{bmatrix}
$$
$$\text{branch} \rightarrow 1 \quad 2 \quad 3 \quad 4 \quad 5 \quad 6 \quad 7 \quad 8$$

More generally, the n linear homogeneous equations in i_m, $m = 1, 2, \ldots, b$, obtained by applying KCL to each of the n fundamental cut-sets constitute a set of n linearly independent equations.

We express KCL for cut-set analysis by

$$\mathbf{Q}\mathbf{i}_b = \mathbf{0} \qquad \text{(KCL)} \qquad\qquad\qquad (7.107)$$

where $\mathbf{Q}$ is an $n \times b$ matrix defined by

$$
q_{jk} = \begin{cases} 1 & \text{if branch } k \text{ belongs to cut-set } j \text{ and has the same reference direction} \\ -1 & \text{if branch } k \text{ belongs to cut-set } j \text{ and has the opposite reference direction} \\ 0 & \text{if branch } k \text{ does not belong to cut-set } j \end{cases}
$$

$\qquad n = $ Number of cut-sets (tree branches) $\qquad b = $ Number of branches

The matrix $\mathbf{Q}$ is called the *fundamental cut-set matrix*. If the columns are arranged so that the links and tree branches are grouped, the cut-set matrix $\mathbf{Q}$ can be written in the form

$$\mathbf{Q} = \underbrace{[\mathbf{I}}_{\substack{n \text{ tree} \\ \text{branches}}} \mid \underbrace{\mathbf{X}]}_{\substack{l \text{ link} \\ \text{branches}}} \qquad n \text{ cut-sets} \qquad\qquad (7.108)$$

where $\mathbf{X}$ is an $n \times l$ matrix identical to the matrix $-\mathbf{H}^\mathsf{T}$ of Eq. (7.105); that is, $\mathbf{Q} = \mathbf{A}_p$. The configuration of $\mathbf{Q}$ illustrated in Eq. (7.108) is important in later developments.

We can express Kirchhoff's voltage law in terms of the cut-set matrix $\mathbf{Q}$. Each branch voltage in the network can be written as a linear combination of the tree-branch voltages. For convenience, we will refer to the tree-branch voltages as $\mathbf{v}_t$ or $v_{t1}, v_{t2}, \ldots, v_{tn}$. The branch voltages for the network in Fig. 7.25 are (using the central node as the zero-potential node)

$$v_{b1} = v_{t1}$$
$$v_{b2} = v_{t2}$$
$$v_{b3} = v_{t3}$$
$$v_{b4} = v_{t4}$$
$$v_{b5} = v_{t4} - v_{t1}$$
$$v_{b6} = v_{t1} - v_{t2} - v_{t3}$$
$$v_{b7} = v_{t2} + v_{t3}$$
$$v_{b8} = v_{t3} - v_{t4}$$

In matrix form

$$\begin{bmatrix} v_{b1} \\ v_{b2} \\ v_{b3} \\ v_{b4} \\ v_{b5} \\ v_{b6} \\ v_{b7} \\ v_{b8} \end{bmatrix} = \begin{bmatrix} 1 & 0 & 0 & 0 \\ 0 & 1 & 0 & 0 \\ 0 & 0 & 1 & 0 \\ 0 & 0 & 0 & 1 \\ -1 & 0 & 0 & 1 \\ 1 & -1 & -1 & 0 \\ 0 & 1 & 1 & 0 \\ 0 & 0 & 1 & -1 \end{bmatrix} \begin{bmatrix} v_{t1} \\ v_{t2} \\ v_{t3} \\ v_{t4} \end{bmatrix}$$

which is exactly the same as

$$\mathbf{v}_b = \mathbf{Q}^\mathsf{T}\mathbf{v}_t\dagger \qquad (\text{KVL}) \qquad\qquad (7.109)$$

That is, the branch voltage vector is obtained by forming the product of the transpose of the fundamental cut-set matrix and the tree-branch voltage vector.

---

†If graph arrows represent either current or voltage, the tip of a voltage arrow must be negative ($-$) and the tail positive ($+$), the opposite of our normal convention. Current arrows are unaffected. (Consider the polarity of the voltage across a resistance relative to the current flow through it).

Cut-Set Analysis Formulation
of Network Equations

In cut-set analysis Kirchhoff's laws are expressed by Eqs. (7.107) and (7.109). These equations are combined with branch equations which relate branch voltages and currents to form network equations with n tree-branch voltages as the network variables. (Consider the relationship between $\mathbf{Q}$ and $\mathbf{v}_t$, and $\mathbf{A}$ and $\mathbf{v}_n$.) For the simple case in which the branch is shown in Fig. 7.5 (p. 237), the branch current is given by

$$\mathbf{i}_b = \mathbf{G}\mathbf{v}_b - \mathbf{J} + \mathbf{GE} \tag{7.110}$$

As earlier in Chapter 7, the branch conductance matrix $\mathbf{G}$ is a diagonal matrix of order b and $\mathbf{E}$ and $\mathbf{J}$ are source vectors of Eq. (7.13). Combining Eqs. (7.107), (7.109), and (7.110), we obtain

$$(\mathbf{Q}\mathbf{G}\mathbf{Q}^\mathsf{T})\mathbf{v}_t = \mathbf{Q}(\mathbf{J} - \mathbf{GE}) \tag{7.111}$$

or

$$\mathbf{G}_q\mathbf{v}_t = \mathbf{J}_q \tag{7.112}$$

where

$$\mathbf{G}_q \triangleq \mathbf{Q}\mathbf{G}\mathbf{Q}^\mathsf{T} \qquad \text{cut-set conductance matrix}$$

and

$$\mathbf{J}_q \triangleq \mathbf{Q}(\mathbf{J} - \mathbf{GE}) \qquad \text{cut-set equivalent current vector}$$

The results shown in Eq. (7.112) for the cut-set analysis method parallel closely the results obtained for the nodal method based upon the incidence matrix $\mathbf{A}$ in Eq. (7.18). By following the same procedure and the branch definition in Fig. 7.9, we similarly obtain the cut-set equations for the general branch

$$\mathbf{Q}\mathbf{G}_2\mathbf{Q}^\mathsf{T}\mathbf{v}_t = \mathbf{Q}(\mathbf{J} - \mathbf{G}_2\mathbf{E}) \tag{7.113}$$

equivalent to Eq. (7.38). Because of the similarity of the results obtained using the $\mathbf{A}$ and $\mathbf{Q}$ matrix methods, we do not pursue cut-set analysis methods further in this chapter. Notice, however, that the solution obtained using the cut-set matrix method yields tree-branch voltages rather than the node voltages obtained from the incidence matrix approach.

The importance of the tree-branch voltage solution is not obvious at this point, but the advantages will become apparent in the state-variable-analysis methods developed in Chapter 11.

Additional Reading

BALABANIAN, N., and T. A. BICKART, *Electrical Network Theory*, John Wiley & Sons, Inc., New York, 1969.

BALABANIAN, N., and S. SESHU, *Linear Network Analysis*, John Wiley & Sons, Inc., New York, 1959.

DESOER, C. A., and E. S. KUH, *Basic Circuit Theory*, McGraw-Hill Book Company, New York, 1969.

SESHU, S., and M. B. REED, *Linear Graphs and Electrical Networks*, Addison-Wesley Publishing Company, Inc., Reading, Mass., 1961.

Problems

7.1. (a) Construct incidence matrices for the networks shown in Fig. P7.1.

   (b) Using the incidence matrices from (a) verify that KVL and KCL are satisfied by

$$\mathbf{v}_b = \mathbf{A}^T \mathbf{v}_n \qquad \text{(KVL)}$$

and

$$\mathbf{A}\mathbf{i}_b = \mathbf{0} \qquad \text{(KCL)}$$

respectively, for each of the networks in Fig. P7.1.

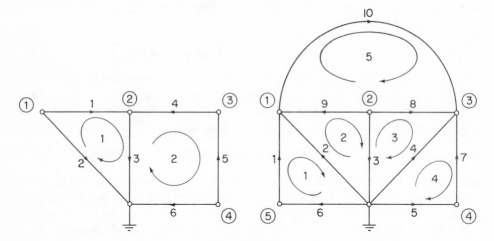

Figure P7.1

7.2. Draw an oriented graph corresponding to the following incidence matrix:

$$\mathbf{A} = \begin{bmatrix} -1 & 1 & 1 & 1 & 0 \\ 0 & 0 & 0 & -1 & 0 \\ 1 & -1 & 0 & 0 & -1 \end{bmatrix}$$

7.3. Draw an oriented graph whose incidence matrix $\mathbf{A}$ is

$$\mathbf{A} = \begin{bmatrix} 1 & 1 & -1 & 0 & 0 & 0 & 0 & 0 & 0 & 0 & 0 & 0 \\ 0 & 0 & 0 & 0 & -1 & -1 & 1 & 0 & 0 & 0 & 0 & 0 \\ -1 & 0 & 0 & 0 & 0 & 0 & 0 & 0 & -1 & 1 & 0 & 0 \\ 0 & 0 & 1 & 1 & 1 & 0 & 0 & 0 & 0 & 0 & 0 & 0 \\ 0 & 0 & 0 & 0 & 0 & 0 & -1 & 1 & 0 & 0 & 0 & 0 \\ 0 & -1 & 0 & -1 & 0 & 1 & 0 & -1 & 1 & 0 & 1 & 1 \end{bmatrix}$$

7.4. Derive the nodal analysis equations using the incidence matrix **A** for the general branch shown in Fig. P7.2.

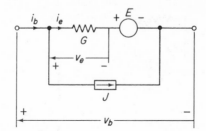

Figure P7.2

7.5. Use the equations developed in Problem 7.4 to write the node equations in matrix form for the circuits shown in Fig. P7.3. The procedure is as follows:
(a) Find the **A** and **G** matrices and the **E** and **J** vectors.
(b) Use the results of (a) to compute G_n.
(c) Use the results of (a) to compute J_n.
(d) Combine the results of (b) and (c) to obtain the node equation.

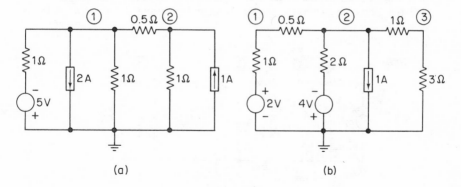

(a)   (b)

Figure P7.3

7.6. Write the node equations for the network shown in Fig. P7.4 using topological methods and the incidence matrix **A**. Verify the equations using the classical methods discussed in Chapter 4.

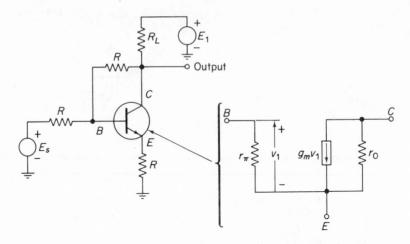

Figure P7.4

7.7. Write the node equations for the network shown in Fig. P7.5 using topological methods and the incidence matrix **A**. Verify the equations using the classical methods discussed in Chapter 4.

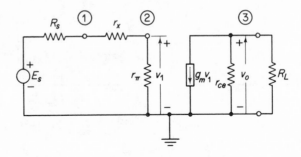

Figure P7.5

7.8. Derive the nodal analysis equations for the general branch shown in Fig. P7.6. The results should be of the form

$$\mathbf{AG_2A^Tv}_n = \pm \mathbf{AG_2E}$$

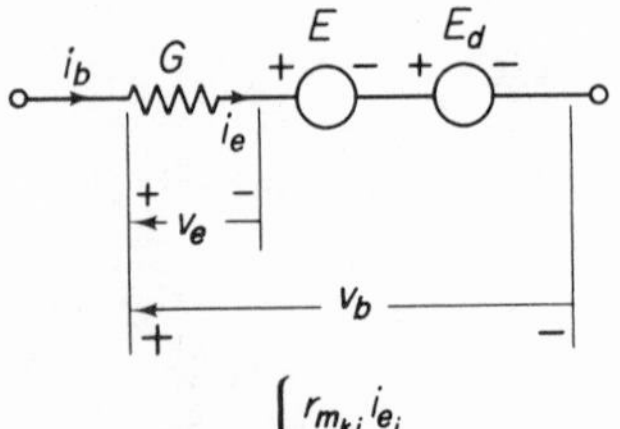

Figure P7.6

$$E_{d_k} = \begin{cases} r_{m_{kj}} i_{e_j} \\ \mu_{k_j} v_{e_j} \end{cases}$$

7.9. Use the results of Problem 7.8 to obtain the node equations for the network shown in Fig. P7.7.

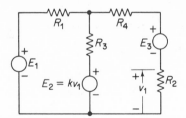

Figure P7.7

7.10. Write the node equations for the network shown in Fig. P7.8 using the general branch and the topological formulation of Problem 7.8. Verify the resulting node equations using the classical methods described in Chapter 4.

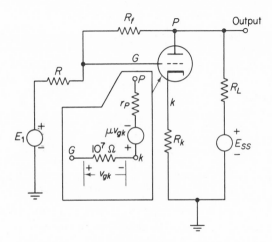

Figure P7.8

7.11. Derive the network equations for nodal analysis of the form

$$(AG_2A^T)v_n = A(\pm J \pm G_2E)$$

for the general branch shown in Fig. P7.9. Are there any restrictions on the use of dependent sources? This branch definition is used as the basic element of the IBM Electronic Circuit Analysis Program (ECAP).

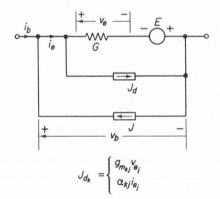

$$J_{d_k} = \begin{cases} g_{m_{kj}} v_{e_j} \\ \alpha_{kj} i_{e_j} \end{cases}$$

Figure P7.9

7.12. Derive the network equations for nodal analysis of the form

$$(\mathbf{A}\mathbf{G}_2\mathbf{A}^\mathsf{T})\mathbf{v}_n = \mathbf{A}(\pm\mathbf{J} \pm \mathbf{G}_2\mathbf{E})$$

for the general branch shown in Fig. P7.10. Are there any restrictions on the use of dependent sources?

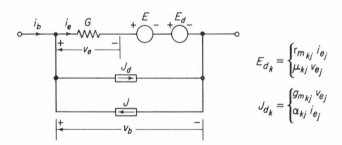

$$E_{d_k} = \begin{cases} r_{m_{kj}} i_{e_j} \\ \mu_{kj} v_{e_j} \end{cases}$$

$$J_{d_k} = \begin{cases} g_{m_{kj}} v_{e_j} \\ \alpha_{kj} i_{e_j} \end{cases}$$

Figure P7.10

7.13. Using the general branch and the network equation formulation of Problem 7.12, find the following for the network in Fig. P7.11.
   (a) The **A**, **G**, **G**₂, **E**, and **J** matrices and vectors.
   (b) The nodal conductance matrix $\mathbf{G}_n$.
   (c) The equivalent current vector $\mathbf{J}_n$.
   (d) Node voltages v_1 and v_2.

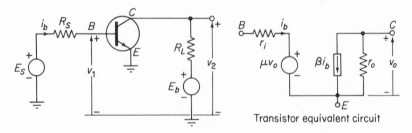

Transistor equivalent circuit

Figure P7.11

(e) The voltage gain $A_v = v_2/E_s$.

Note: Let $E_b = 0$ for voltage-gain calculation.

7.14. Modify the sensitivity coefficient calculation equations [Eqs. (7.52) through (7.55)] to include dependent voltage sources.

7.15. Write the nodal equations representing the network of Fig. P7.12 for an arbitrary driving frequency ω using the topological techniques discussed in this chapter. Verify the equations using the classical methods of Chapter 4.

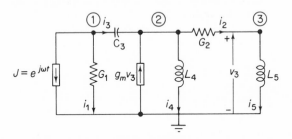

Figure P7.12

7.16. Consider the network shown in Fig. P7.13.
   (a) Obtain the node equations using topological methods for an arbitrary frequency ω.
   (b) Assuming the sinusoidal steady-state frequency $\omega = 1$, calculate the node voltages.

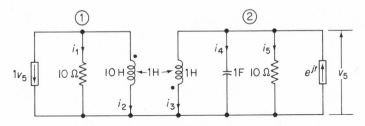

Figure P7.13

7.17. Write a basic dc analysis program based upon the node-voltage formulation

$$(\mathbf{AGA^T})\mathbf{v}_n = \mathbf{A}(\mathbf{J} - \mathbf{GE})$$

and the general branch of Fig. P7.14 without the dependent voltage sources. The program should accept *RLC* networks containing up to 20 nodes, 50 branches, and 50 dependent current sources. (Dependent voltage sources will be added in a later problem.) The input data should contain the following information:

Branch Type	Branch Number	"From" Node	"To" Node	Value	Controlling Branch (for dependent sources)

For example, the data card might read

   1   7   2   3   3400.

indicating branch 7 of type 1 (resistance) connected between nodes 2 and 3 is of value 3400. A dependent source might be specified as

8   9   3   1   4.3   3

indicating that branch 9 (assume type 8 is a transconductance) is connected between nodes 3 and 1 and is equal to 4.3 times the element voltage across branch 3.

The program output should contain in easily readable form:

1. A readable listing of the input data.
2. *The incidence matrix $\mathbf{A}$.
3. *Branch conductance matrix $\mathbf{G}$.
4. *Voltage and current source vectors $\mathbf{E}$ and $\mathbf{J}$.
5. *Nodal conductance matrix $\mathbf{G}_n$.
6. *Equivalent-current-source vector $\mathbf{J}_n$.
7. Node voltages $\mathbf{v}_n$.
8. Branch voltages $\mathbf{v}_b$.
9. Element voltages $\mathbf{v}_e$.
10. Branch currents $\mathbf{i}_b$.
11. Element currents $\mathbf{i}_e$.
12. Solution reliability factor. (KCL states that $\sum i_b = 0$ at each node. Two such factors might be the sum of the absolute values of the nodal current unbalances over the entire network or the largest unbalance obtained at any node.) The outputs marked with an asterisk (*) are useful while debugging the program. These should be suppressible during normal use.

    Since the program is intended for dc analysis applications, the inductances should be automatically replaced by 0.01-Ω resistances and the capacitances should be automatically replaced by 10^7-Ω resistances.

7.18. Modify the dc analysis program developed in Problem 7.17 to include transpotential and transresistance sources as illustrated in Fig. P7.14. Verify the proper function of the program using Example 7.3.

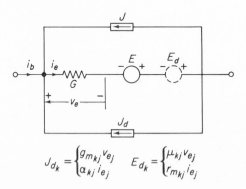

$$J_{d_k} = \begin{cases} g_{m_{kj}} v_{e_j} \\ \alpha_{kj} i_{e_j} \end{cases} \qquad E_{d_k} = \begin{cases} \mu_{kj} v_{e_j} \\ r_{m_{kj}} i_{e_j} \end{cases}$$

Figure P7.14

7.19. Formulate an algorithm to generate the nodal conductance matrix $(\mathbf{AG}_1\mathbf{A}^\mathsf{T})$ and the equivalent current vector $(\mathbf{A}(\mathbf{J} - \mathbf{G}_1\mathbf{E}))$ directly, that is, generate the matrices without performing the implied matrix multiplications for networks containing voltage- and current-dependent current sources. The methods for constructing the matrices described in Chapter 4 should provide some guides.

7.20. Modify the basic ac analysis program shown in Fig. 7.17 to include
(a) Transconductance and transfluence sources.
(b) Transresistance and transpotential sources.
(c) Compute and output on demand node, branch, and element voltages, and branch and element currents.

7.21. Derive the formula for the partial derivatives of the node voltages with respect to the network transconductances as in Eqs. (7.51) through (7.54).

7.22. Repeat Problem 7.19 for ac analysis which includes mutual inductance terms as well as voltage- and current-dependent current sources.

7.23. Implement the sensitivity coefficient calculations described in Eqs. (7.52) through (7.55) in the basic dc analysis program of Fig. 7.8.

7.24. (a) Construct mesh matrices for the networks shown in Fig. P7.1. How does the reversal of the reference direction of mesh current 2 in each of the graphs affect the mesh matrices? What is the counterclockwise reference direction analogous to in node analysis?
(b) Using the mesh matrices from (a) verify that KVL and KCL are satisfied by

$$\mathbf{M}\mathbf{v}_b = 0 \qquad \text{(KVL)}$$

and

$$\mathbf{i}_b = \mathbf{M}^T\mathbf{i}_1 \qquad \text{(KCL)}$$

respectively, for each of the networks in Fig. P7.1.

7.25. Draw oriented graphs corresponding to the mesh matrices given by
(a)
$$\mathbf{M} = \begin{bmatrix} -1 & 1 & 1 & 1 & 0 \\ 0 & 0 & 0 & -1 & 0 \\ 1 & -1 & 0 & 0 & -1 \end{bmatrix}$$
(b)
$$\mathbf{M} = \begin{bmatrix} 1 & 1 & 0 & 0 & 1 \\ -1 & -1 & 1 & 0 & 0 \\ 0 & 0 & -1 & 1 & -1 \\ 0 & 0 & 0 & -1 & 0 \end{bmatrix}$$

7.26. Derive the loop-analysis equations using the loop matrix $\mathbf{B}_f$ for the general branch shown in Fig. P7.2.

7.27. Using the loop equations developed in Problem 7.26 to write the loop equations in matrix form for the circuits shown in Fig. P7.3. Use the following procedure:
(a) Find the incidence matrix $\mathbf{A}$.
(b) Convert the incidence matrix $\mathbf{A}$ to the loop matrix $\mathbf{B}_f$, then rearrange columns of $\mathbf{B}_f$ in ascending branch order.
(c) Find the $\mathbf{G}$ matrix and the $\mathbf{E}$ and $\mathbf{J}$ vectors.
(d) Use the results of (b) and (c) to compute the loop resistance matrix $\mathbf{R}_1$.
(e) Use the results of (b) and (c) to compute the equivalent voltage vector $\mathbf{E}_1$.
(f) Combine the results of (d) and (e) to obtain the loop equations.

7.28. Repeat parts (a) and (b) of Problem 7.27 by first selecting a tree in each of the circuits shown in Fig. P7.3 and then directly constructing the matrix $\mathbf{B}_f$. How does the tree selected compare with the tree selected by manipulation in Problem 7.27(b)? Is the tree selected by inspection unique?

7.29. Write the loop equations for the network shown in Fig. P7.5 using topological methods and the loop matrix $\mathbf{B}_f$. Verify the results using the classical methods described in Chapter 4.

7.30. Derive the loop-analysis equations for the general branch shown in Fig. 7.6. The result should be of the form

$$(\mathbf{B}_f\mathbf{R}_2\mathbf{B}_f^\mathsf{T})\mathbf{i}_l = \mathbf{B}_f(\mathbf{E} - \mathbf{R}_2\mathbf{J})$$

7.31. Use the results of Problem 7.30 to write the loop equations for the network shown in Fig. P7.7.

7.32. Write the loop equations for the network shown in Fig. P7.8 using the general branch and the topological formulation of Problem 7.28. Verify the results using the classical methods described in Chapter 4.

7.33. Derive the network equations for loop analysis of the form

$$(\mathbf{B}_f\mathbf{R}_2\mathbf{B}_f^\mathsf{T})\mathbf{i}_l = \mathbf{B}_f(\pm\mathbf{E} - \mathbf{R}_2\mathbf{J})$$

for the general branch shown in Fig. P7.9. Are there any restrictions on the use of dependent sources?

7.34. Derive the network equations for loop analysis of the form

$$(\mathbf{B}_f\mathbf{R}_2\mathbf{B}_f^\mathsf{T})\mathbf{i}_l = \mathbf{B}_f(\mathbf{E} - \mathbf{R}_2\mathbf{J})$$

for the general branch shown in Fig. P7.10. Are there any restrictions on the use of dependent sources?

7.35. Using the general branch and the network equation formulation of Problem 7.34, find the following for the network in Fig. P7.11.
(a) The $\mathbf{B}_f$, $\mathbf{R}$, $\mathbf{R}_2$, $\mathbf{E}$, and $\mathbf{J}$ matrices and vectors.
(b) The loop resistance matrix $\mathbf{R}_1$.
(c) The equivalent source vector $\mathbf{E}_1$.
(d) The base and collector voltages $\mathbf{v}_1$ and $\mathbf{v}_2$.
(e) The current gain i_c/i_b.
(f) The voltage gain v_2/v_1.

7.36. Derive the loop-sensitivity-coefficient calculation procedure for the loop current definition in Eq. (7.96).
(a) Excluding all dependent sources.
(b) Including dependent voltage sources.
(c) Including both dependent voltage and dependent current sources.

7.37. Write a subroutine to formulate the fundamental loop matrix $\mathbf{B}_f$ from a given nodal incidence matrix $\mathbf{A}$. The subroutine should be capable of converting an incidence matrix for a network with up to 20 nodes and 50 branches. The subroutine call should be of the form

CALL LOOP (A, B, NODES, BRNCHS)

where NODES is the number of nodes, BRNCHS the number of branches, A the input incidence matrix, and B the fundamental loop matrix.

7.38. Write a basic dc analysis program based upon Eq. (7.96) and the general branch of Fig. P7.14 excluding the dependent current sources. The program requirements are identical to those of Problem 7.17 except that output items 2, 3, 5, and 6 should be
2. *The incidence matrix $\mathbf{A}$ and the loop matrix $\mathbf{B}_f$.

3. *Branch resistance matrix **R**.
5. *Loop resistance matrix $\mathbf{R}_1$.
6. *Equivalent-voltage-source vector $\mathbf{E}_1$.
(The asterisks have the some meaning as in Problem 7.17.)

7.39. Modify the dc analysis program developed in Problem 7.39 to include transconductance and transfluence sources as illustrated in Fig. P7.14. Verify the proper function of the program using Example 7.7.

7.40. Write the cut-set matrix **Q** and the incidence matrix **A** for the graph shown in Fig. P7.15. The tree is {5, 6, 7, 8}.

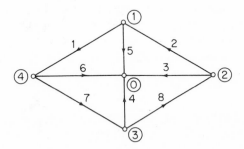

Figure P7.15

7.41. (a) Write the cut-set matrices for the network graphs shown in Figs. 7.2c and d.
(b) Write the fundamental loop matrices for the networks of part (a).

7.42. For the oriented graph shown in Fig. P7.16 and for the tree indicated,
(a) Draw all fundamental loops and fundamental cut-sets.
(b) Construct the fundamental loop and cut-set matrices.
(c) Demonstrate the relationship between the $\mathbf{B}_f$ and **Q** matrices as in Eq. (7.104).
(d) Is there a tree such that all its fundamental loops are meshes?

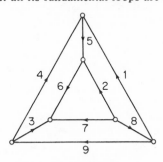

Figure P7.16   Tree = (5, 6, 7, 8, 9)

7.43. The fundamental loop matrix for a connected network is given by

$$\mathbf{B} = \begin{bmatrix} 1 & 0 & 0 & 1 & 0 & 0 \\ 0 & 0 & -1 & 0 & 1 & 0 \\ 1 & -1 & -1 & 0 & 0 & 1 \end{bmatrix}$$

(a) Write, by inspection, the fundamental cut-set matrix which corresponds to the same tree.
(b) Draw an oriented graph of the network.

7.44. Verify the solution obtained in Example 7.3 using cut-set analysis methods. The calculations are minimized by selecting the proper tree.

7.45. Verify the solution obtained in Example 7.7 using loop-analysis methods. The work is minimized by selecting a proper tree.

7.46. Write a subroutine to formulate a cut-set matrix from a given nodal incidence matrix A. The subroutine should be capable of converting an incidence matrix for a network with up to 20 nodes and 50 branches. The subroutine call should be of the form

CALL Q (A, NØDES, BRNCHS)

where NØDES is the number of nodes, BRNCHS the number of branches, and A the input incidence matrix and the resulting cut-set matrix.

Chapter 8

LAPLACE TRANSFORMS

There are more ways than one to skin a cat.
R. A. HABAS, *Morals for Moderns*

8.1 Fourier Transforms and Laplace Transforms

Fourier and Laplace transforms were adopted by engineers because these transforms convert linear differential equations into algebraic equations. The transforms also convert problems in the time domain to problems in the frequency domain and are useful in designing networks in terms of their response as a function of frequency. Since Fourier and Laplace transforms offer a direct relationship between the frequency and time domains, they provide a means of obtaining the time response once the frequency response is known (or vice versa). These transforms may be applied to linear time-varying equations (although they are not usually helpful there), but the transforms cannot be used in the solution of nonlinear differential equations.

Given some function of time $f(t)$, the Fourier transform may be defined as

$$F(\omega) = \int_{-\infty}^{\infty} f(t)e^{-j\omega t}\, dt \tag{8.1}$$

The function $f(t)$ in Eq. (8.1) must be single-valued and satisfy the Dirichlet conditions: (1) the function must possess a finite number of discontinuities in a finite interval, (2) a finite number of points at which it reaches infinity in this interval, and (3) a finite number of maxima and minima in the interval. Most functions of engineering interest meet these conditions. Further, however, we have the rather limiting condition that $f(t)$ must be such that

$$\int_{-\infty}^{\infty} |f(t)|\, dt = K \tag{8.2}$$

where $K < \infty$. This condition is not met by some time functions, as for example the simple step function $u(t)$. If Eq. (8.2) holds, then the transform exists and the inverse Fourier transform is

$$f(t) = \frac{1}{2\pi j} \int_{-\infty}^{\infty} F(\omega)e^{j\omega t}\, d(j\omega) \tag{8.3}$$

Equation (8.3) applies everywhere except at a discontinuity. For a discontinuity in $f(t)$ at a time t_1, Eq. (8.3) gives

$$f(t_1) = \tfrac{1}{2}[f(t_{1+}) + f(t_{1-})] \tag{8.4}$$

Thus the Fourier transform provides a rationale for defining the value of a function at a discontinuity as one half the value as approached from the right plus one half the value as approached from the left. The definition assumes a finite result for Eq. (8.4).

Using Eqs. (8.1) and (8.3), we may make up tables of Fourier transforms useful in various applications. Table 8.1 gives some properties of the Fourier transform, and Table 8.2 gives a few Fourier-transform pairs.

TABLE 8.1 Properties of the Fourier Transform

Function	Transform	
$x(t)$	$X(\omega)$	(a)
$X(t)$	$x(-\omega)$	(b)
$x(at)$	$\dfrac{1}{\|a\|} X\left[\dfrac{\omega}{a}\right]$	(c)
$x(t - t_0)$	$e^{-j\omega t_0} X(\omega)$	(d)
$e^{j\omega_0 t} x(t)$	$X(\omega - \omega_0)$	(e)
$\dfrac{d^n}{dt^n} x(t) \left[\text{if } \displaystyle\sum_{m=0}^{n-1} \dfrac{d^m}{dt^m} x(t) = 0, t \longrightarrow \infty \right]$	$(j\omega)^n X(\omega)$	(f)
$\displaystyle\int_{-\infty}^{t} x_1(\lambda) x_2(t - \lambda)\, d\lambda$	$X_1(\omega) X_2(\omega)$	(g)
$x(t) \cos \omega_0 t$	$\frac{1}{2}[X(\omega - \omega_0) + X(\omega + \omega_0)]$	(h)

TABLE 8.2 Fourier-Transform Pairs

$f(t)$	$F(\omega)$
$\delta(t)$	1
$\displaystyle\sum_{n=-\infty}^{\infty} \delta(t - nT)$	$\dfrac{2\pi}{T} \displaystyle\sum_{n=-\infty}^{\infty} \left(\delta\omega - \dfrac{2\pi n}{T} \right)$
$e^{j\omega_0 t}$	$2\pi\delta(\omega - \omega_0)$
$\cos \omega_0 t$	$\pi[\delta(\omega - \omega_0) + (\omega + \omega_0)]$
$\sin \omega_0 t$	$\dfrac{\pi}{j}[\delta(\omega - \omega_0) - \delta(\omega + \omega_0)]$
	$\dfrac{\tau \sin(\omega\tau/2)}{\omega\tau/2}$
$e^{-a\|t\|}$	$\dfrac{2a}{a^2 + \omega^2}$
$\|t\|$	$-\dfrac{2}{\omega^2}$
$e^{-t^2/2}$	$\sqrt{2\pi}\, e^{-\omega^2/2}$
e^{-at^2}	$\sqrt{\pi/a}\, e^{-\omega^2/4a}$

The Fourier transform is most useful in communication theory, where the time functions are pulses, periodic, decay as e^{-at}, or are random. Perhaps the most important property of Table 8.1 is property (g), which indicates that convolution in the time domain becomes a product in the frequency domain. (Unfortunately, the word "frequency" is often used when strictly angular velocity ω should be.) This property assists design in the frequency domain.

For those familiar with the Fourier series, Eqs. (8.1) and (8.3) may be thought of as the limiting case of the Fourier series as the period T of the fundamental component becomes very long. Thus, any periodic function $f(t)$ of period T may be expressed by a Fourier series

$$f(t) = \sum_{n=-\infty}^{\infty} c_n e^{jn\omega_0 t} \tag{8.5a}$$

with

$$c_n = \frac{1}{T} \int_{-T/2}^{T/2} f(t) e^{-jn\omega_0 t} \, dt \tag{8.5b}$$

and where $\omega_0 = 2\pi f_0 = 2\pi/T$. The term containing $n\omega_0$ with $n = 1$ is known as the *fundamental component*. Although this analogy may be useful as a conceptual tool, we do not pursue it further here.

We overcome the Fourier-transform convergence difficulty posed by the requirement that the left-hand side of Eq. (8.2) converge by replacing $e^{-j\omega t}$ in Eq. (8.1) by $e^{-(\sigma+j\omega)t} = e^{-st}$, with $\sigma > 0$. Thus, Eq. (8.1) becomes

$$F_1(s) = \int_{-\infty}^{\infty} f(t) e^{-\sigma t} e^{-j\omega t} \, dt \tag{8.6}$$

and in effect we have inserted a weighting function $e^{-\sigma t}$ with the function $f(t)$ such that

$$\int_{-\infty}^{\infty} |f(t)| e^{-\sigma t} \, dt$$

will be forced to converge for most functions of engineering interest. (The real number σ may be made as large as necessary.) The modified Fourier transform, called the *Laplace transform*, and its inverse then become

$$F(s) = \int_{-\infty}^{\infty} f(t) e^{-st} \, dt \tag{8.7a}$$

$$f(t) = \frac{1}{2\pi j} \int_{\sigma_a - j\infty}^{\sigma_a + j\infty} F(s) e^{st} \, dt \tag{8.7b}$$

where $s = \sigma + j\omega$ is a complex quantity, $\sigma \geq 0$.

The transform definition given in Eq. (8.7a) is for a two-sided (bilateral) Laplace transform. Usually the history of the function from $t = -\infty$ to some (arbitrary) $t = 0$ is expressed in terms of some initial conditions at $t = 0$. It is thus sufficient to use the one-sided (unilateral) transform and its inverse, defined as

$$F(s) = \int_{0}^{\infty} f(t) e^{-st} \, dt \tag{8.8a}$$

$$f(t) = \frac{1}{2\pi j} \int_{\sigma_a - j\infty}^{\sigma_a + j\infty} F(s) e^{st} \, ds \tag{8.8b}$$

The value of σ_a in Eqs. (8.7b) and (8.8b) must be such that convergence is assured for all $s = \sigma + j\omega$, $\sigma < \sigma_a$. The value of σ_a which assures convergence establishes the region of convergence. The limits on the integral of Eq. (8.8b) indicate that the integration proceeds along a straight line in the complex s-plane σ_a units to the right of the $j\omega$-axis (Fig. 8.1).

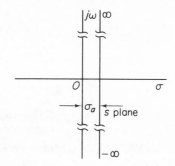

Figure 8.1 Path of integration for the inverse Laplace transform.

8.2 Properties of the Laplace Transform

The definition of the Laplace transform in Eq. (8.8a) can be written more precisely as

$$F(s) = \lim_{\substack{\epsilon \to 0_+ \\ T \to \infty}} \int_{\epsilon}^{T} f(t)e^{-st}\, dt \tag{8.9}$$

The definition in Eq. (8.9) illustrates the fact that the lower limit is to be approached from the right and that the integral is defined by a limit process. These details will be understood but not written in future work. To save space, we define the right-hand side of Eq. (8.9) as $\mathcal{L}[f(t)]$, or Eq. (8.9) becomes

$$F(s) = \mathcal{L}[f(t)] \tag{8.10}$$

Now we can develop two important properties of $\mathcal{L}[f(t)]$, where $f(t)$ is the derivative and the integral of a time function.

From Eq. (8.9), let $u = f(t)$ and $dv = e^{-st}\, dt$ and integrate by parts using $\int u\, dv = uv - \int v\, du$. Then Eq. (8.9) gives

$$\int_{0}^{\infty} f(t)e^{-st}\, dt = -\frac{1}{s} f(t)e^{-st}\Big|_{0+}^{\infty} + \frac{1}{s} \int_{0}^{\infty} \frac{df(t)}{dt} e^{-st}\, dt$$

$$= \frac{f(0_+)}{s} + \frac{1}{s} \int_{0}^{\infty} \frac{df(t)}{dt} e^{-st}\, dt \tag{8.11}$$

Rearranging Eq. (8.11),

$$\int_{0}^{\infty} \frac{df(t)}{dt} e^{-st}\, dt = sF(s) - f(0_+) \tag{8.12}$$

or

$$\mathcal{L}\left[\frac{df(t)}{dt}\right] = sF(s) - f(0_+) \tag{8.13}$$

where $F(s) = \mathcal{L}[f(t)]$ and $f(0_+)$ is the initial value of $f(t)$ at $t = 0_+$. If we continue this process for higher derivatives, we find that

$$\mathcal{L}\left[\frac{d^2 f(t)}{dt^2}\right] = s^2 F(s) - sf(0_+) - \frac{df(t)}{dt}\bigg|_{0+} \tag{8.14}$$

and finally,

$$\mathcal{L}[f^{(n)}(t)] = s^n F(s) - \sum_{k=1}^{n} s^{n-k} f^{(k-1)}(0_+) \tag{8.15}$$

where $f^{(k)} = d^k f(t)/dt^k$ and $f(0_+)$ means that the indicated function is to be evaluated at 0_+.

Reversing the roles of u and v in the integration-by-parts formula; that is, letting $u = e^{-st}$ and $dv = f(t)\, dt$, then

$$\int_0^\infty f(t)e^{-st}\, dt = e^{-st} \int f(t)\, dt \bigg|_0^\infty + s \int_0^\infty \left[\int f(t)\, dt\right] e^{-st}\, dt$$
$$= -\left[\int f(t)\, dt\right](0_+) + s \int_0^\infty \left[\int f(t)\, dt\right] e^{-st}\, dt \tag{8.16}$$

Thus

$$\int_0^\infty \left[\int f(t)\, dt\right] e^{-st}\, dt = \frac{F(s)}{s} + \frac{f^{(-1)}(0_+)}{s}$$

or

$$\mathcal{L}\left[\int f(t)\, dt\right] = \frac{F(s)}{s} + \frac{f^{(-1)}(0_+)}{s} \tag{8.17}$$

where

$$f^{(-1)}(0_+) = \lim_{t \to 0_+} \int_{-\infty}^t f(\lambda)\, d\lambda$$

If we continue the derivation for higher-order integrals, we find in general that

$$\mathcal{L}[f^{(-n)}(t)] = \frac{F(s)}{s^n} + \sum_{k=1}^{n} \frac{f^{(-k)}(0_+)}{s^{n-k+1}} \tag{8.18}$$

where $f^{(-j)} = \int \cdots \int f(t)\, dt^k$ and $f^{(-0)}(t) = f(t)$. Thus, differentiation and integration in the time domain transform into algebraic operations in the (complex) s-domain.

From the integral form of the definition Eq. (8.9), it is clear that

$$\mathcal{L}[af(t)] = aF(s) \tag{8.19}$$

and

$$\mathcal{L}[f_1(t) + f_2(t)] = F_1(s) + F_2(s) \tag{8.20}$$

where a is constant or independent of t or s. Other useful properties of the Laplace transform shown in Table 8.3 may be derived by using the basic definition. We com-

TABLE 8.3 Properties of the Laplace Transform

Function	Transform	
$f(t)$	$F(s)$	(a)
$af(t)$	$aF(s)$	(b)
$f(t-a)u(t-a)$	$e^{-as}F(s)$	(c)
$\dfrac{d}{dt}f(t)$	$sF(s) - f(0_+)$	(d)
$\dfrac{d^2}{dt^2}f(t)$	$s^2F(s) - sf(0_+) - \dfrac{df}{dt}(0_+)$	(e)
$\displaystyle\int_{-\infty}^{t} f(\lambda)\, d\lambda$	$\dfrac{F(s)}{s} + \dfrac{1}{s}\displaystyle\lim_{t\to 0_+}\int_{-\infty}^{t} f(\lambda)\, d\lambda$	(f)
$e^{-at}f(t)$	$F(s+a)$	(g)
$f(t/a)$	$aF(as)$	(h)
$\displaystyle\int_{0}^{t} f_1(t-\lambda)f_2(\lambda)\, d\lambda$	$F_1(s)F_2(s)$	(i)
$\displaystyle\lim_{t\to 0} f(t)$	$\displaystyle\lim_{s\to\infty} sF(s)$	(j)
$\displaystyle\lim_{t\to\infty} f(t)$	$\displaystyle\lim_{s\to 0} sF(s)$	(k)

plete this section by deriving the convolution property, which, together with the derivative and integral properties, forms the cornerstone of Laplace-transform theory.

In Eq. (8.9), let $f(t)=\int_0^t f_1(t-\tau)f_2(\tau)\, d\tau$. Then

$$F(s) = \int_0^\infty \left[\int_0^t f_1(t-\tau)f_2(\tau)\, d\tau\right] e^{-st}\, dt \qquad (8.21)$$

Consider the step function $u(t-\tau)$ along the τ-axis [see inner parentheses of Eq. (8.21)] as shown in Fig. 8.2. Clearly $u(t-\tau)=0$ for $\tau > t$. By multiplying the function $f_1(t-\tau)$ by $u(t-\tau)$ inside Eq. (8.21) we can change the second upper integration limit from t to ∞ and transform Eq. (8.21) to

$$F(s) = \int_0^\infty \int_0^\infty [f_1(t-\tau)u(t-\tau)f_2(\tau)\, d\tau] e^{-st}\, dt \qquad (8.22)$$

Changing the order of integration, we obtain

$$F(s) = \int_0^\infty f_2(\tau)\left[\int_0^\infty f_1(t-\tau)u(t-\tau)e^{-st}\, dt\right] d\tau \qquad (8.23)$$

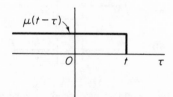

Figure 8.2 The step $u(t-\tau)$ along the τ axis.

Now in the inner integral, along the t-axis, $u(t - \tau) = 0$ for $t < \tau$. Hence

$$\int_0^\infty f_1(t - \tau)u(t - \tau)e^{-st}\,dt = \int_\tau^\infty f_1(t - \tau)e^{-st}\,dt$$

and letting $\lambda = t - \tau$ (or $t = \lambda + \tau$), we obtain

$$\int_\tau^\infty f_1(t - \tau)e^{-st}\,dt = \int_0^\infty f_1(\lambda)e^{-s(\tau+\lambda)}\,d\lambda$$

$$= e^{-s\tau}\int_0^\infty f_1(\lambda)e^{-s\lambda}\,d\lambda$$

Then (8.23) becomes

$$F(s) = \int_0^\infty f_2(\tau)e^{-s\tau}\left[\int_0^\infty f_1(\lambda)e^{-s\lambda}\,d\lambda\right]d\tau$$

$$= \int_0^\infty f_2(\tau)e^{-s\tau}\,d\tau\left[\int_0^\infty f_1(\lambda)e^{-s\lambda}\,dt\right]$$

$$= F_2(s)F_1(s) = F_1(s)F_2(s)$$

or finally,

$$\mathcal{L}\left[\int_0^t f_1(t - \tau)f_2(\tau)\,d\tau\right] = F_1(s)F_2(s)\dagger \qquad (8.24)$$

The integral on the left is termed the *convolution integral* and is abbreviated $f_1(t) * f_2(t)$ to represent convolution between $f_1(t)$ and $f_2(t)$. The integral applies only to linear, time-invariant networks and requires only that $f_1(t)$ and $f_2(t)$ be Laplace transformable. (The more general superposition integral applies to all linear networks.)

The convolution integral enables us to find the response $f(t)$ of a network to some input $f_1(t)$ if the response to another input is known. Specifically, if the zero-state response of a linear fixed network to an impulse $\delta(t)$ is $g(t)$, then the zero-state response $y(t)$ to an input $f_1(t)$ applied at $t = 0$ [that is, $f_1(t) = 0$, $t \leq 0$] is

$$y(t) = \int_0^t f_1(t - \tau)g(\tau)\,d\tau \qquad t \geq 0 \qquad (8.25)$$

or

$$y(t) = f_1(t) * g(t) \qquad t \geq 0 \qquad (8.26)$$

The situation is shown in Fig. 8.3. In the upper figure, we introduce an impulse to the zero-state network G at $t = 0$ and obtain the output response $g(t)$. Now if we apply some other function $f_1(t)$ at $t = 0$ to the same zero-state network G, we obtain the

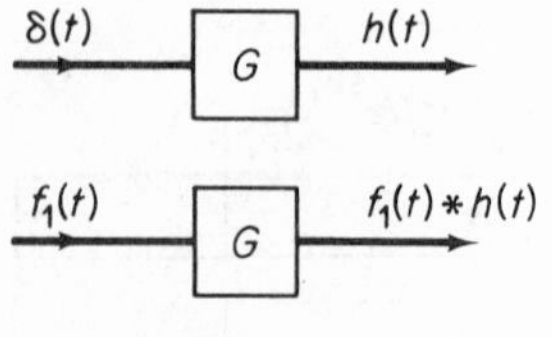

Figure 8.3 Illustration of convolution.

---

†The manipulations in this development depend on linearity assumptions.

response $y(t) = f_1(t) * g(t)$. Clearly $g(t)$ describes the network from a zero-state input–output standpoint.†

If we let $f_1(t) = \delta(t)$, then from Eq. (8.25),

$$g(t) = \int_0^t \delta(t - \tau)g(\tau)\, d\tau$$

and taking the Laplace transform of both sides, we obtain from Eq. (8.24)

$$G(s) = \mathcal{L}[\delta(t)]G(s)$$

from which

$$\mathcal{L}[\delta(t)] = 1 \tag{8.27}$$

Going back to Eq. (8.25) and taking the transform of both sides,

$$Y(s) = F_1(s)G(s) \tag{8.28a}$$
$$= G(s)F_1(s) \tag{8.28b}$$

Equations (8.28) tell us that the output $Y(s)$ to some input $F_1(s)$ in the s-domain is formed by the product of $F_1(s)G(s)$. $G(s)$ is again a characteristic of the zero-state network in the s-domain, and Eqs. (8.28) separate the function $G(s)$, which is dependent only on the network from the input function $F_1(s)$. $G(s)$ is called the *network function*, or more commonly, the *transfer function of the network*. It depends only on the components and interconnections of components in the network. It should be remembered, however, that $G(s)$ is defined by input–output relations. That is, in Fig. 8.3 the network is a black box, and $G(s)$ is determined by the input $\delta(t)$ and the output $g(t)$ with zero initial conditions. These input–output relations may be found by laboratory experiment. The transfer function obtained from the network equations may not result in the same $G(s)$ as obtained from the input–output relationships, owing to possible algebraic cancellations in the numerator and denominator of the transfer function found from the network equations. The fact that the two methods of finding $G(s)$ may give different results leads to an area of study termed "controllability and observability," which we do not discuss further here.‡

We have not proved Eq. (8.25), but if we can prove Eq. (8.27) by an alternative method, we indirectly (but not rigorously) prove Eq. (8.25) by proceeding in the reverse direction. This we will do now. In Eq. (8.9) let $t = (t - b)$, where $b > 0$. Then

$$F(s) = \int_b^\infty f(t - b)e^{-s(t-b)}\, d(t - b)$$

or

$$e^{-bs}F(s) = \int_b^\infty f(t - b)e^{-st}\, dt$$

If $f(t - b) = 0$ for $0 < t < b$, then the lower limit of the last integral may be made

---

†In many texts the impulse response is symbolized as $h(t)$, with a transform $H(s)$.

‡To pursue this point, see, for example, Desoer and Kuh in the Additional Reading.

zero, or under this condition,

$$\int_0^\infty f(t-b)c^{-st}\,dt = e^{-bs}F(s)$$

Hence

$$\mathcal{L}[f(t-b)] = e^{-bs}F(s) \qquad \text{if } f(t-b) = 0 \quad t < b$$

Because $u(t-b) = 0$, $0 < t < b$; this may also be expressed as

$$\mathcal{L}[f(t-b)u(t-b)] = e^{-bs}F(s) \tag{8.29}$$

In Fig. 8.4 we show a pulse of height $1/a$ and breadth a (or area $= 1$). This pulse may be considered as the sum of a step function of magnitude $1/a$ starting at $t = \tau$ and a later negative step function of the same magnitude starting at $t = \tau + a$.

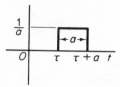

Figure 8.4   A pulse of width a and height $\frac{1}{a}$.

The transform of a step function $u(t)$ (magnitude unity) is given by Eq. (8.9) as

$$F(s) = \lim_{\substack{\epsilon \to 0 \\ T \to \infty}} \int_\epsilon^T (1)e^{-st}\,dt$$

$$= \frac{e^{-st}}{-s}\Big|_0^\infty \tag{8.30}$$

$$= \frac{1}{s}$$

From Eqs. (8.29) and (8.30) the transform of the pulse of Fig. 8.4 is therefore

$$\frac{1}{a}\frac{e^{-\tau s} - e^{-(\tau+a)s}}{s} = \frac{e^{-\tau s}(1 - e^{-as})}{as} \tag{8.31}$$

Now let $a \to 0$ in the right side of Eq. (8.31) [or the pulse becomes a unit impulse $\delta(t - \tau)$]. Then

$$\lim_{a \to 0} \frac{e^{-\tau s}(1 - e^{-as})}{as} = \lim_{a \to 0} e^{-\tau s}\frac{se^{-as}}{s} = e^{-\tau s}$$

by L'Hospital's rule. Thus,

$$\mathcal{L}[\delta(t - \tau)] = e^{-\tau s} \tag{8.32a}$$

and by Eq. (8.29),

$$\mathcal{L}[\delta(t)] = 1 \tag{8.32b}$$

Equation (8.32b) requires some interpretation, since if we examine Fig. 8.5, we

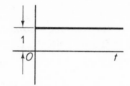

Figure 8.5 A unit step function.

see that the time derivative of a unit step function $u(t)$ is an impulse at $t = 0$ (the slope changes instantaneously at $t = 0$). However, if we let $f(t) = u(t)$ in Eq. (8.13), we run into trouble, since we get

$$\frac{du(t)}{dt} = s\frac{1}{s} - f(0_+)$$

which does not seem to result in Eq. (8.32). We may reconcile this by letting the impulse $\delta(t)$ occur just after $t = 0_+$, or if $f_1(t)$ represents $\delta(t)$, then $\delta(t) = \lim_{\Delta \to 0}[f_1(0 + \Delta)]$, while $f(0_+)$ occurs such that $f(0_+) = \lim_{\epsilon \to 0}[f(0) + \epsilon]$, where $\Delta > \epsilon$ and both Δ and ϵ are positive numbers. Thus $f(0_+) = 0$ if the step to be differentiated occurs an infinitesimal time interval later than $t = 0_+$ and Eq. (8.32) is verified. The practical application of this consists in taking any initial conditions of a network as existing just *prior* to application of an impulse rather than just after. This is necessary because the impulse itself injects energy into the network instantaneously—a physical impossibility but theoretically useful. This rather pragmatic reasoning may be justified by a review of Example 1.1, where we treated the case of the response to a pulse input with the pulse width shrinking.

The unusual care necessary in using the impulse arises from its peculiar characteristics, which we have previously pointed out; the inherent difficulties with correctly evaluating limits; and basic mathematical problems encountered at discontinuities. Some of the conceptual difficulties are alleviated with the bilateral Laplace transform, where integration is performed from $-\infty$ to $+\infty$. However, the concept of the impulse is so useful for both theoretical and experimental work that, similar to the relationships between man and wife, some travail is endured for the greater benefits.

8.3 Application of the Laplace Transform

Since the transient (or homogeneous) solution of the linear differential equation with fixed coefficients is so simple, transform methods are not particularly advantageous here. Conventional methods are even preferable for steady-state (particular integral) solutions if the forcing functions (sources) are constant, sinusoidal, or exponential. The transform technique becomes important for more complicated forcing functions. In addition, the concept of the transfer function becomes very valuable since design in the complex plane has many advantages, and the Laplace and Fourier transforms provide links between the complex frequency design and network operation in the time domain.

To make adequate use of the Laplace transform, we need a table of transform pairs, relating the time function $f(t)$ and the complex frequency function $F(s)$. This table may be prepared by using Eq. (8.9), the definition of the transform, and also the properties of the $\mathcal{L}$ transform. We have already computed some of these transform pairs, and show these and others in Table 8.4. More extensive tables are available in several commonly used handbooks.

TABLE 8.4 A Few Laplace Transforms

$f(t)$ †	$F(s)$	Region of Convergence
$\delta(t)$	1	$\sigma > -\infty$
$u(t)$	$\dfrac{1}{s}$	$\sigma > 0$
t	$\dfrac{1}{s^2}$	$\sigma > 0$
$\dfrac{t^{n-1}}{(n-1)!}$	$\dfrac{1}{s^n}$	$\sigma > 0$
e^{-at}	$\dfrac{1}{s+a}$	$\sigma > -a$
te^{-at}	$\dfrac{1}{(s+a)^2}$	$\sigma > -a$
$\sin \alpha t$	$\dfrac{\alpha}{s^2+\alpha^2}$	$\sigma > 0$
$\cos \alpha t$	$\dfrac{s}{s^2+\alpha^2}$	$\sigma > 0$
$e^{-at}\sin \alpha t$	$\dfrac{\alpha}{(s+a)^2+\alpha^2}$	$\sigma > -a$
$e^{-at}\cos \alpha t$	$\dfrac{s+a}{(s+a)^2+\alpha^2}$	$\sigma > -a$

†Valid for $t \geq 0$.

EXAMPLE 8.1

As the first example, look at Fig. 8.6a which shows an RC circuit with a capacitance initially charged to a voltage v_{c0} (polarity shown) and an impulse source $A(t)$. The objective is to find the voltage $v_c(t)$ on the capacitance.

Solution. The differential equation is

$$RC\frac{dv_c}{dt} + v_c = A\delta(t) \tag{8.33}$$

where $v_c = v_c(t)$. We can apply Eqs. (8.13) and (8.32), or use Table 8.4, to obtain the transformed equation

$$(RCs + 1)V_c(s) - (-v_{c0}RC) = A(1) \tag{8.34}$$

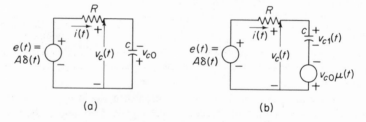

(a)   (b)

Figure 8.6 *RC* circuit for Example 8.1.

We follow the convention of using lower case letters for time variables and capital letters for transformed variables in the frequency domain.

From Eq. (8.34),

$$V_c(s) = \left[\frac{1}{RCs + 1}\right][A - v_{c0}RC] \tag{8.35}$$

The response of the capacitance voltage $v_c(t)$ could have been calculated by using an equivalent voltage source in series with the capacitance with a value equal to the initial voltage v_{c0}, as shown in Fig. 8.6b. The idea of using equivalent sources, introduced in Chapter 2 in the definitions of the capacitance and inductance models, is generally quite useful, although in this particular example, the method has questionable advantage. The first step in the application of the equivalent sources method consists of converting all initial conditions to equivalent voltage and current sources (see Fig. 8.6b).

The transformed differential equation describing the RC network is then

$$(RCs + 1)V_{c1}(s) = A + \frac{v_{c0}}{s} \tag{8.36}$$

where v_{c0}/s represents the capacitance initial voltage as an equivalent source. Solving Eq. (8.36) for $V_{c1}(s)$, we find

$$V_{c1}(s) = \frac{A}{RC}\frac{1}{s + 1/RC} + \frac{v_{c0}}{RC}\frac{1}{s(s + 1/RC)} \tag{8.37}$$

The capacitance voltage $v_c(t)$ is given by

$$v_c(t) = v_{c1}(t) - v_{c0}$$

or

$$V_c(s) = V_{c1}(s) - \frac{v_{c0}}{s} \tag{8.38}$$

Combining Eqs. (8.37) and (8.38) the capacitance voltage becomes

$$V_c(s) = \frac{A}{RC}\left[\frac{1}{s + 1/RC}\right] + \frac{v_{c0}}{RC}\left[\frac{1}{s(s + 1/RC)}\right] - \frac{v_{c0}}{s}$$

$$= \frac{A - v_{c0}RC}{RCs + 1}$$

which is identical to the result obtained in Eq. (8.35).

In Eq. (8.35) we separate the parentheses since this places in evidence the relationship mentioned in connection with Eq. (8.28). Thus the transfer function is

$$G(s) = \frac{1}{RCs + 1}$$

and the input function is

$$F_1(s) = [A - v_{c0}RC]$$

The justification for algebraically manipulating Eq. (8.34) into Eq. (8.35) is contained in the discussion of the convolution integral of the preceding section. Note that the initial condition (with a multiplier) becomes part of the source term and is included in the original equation rather than being considered separately as in the conventional method.

The transfer function $G(s)$ is said to have a real pole at $s = -1/RC$, as when $s = -1/RC$ the magnitude of the transfer function becomes infinite. This value of the pole is obtained by setting the denominator polynomial in the transfer function to zero, which forms the characteristic equation of the network. Thus, the characteristic equation is

$$RCs + 1 = 0 \qquad (8.39)$$

We see that this is the same equation previously obtained in solving the homogeneous equation by conventional methods. Thus, $\lambda = -1/RC$ is an eigenvalue of the network and "poles" and "eigenvalues" of the network are (virtually) synonymous. (Exceptions need not concern us here.) Dividing the numerator and denominator of Eq. (8.35) by RC we obtain

$$V_c(s) = \left[\frac{1}{s + 1/RC}\right]\left[\frac{A}{RC} - v_{c0}\right] \qquad (8.40)$$

Taking the inverse Laplace transform of both sides of Eq. (8.40) (using Table 8.4), we find

$$v_c(t) = \left[\frac{A}{RC} - v_{c0}\right]e^{-(1/RC)t} \qquad (8.41)$$

From Eq. (8.41) we see that at $t = 0_+$,

$$v_c(0_+) = \left[\frac{A}{RC} - v_{c0}\right] \qquad (8.42)$$

Thus, an impulse of magnitude A instantaneously puts a voltage of A/RC or a charge $Q = A/R$ on the capacitance. Practically, as we discussed previously, this is not possible, but can be approached as a limit.

Note from the example that if $v_{co} = 0$ and the impulse has unit value ($A = 1$), the zero-state network response to a unit impulse is just the inverse transform of the transfer function $G(s)$, or $g(t) = \mathcal{L}^{-1}[G(s)]$. This is an important property of the impulse.

The example suggests a general procedure in applying the Laplace transformation to the solution of integrodifferential equations. A differential equation of the form

$$a_0\frac{d^ny}{dt^n} + a_1\frac{d^{n-1}y}{dt^{n-1}} + \cdots + a_{n-1}\frac{dy}{dt} + a_ny = m(t) \qquad (8.43)$$

becomes, as a result of the Laplace transformation, an algebraic equation which can be solved for $Y(s)$, or

$$Y(s) = \frac{\mathcal{L}[m(t)] + \text{initial condition terms}}{a_0s^n + a_1s^{n-1} + \cdots + a_{n-1}s + a_n} \qquad (8.44)$$

The right-hand side of Eq. (8.44) reduces to a quotient of polynomials in s if $\mathcal{L}[m(t)] = 0$, or if $\mathcal{L}[m(t)]$ is also a quotient of polynomials. $\mathcal{L}[m(t)]$ often satisfies these requirements; for example, the functions $\delta(t)$, $u(t)$, $\sin at$, $\cos bt$, $e^{\alpha t}$ and so on, transform into a quotient of polynomials. If we designate the numerator and denominator polynomials as $P(s)$ and $Q(s)$, respectively, then

$$Y(s) = \frac{P(s)}{Q(s)} \qquad (8.45)$$

If the transform pair for $P(s)/Q(s)$ can be found in a transform table, we may write the solution $y(t)$ directly. However, the transformed expression for $Y(s)$ generally must be broken into simpler terms before using any practical transform table. The method of partial-fraction expansion is a way of accomplishing this simplification.

The first step in the expansion of the quotient of polynomials $P(s)/Q(s)$ checks the order of $P(s)$ relative to that of $Q(s)$. If the order of $P(s)$ is greater than that of $Q(s)$, divide $Q(s)$ into $P(s)$ to obtain

$$\frac{P(s)}{Q(s)} = B_0 + B_1 s + \cdots + B_{m-n} s^{m-n} + \frac{P_1(s)}{Q(s)} \tag{8.46}$$

where m is the order of $P(s)$ and n is the order of the $Q(s)$. Now $P_1(s)$ is of lower order than $Q(s)$ and we can proceed with the next step in the expansion.

Second, we factor the denominator polynomial, $Q(s)$, into the form

$$\begin{aligned} Q(s) &= a_0 s^n + a_1 s^{n-1} + \cdots + a_n \\ &= a_0 (s - s_1) \cdots (s - s_n) \end{aligned}$$

or

$$Q(s) = a_0 \prod_{j=1}^{n} (s - s_j) \tag{8.47}$$

The roots of $Q(s)$ can be (1) real and simple (distinct), (2) multiple, (3) conjugate complex, or (4) a combination of these. We consider each of the classifications of roots separately.

If $Q(s)$ has n simple roots at $s = s_k$, where $k = 1, 2, \ldots, n$, $P_1(s)/Q(s)$ may be expressed in a partial-fraction expansion as

$$\frac{P_1(s)}{a_0(s - s_1) \cdots (s - s_n)} = \frac{K_1}{s - s_2} + \frac{K_2}{s - s_2} + \cdots + \frac{K_n}{s - s_n} \tag{8.48}$$

where the K's are real constants.

The constants K_j are determined from the relationship

$$K_j = (s - s_j) \frac{P_1(s)}{Q(s)} \Big|_{s=s_j} \qquad j = 1, 2, \ldots, n \tag{8.49}$$

EXAMPLE 8.2

Let

$$G(s) = \frac{P(s)}{Q(s)} = \frac{s^3 + 4s^2 + 4s + 5}{s^2 + 3s + 2} \tag{8.50}$$

Applying the first expansion step to $G(s)$ we have

$$G(s) = s + 1 + \frac{-s + 3}{(s + 1)(s + 2)}$$

$$= s + 1 + \frac{K_1}{s + 1} + \frac{K_2}{s + 2}$$

Then, according to Eq. (8.49),

$$K_1 = (s + 1)P_1(s)|_{s=-1} = \frac{-s + 3}{s + 2}\bigg|_{s=-1} = 4$$

and

$$K_2 = (s + 2)P_1(s)|_{s=-2} = \frac{-s + 3}{s + 1}\bigg|_{s=-2} = -5$$

Therefore, the complete partial-fraction expansion of Eq. (8.49) is

$$G(s) = s + 1 + \frac{4}{s + 1} - \frac{5}{s + 2} \tag{8.51}$$

Justification of Eq. (8.49) consists in multiplying both sides of Eq. (8.48) by $(s - s_j)$ and then letting $s = s_j$.

If a root of $Q(s) = 0$ if of multiplicity r, the partial-fraction expansion of the repeated root is given by

$$\frac{P_1(s)}{(s - s_j)r} = \frac{K_{j1}}{s - s_j} + \frac{K_{j2}}{(s - s_j)^2} + \cdots + \frac{K_{jr}}{(s - s_j)r} \tag{8.52}$$

Multiplying Eq. (8.52) by $(s - s_j)^r$, we have

$$P_1(s) = K_{j1}(s - s_j)^{r-1} + K_{j2}(s - s_j)^{r-2} + \cdots + K_{jr} \tag{8.53}$$

If we let $s = s_j$, in Eq. (8.53), all terms in the equation disappear except K_{jr}, which can be evaluated. Next, differentiate Eq. (8.53) once with respect to s. The constant K_{jr} will vanish, but K_{jr-1} will remain without a multiplying function of s. Thus, K_{jr-1} can be evaluated as before by letting $s = s_j$. To find the general term K_{jn}, we differentiate Eq. (8.53) $(r - n)$ times and let $s = s_j$; then

$$K_{jn} = \frac{1}{(r - n)!} \frac{d^{(r-n)}}{ds^{(r-n)}} P_1(s)\bigg|_{s=s_j} \tag{8.54}$$

The use of Eq. (8.54) is easier than might appear.

EXAMPLE 8.3

Consider

$$\frac{2s^2 + 3s + 2}{(s + 1)^3} = \frac{K_{11}}{s + 1} + \frac{K_{12}}{(s + 1)^2} + \frac{K_{13}}{(s + 1)^3} \tag{8.55}$$

Multiplying Eq. (8.55) by $(s + 1)^3$, we have

$$2s^2 + 3s + 2 = K_{11}(s + 1)^2 + K_{12}(s + 1) + K_{13} \tag{8.56}$$

From this equation

$$K_{13} = 2s^2 + 3s + 2|_{s=-1} = 1$$

Next, differentiate Eq. (8.56) with respect to s to obtain

$$4s + 3 = 2K_{11}(s + 1) + K_{12} \tag{8.57}$$

from which

$$K_{12} = 4s + 3|_{s=-1} = -1$$

Again differentiate the last equation to get

$$4 = 2K_{11} \tag{8.58}$$

or

$$K_{11} = 2$$

The final partial-fraction expansion is then

$$\frac{2s^2 + 3s + 2}{(s+1)^2} = \frac{2}{s+1} + \frac{-1}{(s+1)^2} + \frac{1}{(s+1)^3} \tag{8.59}$$

EXAMPLE 8.4

If $Q(s)$ contains both simple and repeated roots, a combination of both methods may be used. As an example, let

$$F(s) = \frac{P(s)}{Q(s)} = \frac{s+2}{(s+1)^2(s+3)} \tag{8.60}$$

The form of the partial-fraction expansion of Eq. (8.60) is

$$\frac{s+2}{(s+1)^2(s+3)} = \frac{K_{11}}{s+1} + \frac{K_{12}}{(s+1)^2} + \frac{K_2}{s+3} \tag{8.61}$$

The first coefficient K_2 can be evaluated by Eq. (8.49) and coefficients K_{11} and K_{12} can be evaluated by applying Eq. (8.54). Then

$$K_2 = \frac{s+2}{(s+1)^2}\bigg|_{s=-3} = -\frac{1}{4}$$

Multiplying Eq. (8.61) by $(s+1)^2$, we have

$$\frac{s+2}{s+3} = K_{11}(s+1) + K_{12} + \frac{K_2(s+1)^2}{s+3} \tag{8.62}$$

Letting $s = -1$, the constant K_{12} is

$$K_{12} = \frac{s+2}{s+3}\bigg|_{s=-1} = \frac{1}{2}$$

Differentiating Eq. (8.62) and letting $s = -1$ we have

$$\frac{(s+3) - (s+2)}{(s+3)^2} = K_{11} + K_2 \frac{d}{ds}\left[\frac{(s+1)^2}{s+3}\right]$$

from which

$$K_{11} = \frac{1}{(s+3)^2}\bigg|_{s=-1} = \frac{1}{4}$$

The resulting expansion is finally

$$\frac{s+2}{(s+1)^2(s+3)} = \frac{\frac{1}{4}}{s+1} + \frac{\frac{1}{2}}{(s+1)^2} - \frac{\frac{1}{4}}{s+3} \tag{8.63}$$

The partial-fraction expansion for a function containing two roots which form a complex-conjugate pair is

$$\frac{P_1(s)}{Q(s)} = \frac{P_1(s)}{Q_1(s)(s + \alpha + j\omega)(s + \alpha - j\omega)}$$

$$= \frac{K_1}{(s + \alpha + j\omega)} + \frac{K_1^*}{(s + \alpha - j\omega)} + \cdots \qquad (8.64a)$$

$$= \frac{As + B}{(s^2 + 2\alpha s + \alpha^2 + \omega^2)} + \cdots \qquad (8.64b)$$

where K_1^* is the complex conjugate of K_1 but A and B are real. An expansion of this type is necessary for each pair of complex-conjugate roots. The constant K_1 is determined from the relationship

$$K_1 = \left| (s + \alpha + j\omega)\frac{P_1(s)}{Q(s)} \right|_{s = -\alpha - j\omega} \qquad (8.65)$$

as in the case of real and simple roots. With K_1 known, we can immediately write down K_1^*, the conjugate of K_1, without going through another application of Eq. (8.49).

EXAMPLE 8.5

Let $e(t)$ in Fig. 8.6a (p. 314) be $A \sin \omega t$ (applied at $t = 0$) and $v_{c0} = 0$. The differential equation describing the network with the new driving function is

$$RC\frac{dv_c}{dt} + v_c = A \sin \omega t \qquad (8.66)$$

The Laplace-transformed network equation is

$$\mathcal{L}\left[RC\frac{dv_c}{dt} + v_c \right] = \mathcal{L}[A \sin \omega t]$$

or

$$sRCV_c(s) + V_c(s) = A \frac{\omega}{s^2 + \omega^2} \qquad (8.67)$$

Then

$$V_c(s) = \left[\frac{1}{RCs + 1} \right] \frac{A\omega}{s^2 + \omega^2} \qquad (8.68)$$

Note that the transfer function $G(s)$ is exactly as before, since it depends only on the network configuration.

From Eq. (8.75),

$$V_c(s) = \frac{A\omega}{RC}\left[\frac{1}{(s + 1/RC)(s^2 + \omega^2)} \right] \qquad (8.69)$$

If we have an extensive table, we may immediately look up the inverse transform. If not, again the quantity inside the brackets must be expanded into partial-fraction form. Thus,

$$\frac{1}{(s + 1/RC)(s^2 + \omega^2)} = \frac{K_1}{s + 1/RC} + \frac{K_2}{s + j\omega} + \frac{K_3}{s - j\omega} + K_4 \qquad (8.70)$$

Here the numerator of the function $V_c(s)$ is already of lower degree than the denominator, so in Eq. (8.70) the constant $K_4 = 0$.

Multiply both sides of Eq. (8.70) by $(s + 1/RC)$, or

$$\frac{1}{s^2 + \omega^2} = K_1 + \left[\frac{K_2}{s + j\omega} + \frac{K_3}{s - j\omega}\right]\left(s + \frac{1}{RC}\right)$$

Then let $s = -1/RC$. Thus,

$$K_1 = \frac{1}{s^2 + \omega^2}\bigg|_{s = -1/RC}$$

$$= \frac{1}{\omega^2 + (1/RC)^2}$$

Multiply both sides of Eq. (8.70) by $s + j\omega$ and let $s = -j\omega$ to get

$$K_2 = \frac{1}{(s + 1/RC)(s - j\omega)}\bigg|_{s = -j\omega}$$

$$= \frac{1}{[(1/RC) - j\omega](-2j\omega)} = \frac{-(1/RC + j\omega)}{[\omega^2 + (1/RC)^2]2j\omega}$$

and similarly,

$$K_3 = \frac{1}{(s + 1/RC)(s + j\omega)}\bigg|_{s = j\omega}$$

$$= \frac{1}{[(1/RC) + j\omega](2j\omega)} = \frac{(1/RC) - j\omega}{[\omega^2 + (1/RC)^2]2j\omega}$$

Since K_2 is complex, K_3 will be its complex conjugate. The partial-fraction expansion of the term inside the brackets in Eq. (8.69) becomes

$$\left[\frac{1}{\omega^2 + (1/RC)^2}\right]\left[\frac{1}{s + 1/RC} - \frac{\sqrt{\omega^2 + (1/RC)^2}\, e^{j\theta}}{2j(s + j\omega)\omega} + \frac{\sqrt{\omega^2 + (1/RC)^2}\, e^{-j\theta}}{2j(s - j\omega)\omega}\right]$$

where $\theta = \tan^{-1} \omega RC$. From Table 8.4 the transform of this is (changing the order of the last two terms)

$$\left[\frac{1}{\omega^2 + (1/RC)^2}\right]\left[e^{-(1/RC)t} + \frac{\sqrt{\omega^2 + (1/RC)^2}\, e^{-j\theta}e^{j\omega t}}{2j\omega} - \frac{\sqrt{\omega^2 + (1/RC)^2}\, e^{j\theta}e^{-j\omega t}}{2j\omega}\right]$$

or

$$\left[\frac{1}{\omega^2 + (1/RC)^2}\right]\left[e^{-(1/RC)t} + \frac{\sqrt{\omega^2 + (1/RC)^2}}{\omega}\frac{e^{j(\omega t - \theta)} - e^{-j(\omega t - \theta)}}{2j}\right]$$

Thus, from Eq. (8.69),

$$v_c(t) = \frac{A\omega}{RC}\left[\frac{1}{\omega^2 + (1/RC)^2}e^{-(1/RC)t} + \frac{1}{\omega\sqrt{\omega^2 + (1/RC)^2}}\sin(\omega t - \theta)\right] \qquad (8.71)$$

It is clear from this example that the labor in the Laplace transform method resides in taking the inverse transform. Fortunately, the engineer can deduce many characteristics of the network by examination of the location of the poles of the transfer function. Note that in this example additional poles of the response not belonging to the transfer function are introduced by the forcing function $A \sin \omega t$.

EXAMPLE 8.6

As another example, let us examine the circuit of Fig. 8.7, in which the switch sw closes at $t = 0$. Find the current i.

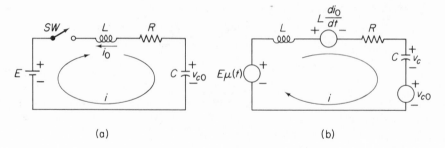

Figure 8.7 *RLC* network for Example 8.6.

Solution. The initial current in the inductance is $-i_0$ and the initial voltage across the capacitance is $-v_{c0}$. The differential equation after sw closes is

$$L\frac{di}{dt} + Ri + \frac{1}{C}\int i\,dt = E \tag{8.72}$$

Applying Eqs. (8.13) and (8.17) to the transformed network in Fig. 8.7b we get

$$LsI(s) + RI(s) + \frac{1}{Cs}I(s) = \frac{E}{s} + \frac{v_{c0}}{s} - Li_0$$

Collecting terms and multiplying through by s we get

$$\left(Ls^2 + Rs + \frac{1}{C}\right)I(s) = -sLi_0 + v_{c0} + E$$

or

$$I(s) = \frac{-si_0 + (v_{c0} + E)/L}{s^2 + (R/L)s + 1/LC} \tag{8.73}$$

The transfer function here is

$$G(s) = \frac{1}{L(s^2 + (R/L)s + (1/LC))}$$

The poles (and the network eigenvalues) are the roots of the characteristic equation

$$s^2 + \frac{R}{L}s + \frac{1}{LC} = 0$$

The inverse transformation will depend on whether the roots are real and unequal, real and equal, or complex, as in the conventional case. If they are real, the inverse transform gives two exponentials. Thus assume that

$$s^2 + \frac{R}{L}s + \frac{1}{LC} = (s + \alpha_1)(s + \alpha_2), \qquad \alpha_1, \alpha_2 > 0$$

Then

$$I(s) = \frac{-si_0 + (v_{c0} + E)/L}{(s + \alpha_1)(s + \alpha_2)} \tag{8.74}$$

The partial-fraction expansion of Eq. (8.74) is

$$I(s) = \frac{K_1}{s + \alpha_1} + \frac{K_2}{s + \alpha_2} \tag{8.75}$$

where

$$K_1 = \frac{-si_0 + (v_{c0} + E)/L}{s + \alpha_2}\bigg|_{s=-\alpha_1} = \frac{\alpha_1 i_0 + (v_{c0} + E)/L}{\alpha_2 - \alpha_1}$$

and

$$K_2 = \frac{-si_0 + (v_{c0} + E)/L}{s + \alpha_1}\bigg|_{s=-\alpha_2} = \frac{\alpha_2 i_0 + (v_{c0} + E)/L}{\alpha_1 - \alpha_2}$$

Hence,

$$i(t) = \frac{\alpha_1 i_0 + (v_{c0} + E)/L}{\alpha_2 - \alpha_1} e^{\alpha_1 t} + \frac{-\alpha_2 i_0 + (v_{c0} + E)/L}{\alpha_1 - \alpha_2} e^{-\alpha_2 t} \tag{8.76}$$

After finding K_1 and K_2, it is a good practice to put the right-hand side of Eq. (8.75) back over a common denominator to see if Eq. (8.74) results. This is arduous, but reveals any computational errors.

If the roots of the characteristic equation are complex, then

$$s^2 + \frac{R}{L}s + \frac{1}{LC} = s^2 + 2\zeta\omega_n s + \omega_n^2$$

$$= (s + \zeta\omega_n)^2 + \omega_d^2 \tag{8.77}$$

where $\omega_d = \omega_n\sqrt{1 - \zeta^2}$, and ζ and ω_n have the usual meanings. The last operation is called *completing the square* and is necessary because most tables give the function in this form. Then

$$I(s) = \frac{-i_0 s + (v_{c0} + E)/L}{(s + \zeta\omega_n)^2 + \omega_d^2} \tag{8.78}$$

The second term on the right-hand side may be written

$$\frac{v_{c0} + E}{L\omega_d}\left[\frac{\omega_d}{(s + \zeta\omega_n)^2 + \omega_d^2}\right]$$

and the transform from the table is

$$\frac{v_{c0} + E}{L\omega_d}[e^{-\zeta\omega_n t} \sin \omega_d t] \tag{8.79}$$

The first term on the right-hand side of Eq. (8.78) may be written

$$-\frac{i_0}{\omega_d} s \frac{\omega_d}{(s + \zeta\omega_n)^2 + \omega_d^2}$$

and since s represents the derivative in the time domain, the inverse is

$$-\frac{i_0}{\omega_d}\frac{d}{dt}(e^{-\zeta\omega_n t}\sin\omega_d t) = -\frac{i_0}{\omega_d}(-\zeta\omega_n e^{-\zeta\omega_n t}\sin\omega_d t + \omega_d e^{-\zeta\omega_n t}\cos\omega_d t)$$

$$= -\frac{i_0}{\omega_d}e^{-\zeta\omega_n t}(-\zeta\sin\omega_d t + \omega_d\cos\omega_d t) \tag{8.80}$$

The complete current is the sum of the two inverses in Eqs. (8.79) and (8.80). Usually it is much easier to deal directly with numerical values than to develop general algebraic equations as we have here. Also, it is better to find a transform pair which corresponds to the one desired than to use the methods shown previously, since the likelihood of errors is reduced. However all-inclusive tables are seldom available.

If the characteristic equation is of third order, it will have at least one real eigenvalue. If the equation is of fourth order, we have several possible root combinations. If all roots are real except for one pair, the arduous task of using complex quantities may be avoided by a technique now shown.

EXAMPLE 8.7

Suppose that we have a function $F(s) = P(s)/Q(s)$ in which $Q(s)$ is factored as

$$F(s) = \frac{10s + 3}{(s + 1)(s^2 + 5s + 25)} \tag{8.81}$$

The partial-fraction expansion of $F(s)$ can be written

$$F(s) = \frac{K_1}{s + 1} + \frac{K_2 s + K_3}{s^2 + 5s + 25} \tag{8.82}$$

where K_1, K_2, and K_3 are real constants. The coefficients for the real roots, such as K_1, are found using Eq. (8.49). Thus, as before,

$$K_1 = \frac{10s + 3}{s^2 + 5s + 25}\bigg|_{s=-1} = -\frac{7}{21} = -\frac{1}{3}$$

Then

$$F(s) = \frac{-\frac{1}{3}(s^2 + 5s + 25) + (K_2 s + K_3)(s + 1)}{(s + 1)(s^2 + 5s + 25)} \tag{8.83}$$

Collecting terms in the numerator of Eq. (8.83) and comparing the terms with the numerator of Eq. (8.81), we find that

$$10s + 3 = (K_2 - \tfrac{1}{3})s^2 + (K_2 + K_3 - \tfrac{5}{3})s + (K_3 - \tfrac{25}{3})$$

The constants K_2 and K_3 which satisfy Eq. (8.81) are thus

$$K_2 = \tfrac{1}{3} \quad (s^2 \text{ coeff.})$$

$$K_3 - \tfrac{25}{3} = 3 \quad \text{or} \quad K_3 = \tfrac{34}{3} \quad (s^0 \text{ coeff.})$$

We can verify the validity of the coefficients by inserting them in the equation for the s^1 term. Thus,

$$(\tfrac{1}{3} + \tfrac{34}{3} - \tfrac{5}{3}) = \tfrac{30}{3} = 10 \quad (s^1 \text{ coeff.})$$

The method becomes more difficult with the number of roots and does not allow the check previously advocated, but the method may be less frustrating than finding the complex coefficients K_4 and K_4^* if the last term is expressed as

$$\frac{K_2 s + K_3}{s^2 + 5s + 25} = \frac{K_4}{s + 2.5 + j(\tfrac{75}{4})^{1/2}} + \frac{K_4^*}{s + 2.5 - j(\tfrac{75}{4})^{1/2}}$$

and then combining the results of the complex exponential inverses. Unfortunately, complex roots become very prevalent in high-order polynomials.

It should be noted with respect to repeated roots that if the characteristic equation has nonintegral coefficients which are rather randomly obtained, repeated roots will be extremely unlikely. Furthermore, slight variation in the polynomial coefficients (such as occurs in computer roundoff) results in large root changes. Hence, except for integral coefficients (such as in some quantum mechanics problems), repeated roots in engineering problems (or repeated eigenvalues) are quite unlikely.

An alternative method of finding a partial-fraction expansion which can be implemented on a digital computer, and the associated computer program based on this method, are presented in Chapter 10. This program obtains the partial-fraction expansion of a ratio of two polynomials where the numerator polynomial is of lower degree than the denominator, and where the denominator roots are known and nonrepeated.

Nodal or loop equations can be written using the same methods as illustrated in Chapter 4.

If we consider the node equations for the circuit in Fig. 8.8, we have

$$\begin{bmatrix} C_1 \dfrac{d}{dt} + G_1 + G_3 + \Gamma_3 \int dt & -G_3 - \Gamma_3 \int dt \\[2ex] -G_3 - \Gamma_3 \int dt & G_2 + G_3 + (\Gamma_2 + \Gamma_3) \int dt \end{bmatrix} \begin{bmatrix} v_1 \\[2ex] v_2 \end{bmatrix} = \begin{bmatrix} J_{s1}(t) \\[2ex] J_{s2}(t) \end{bmatrix}$$

(8.84)

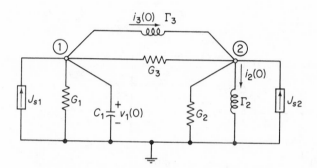

Figure 8.8 Network for simultaneous equations.

Equation (8.84) transforms to a function of s by taking the Laplace transform of the equations for $i_1(t)$ and $i_2(t)$, which produces

$$\begin{bmatrix} sC_1 + G_1 + G_3 + \dfrac{\Gamma_3}{s} & -G_3 - \dfrac{\Gamma_3}{s} \\[2ex] -G_3 - \dfrac{\Gamma_3}{s} & G_2 + G_3 + \dfrac{\Gamma_2 + \Gamma_3}{s} \end{bmatrix} \begin{bmatrix} V_1(s) \\[2ex] V_2(s) \end{bmatrix} = \begin{bmatrix} J_{s1}(s) \\[2ex] J_{s2}(s) \end{bmatrix}$$

(8.85)

The sources representing initial conditions have been omitted in Eq. (8.85). The network in Fig. 8.8 is equivalent to the network in Fig. 8.9, where the initial conditions for the reactive elements are represented by equivalent sources. The addition of the initial conditions modify the equivalent current vector so that the nodal equations now appear as

$$\begin{bmatrix} sC_1 + G_1 + G_3 + \dfrac{\Gamma_3}{s} & -G_3 - \dfrac{\Gamma_3}{s} \\[2ex] -G_3 - \dfrac{\Gamma_3}{s} & G_2 + G_3 + \dfrac{\Gamma_2 + \Gamma_3}{s} \end{bmatrix} \begin{bmatrix} V_2(s) \\[2ex] V_2(s) \end{bmatrix} = \begin{bmatrix} J_{s1}(s) + C_1 v_1(0) - \dfrac{i_3(0)}{s} \\[2ex] J_{s2}(s) + \dfrac{i_3(0)}{s} - \dfrac{i_2(0)}{s} \end{bmatrix}$$

(8.86)

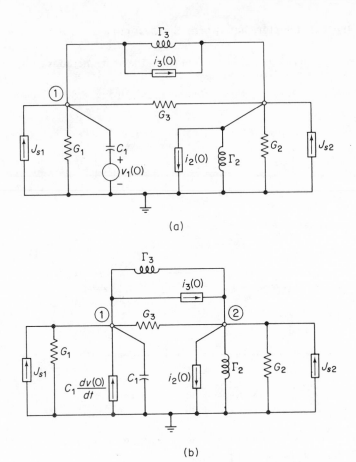

Figure 8.9 Equivalent representation of Fig. 8.8.

The term $C_1 v_1(0)$ is derived from $sC_1[v_1(0)/s]$ and is equal to the initial charge on the capacitance. (The initial conditions are all applied as step inputs applied at $t = 0$.) The nodal equations can also be written

$$\mathbf{Y}(s)\mathbf{V}(s) = \mathbf{J}(s) \tag{8.87}$$

where $\mathbf{J}(s)$ contains all the network driving sources as well as the initial conditions. We apply loop analysis similarly, as shown in the following example.

EXAMPLE 8.8

In the network shown in Fig. 8.10, the switch is closed at $t = 0$. With the network parameter values given, the network equations are

$$\begin{bmatrix} s + 20 & -10 \\ -10 & s + 20 \end{bmatrix} \begin{bmatrix} I_1(s) \\ I_2(s) \end{bmatrix} = \begin{bmatrix} \dfrac{100}{s} \\ 0 \end{bmatrix}$$

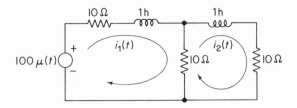

Figure 8.10 Network for Example 8.8.

if the initial values of i_1 and i_2 at $t = 0_-$ are zero. Suppose that we are required to find $i_2(t)$.

Solution. The transformed current $I_2(s)$ may be found from the network equations by Cramer's rule as

$$I_2(s) = \frac{\begin{vmatrix} s + 20 & \dfrac{100}{s} \\ -10 & 0 \end{vmatrix}}{\begin{vmatrix} s + 20 & -10 \\ -10 & s + 20 \end{vmatrix}}$$

$$= \frac{1000}{s(s^2 + 40s + 300)}$$

The partial-fraction expansion of $I_2(s)$ is

$$\frac{1000}{s(s + 10)(s + 30)} = \frac{3.33}{s} - \frac{5}{s + 10} + \frac{1.67}{s + 30}$$

The inverse Laplace transform of $I_2(s)$ is

$$i_2(t) = 3.33 - 5e^{-10t} + 1.67e^{-30t} \qquad t \geq 0$$

which is the desired solution.

We have considered several network elements, which we called *trans*conductance, *trans*potential, and so on. These "trans" terms could also be considered to be *transfer functions;* that is, the output (either voltage or current) becomes a function of an input voltage or current. These transfer functions are of the form

$$y = hx \tag{8.88}$$

where h is a constant, y the output of the element, and x the input to the element. Extending this concept to networks with multiple inputs and outputs, we arrive at the same representation, except written in matrix form as

$$\mathbf{y} = \mathbf{Hx} \tag{8.89}$$

In terms of Laplace-transformed networks, the relationship becomes

$$\mathbf{Y}(s) = \mathbf{H}(s)\mathbf{X}(s) \tag{8.90}$$

As discussed briefly previously, $\mathbf{H}(s)$ is known as the transfer function of the network.†
For example, Eq. (8.87) can be written as

$$\mathbf{V}(s) = \mathbf{Y}(s)^{-1}\mathbf{J}(s) \tag{8.91}$$

from which we deduce that the transfer function for the network is

$$\mathbf{H}(s) = \mathbf{Y}(s)^{-1} \tag{8.92}$$

The transfer function is not a function of the inputs or time in linear time-invariant
networks but is a function of the network parameters only.

The transfer-function matrix $\mathbf{H}(s)$ is given by

$$\mathbf{H}(s) = \mathbf{Y}(s)^{-1}$$

$$= \frac{1}{\Delta_r(s)}\begin{bmatrix} Y_{11}(s) & Y_{21}(s) & \cdots & Y_{n1}(s) \\ Y_{12}(s) & Y_{22}(s) & \cdots & Y_{n2}(s) \\ \cdot & \cdot \cdot \cdot \cdot \cdot \cdot \cdot \cdot & \cdot \cdot \\ Y_{1n}(s) & Y_{2n}(s) & \cdots & Y_{nn}(s) \end{bmatrix} \tag{8.93}$$

Note that each term coupling $J(s)$ to the output $V(s)$ has the denominator $\Delta_r(s)$. This
is again the characteristic polynomial of the network. Or the roots of

$$\Delta_r(s) = 0 \tag{8.94}$$

represent the poles of the network.

In the case of relatively simple networks, the transfer function may be found by
defining the general impedances $Z(s) = Ls$ for an inductance, $Z(s) = 1/Cs$ for a
capacitance, and $Z(s) = R$ for a resistor. That is, we make the substitution of s for
the $j\omega$ of the sinusoidal case (see Section 6.3). We then combine the impedances just
as in the sinusoidal case.

EXAMPLE 8.9

Find the transfer function $V_c(s)/E(s)$ in Fig. 8.7.

Solution. Since the current $i(t)$ is common to all elements, the desired voltage ratio is the
same as the ratio of the impedance of the capacitance to the total impedance. Hence,

$$\frac{V_c(s)}{E(s)} = \frac{1/Cs}{Ls + R + 1/Cs}$$

or

$$\frac{V_c(s)}{E(s)} = \frac{1/LC}{s^2 + (R/L)s + 1/LC}$$

This method may prove simpler than writing the Laplace-transformed equations,
particularly for single input–output networks, where the transfer-function matrix
consists of a single element.

---

†We adopt $\mathbf{H}(s)$ as the symbol for a transfer-function matrix.

From the examples given we see that the difficult computational problems with the Laplace-transform methods arise in the inverse transformation process. Computer programs become necessary for the solution of problems of any magnitude. In many cases, however, the designer can judge the performance of the network, at least in a general way, without taking the inverse transform. Hence, we will proceed no further with inverse transformation theory but include a list of excellent texts on the subject. We also remark that we have not proved that the integral in Eq. (8.8b) actually converges to $f(t)$, or the restrictions on $F(s)$ for it to do so. Appendix A presents some discussion and gives references for rigorous proofs. Excusing our indolence, we plea that now the solution of differential equations by the digital (or analog) computer in the time domain is now so well established that inverse Laplace transformation has considerably less importance than previously. We do remark that at points of discontinuity, Eq. (8.4) holds for the Laplace transform for reasonably well-behaved functions.

Additional Reading

ASELTINE, J. A., *Transform Methods in Linear System Analysis*, McGraw-Hill Book Company, New York, 1958.

CHURCHILL, R. V., *Modern Operational Mathematics in Engineering*, McGraw-Hill Book Company, New York, 1944.

DESOER, C. A., and E. S. KUH, *Basic Circuit Theory*, McGraw-Hill Book Company, New York, 1969.

GARDNER, M. F., and J. L. BARNES, *Transients in Linear Systems*, John Wiley & Sons, Inc., New York, 1942.

KUO, B. C., *Linear Networks and Systems*, McGraw-Hill Book Company, New York, 1967.

PAPOULIS, A., *The Fourier Integral and Its Applications*, McGraw-Hill Book Company, New York, 1962.

SANFORD, R. S., *Physical Networks*, Prentice-Hall, Inc., Englewood Cliffs, N.J., 1965.

SPIEGEL, M. R., *Laplace Transforms*, Schaum's Outline Series, McGraw-Hill Book Company, New York, 1965.

WATKINS, B. O., *Introduction to Control Systems*, The Macmillan Company, New York, 1969.

WYLIE, C. R., Jr., *Advanced Engineering Mathematics*, McGraw-Hill Book Company, New York, 1960.

Problems

8.1. Demonstrate that the Laplace-transform pairs of Table 8.4 are correct by using Eq. (8.8a), Eq. (8.8b), or any previously demonstrated pair.

8.2. Find the partial-fraction expansion of the following functions. If a quadratic deno-
minator of a fraction has complex roots, find it in the form $(As + B)/(s^2 + 2\zeta\omega_n s + \omega_n^2)$. Verify the results by recombining the fractions over a common denominator.

(a) $\dfrac{3}{s^2 + 5s + 6}$

(f) $\dfrac{s^2 + 5}{(s^2 + 2s + 26)(s + 1)(s + 2)}$

(b) $\dfrac{s + 4}{s^2 + 5s + 6}$

(g) $\dfrac{s + 1}{s^3 + 6s^2 + 11s + 6}$

(c) $\dfrac{2s^2 + 5s + 3}{s^2 + 5s + 6}$

(h) $\dfrac{s + 2}{(s^2 + 2s + 1)(s + 3)}$

(d) $\dfrac{5s + 3}{(s^2 + 7s + 12)(s + 1)}$

(i) $\dfrac{s + 4}{s(s^2 + 5s + 6)}$

(e) $\dfrac{5s + 3}{(s^2 + 4s + 13)(s + 1)}$

(j) $\dfrac{5s + 3}{s(s^2 + 4s + 13)(s + 1)}$

8.3. Note in the last two lines of Table 8.4 that the denominator of the transform must be
put into the form $[(s + a)^2 + \alpha^2]$ if the time function is the product of an exponential
and a sine or cosine function [see Eq. (8.77)]. This always occurs if the roots of the
characteristic equation are complex. Put the following fractions in this form:

(a) $\dfrac{s}{s^2 + 4s + 13}$

(b) $\dfrac{3}{s^2 + 2s + 26}$

(c) $\dfrac{4s + 2}{s^2 + 2s + 10}$

8.4. If the s-domain functions of Problem 8.2 represent Laplace transforms of time functions, find these functions by taking the inverse Laplace transform.

8.5. In Example 8.1 let $C = 0.05$ μF, $R = 0.5$ MΩ, $A = 5$, and $v_{c0} = +10$. Find the
response $v_c(t)$, $t \geq 0$.

8.6. In Example 8.5 let $C = 0.1$ μF, $R = 10 \times 10^6$ Ω, and $A = 10$. Find the response
$v_c(t)$, $t \geq 0$.

8.7. In Fig. 8.8 assume that J_{s1} and J_{s2} are time step functions of value 2 and 3 A, respectively, and that $v_1(0) = 5$ V, $i_2(0) = 2$ A, and $i_3(0) = 1$ A. Let $G_1 = 0.5$, $G_2 = 1$,
and $G_3 = 0.1$ (all in mhos), $\Gamma_2 = 0.5$ and $\Gamma_3 = 0.2$ in inverse henries, and $C = 0.1$
μF. Develop Eq. (8.86) with these values. Obtain equations for $V_1(s)$ and $V_2(s)$,
but do not find the inverse transformations. What are the roots of the characteristic
equation (the eigenvalues of the admittance matrix)?

8.8. Find the transfer function $I(s)/E(s)$ in Fig. P8.1 if $R = L/C = 1$.

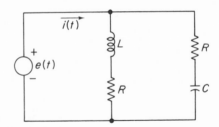

Figure P8.1

8.9. Assume that a transfer function is given by

$$G(s) = \frac{5(s - a)}{(s - 1)(s + 2)(s + 3)}$$

What time-response terms might we expect with an impulse input if $a = 1.0$? What terms might we expect if $a = 1.00001$?

8.10. Solve the following differential equations using the Laplace-transform method:

(a) $\dfrac{d^2i}{dt^2} - i = 25 + e^{2t}$ $i(0) = 2$ $\dfrac{di(0)}{dt} = 0$

(b) $\dfrac{d^2i}{dt^2} + \dfrac{di}{dt} = t^2 + 2t$ $i(0) = 4$ $\dfrac{di(0)}{dt} = -2$

8.11. Find the required output in the following problems in Chapter 5 using Laplace-transform methods (if an output is not given, select an element current or voltage): (a) 5.6, (b) 5.7, (c) 5.13, (d) 5.14, (e) 5.15, (f) 5.17, (g) 5.18, (h) 5.19.

Chapter 9

ANALYSIS IN THE COMPLEX PLANE

When a response becomes stipulated,
poles and zeros must be manipulated.

9.1 Sinusoidal Response in Terms of Poles and Zeros

In Chapter 6 we saw that if the input to a linear network varies sinusoidally with time, the steady-state output is also sinusoidal. The output differs from the input in (1) magnitude and (2) phase angle; both of these quantities vary with the angular velocity ω. For these steady-state conditions, we define the ratio of the output to the input as the transfer function. Thus, if $v_1(t)$ in Fig. 9.1 represents an input and $v_2(t)$

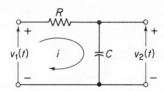

Figure 9.1 *RC* network.

represents an output, and if $v_2(t) = V_1 e^{j\omega t}$, then the input impedance Z is given by

$$Z = R + \frac{1}{j\omega C} = R - jX_c$$

where

$$j = \sqrt{-1}$$

Hence

$$\hat{I} = \frac{\hat{V}_1}{Z}$$

and

$$\hat{V}_2 = \hat{I}\left(\frac{1}{j\omega C}\right)$$

or

$$\hat{V}_2 = \frac{\hat{V}_1}{Z} X_c$$

$$= \hat{V}_1\left[\frac{1}{R + 1/j\omega C}\right]\left[\frac{1}{j\omega C}\right]$$

$$= \hat{V}_1\left[\frac{1}{j\omega RC + 1}\right] \tag{9.1a}$$

$$= \hat{V}_1\left[\frac{1/RC}{j\omega + 1/RC}\right] \tag{9.1b}$$

The quantity in the brackets on the right-hand side of Eq. (9.1a) or Eq. (9.1b) is the transfer function $G(j\omega)$ and gives us the ratio of the magnitudes of V_1 and V_2, and their phase difference. Thus,

$$G(j\omega) = \frac{\hat{V}_2}{\hat{V}_1}$$

$$= \frac{1}{[(\omega RC)^2 + 1]^{1/2}} e^{j\theta(\omega)}$$

$$= M(\omega)e^{j\theta(\omega)}$$

where

$$\theta = -\tan^{-1}\omega RC$$

or

$$v_2(t) = V_1 | G(j\omega) | e^{j\omega t} e^{j\theta(\omega)}$$

The differential equation describing the network in Fig. 9.1 is

$$RC\frac{dv_2}{dt} + v_2 = v_1(t) \tag{9.2}$$

Using the Laplace transform with zero initial conditions, we get

$$(RCs + 1)V_2(s) = V_1(s)$$

or

$$V_2(s) = \left[\frac{1}{RCs + 1}\right]V_1(s) \tag{9.3a}$$

$$= \left[\frac{1/RC}{s + 1/RC}\right]V_1(s) \tag{9.3b}$$

In Chapter 8 we defined the quantity in the brackets in Eq. (9.3a) or Eq. (9.3b) as the transfer function in the s-domain. This quantity expresses the network characteristics, as opposed to the signal characteristics. We observe that if we replace the variable s in the denominator of the transfer function in Eq. (9.3a) by $j\omega$, we obtain the transfer function $G(j\omega)$ of Eq. (9.1a). Thus, if we know the transfer function in the s-domain, we can obtain the transfer function for the steady-state sinusoidal case by simply replacing s by $j\omega$. Alternatively, to obtain transfer functions in the s-domain, we change the inductive impedance $j\omega L$ to sL and the capacitive impedance $1/j\omega C$ to $1/sC$. By combining the "impedances" in series and parallel, we may rapidly obtain transfer functions in the s-domain.

EXAMPLE 9.1

Find the transfer function $G(s) = Z(s) = V(s)/I(s)$ and the differential equation relating $i(t)$ and $v(t)$ in Fig. 9.2.

Solution

$$Z(j\omega) = R_2 + j\omega L + \frac{R_1(1/j\omega C)}{R_1 + 1/j\omega C}$$

$$= R_2 + j\omega L + \frac{R_1}{j\omega R_1 C + 1}$$

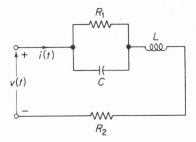

Figure 9.2 *RLC* network for Example 9.1.

Then

$$Z(s) = \frac{V(s)}{I(s)}$$

$$= R_2 + sL + \frac{R_1}{R_1 Cs + 1}$$

$$= \frac{R_2 R_1 Cs + R_2 + R_1 LCs^2 + sL + R_1}{R_1 Cs + 1}$$

$$= \frac{Ls^2 + (R_2 + L/R_1 C)s + (R_1 + R_2)/R_1 C}{s + 1/R_1 C}$$

Solving for $I(s)$ we find that $I(s) = V(s)/Z(s)$, which then provides the differential equation relating i and v as follows:

$$L\frac{d^2 i}{dt^2} + \left(R_2 + \frac{L}{R_1 C}\right)\frac{di}{dt} + \frac{R_1 + R_2}{R_1 C}i = \frac{dv}{dt} + \frac{1}{R_1 C}v$$

Note that in general the solution of the differential equation involves transient terms as well as steady-state terms, even though we used only the transfer function to obtain it.

We also observed in Chapter 8 that for linear networks the transfer function is always a ratio of polynomials in s. In general, if $G(s)$ is a transfer function between input and output, then

$$G(s) = \frac{KP(s)}{Q(s)} \tag{9.4a}$$

$$= \frac{a_0 s^m + a_1 s^{m-1} + \ldots + a_m}{b_0 s^n + b_1 s^{n-1} + \ldots + b_n} \tag{9.4b}$$

where $K = a_0/b_0$. We may factor the numerator $P(s)$ and denominator $Q(s)$ of Eq. (9.4) to obtain

$$G(s) = \frac{a_0(s - z_1)(s - z_2)\ldots(s - z_m)}{b_0(s - p_1)(s - p_2)\ldots(s - p_m)} \tag{9.5}$$

In Eq. (9.5), the denominator roots p_i, $i = 1, 2, \ldots, n$, are called *poles* since when $s = p_i$, $G(s)$ becomes infinite. The numerator roots z_i, $i = 1, 2, \ldots, m$, are called *zeros* by similar reasoning. We have already discussed these terms.

Using the results of our previous analysis, the sinusoidal transfer function then becomes

$$G(j\omega) = \frac{K(j\omega - z_1)(j\omega - z_2)\ldots(j\omega - z_m)}{(j\omega - p_1)(j\omega - p_2)\ldots(j\omega - p_n)} \tag{9.6}$$

where $K = a_0/b_0$.

Assuming for the moment that all roots p_i and z_i in Eq. (9.6) are real, the p_i may be considered as inverse time constants. Thus, in Eq. (9.3a) if $V_1(s) = 1$ $[v_1(t) = \delta(t)]$, then

$$v_2(t) = \mathcal{L}^{-1}\left[\frac{1}{RCs + 1}\right]$$

$$= \frac{1}{RC}\mathcal{L}^{-1}\left[\frac{1}{s + 1/RC}\right]$$

$$= \frac{1}{RC}e^{-t/RC}$$

$$= \frac{1}{\tau}e^{-t/\tau}$$

where $\tau = RC$ is the time constant for the circuit (see Chapter 5). Now let $p_1 = -1/\tau_1$, $p_2 = -1/\tau_2, \ldots$, and $z_1 = -1/\tau_a$, $z_2 = -1/\tau_b, \ldots$, so that we may write Eq. (9.6) as

$$G(j\omega) = \frac{a_0\tau_1\tau_2\ldots\tau_n(1 + j\omega\tau_a)(1 + j\omega\tau_b)\ldots(1 + j\omega\tau_m)}{b_0\tau_a\tau_b\ldots\tau_m(1 + j\omega\tau_1)(1 + j\omega\tau_2)\ldots(1 + j\omega\tau_n)} \tag{9.7}$$

We may use Eq. (9.6) or Eq. (9.7), whichever is convenient, as they are identical algebraically and differ only in form.

Equation (9.5) may be depicted graphically in the s-plane by drawing vectors form to poles and zeros to a particular point s_1. To illustrate, consider the transfer function

$$G(s) = \frac{k(s - z_1)}{(s - p_1)(s - p_2)}$$

where p_1 and p_2 are each shown by a cross ($\times$) in the s-plane and z_1 by a circle ($\bigcirc$), as shown in Fig. 9.3. We now select a point $s = s_1$ and draw vectors from p_1 to s_1, p_2 to

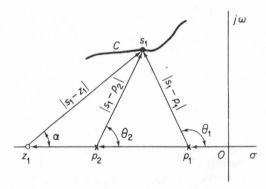

Figure 9.3 Vectors in the s-plane.

s_1, and z_1 to s_1, as shown. These vectors are then, respectively, $s_1 - p_1$, $s_1 - p_2$, and $s_1 - z_1$. Hence, $G(s_1)$ may be considered as a vector represented by Eq. (9.7); that is, $|G|$, the magnitude of $G(s_1)$ is

$$|G(s_1)| = \frac{k|s_1 - z_1|}{|s_1 - p_1||s_1 - p_2|}$$

where $|s_1 - z_1|$ is the magnitude of the vector from z_1 to s_1, $|s_1 - p_1|$ is the magnitude of the vector from p_1 to s_1, and so on. The phase angle of $G(s_1)$ is

$$\arg G(s_1) = \alpha - \theta_1 - \theta_2$$

where α, θ_1, and θ_2 are shown in Fig. 9.3. Thus, $G(s_1)$ might appear in a (different) complex plane as shown in Fig. 9.4. As s varies along a curve C in Fig. 9.3, $G(s)$ will

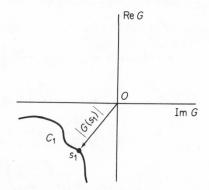

Figure 9.4 The function $G(s)$ in a complex u-v plane.

vary along a corresponding curve C_1 in Fig. 9.4. We will utilize these ideas later in the chapter when we discuss the root locus and its application.

Carrying the vector concept on to the sinusoidal case, the position of the point s in the complex plane is now restricted to the $j\omega$-axis, and our previous illustration now becomes as shown in Fig. 9.5 for some particular $\omega = \omega_1$. In this case,

$$G(j\omega) = \frac{k(j\omega - z_1)}{(j\omega - p_1)(j\omega - p_2)}$$

$$= \frac{k|j\omega - z_1|}{|j\omega - p_1||j\omega - p_2|} e^{j(\alpha - \theta_1 - \theta_2)} \qquad (9.8)$$

where the bars indicated magnitudes.

EXAMPLE 9.2

The transfer function representing the impedance of a series *RLC* circuit is

$$Z(s) = Ls + R + \frac{1}{Cs}$$

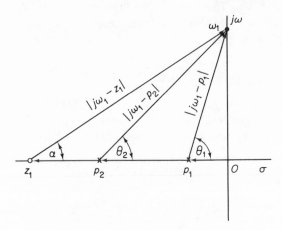

Figure 9.5 Point s_1 confined to the $j\omega$ axis.

or the admittance is

$$Y(s) = \frac{1}{L} \frac{s}{s^2 + sR/L + 1/LC}$$

$$= \frac{1}{L} \frac{s}{s^2 + 2\zeta\omega_n s + \omega_n^2}$$

Then

$$Y(j\omega) = \frac{1}{L} \frac{j\omega}{(j\omega - p_1)(j\omega - p_1^*)}$$

where

$$p_1 = -\frac{R}{2L} + j\sqrt{\frac{1}{LC} - \left(\frac{R}{2L}\right)^2}$$

$$= -\zeta\omega_n + j\omega_n\sqrt{1 - \zeta^2}$$

and

$$p_1^* = -\frac{R}{2L} - j\sqrt{\frac{1}{LC} - \left(\frac{R}{2L}\right)^2}$$

(We assume the poles to be complex.) Figure 9.6 shows the vectors $(j\omega - p_1)$, $(j\omega - p_1^*)$, and $(j\omega - 0)$ drawn for some $\omega = \omega_1$. The magnitude of $Y(j\omega)$ for any ω can be found by dividing the product of the distances from the zeros to ω_1 by the product of the distances from the poles to ω_1. The magnitude of the admittance $|Y(j\omega_1)|$ for the con-figuration shown in Fig. 9.6 is

$$|Y(j\omega)| = \frac{1}{L} \frac{|j\omega|}{|j\omega - p_1||j\omega - p_1^*|}$$

while the phase is the sum of the angles of the vectors from the zeros to ω_1 minus the sum of the angles of the vectors from the poles to ω_1. In Fig. 9.6 the phase angle becomes $90° - \theta_1 - \theta_2$. The magnitude result is the familiar resonance curve shown in Fig. 9.7. If we let the circuit $Q = \omega_n L/R$, then in terms of ζ, $Q = 1/2\zeta$. Thus, the smaller ζ, the larger the Q, and the sharper the resonance curve. If we define resonance as the point of unity power factor ($\omega L = 1/\omega C$), then the maximum of the curve of $|Y(j\omega)|$

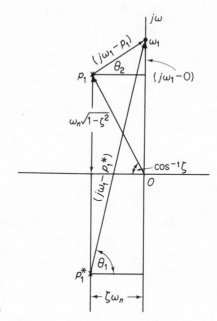

Figure 9.6 Vectors for Example 9.2.

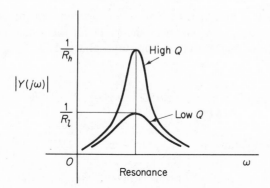

Figure 9.7 Magnitude curve for Example 9.2.

$= 1/R$ and occurs at $\omega = \omega_n = \sqrt{1/LC}$. The circuit current at this point is $|I| = |V|/R$. When the current has the value

$$|I| = \frac{|V|}{\sqrt{2}\,R}$$

the circuit power will be half the power at resonance. If we define ω_1 and ω_2 as the angular velocities at which this occurs (half-power angular velocities), then we may define the half-power bandwidth BW as

$$\mathrm{BW} = \omega_2 - \omega_1$$

It is easy to show that $\mathrm{BW} = 2\zeta\omega_n = \omega_n/Q$ (see the Problems). To be mathematically correct, we should continue the process for all negative ω in Fig. 9.6. This would result in the same resonance curves along the negative axis in Fig. 9.7.

Vector representation as illustrated by the previous discussion and Example 9.2 is an interesting way to visualize the transfer function, but perhaps not too useful in computational work. H. W. Bode† suggested a method of obtaining the magnitude and phase curves by working with the logarithm of $G(s)$. If we take the logarithm of $G(j\omega)$ in Eq. (9.7), we obtain

$$\log G(j\omega) = \log \frac{a_0 \tau_1 \tau_2 \cdots \tau_n}{b_0 \tau_a \tau_b \cdots \tau_m} + \log(1 + j\omega\tau_a) + \log(1 + j\omega\tau_b)$$
$$+ \ldots + \log(1 + j\omega\tau_m) - \log(1 + j\omega\tau_1) - \log(1 + j\omega\tau_2) \quad (9.9)$$
$$- \ldots - \log(1 + j\omega\tau_n)$$

Multiplication and division of the factors has now been simplified to addition and subtraction, and if we can determine the response of each factor, we may combine them by addition (with proper signs) to find the response of the complete transfer function. The first factor on the right-hand side of Eq. (9.9) is a constant; the remaining factors are functions of ω and are all of the same form. Let us examine the characteristics of one of these factors.

$$\log(1 + j\omega\tau_k) = \log|1 + j\omega\tau_k| + j \tan^{-1} \omega\tau_k$$
$$= \tfrac{1}{2}\log(1 + \omega^2\tau_k^2) + j \tan^{-1} \omega\tau_k \quad (9.10)$$

Notice that the logarithm of a factor becomes the logarithm of the factor magnitude plus an associated phase angle. We may then add all the magnitudes and phases separately to obtain the total magnitude and total phase. Thus,

$$\log|G(j\omega)| = \log \frac{a_0 \prod\limits_{j=1}^{n} \tau_j}{b_0 \prod\limits_{i=1}^{m} \tau_i} + \sum_{i=1}^{m} \log|1 + j\omega\tau_i| - \sum_{j=1}^{n} \log|1 + j\omega\tau_j| \quad (9.11a)$$

and

$$\arg G(j\omega) = \sum_{i=1}^{m} \tan^{-1} \omega\tau_i - \sum_{j=1}^{n} \tan^{-1} \omega\tau_j \quad (9.11b)$$

where the τ_i represents the numerator time constants and the τ_j the denominator time constants. Let us now look at each factor in greater detail and see how to combine factors.

9.2 Bode Gain and Phase Curves

In Eq. (9.10), if the log magnitude of a factor is defined as $\log|G_k|$, then

$$\log|G_k| = \log|1 + j\omega\tau_k|$$
$$= \log(1 + \omega^2\tau_k^2)^{1/2} \quad (9.12)$$
$$= \tfrac{1}{2}\log(1 + \omega^2\tau_k^2)$$

---

†H. W. Bode, *Network Analysis and Feedback Amplifier Design*, Van Nostrand Reinhold Company, New York, 1945.

If $\omega\tau_k \ll 1$, then the log magnitude is

$$\log|G_k| = 0 \tag{9.13}$$

Alternatively, if $\omega\tau_k \gg 1$, then

$$\log|G_k| = \log\omega\tau_k \tag{9.14}$$

Let us plot these equations on a coordinate system of $\log|G_k|$ versus $\log\omega$, as in Fig. 9.8a. Equation (9.13) is just the $\log|G_k| = 0$ axis ($|G_k| = 1$). Then Eq. (9.14) is a

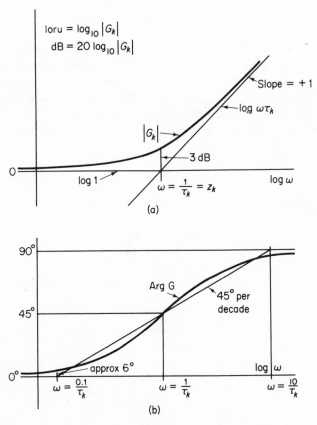

Figure 9.8 Magnitude and phase curves for the factor $(1 + j\omega\tau_k)$.

straight line with a slope of $+1$ loru per decade. A 1-loru increase in $\log|G_k|$ represents a multiplication of 10 by $|G_k|$ if we use logarithms to the base ten. That is, $\log_{10}|10\,G_k| = 1.0 + \log_{10}|G_k|$. (We could also use logarithms to the base two, and so on.) This slope results because if we increase the angular velocity ω in Eq. (9.14) by a factor of 10 (a decade), then we multiply $|G_k|$ by 10, or $\log|G_k|$ increases by 1.0 loru.

We may also measure the magnitude in decibels where (dB) $= 20\log|G_k|$. The slope in these units is $+20$ dB/decade. We indicate this slope by $+1$ in Fig. 9.8a. In subsequent discussions we will use the abbreviated designation of $+1$ for a slope of $+20$ dB/decade, -1 for a slope of -20 dB/decade, -2 for a slope of -40 dB/decade, and so on. The line of Eq. (9.14) intersects the line of Eq. (9.13) at $\omega = 1/\tau_k = z_k$, since at this point $\omega\tau_k = 1$ and $\log 1 = 0$.

We now use Eq. (9.12) to calculate $\log |G_k|$ at $\omega = 1/\tau_k = z_k$. Thus,

$$\log|G_k| = \tfrac{1}{2}\log 2$$
$$= 0.15 \text{ loru}$$

or

$$20\log|G_k| = 20(0.15) = 3 \text{ dB}$$

The curve of $\log |G_k|$ therefore is asymptotic to the two straight-line segments at $\omega = 0$ and $\omega = \infty$, while at $\omega = 1/\tau_k$ it is 3 dB above the intersection of the lines. This intersection is termed a *break point* or *corner point*. If we calculate additional points from Eq. (9.12), we find that 3 dB is the farthest the $\log|G_k|$ curve departs from one or the other straight-line asymptote. Thus, having drawn the straight-line segments, the magnitude may be quickly sketched with an error of less than 1 dB.

The phase angle of a factor is given by

$$\arg G_k = \tan^{-1} \omega\tau_k \tag{9.15}$$

For $\omega = 0$, $\arg G_k = 0$, while for $\omega = \infty$, $\arg G_k = 90°$. When $\omega = 1/\tau_k$,

$$\arg G_k = \tan^{-1} 1$$
$$= 45°$$

We may calculate additional points, obtaining Table 9.1 and the phase curve of Fig. 9.8b

TABLE 9.1   Phase Angle of $(1 + j\omega\tau_k)$

ω	$\arg G_k$
$0.1/\tau_k$	6.0°
$0.2/\tau_k$	11.5°
$0.5/\tau_k$	26.5°
$1/\tau_k$	45.0°
$2/\tau_k$	63.5°
$5/\tau_k$	78.5°
$10/\tau_k$	84.0°

If we draw a straight line through the 45° phase point with a slope of 45° per decade, we find that the phase curve departs no more than 6° from this line in the 2-decade region from $\omega = 0.1/\tau_k$ to $\omega = 10/\tau_k$. Hence, the phase curve may be crudely approximated by the three straight lines shown in Fig. 9.8. This straight-line approximation is not very useful unless the total phase is monotonic in ω, as otherwise the phase error becomes too great. For a single term or factor, it is helpful in visualization.

For the negative factors in Eq. (9.9), the curves of Fig. 9.8 are each simply folded around the horizontal axis; that is, we change the signs of $\log|G_k|$ and $\arg G_k$, and the corner point occurs at $\omega = 1/\tau_k = p_k$. Because of the importance of the gain-phase curves for the single factor, it is helpful to make up curves with a normalized frequency scale. Thus, let $p_k = 1/\tau_k = \omega_n$. Then the transfer function for a single factor in the denominator of Eq. (9.7) is

$$
\begin{aligned}
G_k(j\omega) &= \frac{1}{1 + j\omega\tau_k} \\
&= \frac{\omega_n}{\omega_n + j\omega} \qquad (9.16) \\
&= \frac{1}{1 + j\omega/\omega_n}
\end{aligned}
$$

If we make ω/ω_n the independent variable, then we may plot $\log|G_k|$ and $\arg G_k$ for all factors in Eq. (9.11), with the understanding that for factors associated with the original transfer-function numerator, we change the dependent variable signs. We show these normalized curves in Fig. 9.9. From Fig. 9.9 we note that the curve is down

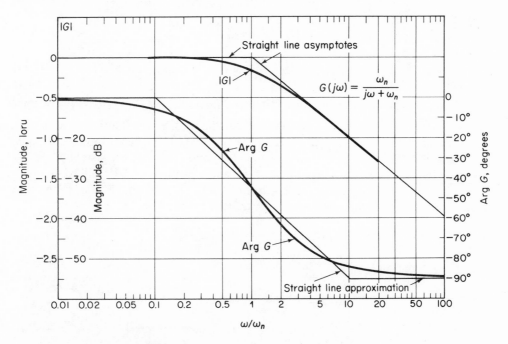

Figure 9.9 Magnitude and phase curves for the

factor $\dfrac{1}{1 + j\omega/\omega_n}$.

1 dB at $f_c/2$, 3 dB at f_c, and 7 dB at $2f_c$, where f_c is the break or corner frequency. These values are often useful in design work.

If we have a multiple pole or zero of order 2 in the transfer function, then in Eq. (9.9) we have factors of the type

$$\log(1 + j\omega\tau) + \log(1 + j\omega\tau) = 2\log(1 + j\omega\tau)$$

The resulting log-magnitude curve has the same corner point as in Fig. 9.8 or Fig. 9.9, but one asymptotic line slope is $+2$, or $+40$ dB/decade, and the actual curve is 6 dB above the corner point. All phase angles double, or the phase at $\omega = \infty$ is $180°$. Similar statements (with reverse signs) apply to factors representing denominator terms.

Often transfer functions have a pole or zero at the s-plane origin, so in Eq. (9.11) some $\tau_i = 1/z_i$ or some $\tau_j = 1/p_j$ becomes infinite at $\omega = 0$. Terms of this type are handled in the following manner: First, assume an arbitrary $\tau_i \rightarrow \infty$ in Eq. (9.11). This τ_i must be removed from the first term on the right-hand side (constant term) of Eq. (9.11a) and combined with its corresponding factor in the second term so that the combined factor becomes

$$|G_i| = \log\frac{1}{\tau_i}|1 + j\omega\tau_i|$$

$$= \log|\frac{1}{\tau_i} + j\omega|$$

As $\tau_i \rightarrow \infty$, then

$$|G_i| = \log|0 + j\omega|$$

or

$$|G_i| = \log\omega \tag{9.17}$$

From Eq. (9.11),

$$\arg G_i = 90° \tag{9.18}$$

The log-magnitude curve is a straight line passing through the 0-loru axis at $\omega = 1$, and with a slope of $+1$. The phase is a constant $+90°$. If the factor represents a pole at the origin, the slope of the line becomes -1, and the phase is constantly $-90°$, a double zero at the origin gives a slope of $+2$ and a phase of $+180°$; and so on.

The total log-magnitude curve is, from Eq. (9.11a), the sum of terms of the type discussed. Thus, suppose that we have a transfer function

$$G(j\omega) = \frac{a_0\tau_1\tau_2(1 + j\omega\tau_3)}{b_0\tau_3(j\omega)(1 + j\omega\tau_1)(1 + j\omega\tau_2)} \tag{9.19}$$

Then Eq. (9.11a) becomes

$$\log|G(j\omega)| = \log\frac{a_0\tau_1\tau_2}{b_0\tau_3} + \log|1 + j\omega\tau_3| - \log\omega$$
$$- \log|1 + j\omega\tau_1| - \log|1 + j\omega\tau_2| \tag{9.20a}$$

while Eq. (9.11b) becomes

$$\arg G(j\omega) = \tan^{-1}\omega\tau_3 - \tan^{-1}\omega\tau_1 - \tan^{-1}\omega\tau_2 - 90° \tag{9.20b}$$

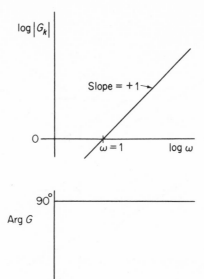

Figure 9.10 Magnitude and phase for the factor $j\omega$.

We may draw the straight-line segments for each of the terms on the right-hand side of Eq. (9.20a) as in Fig. 9.11a (assume that $1 < 1/\tau_1 < 1/\tau_3 < 1/\tau_2$). The magnitude characteristic expressed in logarithmic form is the arithmetic sum of the line segments. Notice that once the angular velocity becomes greater than the break-point value, the sloping asymptote for each term continues on forever, so we can draw in the sum of the straight-line asymptotes by changing the slope of the magnitude curve as we come to each break point. Starting from the first line segment at a slope of -1, the slope decreases to -2 at $1/\tau_1$, becomes more positive at $1/\tau_3$, and then decreases again at $1/\tau_2$. The result is shown in Fig. 9.11b, with the associated slopes as labeled. We must now take into account the first term on the right-hand side of Eq. (9.20a). This term simply shifts the magnitude characteristic up or down on the magnitude axis depending on whether the factor $(a_0\tau_1\tau_2/b_0\tau_3)$ is greater or less than 1, respectively. Let $a_0\tau_1\tau_2/b_0\tau_3 = 22$. Then log 22 = 1.342. Thus, the lines are shifted up by 1.342 loru or 26.84 dB as in Fig. 9.11b. The actual curve may now be drawn from this approximation with an error of less than 1 dB. Equation (9.20b) gives the total phase angle, which is therefore the sum of the phase curves drawn in Fig. 9.12. This summation may be made graphically or algebraically. At low angular velocities, the argument or phase angle approaches 0, $\pm90°$, $\pm180°$, . . . , depending on whether or not there is a pole or zero at the origin, and the order of this pole or zero, if it exists. In this example, we have a pole of order 1 at the origin in Eq. (9.19), and hence the phase angle approaches $-90°$ at low angular velocities. It is important to remember that this phase is present for all ω. At large angular velocities, the phase angle approaches an angle of $-(n - m)90°$, where n and m are the orders of the denominator and numerator, respectively. A quick check of the phase angle at low and high ω values is definitely of value in verifying calculations.

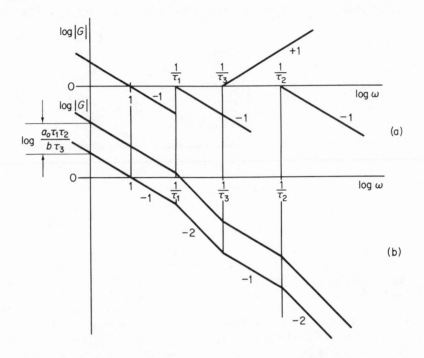

Figure 9.11 Magnitude curves for $\dfrac{K(1 + j\omega\tau_3)}{j\omega(1 + j\omega\tau_1)(1 + j\omega\tau_2)}$.

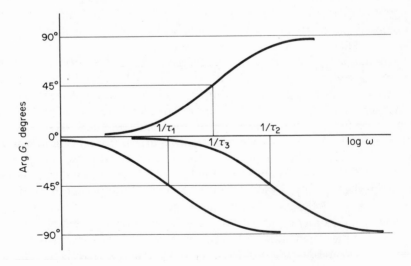

Figure 9.12 Phase curves for $\dfrac{K(1 + j\omega\tau_3)}{j\omega(1 + j\omega\tau_1)(1 + j\omega\tau_2)}$.

The log-magnitude and phase curves for transfer functions with sinusoidal signals are often called *Bode plots*. We now give an example for a case with real poles and zeros.

EXAMPLE 9.3

Let

$$G(s) = \frac{1108(s + 2)(s + 6.67)}{s(s + 0.333)(s + 0.667)(s + 33.3)(s + 200)} \tag{9.21}$$

Draw the magnitude and argument curves for $G(j\omega)$ as ω varies from 0 to ∞.

Solution. Choose four-cycle semilog paper. Since the lowest break-point angular velocity is $\omega_1 = 0.333$ and the highest is $\omega_2 = 200$, we choose a beginning angular velocity of $\omega = 0.1$ so as to cover the range $0.1 \leq \omega \leq 1000$.

The asymptotic line slope at $\omega = 0.1$ is -1, as a result of the single s-term in the denominator of Eq. (9.21). We then have denominator break or corner points at $\omega = 0.333, 0.667, 33.3,$ and 200 and numerator break points at $\omega = 2$ and 6.67. The straight-line asymptotes have decreasing slopes at the denominator points and increasing slopes at the numerator points. Thus, the overall line slopes at 0.333 and 0.667 become -2 and -3, respectively. At $\omega = 2$ and $\omega = 6.67$ we have increasing slopes, so the overall line slopes become -2 and -1, respectively, at these points. At $\omega = 33.3$ and $\omega = 200$, the overall line slopes decrease to -2, and -3, respectively, and the final slope is thus -3 loru (or -60 dB) per decade. Adding all the straight-line characteristics together as we did in Fig. 9.11b, we first have an asymptote with a slope of -1 which goes through the point log (1) or 0 loru at $\omega = 1$, and also the point 1 loru or 20 dB at $\omega = 0.1$. As a result of the nearby break point at $\omega = 0.333$, the total magnitude asymptote has a slope of -2 beyond $\omega = 0.333$ (and a slope of -3 beyond $\omega = 0.667$), hence the line with a slope of -1 at $\omega = 0.1$ is not effective beyond $\omega = 0.333$. Continuing as before, we get the dashed-line asymptotes in Fig. 9.13. Now, rearranging Eq. (9.21) we obtain

$$G(s) = \frac{1108(2)(6.67)}{(0.333)(0.667)(33.3)(200)} \frac{(1 + 0.5s)(1 + 0.15s)}{s(1 + 3s)(1 + 1.5s)(1 + 0.03s)(1 + 0.005s)}$$

$$= (10)\frac{(1 + 0.5s)(1 + 0.15s)}{s(1 + 3s)(1 + 1.5s)(1 + 0.03s)(1 + 0.005s)}$$

We see from this that the constant multiplier is 10. Hence the dashed asymptotes must be raised 1 loru or 20 dB, giving the magnitude asymptotes shown by the solid lines in Fig. 9.13.

We can avoid the rearrangement of Eq. (9.21) to find the constant by letting $s = j\omega$ and making ω less than 0.333 in Eq. (9.21) directly. That is, if $s = j\omega$ and $\omega = 0.1$ in Eq. (9.21),

$$G(j0.1) = \frac{1108(j0.1 + 2)(j0.1 + 6.67)}{0.1(j0.1 + 0.333)(j0.1 + 0.667)(j0.1 + 33.3)(j0.1 + 200)}$$

Clearly in computing $|G(j0.1)|$ we may ignore any imaginary component in a factor which is less than $\frac{1}{10}$ that of the real component. For example, $|j0.1 + 2| = (0.01 + 4)^{1/2} \simeq 2$. Hence at $\omega = 0.1$ the straight-line asymptote passes through the point given by

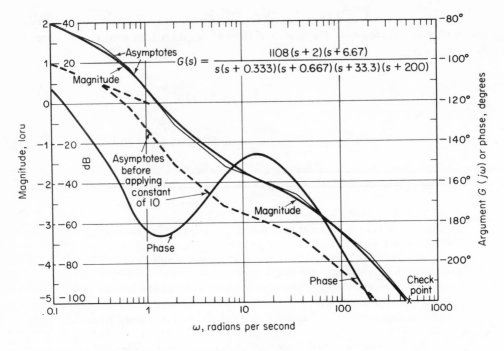

Figure 9.13

the computation

$$|G(j0.1)| \simeq \frac{1108(2)(6.67)}{0.1(0.333)(0.667)(33.3)(200)}$$

$$\simeq 100 = 2 \text{ loru or } 40 \text{ dB}$$

The actual magnitude curve at $\omega = 0.1$ will be about 1 dB below the asymptotic line point of 40 dB due to the fact that we neglected the relative closeness of the nearby break point at $\omega = 0.333$ to the chosen angular velocity of 0.1.

We now need a check point at large angular velocity ω_z to see if we drew the lines in properly. Here let $\omega_z = 500$. Now we can disregard any real terms in any factor $(j\omega + \omega_n)$, where $\omega_n < \omega_z$, and at $\omega = 500$, from Eq. (9.21), we obtain

$$|G(j500)| \simeq \frac{1108(500)^2}{(500)^5}$$

$$= 8.86 \times 10^{-6} \quad \text{or} \quad -6 + 0.948 \text{ loru}$$

$$= -5.052 \text{ loru} \quad \text{or} \quad -101.4 \text{ dB}$$

This point should be on the straight-line summation previously made using break points and slopes, if we made no errors. This check point is shown in Fig. 9.13.

The phase curve can be calculated from the formula

$$\arg G = -90° - \tan^{-1}\frac{\omega}{0.333} - \tan^{-1}\frac{\omega}{0.667} - \tan^{-1}\frac{\omega}{33.3} - \tan^{-1}\frac{\omega}{200}$$
$$+ \tan^{-1}\frac{\omega}{2} + \tan^{-1}\frac{\omega}{6.67} \tag{9.22}$$

and is shown in Fig. 9.13 (right-side scale); or by graphically summing the phase-response curves for the individual transfer-function terms. As a check on the phase angle, it should approach $-90°$ at low angular velocities and $-(5 - 2)90° = -270°$ at high angular velocities, which it does.

Returning to Eq. (9.5), some of the polynomial factors may be quadratic with complex roots in conjugate pairs. For example, if p_1 is complex, then $p_2 = p_1^*$ and

$$(s - p_1)(s - p_1^*) = (s^2 - 2p_1s + p_1p_1^*)$$
$$= s^2 + 2\zeta\omega_n s + \omega_n^2 \tag{9.23}$$

In this instance the straight-line segments for the log-magnitude curve correspond to those for a multiple root of order 2; that is, the sloping segment is at a slope of -2 or -40 dB/decade, and the intersection of the segments is at $\omega = \omega_n$. The phase curve also starts at $0°$ at $\omega = 0$ and arrives at $-180°$ at $\omega = \infty$, as for the multiple root of order 2. However the behavior of the magnitude curve in the vicinity of $\omega = \omega_n$ depends on ζ, the damping factor, while the phase curve also depends on ζ.

The transfer function of a network with a complex-conjugate pole pair from Eq. (9.23) is

$$G(s) = \frac{\omega_n^2}{s^2 + 2\zeta\omega_n s + \omega_n^2} \tag{9.24}$$

where the ω_n^2 term in the numerator ensures that $|G| = 1$ when $s = 0$. Substituting $s = j\omega$ to obtain the sinusoidal case, we obtain

$$G(j\omega) = \frac{\omega_n^2}{\omega_n^2 - \omega^2 + j2\zeta\omega_n\omega} \tag{9.25}$$

Dividing numerator and denominator of the right-hand side of Eq. (9.25) by ω_n^2, we then find that

$$|G(j\omega)| = \frac{1}{\{[1 - (\omega/\omega_n)^2]^2 + 4\zeta^2(\omega/\omega_n)^2\}^{1/2}} \tag{9.26a}$$

and

$$\arg G(j\omega) = \frac{2\zeta\omega/\omega_n}{1 - (\omega/\omega_n)^2} \tag{9.26b}$$

If we plot $\log|G(j\omega)|$ and $\arg G(j\omega)$ as functions of $\log \omega/\omega_n$ we obtain the (normalized) Bode plots for a second-order factor with complex roots. These are shown in Figs. 9.14 and 9.15, respectively.

EXAMPLE 9.4

Make Bode plots for the transfer function

$$G(s) = \frac{2000(s + 2)}{s(s^2 + 4s + 20)}$$

Solution. There is a zero at $s = -2$, and ω_n for the quadratic factor is $\sqrt{20}$. We use the alternative technique of Example 9.3, and evaluate $|G(j\omega)|$ directly at a value of ω less

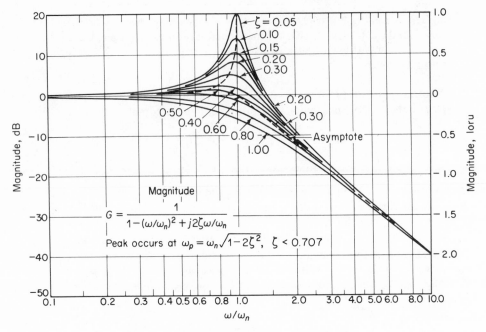

Figure 9.14 Sinusoidal response, second-order
network, magnitude.

than 2/10. Let $s = j\omega$ and $\omega = 0.1$. Then

$$|G(j\omega)| = \frac{2000|j0.1 + 2|}{0.1|0.01 + 20 + j0.4|} \simeq \frac{2000(2)}{0.1(20)}$$

$$= 2000$$

$$\log|G(j\omega)| = 3.301 \text{ loru}$$

$$= 66.020 \text{ dB}$$

The magnitude curve thus goes through the point 66.02 dB at $\omega = 0.1$, and the asymptote has a slope of -1 at this point. At $\omega = 2$, the asymptote slope becomes unity due to the numerator factor. The natural angular velocity $\omega_n = \sqrt{20} = 4.47$ rad/s. Starting here the asymptote has a slope of -2, which continues indefinitely. As a check point, let $\omega = 500$. At this point

$$|G(j\omega)| \simeq \frac{2000(500)}{500(500)^2}$$

$$\log|G(j\omega)| = -\log\frac{1}{|G(j\omega)|} = -\log\frac{500(500)}{2000}$$

$$= -2.096 \text{ loru}$$

$$= -41.920 \text{ dB}$$

Since $2\zeta\omega_n = 4$, $\zeta = 4/(12)(4.47) = 0.448$. From Fig. 9.14, the magnitude curve for this ζ "peaks up" about 1.5 dB when $\omega/\omega_n \simeq 0.85$, or in this case, when $\omega \simeq 0.85(4.47)$ $= 3.8$. Because the actual magnitude curve is 3 dB above the break point at $\omega = 2$ due to the zero, the actual magnitude "peak" will be raised about 2 dB more and

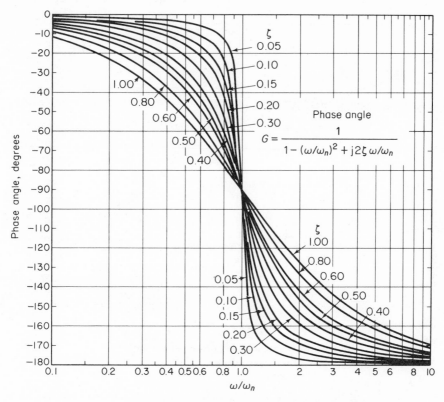

Figure 9.15 Sinusoidal response, second-order
network, argument or phase angle.

the peak of the curve will be approximately 3.5 dB above the asymptote line with zero slope. The magnitude curve may now be sketched as in Fig. 9.16. This curve is within 1 dB of the calculated magnitude at $\omega = 4$, which is 42.7 dB. To calculate the phase angle, we prepare the following table:

ω	$\omega/5$	θ_1	ω/ω_n	θ_2	θ_T
1	0.2	11.3	0.224	-12	-90.7
2	0.4	21.8	0.447	-26.5	-94.7
4	0.8	38.6	0.894	-76	-127.4
8	1.6	58.0	1.79	-144	-176.0
10	2.0	63.5	2.24	-153	-179.5
20	4.0	76.0	4.47	-168	-182
40	8.0	82.9	8.94	-174	-182
80	16.0	86.4	17.90	-177	-181
100	20.0	87.1	22.40	-178	-181
200	40.0	88.6	44.70	-179	-180

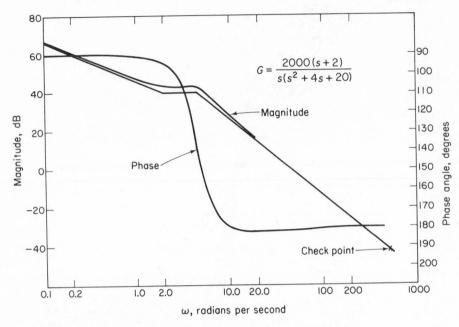

Figure 9.16 Magnitude and phase curves for
Example 9.4.

In the table $\theta_1 = \tan^{-1} \omega/5$. θ_2 is read directly from Fig. 9.15 or calculated from Eq. (9.26b). The total phase angle is $\theta_1 + \theta_2 - 90°$. We observe that the phase is $-90°$ at low frequencies, and $-180°$ at high frequencies, as it should be.

Since we extol the use of the computer, it is natural to provide a program which obtains the magnitude and phase relations from the transfer function. The program BODE which serves this purpose, and a discussion of its algorithm and use, is contained in Appendix B.

9.3 Bode Diagrams—Interpretation

The Bode plots show the sinusoidal response of a network in magnitude and phase as a function of angular velocity $\omega = 2\pi f$. A change of the factor K moves the magnitude curve up or down the log-magnitude axis. Thus, if we decrease K in Eq. (9.4a) by a factor of 10, the entire magnitude curve moves down 1 loru or 20 dB; and vice versa, if we increase the gain K, the magnitude curve moves up. Thus, the effect of "pure" gain becomes readily apparent, and in fact we see that the magnitude curve may be plotted for any arbitrary but convenient gain K. Naturally we cannot construct an amplifier with complete frequency-independent gain; however, we may include its frequency characteristics in the transfer function with which it is associated.

Specifications for electronic amplifiers are often written in terms of frequency

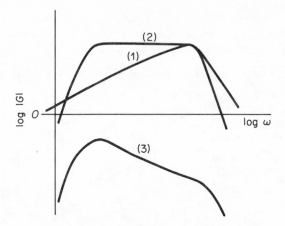

Figure 9.17 Magnitude curves for an amplifier.

response. For amplifiers in the audio range 20 to 20,000 Hz, the magnitude curve is important. For example, we may be given a device (such as a transducer) with a frequency-response curve (1) as shown in Fig. 9.17, to build a network and amplifier so as to obtain a frequency response (2). Obviously, we must construct a network which has a response shape (3) which is the difference between curves (1) and (2). Then if we add (or subtract) sufficient frequency-independent gain K, we may combine vertically displaced curve (3) with curve (1) to achieve response curve (2). The eye, on the other hand, responds to phase angle, and it becomes desirable in television circuits to have the phase angle change linearly with frequency in the frequency range of interest. To comply with both magnitude and phase specifications may often be difficult.

Obviously frequency measurements on a device determine its transfer function. That is, if we make tests on a system by applying a sinusoidal input and observing the output over a frequency range, we can obtain magnitude and phase plots. We can proceed backward in our analysis process by drawing straight-line asymptotes to the magnitude curve at slopes of $20n$ dB/decade, where $n = 0, \pm 1, \pm 2, \ldots$. (*Important:* The slopes *cannot* be other than this, so a simple tangent to the magnitude curve does not qualify unless it meets this specification.) The intersection of these line asymptotes roughly determines the poles and zeros of the system, whereas the frequency-independent gain may be calculated from one point on the curve. Using the break points, a phase curve may now be calculated for the assumed transfer function and compared with that obtained in the experiment. The closeness of agreement between the two phase curves comprises a measure of the astuteness in choosing the straight-line asymptotes of the magnitude curve. If agreement seems poor, new choices of factors are involved, as we have some difficulty in estimating the damping factor of quadratic factors. Noise also obscures the data. Nevertheless, steady-state sinusoidal frequency identification may be one of the easiest and best methods for network analysis.

EXAMPLE 9.5

Tests in the laboratory with a sinusoidal input gives the output/input ratio in magnitude and phase as shown by the small circles in Fig. 9.18. Find the transfer function in terms of the complex variable s.

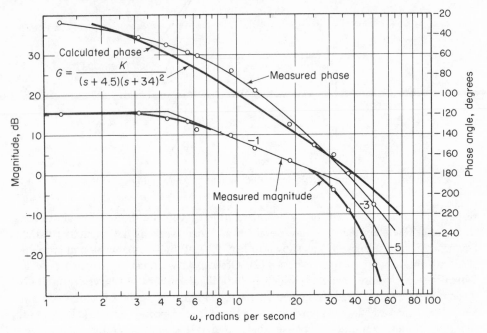

Figure 9.18 Magnitude and phase curves for Example 9.5.

Solution. We first draw in straight-line asymptotes at slopes of $-20n$ dB/decade, where n is an integer, to obtain the lines shown in Fig. 9.18. (Recall that here the actual curve lies below the asymptotes, particularly with a rapidly changing magnitude.) We obtain corner points at $\omega = 4.5$, 34, and 49. We arbitrarily neglect distant poles at $\omega = 49$ and assume a transfer function

$$G = \frac{K}{(s + 4.5)(s + 34)^2}$$

We then plot a phase curve for this function for $s = j\omega$. The calculated phase curve, shown as the heavy line in Fig. 9.18, seems in error, as the calculated phase is too large in the $\omega = 10$ rad/s region. We surmise that the first corner point probably occurs at a larger ω, of say $\omega = 5$. We may obtain this in Fig. 9.18 by lowering the low-frequency asymptote slightly and shifting the -1 slope line to a higher ω. The next corner point will occur at about $\omega = 35$. Obviously, now the asymptote slope must be -4, or else we must take the succeeding corner point around $\omega = 50$ into consideration, since our calculated phase curve is not low enough in the $\omega = 50$ rad/s region. We then deduce that the transfer function is approximated reasonably by the function

$$G(s) = \frac{K}{(s + 5)(s + 35)^2(s + 50)}$$

If desired, we may plot a new phase curve for this function. Since the measured values at high frequencies (both magnitude and phase) are obscured by noise and hence inaccurate, the value of the final corner point and the final slope are rather arbitrary. However, as we will see, distant poles have little effect on the time response. The value of K may now be found by taking $\omega = 1$ in Fig. 9.18.

We have not necessarily arrived at the best possible approximation to the measured response in Fig. 9.18. We might have chosen a function like

$$G(s) = \frac{K(s + 6)}{(s + 3.5)(s + 9)(s + 35)^2}$$

as an approximation and assumed a greater amount of noise in the low-frequency magnitude measurements.

9.4 Second-Order Network

The relationships between the time response of a second-order network and its pole locations in the complex s-plane are very useful in circuit design. If we understand the response of the second-order network to simple inputs such as the unit step function $u(t)$ and the sinusoidal function, $\sin \omega t$, we can extend this comprehension to higher order and more complicated networks. We therefore examine this type of network in detail.

A second-order network is a special case in which the highest-order derivative of the output is of second degree. The general response of the second-order network to a step function input is described by

$$\frac{d^2y}{dt^2} + 2\zeta\omega\frac{dy}{dt} + \omega_n^2 y = Au(t) \tag{9.27}$$

where $y = y(t)$. Transforming, with all initial conditions equal to zero, we obtain

$$(s^2 + 2\zeta\omega_n s + \omega_n^2)\,Y(s) = \frac{A}{s}$$

Thus,

$$Y(s) = \frac{A}{s(s^2 + 2\zeta\omega_n s + \omega_n^2)} \tag{9.28}$$

It is convenient to normalize the step forcing function such that we obtain a final value of unity. Thus, we let the normalized response function $Y_1(s) = Y(s)\omega_n^2/A$, or

$$Y_1(s) = \frac{1}{s}\,\frac{\omega_n^2}{s^2 + 2\zeta\omega_n s + \omega_n^2} \tag{9.29}$$

Our primary interest concerns the performance of the network with ζ and ω_n varying. Figure 9.19 shows the location of the poles of $Y_1(s)$ in the complex plane as related to the parameters ζ and ω_n.

If $0 \leq \zeta < 1$ in Fig. 9.19, the response is oscillatory and the solution may be found from Eq. (9.29) by expanding $Y_1(s)$ into partial-fraction form, or directly from

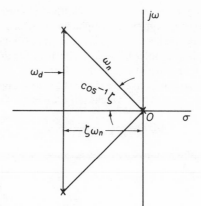

Figure 9.19 Second-order
poles in the complex plane.

a Laplace transform table. The response is given by

$$y_1(t) = 1 - \frac{1}{\sqrt{1 - \zeta^2}} e^{-\zeta\omega_n t} \sin(\omega_n\sqrt{1 - \zeta^2}\, t - \phi) \tag{9.30}$$

where

$$\phi = \tan^{-1}\frac{\sqrt{1 - \zeta^2}}{-\zeta}$$

Figure 9.20 shows the general time response of the network. The important characteristics of the step response are:

1. Per cent overshoot on the first reversal of the output, or per cent maximum overshoot, usually abbreviated "per cent overshoot."

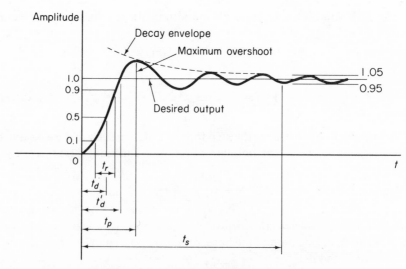

Figure 9.20 General response of a step function, second-order network.

2. Ten to 90 per cent rise time t_r.

3. Fifty per cent rise time (delay time) t_d.

4. Delay time t_d', where output first equals final value.

5. Settling time t_s, or time required for the response to reach and remain within a certain specified tolerance band (shown as ± 5 per cent in Fig. 9.20).

6. Frequency of oscillation during settling.

7. Decay rate during settling.

If we plot Eq. (9.30) on a normalized time scale $\omega_n t$, we eliminate the variable ω_n and obtain Fig. 9.21. From this figure you see that the overshoot above the final value depends only on ζ. The peak overshoot may be found by differentiating Eq. (9.30) [or more simply, by multiplying Eq. (9.29) by s and transforming] to get

$$\frac{dy_1(s)}{dt} = \frac{\omega_n}{\sqrt{1-\zeta^2}} e^{-\zeta\omega_n t} \sin(\omega_n\sqrt{1-\zeta^2}\,t) \qquad (9.31)$$

The first (and peak) overshoot occurs when $(\omega_n\sqrt{1-\zeta^2})t = \pi$, or

$$\omega_n t_p = \frac{\pi}{\sqrt{1-\zeta^2}} \qquad (9.32)$$

Substituting Eq. (9.32) into Eq. (9.30) gives the maximum per cent overshoot as

$$\% \text{ overshoot } 100e^{-\zeta\pi/\sqrt{1-\zeta^2}} \qquad (9.33)$$

This is plotted in Fig. 9.22. The maximum overshoot is less than about 25 per cent for $\zeta > 0.4$, but increases rapidly for smaller ζ.

It is more difficult to obtain explicit relations for rise times. If we tabulate the approximate 10 to 90 per cent rise time t_r in terms of ω_n and ζ, we obtain Table 9.2.

From Table 9.2 the rise time varies almost inversely with ω_n for $\zeta < 0.8$.

EXAMPLE 9.6

Given $G(s) = 1/(s^2 + 5s + 25)$ find the 10–90 per cent rise time on a step function input.

Solution

$$\omega_n = \sqrt{25} = 5$$
$$\zeta = 0.5$$

Then $\omega_n t_r = 1.58$ from Table 9.2, or $t_r = 1.58/5 = 0.316$ s.

The frequency of oscillation and the decay rate arise directly from Eq. (9.30). If $\zeta > 1$, then the network will not oscillate, has no overshoot, and becomes sluggish, as shown in Fig. 9.21.

We find the response to an impulse as a function of time by taking the derivative of the response to a step function. Since time differentiation transforms into multiplica-

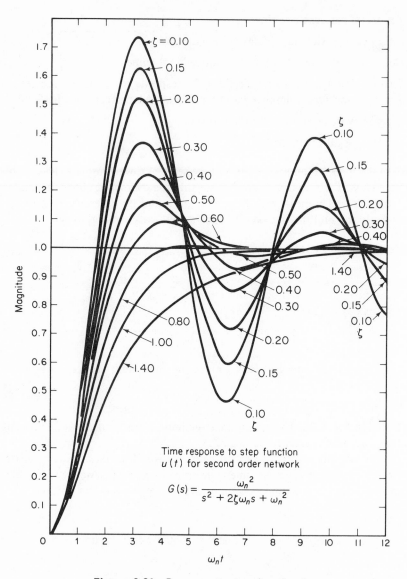

Figure 9.21 Response to step function for a
second-order network.

tion by s in the complex s-domain, we observe that this operation on Eq. (9.29) results
in the network transfer function itself. Hence, as previously stated in Chapter 8, the
time response $g(t)$ of a zero-state network to a unit impulse becomes simply the inverse
transform of the transfer function $G(s)$.

We have already studied to some extent the sinusoidal steady-state solution of the
second-order network in Section 9.4. If we let the right-hand side of Eq. (9.27) become
$A \sin \omega t$, we know that the zero-initial-condition output is $B \sin(\omega t + \theta)$. The ratio of

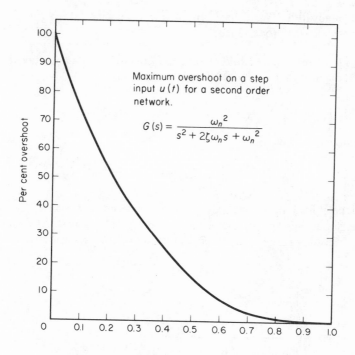

Figure 9.22 Maximum overshoot as a function of ζ, second-order network.

TABLE 9.2 Approximate 10 to 90 Per Cent Rise Time t_r (in seconds) in Terms of ω_n

ζ	$\omega_n t_r$
0.4	1.45
0.5	1.58
0.6	1.86
0.7	2.16
0.8	2.55
1.0	3.47

output magnitude B to the input magnitude A is

$$\frac{B}{A} = |G(j\omega)|$$

as given by Eq. (9.26a), while $\theta = \arg G(j\omega)$, as given by Eq. (9.26b). The magnitude and phase responses to a sinusoidal input are shown in Figs. 9.14 and 9.15.

Performance indices based on the sinusoidal function are

1. The peak response, M_P.

2. The angular velocity ω_p at which M_P occurs.

3. The bandwidth BW, usually the angular velocity (or frequency) at which $\log|G| = -0.15$ (or -3 dB).

Since a decade represents a tenfold change in frequency, while an octave is a twofold change in frequency, the number of decades between ω_2 and ω_1 is

$$N_1 = \log_{10}\frac{\omega_2}{\omega_1} \text{ decades} \qquad \omega_2 > \omega_1 \tag{9.34}$$

and the number of octaves is

$$N_2 = \log_2\frac{\omega_2}{\omega_1} \text{ octaves} \qquad \omega_2 > \omega_1 \tag{9.35}$$

The final slope at high ω for $\log|G|$ is -2 loru per decade (or -40 dB/decade). Thus, this slope is about -12.04 dB/octave.

Figure 9.23 shows a general $|G|$ curve for $\zeta < 0.707$. The peak value M_P and the

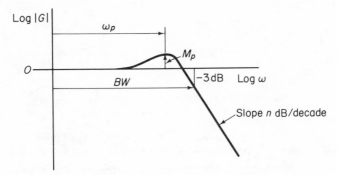

Figure 9.23 General sinusoidal magnitude response, second-order network.

angular velocity ω_p may be found by differentiating Eq. (9.26a) with respect to ω and setting this to zero. The result is

$$\omega_p = \omega_n\sqrt{1 - 2\zeta^2} \qquad 0 \leq \zeta < \tfrac{1}{2}\sqrt{2} \tag{9.36}$$

Substituting this back into Eq. (9.26a), we obtain

$$M_P = \frac{1}{2\zeta\sqrt{1 - \zeta^2}} \qquad 0 \leq \zeta < \tfrac{1}{2}\sqrt{2} \tag{9.37}$$

The bandwidth BW becomes greater with decreasing ζ, varying from about $1.2\omega_n$ at $\zeta = 0.6$ to about $1.5\omega_n$ at $\zeta = 0.1$. For $\zeta > 0.7$, the bandwidth is less than ω_n. Note from Eq. (9.36) that $\omega_p \neq \omega_n$ and $\omega_p \neq \omega_d$, except when $\zeta = 0$. The curve does not peak up at all (M_P does not exist) unless $\zeta < 0.707$, but goes to infinity at $\omega = \omega_n$ for $\zeta = 0$ (no damping). The curve is very similar to the resonance curve previously discussed in Example 9.2, except that here there is no zero at the origin.

As previously stated, the relation between the frequency domain and the time domain enables us to predict time-domain performance from measured frequency characteristics. Figure 9.24 contains plots of Eqs. (9.36) and (9.37) which enables us to find ζ if M_P is known, and from this to find ω_n if ω_p is known. Once ω_n and ζ are found, the time response to a step function is clear from the preceding curves.

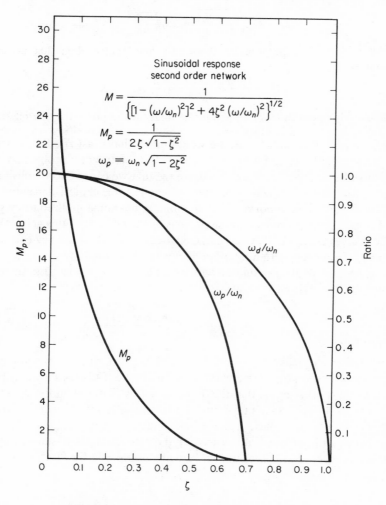

Sinusoidal response
second order network

$$M = \frac{1}{\left\{[1-(\omega/\omega_n)^2]^2 + 4\zeta^2\,(\omega/\omega_n)^2\right\}^{1/2}}$$

$$M_p = \frac{1}{2\zeta\sqrt{1-\zeta^2}}$$

$$\omega_p = \omega_n\sqrt{1-2\zeta^2}$$

Figure 9.24 Dependence of M_P and ω_p on the damping factor ζ.

EXAMPLE 9.7

Given a network which has an $M_P = 1.5$ (3.5 dB) at $\omega_p = 5$ rad/s. The network response magnitude falls off at 40 dB/decade at high frequencies. What will be the peak overshoot and frequency of oscillation of the same network to a step input?

Solution. The falloff at high frequencies indicates a second-order network, or one approximating this. From Fig. 9.14 or Fig. 9.24, $\zeta = 0.354$. Since

$$\omega_p = \omega_n\sqrt{1 - 2\zeta^2}$$

and

$$\omega_d = \omega_n\sqrt{1 - \zeta^2}$$

then

$$\omega_d = \omega_p\sqrt{\frac{1 - \zeta^2}{1 - 2\zeta^2}} = (5)\frac{0.936}{0.866} = 5.4 \text{ rad/s}$$

From Fig. 9.22, the peak overshoot on a step input is about 30.6 per cent. The time to peak is

$$t_p = \frac{\pi}{\omega_n\sqrt{1 - \zeta^2}} = \frac{\pi}{5.76(0.936)} = 0.582 \text{ s}$$

Alternatively, given the time response, the frequency response may be inferred.

In designing an amplifier, say, we would like to have fast rise time, small overshoot, fast settling time, and a narrow bandwidth (due to noise problems with large bandwidths). Since these are interrelated, we see that we cannot accomplish all these simultaneously but must make some sort of engineering choice, depending on the factors considered most important. If the amplifier has some midfrequency gain, the curve of Fig. 9.23 merely moves up the $|G|$ axis by the amount of this gain. Amplifier-frequency curves usually also have characteristics at the low-frequency end similar to those at high frequency, with a falloff in gain with reduction in frequency.

Now, let us look at the situation with an added zero at $s = z_1$. In this case for a step input the (normalized) output is

$$y_1(t) = \mathcal{L}^{-1}\frac{(1/z_1)\omega_n^2(s + z_1)}{s[(s + \zeta\omega_n)^2 + \omega_n^2(1 - \zeta^2)]} \tag{9.38}$$

The $(1/z_1)$ factor is to make $y_1(t) = 1$ at low s, to compare with the previous case. The pole-zero configuration for the network described by Eq. (9.38) is shown in Fig. 9.25. The inverse transform reveals that the frequency of oscillation and the decay rate are the same as before. The added numerator term (sometimes called *numerator dynamics*) does, however, alter the phase angle of the sinusoidal term and provides a larger overshoot than with the poles alone, making the network more unstable. Figure 9.26 illustrates this effect. The curves in Fig. 9.26 are asymptotic for large values of z_1 to

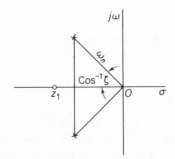

Figure 9.25 Zero added to second-order network.

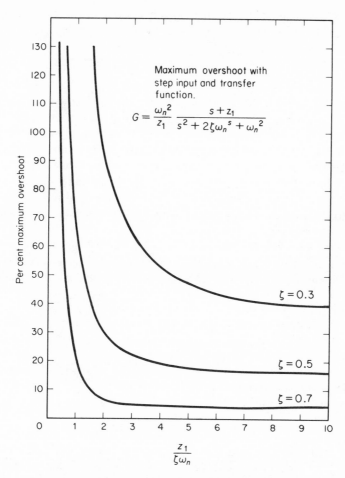

Figure 9.26 The effect of an added zero at the peak overshoot for a step input second-order network.

the situation without the zero shown in Fig. 9.22, and as the zero comes closer to the origin, it increases the overshoot and the settling time. The effect of the zero may be visualized qualitatively by analogy to an electrostatic field in two dimensions, where the poles represent unit positive charges and the zero a unit negative charge. The field in the region of the origin, as measured by the force on a unit test charge, is clearly changed as z_1 comes closer to the origin. Electrostatic fields obey Laplace's partial differential equation, and these relations have been exploited to obtain quantitative data.

In the sinusoidal response, the zero will increase the peak response M_P, increase the bandwidth, and reduce the final phase angle to $-90°$ rather than $-180°$.

An added real pole stabilizes the second-order network or makes it more sluggish. It adds an exponentially decaying term, reduces the overshoot, and increases the rise time. The step-response equation for the stabilized network is

$$y_1(t) = \mathcal{L}^{-1} \frac{p_1 \omega_n^2}{s(s + p_1)[(s + \zeta\omega_n)^2 + \omega_n^2(1 - \zeta^2)]} \qquad (9.39)$$

By evaluating the slope of this response, you can show that the slope never becomes negative if $p_1 \leq \zeta\omega_n$. Hence for this case (pole inside real part of complex poles), there can be no response overshoot. If $p_1 \gg \omega_n$, the effect of the real pole becomes very small. Thus, if $p_1 = 5\omega_n$, the initial exponential value is only about $\frac{1}{16}$ or less of the final value, or distant poles have little effect on the network response. The same conclusion follows if we use the potential analogy by viewing the effect in the region of the origin due to a positive charge on the $-\sigma$ axis as it moves out from the origin. In the sinusoidal domain, the real pole lowers M_P, reduces bandwidth, and increases the cutoff rate at high frequencies.

A pole and zero close together is called a "dipole," and again from potential analogy you see that the field at a distance from the dipole is disturbed only slightly. Thus, the configuration of Fig. 9.27, although it looks complicated, gives a response to

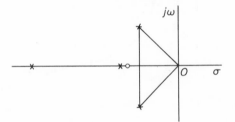

Figure 9.27   Second-order network with added real poles and dipole.

a step input essentially the same as would result due to a pair of complex poles alone. Thus, the second-order network forms a basis for design for higher-order networks.

If poles of the transfer function lie in the right-hand s-plane, then terms of the form $e^{\alpha t}$ appear in the response, which means that we have instability in the network response. Such terms will appear even with zero sources, since any slight disturbance of the network state, such as movement of an electron, will start the exponential term growing. Clearly poles cannot be allowed in the right half of the s-plane. You might ask whether or not we could cancel a right-half-plane pole with a zero. This has theoretical possibilities, but is not practical, since a slight change in pole or zero position (due to an element change) results in a dipole in the right half plane. With this combination we have an exponentially growing term in the response and the network is unstable regardless of the nearby zero.

We use the term *relative stability* to indicate the degree of damping. Obviously, the closer the poles move toward the $j\omega$ axis, the less the relative stability. If the network has real poles only, it will not oscillate at all, and the relative stability is high. In design the degree of stability depends on the required performance indices. Since the relative stability and the network performance are intimately related to the location of the poles, which are the roots of the characteristic equation, the key question with respect to linear networks becomes: Where are the roots of the characteristic equation? We discuss this question in the next section.

9.5 Root Locus

Since the pole positions assume such importance in design, we would like to know the location of the poles as we change some network element value. The movement of a root of the characteristic equation in the complex s-plane with the variation of a parameter generates a curve known as the *root locus*. In general, there will be n such loci, where n is the degree of the characteristic polynomial.

For example, suppose that we plot the roots of the polynomial $(s^2 + as + b)$ as b varies (but $b \geq 0$). Then

$$s^2 + as + b = 0$$

or

$$s^2 + as = -b \tag{9.40}$$

Divide both sides of Eq. (9.40) by $(s^2 + as)$. Then

$$\frac{b}{s(s + a)} = -1 \tag{9.41}$$

Now plot the poles of the left-hand side of Eq. (9.41) as in Fig. 9.28 and take some

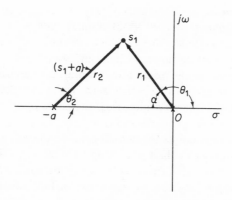

Figure 9.28 Vectors in the s plane for $G(s) = \dfrac{b}{s(s + a)}$.

point in the complex plane such as s_1 in that figure. The vector from the origin to s_1 represents the term s in the denominator of Eq. (9.41) when $s = s_1$, while the vector from $-a$ to s_1 represents the term $(s + a)$ when $s = s_1$. Changing Eq. (9.41) to polar form, we obtain

$$\frac{b}{r_1 e^{j\theta_1} r_2 e^{j\theta_2}} = 1 \cdot e^{j(2k+1)\pi} \qquad k = \text{integer} \tag{9.42}$$

where r_1 and r_2 are the magnitudes of the vectors, and θ_1 and θ_2 their arguments, as shown in Fig. 9.28. From Eq. (9.42) we find two conditions necessary, as follows:

$$b = r_1 r_2 \tag{9.43a}$$

$$-(\theta_1 + \theta_2) = (2k + 1)\pi \qquad k = \text{integer} \tag{9.43b}$$

These equations follow from the required identity of magnitudes and angles on both sides of Eq. (9.42).

From Fig. 9.28, $\theta_1 + \alpha = 180° = \pi$. Now $\theta_2 = \alpha$ if the triangle is isosceles or if the point s_1 lies on a straight line parallel to the $j\omega$ axis and $a/2$ distance to the left of the origin, as in Fig. 9.29. Any point s lying on this line satisfies Eq. (9.43b), and thus

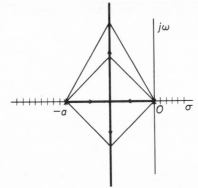

Figure 9.29 Root locus for Eq.
(9.41), b varying. Crossed lines
        are for b negative.

this line is part of the locus of roots of $(s^2 + as + b)$, with b varying ($b \geq 0$). For each point on the locus above the σ-axis, there must be a corresponding point below the σ-axis, since complex roots of a polynomial with real coefficients must always occur in conjugate pairs. Hence, the root locus in the lower half of the s-plane must be the mirror image of that in the upper half, and we only need to draw the upper half loci. The real axis between the origin and $-a$ must also be part of a locus, since the vector to the right from $s = -a$ has angle zero, while the vector from the left has angle π.

Following the loci as b increases, the roots of $(s^2 + as + b)$ are at $s = 0$, and $s = -a$, or the poles of the left-hand side of Eq. (9.41), when $b = 0$. As b increases positively, one root moves from the origin to the left along the real axis; the other moves from $s = -a$ to the right along the real axis. When $s = -a/2$, the two real roots coincide ($b = a^2/4$). For $b > a^2/4$, the roots become complex. In this region the value of b at any point s_1 on the locus may be found from Eq. (9.46a), or in this case, since $r_1 = r_2$,

$$b = r_1^2$$

$$= \left(\frac{a}{2}\right)^2 + \omega_1^2$$

where $s_1 = a/2 + j\omega_1$. Along the negative real axis,

$$s^2 + as + b = (s + \sigma_1)(s + \sigma_2)$$

or

$$\sigma_2 = a - \sigma_1 \tag{9.44a}$$

and

$$b = \sigma_1 \sigma_2$$

$$= \sigma_1(a - \sigma_1) \tag{9.44b}$$

The arrows in Fig. 9.29 show the direction the roots move as b increases. From this figure a network with this characteristic equation is never unstable if $b > 0$ since no root crosses into the right half plane.

Now let us look at the case for $b < 0$. The characteristic equation may now be written

$$s^2 + as - b_1 = 0$$

where

$$b_1 \geq 0$$

Then Eq. (9.41) becomes

$$\frac{b_1}{s(s+a)} = 1 \qquad b_1 \geq 0$$

and Eqs. (9.43) become

$$b_1 = r_1 r_2$$

$$-(\theta_1 - \theta_2) = 2k\pi \qquad k \text{ integer}$$

In our example the loci for b negative lies along the real axis to the right of the origin and also along the real axis to the left of $s = -a$ as shown by the crossed lines in Fig. 9.29. A network with this polynomial would be unstable for all $b_1 > 0$.

Let us pursue the polynomial $(s^2 + as + b)$, but this time plot the root locus for varying a. As before,

$$s^2 + as + b = 0$$

or

$$s^2 + b = -as$$

and

$$\frac{as}{s^2 + b} = -1 \tag{9.45}$$

Plotting the poles and zeros of the left-hand side of Eq. (9.45) in Fig. 9.30, we see pure

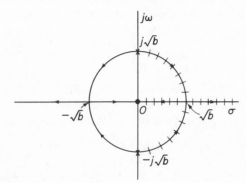

Figure 9.30 Root locus for Eq. (9.45), varying a. Crossed lines are for negative a.

imaginary poles at $\pm j\sqrt{b}$ and a zero at the origin. As a increases, one locus proceeds to the left from one pole and the other proceeds to the left from the other until they meet at $s = -\sqrt{b}$ as in Fig. 9.30. Then with continued increase in a, one locus goes to zero along the negative real axis.

For the case $a < 0$, the locus is the mirror image of the other in the right half plane, as shown in Fig. 9.30. For this particular locus plot, we obtain a circle of radius $\sqrt{b}$ around the origin.

The root locus is particularly useful in feedback design. In Fig. 9.31 we are given

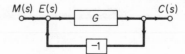

Figure 9.31 Feedback network.

an input $M(s)$ into an overall network with a forward transfer function $G(s) = KP(s)/Q(s)$. The output $C(s)$ is fed back into the summing point $E(s)$ such that

$$E(s) = M(s) - C(s) \tag{9.46}$$

To find the overall transfer function $C(s)/M(s)$, we proceed as follows:

$$C(s) = G(s)E(s)$$

but using Eq. (9.46),

$$C(s) = G(s)[M(s) - C(s)]$$

Collecting terms,

$$C(s)[1 + G(s)] = G(s)M(s)$$

or

$$\frac{C(s)}{M(s)} = \frac{G(s)}{1 + G(s)} \tag{9.47}$$

From Eq. (9.47) the poles of $C(s)/M(s)$ occur when

$$G(s) = -1 \tag{9.48}$$

Substituting in the definition of $G(s)$, we then get

$$\frac{KP(s)}{Q(s)} = -1 \tag{9.49}$$

which is precisely the correct form for a plot of the locus of the poles of $C(s)/M(s)$ with varying K. Thus, in this case the root locus determines what constant gain $K > 0$ we should use in the given "forward" network $G(s)$ to obtain the overall performance we want. From Eq. (9.47),

$$\frac{C(s)}{M(s)} = \frac{KP(s)/Q(s)}{1 + KP(s)/Q(s)}$$

or

$$\frac{C(s)}{M(s)} = \frac{KP(s)}{Q(s) + KP(s)} \tag{9.50}$$

The root-locus plot using Eq. (9.49) thus finds the roots of the polynomial $Q(s) +$ $KP(s)$. We present in Appendix B a computer program LOCUS, which produces a root locus given the form of Eq. (9.49).

Additional Reading

BODE, H. W., *Network Analysis and Feedback Amplifier Design*, Van Nostrand Reinhold, New York, 1945.

BROWN, G. S., and D. A. CAMPBELL, *Principles of Servomechanisms*, John Wiley & Sons, Inc., New York, 1948.

CLARK, R. N., *Introduction to Automatic Control Systems*, John Wiley & Sons, Inc., New York, 1962.

TERMAN, F. E., *Electronic and Radio Engineering*, McGraw-Hill Book Company, 1955.

THALER, G. J., and R. G. BROWN, *Servomechanism Analysis*, McGraw-Hill Book Company, New York, 1953.

TRUXAL, J. G., *Automatic Feedback Control System Synthesis*, McGraw-Hill Book Company, New York, 1955.

VAN VALKENBURG, M. E., *Network Analysis*, Prentice-Hall, Inc., Englewood Cliffs, N.J., 1964.

Problems

9.1. Given the series RLC circuit of Example 9.2, define $Q = \omega_n L/R$, where $\omega_n^2 = 1/LC$. Show that $Q = \frac{1}{2}\zeta$. Also show that the half-power bandwidth BW $= 2\zeta\omega_n = \omega_n/Q$.

9.2. For the following network functions, plot the straight-line asymptotic magnitude response and the phase response. Use four-cycle semilog paper.

(a) $G(s) = \dfrac{100}{s(0.01s + 1)(0.001s + 1)}$

(b) $G(s) = \dfrac{(0.1s + 1)(0.01s + 1)}{(s + 1)(0.001s + 1)}$

(c) $G(s) = \dfrac{100_s^2}{(0.17s + 1)(0.53s + 1)}$

(d) $G(s) = \dfrac{100(10s + 1)}{(s + 1)(s + 100)}$

(e) $G(s) = \dfrac{1250s}{(s + 0.4)(s + 50)^2}$

(f) $G(s) = \dfrac{s^2(20 - s)}{(50s + 1)(s + 1)^2}$

9.3. Determine the true response (magnitude and phase) for the functions of Problem 9.2 at the following frequencies:

(a) $\omega = 100, 500, 1000$ rad/s
(b) $\omega = 0, 5, 50, 100, 10^5$ rad/s

Use the results to verify the straight-line approximations of Problem 9.2.

9.4. Plot the straight-line asymptotic magnitude response, and determine the actual response for the network functions:

(a) $G(s) = 1000 \dfrac{(0.25s + 1)(0.1s + 1)}{(s + 1)(0.025s + 1)}$

(b) $G(s) = 180 \dfrac{s(0.01s + 1)}{(0.05s + 1)(0.001s + 1)}$

(c) $G(s) = 120 \dfrac{(0.2s + 1)}{s(s^2 + 2s + 10)}$

On the same coordinate system, plot the phase response for each function. Use four-cycle semilog paper for the plot.

9.5. Plot the magnitude and phase-response curves for the following transfer functions:

(a) $G(s) = \dfrac{s}{(s + 1)(s + 2)}$

(b) $G(s) = \dfrac{100s}{s^2 + 2s + 100}$

(c) $G(s) = \dfrac{10^8 s}{(s + 10)(s + 10^5)}$

9.6. Over what frequency range is the phase shift of the transfer function less than 5 degrees in the amplifier in Fig. P9.1 if

$$G(s) = \frac{E_0(s)}{E_i(s)} = \frac{10^8 s}{(s + 10)(s + 10^5)}$$

Phase shift in the amplifier corresponds to phase distortion.

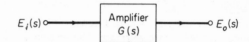

$E_i(s)$ ○——▶ | Amplifier $G(s)$ | ——○ $E_o(s)$

Figure P9.1

9.7. The transfer function of an amplifier is given by

$$G(s) = \frac{E_0(s)}{E_i(s)} = \frac{100}{(1 + 100/s)(10^{-5}s + 1)}$$

(a) Sketch the magnitude and phase-response curves for the amplifier.
(b) Determine the upper and lower half-power points of the magnitude response.

(c) Determine the midband gain of the amplifier.

9.8. Given the network in Fig. P9.2, sketch the Bode diagrams (magnitude and phase-response curves) for v_c when $R = 21\ \Omega$, $L = 1$ H, and $C = 0.05$ F. Also sketch the pole and zero locations in the complex plane.

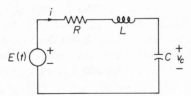

Figure P9.2

9.9. Repeat Problem 9.8 for $R = 8\ \Omega$ and $R = 2\ \Omega$.

9.10. Plot the poles and zeros for the following network functions:

(a) $F(s) = \dfrac{s^2 + 3s + 2}{s^2 + 3s}$

(b) $F(s) = \dfrac{1}{s^2 + s + 1}$

(c) $F(s) = \dfrac{100(s + 1)}{s^2 + 2s + 400}$

(d) $F(s) = \dfrac{100s^2}{(s^2 + 2s + 100)^2}$

(e) $F(s) = \dfrac{100s^2}{(s^2 + 2s + 100)(s^2 + 2s + 400)}$

(f) $F(s) = \dfrac{s^2 + 2s + 100}{s(s + 1)}$

Find numerical values for the bandwidth in parts (c) and (d).

9.11. The straight-line approximation to the gain curve is given for a certain amplifier in Fig. P9.3.

(a) What is the transfer function of the amplifier?
(b) What is the bandwidth of the amplifier?
(c) Suppose that two amplifiers with the above gain characteristic are connected in cascade and that the second amplifier does not load the first. Sketch the resulting gain characteristic and determine the bandwidth for the combination.

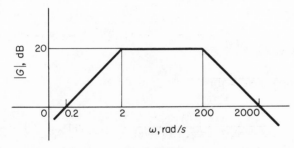

Figure P9.3

9.12. A phonograph pickup has a magnitude characteristic G approximated by the following expression:

$$G(j\omega) = \frac{10^6(j\omega)(j\omega + 3000)}{(j\omega + 600)(j\omega + 15,000)(j\omega + 120,000)}$$

A characteristic which is flat to within 3 dB is required from $\omega = 600$ to $\omega = 120,000$ and with a magnitude of 10 dB at $\omega = 60,000$. At ω values less than 600 and more than 120,000, the characteristic should drop off by at least 6 dB/octave or more. Draw $|G(j\omega)|$ on semilog paper. What should be the characteristic of a compensator in tandem with G to give the required results? (Note that angular velocity $\omega = 2\pi f$, where f is the frequency.)

9.13. A step input $x(t) = 10u(t)$ occurs on a network of second order. It is observed that the output peak overshoot of the final value of the output $y(t)$ is about 50 per cent and that the output has an oscillation frequency of 3 Hz during the transient. The final value of $y(t)$ at $t \longrightarrow \infty$ is 20. Write the differential equation relating $x(t)$ and $y(t)$ and their derivatives.

9.14. Using a sinusoidal generator, the ratio of the output to input as a function of angular velocity is determined for a network. This shows a ratio of 1 (0 dB) at low frequencies, a peak ratio M_P of about 6 dB at $\omega_p = 10$ rad/s, and a rolloff of about 40 dB/decade at high frequencies. Describe the time response of this same network to a step input as to per cent peak overshoot, oscillation frequency, and rise time. Write the response to an impulse $\delta(t)$.

9.15. It is desired to design a second-order network to have a 10 to 90 per cent rise time of 0.1 s, and a maximum overshoot of 20 per cent on a step input. Write the governing differential equation if the output equals the input as $t \longrightarrow \infty$.

9.16. A second-order network shows a peak overshoot of 60 per cent and an oscillation frequency of 6 Hz when a step input is applied. About what will M_P and ω_p be for a sinusoidal test? Find the phase angle between sinusoidal output and input at 1 and 15 Hz.

9.17. A term $(\frac{1}{3}s + 1)$ is added to the numerator of the transfer function of the network of Problem 9.15. About what is the per cent peak overshoot on a step function now? What is the oscillation frequency?

9.18. A term $(\frac{1}{3}s + 1)$ is added to the denominator of the transfer function of the network of Problem 9.15. What effect does this new term have on step response?

9.19. Find the root locus for the following functions with $K > 0$:

(a) $\dfrac{K(s + 5)}{s(s + 3)} = -1$

(b) $\dfrac{K(s + 5)}{s^2 + 4s + 20} = -1$

(c) $\dfrac{K}{s(s + 3)(s^2 + 3s + 10)} = -1$

9.20. Given

$$\frac{K(s + 4)}{s(s + 6)(s^2 + 2\zeta\sqrt{8}\,s + 8)} = -1 \quad \text{and} \quad 2\zeta\sqrt{8} = 4$$

Find the root locus for $K > 1$ and the roots of the characteristic equation $s(s + 6)$

$(s^2 + 4s + 8) + K(s + 4) = 0$, $K = 12.2$. With this value of K, plot the root locus with ζ varying both positively and negatively from its nominal value of $1/\sqrt{2}$.

9.21. Plot the root locus for

$$G = \frac{K}{(s + 5)(s^2 - 4)} = -1$$

Now let

$$G = \frac{K[(1/z)s + 1]}{(s + 5)(s^2 - 4)} = -1$$

Find the value of z such that the characteristic equation $(s + 5)(s^2 - 4) + K[(1/z)s + 1] = 0$ has no roots in the right half plane when $K = 50$. Repeat for $K/z = 50$.

Chapter 10

NETWORKS AND SYSTEMS IN STATE SPACE

These are the times that try men's souls.

THOMAS PAINE, *The American Crisis*,
No. 1, Dec. 19, 1776

10.1 Introduction

The differential equations of a lumped linear network can be written in the form

$$\dot{\mathbf{x}} = f(\mathbf{x}, \mathbf{m}, t)$$

(where $\mathbf{x}$ is a vector, $\mathbf{m}$ is a vector of inputs, and t represents time) as well as in the classical form discussed in Chapters 5 and 6. This system of first-order differential equations is known as the state equations of the network and $\mathbf{x}$ is the state vector. There are three basic reasons for writing the network equations in the state form: (1) the form lends itself easily to digital- and/or analog-computer methods of solution, (2) the method can be easily extended to the analysis of nonlinear networks, and (3) a number of systems concepts can be readily applied to the analysis of networks.

Beginning with this chapter we will investigate many important concepts based upon state-space theory which are useful in the time-domain analysis of networks.

10.2 Vectors in State Space

We discussed in Chapter 3 the notion of an array of n independent quantities as representing a vector in an n-dimensional space. We observed that if such a vector were depicted as

$$\mathbf{x} = \begin{bmatrix} x_1 \\ x_2 \\ \cdots \\ x_n \end{bmatrix} \tag{10.1}$$

then its "length" might be defined as

$$||\mathbf{x}|| = \langle \mathbf{x}, \mathbf{x} \rangle^{1/2} \tag{10.2a}$$

$$||\mathbf{x}|| = \left([x_1\, x_2 \ldots x_n] \begin{bmatrix} x_1 \\ x_2 \\ \cdots \\ x_n \end{bmatrix}\right)^{1/2} \tag{10.2b}$$

375

and the angle θ between two vectors $\mathbf{x}$ and $\mathbf{y}$ could be defined by the equation

$$\cos \theta = \frac{\langle x, y \rangle}{\|\mathbf{x}\| \, \|\mathbf{y}\|} \tag{10.3}$$

It is revealing to study the excursions of $\mathbf{x}$ as a function of time, where $\mathbf{x}$ may represent some quantity in a network (for example, the node voltages). The condition or state of the network at any time may be described by the location of this vector at that time, and the vector is said to exist in state space.

We should emphasize here that the analysis of networks constitutes a subset of the study of more general systems which may be described by differential equations. Usually such systems may contain mechanical, hydraulic, pneumatic, or other devices, as well as electrical elements. The theory presented here applies to such systems, as well as to electrical networks.

If a set of differential equations defines the network, as is usually the case, then in order to obtain a solution in terms of a state vector, the differential equations must be written in a form such that only the first derivative of the state vector is involved. That is, we must formulate the network by a set of first-order differential equations. Suppose that the network is linear, and $\mathbf{x} = \mathbf{x}(t)$ is the state variable. Let $\mathbf{m} = \mathbf{m}(t)$ be the input vector and $\mathbf{y} = \mathbf{y}(t)$ be an output vector. Then we have a situation as shown in Fig. 10.1, where N represents the network. We imply that there may be more than one

Figure 10.1 Representation of a network N in state space.

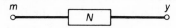

input, or that

$$\mathbf{m} = \begin{bmatrix} m_1 \\ m_2 \\ \cdots \\ m_r \end{bmatrix}$$

and more than one output given by

$$\mathbf{y} = \begin{bmatrix} y_1 \\ y_2 \\ \cdots \\ y_p \end{bmatrix}$$

We further assume the network state vector $\mathbf{x}$ to be n dimensional, or

$$\mathbf{x} = \begin{bmatrix} x_1 \\ x_2 \\ \cdots \\ x_n \end{bmatrix}$$

We then require the network be characterized by the equations

$$\dot{\mathbf{x}} = \mathbf{Ax} + \mathbf{Bm} \qquad (10.4a)$$

$$\mathbf{y} = \mathbf{Cx} + \mathbf{Dm} \qquad (10.4b)$$

where $\dot{\mathbf{x}} = d(\mathbf{x})/dt$ by definition.

If there are r inputs, p outputs, and n components for $\mathbf{x}$, then $\mathbf{A}$ in Eq. (10.4) is a $n \times n$ matrix, $\mathbf{B}$ is $n \times r$, $\mathbf{C}$ is $p \times n$, and $\mathbf{D}$ is $p \times r$. If the network is time varying, some elements (particularly of $\mathbf{A}$) will be time dependent.

If the input $\mathbf{m} = \mathbf{0}$, then Eqs. (10.4) become

$$\dot{\mathbf{x}} = \mathbf{Ax} \qquad (10.5a)$$

$$\mathbf{y} = \mathbf{Cx} \qquad (10.5b)$$

or we have a free network where the output $\mathbf{y}$ depends only on the initial condition of $\mathbf{x}$, which we symbolize as $\mathbf{x}(0)$. Since the output $\mathbf{y}$ in Eq. (10.5) consists of a linear combination of the components of $\mathbf{x}$, it is clear that the matrix $\mathbf{A}$ determines the network operation and this matrix thus becomes the fundamental specification of the network itself.

A powerful motivation for putting the network in terms of Eqs. (10.4) is that solutions on the digital computer require this form; that is, simultaneous equations with only first derivatives are easily solved in the time domain on the digital computer, whereas higher-order differential equations become quite difficult. In fact, a nonlinear network may be described by the equations

$$\dot{\mathbf{x}} = f(\mathbf{x}, \mathbf{m}, t) \qquad (10.6a)$$

$$\mathbf{y} = g(\mathbf{x}, \mathbf{m}, t) \qquad (10.6b)$$

and can also be similarly solved by the digital computer if the functions $f(\mathbf{x}, \mathbf{m}, t)$ and $g(\mathbf{x}, \mathbf{m}, t)$ are known. In Eqs. (10.6) we may still consider $\mathbf{x}$ to be in some sort of warped state space, but the notions of vector "length" and "angle" become rather vague.

10.3 Formulation in State Space

Accepting the fact that for digital-computer solutions it is necessary to describe the network in terms of Eqs. (10.4) or (10.6), how must we proceed to obtain the system of first-order differential equations? Assuming a linear, fixed network with a single input and output for the present, we may have the network description in terms of a differential equation of high order, for example

$$\frac{d^3y}{dt^3} + P\frac{d^2y}{dt^2} + Q\frac{dy}{dt} + Ry = Tm(t) \qquad (10.7)$$

where $y = y(t)$. If the right-hand side of Eq. (10.7) does not involve derivatives of the input $m(t)$, as here, then we may proceed as follows: Let the state variables be

$$x_1 = \frac{y}{T}$$

$$x_2 = \frac{\dot{y}}{T}$$

$$x_3 = \frac{\ddot{y}}{T}$$

where the dots over the variable indicate time derivatives. Then from Eq. (10.7) we obtain the set of equations

$$\dot{x}_1 = x_2$$
$$\dot{x}_2 = x_3$$
$$\dot{x}_3 = -Rx_1 - Qx_2 - Px_3 + m(t)$$
$$y = Tx_1$$

(10.8)

or

$$\dot{\mathbf{x}} = \mathbf{Ax} + \mathbf{Bm}$$
$$y = \mathbf{Cx} + \mathbf{Dm}$$

(10.9)

where

$$\mathbf{A} = \begin{bmatrix} 0 & 1 & 0 \\ 0 & 0 & 1 \\ -R & -Q & -P \end{bmatrix} \qquad \mathbf{B} = \begin{bmatrix} 0 \\ 0 \\ 1 \end{bmatrix}$$

$$\mathbf{C} = [T \quad 0 \quad 0] \qquad \mathbf{D} = 0$$

and

$$\mathbf{x} = [x_1 \; x_2 \; x_3]^\mathsf{T}$$

The choice of $x_1 = y/T$ is arbitrary. We could equally well choose the state vector as $x_1 = y$, $x_2 = \dot{y}$, and $x_3 = \ddot{y}$. The coefficients P, Q, R, and T in this case could also be functions of time without complicating the formulation.

Alternatively, we may select the state vector as follows: Let

$$x_3 = y$$
$$x_2 = \dot{y} + Py$$
$$x_1 = \ddot{y} + P\dot{y} + Qy$$

Then from Eq. (10.7) we obtain the set of equations

$$\dot{x}_1 = -Rx_3 + Tm(t)$$
$$\dot{x}_2 = -Qx_3 + x_1$$
$$\dot{x}_3 = -Px_3 + x_2$$
$$y = x_3$$

(10.10)

or

$$\dot{\mathbf{x}} = \mathbf{Ax} + \mathbf{Bm}$$
$$\mathbf{y} = \mathbf{Cx} + \mathbf{Dm} \qquad (10.11)$$

where

$$\mathbf{A} = \begin{bmatrix} 0 & 0 & -R \\ 1 & 0 & -Q \\ 0 & 1 & -P \end{bmatrix} \qquad \mathbf{B} = \begin{bmatrix} T \\ 0 \\ 0 \end{bmatrix}$$

$$\mathbf{C} = [0 \quad 0 \quad 1] \qquad \mathbf{D} = 0$$

We see that the state vector $\mathbf{x}$ is not unique. In fact, we could find a large number of specifications for Eq. (10.7) in other state-space forms. For any particular state-space form, we must find the initial conditions on the state variable in terms of the original variables. In the first case previously shown, the determination of initial conditions is simple, due to the choice of state variables, or

$$x_1(0) = \frac{1}{T} y(0)$$

$$x_2(0) = \frac{1}{T} \dot{y}(0) \qquad (10.12)$$

$$x_3(0) = \frac{1}{T} \ddot{y}(0)$$

In the second case, the initial state is

$$x_1(0) = \ddot{y}(0) + P\dot{y}(0) + Qy(0)$$
$$x_2(0) = \dot{y}(0) + Py(0) \qquad (10.13)$$
$$x_3(0) = y(0)$$

In other cases, finding the initial conditions becomes quite a problem in itself, as we will illustrate.

If derivatives of the input m are involved, then expression in the form of Eq. (10.4) involves more difficulty. Suppose we are given

$$y^{(n)} + \alpha_1 y^{(n-1)} + \cdots + \alpha_n y^{(0)} = \beta_0 m^{(n)} + \beta_1 m^{(n-1)} + \cdots + \beta_n m^{(0)}$$

where

$$y^{(n)} = \frac{d^n y}{dt^n} \quad \text{and} \quad y^{(0)} = y \qquad (10.14)$$

We need to find a form

$$\dot{x}_i = \sum_{r=1}^{n} a_{ir} x_r + b_i m \qquad i = 1, 2, \ldots, n \qquad (10.15a)$$

$$y = \sum_{j=1}^{n} c_j x_j + dm \qquad (10.15b)$$

In Eqs. (10.15) there are $n^2 + 2n + 1$ unknown constants. If we substitute Eq. (10.15) into Eq. (10.14), we find $2n + 1$ equations. Hence, n^2 of the constants may be selected arbitrarily. For simplicity, let $c_1 = 1$ and $c_2 = c_3 = \ldots = c_n = 0$. (We cannot make all c's $= 0$.) We now have $n^2 - n$ constants to select. Let

$$a_{ir} = \begin{cases} 0 & \text{for } r = 1, 2, \ldots, n \quad r \neq i + 1 \\ 1 & \text{for } r = i + 1 \end{cases} \quad i = 1, 2, \ldots, n - 1$$

With these choices, Eqs. (10.15a) and (10.15b) become

$$y = x_1 + dm$$
$$\dot{x}_1 = x_2 + b_1 m$$
$$\dot{x}_2 = x_3 + b_2 m$$
$$\cdots \cdots \cdots \cdots$$
$$\dot{x}_{n-1} = x_n + b_{n-1} m$$
$$\dot{x}_n = a_{n1} x_1 + a_{n2} x_2 + \cdots + a_{nn} x_n + b_n m \tag{10.16}$$

Then

$$y^{(1)} = \dot{x}_1 + d\dot{m} = x_2 + b_1 m + d\dot{m}$$
$$y^{(2)} = \ddot{x}_1 + dm = \dot{x}_2 + b_1 \dot{m} + dm = x_3 + b_2 m + b_1 \dot{m} + d\ddot{m}$$
$$\cdot \cdot \cdot \cdot \cdot \cdot \cdot \cdot \cdot \cdot \cdot \cdot \cdot \cdot \cdot \cdot \cdot \cdot \cdot \cdot \cdot \cdot \cdot \cdot \cdot \cdot \cdot$$
$$y^{(n)} = a_{n1} x_1 + a_{n2} x_n + \cdots + a_{nn} x_n + b_n m + b_{n-1} \dot{m} + \cdots + dm^{(n)}$$

where the superscript indicates a derivative with respect to time. Substituting these equations in the left-hand side of Eq. (10.14), we equate coefficients to find that

$$a_{nr} = -\alpha_{n-r+1} \quad r = 1, 2, \ldots, n \tag{10.17a}$$

and

$$d = \beta_0 \tag{10.17b}$$

From the previous choices, the other a_{ir} are given by

$$a_{ir} = \begin{cases} 0 & r = 1, 2, \ldots, n \quad r \neq i + 1 \\ 1 & r = i + 1 \qquad i = 1, 2, \ldots, n - 1 \end{cases} \tag{10.18}$$

For example, if $n = 4$, $a_{12} = a_{23} = a_{34} \ldots = 1$. The b values may be found by solving the matrix equation

$$\begin{bmatrix} 1 & 0 & 0 & \cdots & 0 \\ \alpha_1 & 1 & 0 & \cdots & 0 \\ \alpha_2 & \alpha_1 & 1 & \cdots & 0 \\ \cdot & \cdot & \cdot & \cdots & \cdot \\ \alpha_n & \alpha_{n-1} & \alpha_{n-2} & \cdots & 1 \end{bmatrix} \begin{bmatrix} d \\ b_1 \\ b_2 \\ \cdots \\ b_n \end{bmatrix} = \begin{bmatrix} \beta_0 \\ \beta_1 \\ \beta_2 \\ \cdots \\ \beta_n \end{bmatrix} \tag{10.19}$$

EXAMPLE 10.1

Given the fourth-order differential equation

$$\frac{d^4y}{dt} + 5\frac{d^3y}{dt^3} + 6\frac{d^2y}{dt^2} + 4\frac{dy}{dt} + 10y = 2\frac{d^2m}{dt^2} + 3\frac{dm}{dt} + 4m$$

Find a state-space representation of the equation.

Solution. Using Eq. (10.19) we write

$$\begin{bmatrix} 1 & 0 & 0 & 0 & 0 \\ 5 & 1 & 0 & 0 & 0 \\ 6 & 5 & 1 & 0 & 0 \\ 4 & 6 & 5 & 1 & 0 \\ 10 & 4 & 6 & 5 & 1 \end{bmatrix} \begin{bmatrix} d \\ b_1 \\ b_2 \\ b_3 \\ b_4 \end{bmatrix} = \begin{bmatrix} 0 \\ 0 \\ 2 \\ 3 \\ 4 \end{bmatrix}$$

Solving the matrix equation for the d and b_i coefficients, we find

$$d = 0 \qquad b_1 = 0 \qquad b_2 = 2$$
$$5b_2 + b_3 = 3 \quad \text{or} \quad b_3 = 3 - 10 = -7$$
$$6b_2 + 5b_3 + b_4 = 4 \quad \text{or} \quad b_4 = 4 + 35 - 12 = 27$$

The state-space representation becomes

$$\dot{\mathbf{x}} = \mathbf{A}\mathbf{x} + \mathbf{B}m$$
$$y = \mathbf{C}\mathbf{x} + \mathbf{D}m$$

The matrices **A, B, C**, and **D** may be found from Eqs. (10.16), (10.17), and (10.18) [apply Eq. (10.16) in reverse order]. They are

$$\mathbf{A} = \begin{bmatrix} 0 & 1 & 0 & 0 \\ 0 & 0 & 1 & 0 \\ 0 & 0 & 0 & 1 \\ -10 & -4 & -6 & -5 \end{bmatrix} \qquad \mathbf{B} = \begin{bmatrix} 0 \\ 2 \\ -7 \\ 27 \end{bmatrix}$$

$$\mathbf{C} = \begin{bmatrix} 1 & 0 & 0 & 0 \end{bmatrix} \qquad \mathbf{D} = 0$$

These state equations can also be written in an expanded form as

$$\dot{x}_1 = x_2$$
$$\dot{x}_2 = x_3 + 2m$$
$$\dot{x}_3 = x_4 - 7m$$
$$\dot{x}_4 = -10x_1 - 4x_2 - 6x_3 - 5x_4 + 27m$$
$$y = x_1$$

The initial conditions of the state vector are obtained by taking the derivative of the last equation. Thus,

$$\dot{y} = \dot{x}_1 = x_2$$
$$\ddot{y} = \ddot{x}_1 = \dot{x}_2 = x_3 + 2m$$
$$y^{(3)} = x_1^{(3)} = \ddot{x}_2 = \dot{x}_3 + 2\dot{m} + x_4 - 7m + 2\dot{m}$$

or rearranging,

$$x_1 = y$$
$$x_2 = \dot{y}$$
$$x_3 = \ddot{y} - 2m$$
$$x_4 = y^{(3)} + 7m - 2\dot{m}$$

which, since they hold for all times, apply also at $t = 0$, the initial time. The initial conditions on the state variables depend on the initial values of m and $\dot{m}$ as well as on the initial values of the original variables y, $\dot{y}$, $\ddot{y}$, and $y^{(3)}$.

The form in which the matrix $\mathbf{A}$ occurs in this development is sometimes called the *standard form*. The diagonalizing matrix $\mathbf{T}$, if the eigenvalues $\lambda_1, \lambda_2, \ldots, \lambda_n$ are distinct, is given in Section 3.6.

If a network consists of more than a simple node or loop, a system of simultaneous differential equations describes the network. We handle the system of equations in a similar manner as we did a single equation. The number of components of the state vector increases so that the order of the vector equals the sum of the orders of the highest derivative of each variable in the system of equations. The order of the output vector is determined by the number of outputs desired.

EXAMPLE 10.2

Given the system of differential equations

$$\frac{d^3y}{dt^3} + 6\frac{d^2y}{dt^2} + 4\frac{dy}{dt} + 10y + 6\frac{dz}{dt} = 10\dot{m} + 5m$$

$$\frac{d^2z}{dt^2} + 4\frac{dz}{dt} + 6z + 3\frac{dy}{dt} + 4y = 5\dot{n} + 2n$$

Find a state-space representation.

Solution. Since the order of the highest derivative in y and z is 3 and 2, respectively, the state vector must have $3 + 2 = 5$ components.

Let the components be†

$$x_1 = y$$
$$x_2 = \dot{y}$$
$$x_3 = \ddot{y} - 10m$$
$$x_4 = z$$
$$x_5 = \dot{z} - 5n$$

Substituting the state variables into the original equations, we get

$$\dot{x}_3 = -6(x_3 + 10m) - 4x_2 - 10x_1 - 6(x_5 + 5n) + 5m$$
$$\dot{x}_5 = -4(x_5 + 5n) - 6x_4 - 3x_2 - 4x_1 + 2n$$

---

†If derivatives of the input are not found in the original equations, the input terms may be omitted in selecting the state-space variables.

One set of state space equations is thus

$$\dot{\mathbf{x}} = \mathbf{Ax} + \mathbf{Br}$$

$$\boldsymbol{\omega} = \mathbf{Cx} + \mathbf{Dr}$$

where

$$\mathbf{A} = \begin{bmatrix} 0 & 1 & 0 & 0 & 0 \\ 0 & 0 & 1 & 0 & 0 \\ -10 & -4 & -6 & 0 & -6 \\ 0 & 0 & 0 & 0 & 1 \\ -4 & -3 & 0 & -6 & -4 \end{bmatrix} \quad \mathbf{B} = \begin{bmatrix} 0 & 0 \\ 10 & 0 \\ -55 & -30 \\ 0 & 5 \\ 0 & -18 \end{bmatrix}$$

$$\mathbf{C} = \begin{bmatrix} 1 & 0 & 0 & 0 & 0 \\ 0 & 0 & 0 & 1 & 0 \end{bmatrix} \quad \mathbf{D} = [0 \quad 0]$$

and

$$\mathbf{x} = \begin{bmatrix} x_1 \\ x_2 \\ x_3 \\ x_4 \\ x_5 \end{bmatrix} \quad \mathbf{r} = \begin{bmatrix} m \\ n \end{bmatrix} \quad \boldsymbol{\omega} = \begin{bmatrix} y \\ z \end{bmatrix}$$

This system of equations has five state variables, two inputs, and two outputs. Note that the number of inputs and outputs does not depend on the number of state variables in the system.

Other state-space forms may be found by using topological methods, as we will discuss in the succeeding chapters or by using flow diagrams.[†]

10.4 Laplace-Transform Solutions, Free Network

Assume that we have a network described in the state-space form

$$\dot{\mathbf{x}} = \mathbf{Ax} + \mathbf{Bm} \tag{10.20a}$$

$$\mathbf{y} = \mathbf{Cx} + \mathbf{Dm} \tag{10.20b}$$

How can we solve this set of equations? Since in Eq. (10.20b) $\mathbf{y}$ is a linear combination of $\mathbf{x}$ and $\mathbf{m}$, our primary concern is Eq. (10.20a). We may solve the set of equations represented by Eq. (10.20a) numerically on the digital computer by using the differential-equation algorithms discussed in Chapter 12 which are valid even for nonlinear equations. However, if $\mathbf{A}$ has constant elements, analytic solutions are possible. Let us first investigate the undriven network described by

$$\dot{\mathbf{x}}(t) = \mathbf{Ax}(t) \tag{10.21a}$$

$$\mathbf{x}(0) = \mathbf{x}_0 \tag{10.21b}$$

†B. O. Watkins, *Introduction to Control Systems*, The Macmillan Company, New York, 1969, pp. 139–176.

Since in the scalar case the exponential function is a solution, we try

$$\mathbf{x}(t) = e^{\mathbf{A}t}\mathbf{k} \tag{10.22}$$

where

$$e^{\mathbf{A}t} = \mathbf{I} + \mathbf{A}t + \mathbf{A}^2 t^2/2! + \ldots + \mathbf{A}^k t^k/k! + \ldots \tag{10.23}$$

and $\mathbf{k}$ is a constant vector. Then

$$\frac{d\mathbf{x}(t)}{dt} = \mathbf{A}e^{\mathbf{A}t}\mathbf{k} \tag{10.24}$$

as can be demonstrated by differentiating $e^{\mathbf{A}t}$. Substituting Eqs. (10.22) and (10.24) in Eqs. (10.21), we see that Eq. (10.22) is a solution and that

$$\mathbf{k} = \mathbf{x}_0 \tag{10.25}$$

Hence, the undriven solution is

$$\mathbf{x}(t) = e^{\mathbf{A}t}\mathbf{x}_0 \tag{10.26}$$

where $\mathbf{x}_0$ is the initial-condition vector. The matrix $e^{\mathbf{A}t} = \exp(\mathbf{A}t)$ is often called the *state transition matrix*, $\boldsymbol{\phi}(t)$, or

$$\boldsymbol{\phi}(t) = e^{\mathbf{A}t} \tag{10.27}$$

by definition.

There are several methods of finding $\boldsymbol{\phi}(t)$, one of which is to expand the exponential as in Eq. (10.23), which converges due to the factorial denominator. A closed-form solution for $\mathbf{x}(t)$ is also possible. One of the most useful methods applies the Laplace-transform technique. The method can be best illustrated by transforming Eq. (10.21a) to obtain

$$(s\mathbf{I} - \mathbf{A})\mathbf{X}(s) = \mathbf{x}_0 \tag{10.28}$$

Solving for $\mathbf{X}(s)$, we get

$$\mathbf{X}(s) = (s\mathbf{I} - \mathbf{A})^{-1}\mathbf{x}_0 \tag{10.29}$$

Taking the inverse Laplace transform of Eq. (10.29) and comparing the result to Eq. (10.26), we see that

$$\boldsymbol{\phi}(t) = \mathcal{L}^{-1}[(s\mathbf{I} - \mathbf{A})^{-1}] \tag{10.30}$$

EXAMPLE 10.3

Let

$$\mathbf{A} = \begin{bmatrix} 5 & \sqrt{2} \\ \sqrt{2} & 4 \end{bmatrix}$$

Then

$$(s\mathbf{I} - \mathbf{A}) = \begin{bmatrix} s & 0 \\ 0 & s \end{bmatrix} - \begin{bmatrix} 5 & \sqrt{2} \\ \sqrt{2} & 4 \end{bmatrix} = \begin{bmatrix} s-5 & -\sqrt{2} \\ -\sqrt{2} & s-4 \end{bmatrix}$$

and

$$(s\mathbf{I} - \mathbf{A})^{-1} = \frac{1}{\det(s\mathbf{I} - \mathbf{A})}\begin{bmatrix} s - 4 & \sqrt{2} \\ \sqrt{2} & s - 5 \end{bmatrix}$$

$$\det(s\mathbf{I} - \mathbf{A}) = (s^2 - 9s + 18) = (s - 6)(s - 3)$$

Thus,

$$\boldsymbol{\phi}(t) = \mathcal{L}^{-1}\left\{\frac{1}{(s - 6)(s - 3)}\begin{bmatrix} s - 4 & \sqrt{2} \\ \sqrt{2} & s - 5 \end{bmatrix}\right\}$$

or

$$\boldsymbol{\phi}(t) = \frac{1}{3}\begin{bmatrix} 2e^{6t} + e^{3t} & \sqrt{2}(e^{6t} - e^{3t}) \\ \sqrt{2}(e^{6t} - e^{3t}) & e^{6t} + 2e^{3t} \end{bmatrix}$$

We now develop a general algorithm for obtaining $\boldsymbol{\Phi}(s) = (s\mathbf{I} - \mathbf{A})^{-1} = \mathcal{L}\boldsymbol{\phi}(t)$, which we call the *Faddeeva algorithm*. Let

$$(s\mathbf{I} - \mathbf{A})^{-1} = \frac{s^{n-1}\mathbf{B}_0 + s^{n-2}\mathbf{B}_1 + \cdots + s\mathbf{B}_{n-2} + \mathbf{B}_{n-1}}{s^n + d_1 s^{n-1} + d_2 s^{n-2} + \cdots + d_n} \qquad (10.31)$$

where the denominator is the characteristic polynomial and the **B** matrices have constant elements. Then

$$\mathbf{I}(s^n + d_1 s^{n-1} + d_2 s^{n-2} + \cdots + d_n) = (s^{n-1}\mathbf{B}_0 + s^{n-2}\mathbf{B}_1 + \cdots s\mathbf{B}_{n-2} + \mathbf{B}_{n-1})(s\mathbf{I} - \mathbf{A})$$

Comparing coefficient, we find that

$$\mathbf{B}_0 = \mathbf{I}$$
$$\mathbf{B}_1 = \mathbf{A} + d_1\mathbf{I}$$
$$\mathbf{B}_2 = \mathbf{A}\mathbf{B}_1 + d_2\mathbf{I}$$
$$\cdot \quad \cdot \quad \cdot \quad \cdot \quad \cdot \quad \cdot \quad \cdot$$
$$\mathbf{B}_k = \mathbf{A}\mathbf{B}_{k-1} + d_k\mathbf{I} \qquad (10.32)$$
$$\cdot \quad \cdot \quad \cdot \quad \cdot \quad \cdot \quad \cdot \quad \cdot \quad \cdot$$
$$\mathbf{B}_{n-1} = \mathbf{A}\mathbf{B}_{n-2} + d_{n-1}\mathbf{I}$$
$$\mathbf{O} = \mathbf{A}\mathbf{B}_{n-1} + d_n\mathbf{I}$$

For the 2×2 matrix,

$$\mathbf{A} = \begin{bmatrix} a_{11} & a_{12} \\ a_{21} & a_{22} \end{bmatrix}$$

$$\det[\lambda\mathbf{I} - \mathbf{A}] = (\lambda - a_{11})(\lambda - a_{22}) - a_{21}a_{12}$$
$$= \lambda^2 - \lambda(a_{11} + a_{22}) + a_{11}a_{22} - a_{21}a_{12}$$

The characteristic equation in this case is

$$\lambda^2 + d_1\lambda + d_2 = 0$$

Comparing coefficients,

$$-d_1 = a_{11} + a_{22}$$

Using the same technique, if $\mathbf{A}$ is $n \times n$,

$$-d_1 = a_{11} + a_{22} + \cdots + a_{nn}$$

Now define the trace of $\mathbf{A}$ (tr $\mathbf{A}$) by the equation

$$\text{tr } \mathbf{A} = a_{11} + a_{22} + \cdots + a_{nn}$$

or

$$\text{tr } \mathbf{A} = \sum_{i=1}^{n} a_{ii} \tag{10.33}$$

Then

$$-d_1 = \text{tr } \mathbf{A}$$

Using these ideas and following an inductive method Faddeeva shows that[†]

$$
\begin{aligned}
-d_1 &= \text{tr } \mathbf{A} \\
-d_2 &= \frac{1}{2} \text{tr } \mathbf{AB}_1 \\
&\cdot \cdot \cdot \cdot \cdot \cdot \\
-d_k &= \frac{1}{k} \text{tr } \mathbf{AB}_{k-1} \\
&\cdot \cdot \cdot \cdot \cdot \cdot \cdot \\
-d_n &= \frac{1}{n} \text{tr } \mathbf{AB}_{n-1}
\end{aligned}
\tag{10.34}
$$

The algorithm of Eqs. (10.32) and (10.34) is easily programmed on the computer (see Fig. 10.2).

EXAMPLE 10.4

     Use the Faddeeva algorithm to find $(s\mathbf{I} - \mathbf{A})^{-1}$ for the $\mathbf{A}$ of Example 10.3.

     Solution.

$$\mathbf{B}_0 = \begin{bmatrix} 1 & 0 \\ 0 & 1 \end{bmatrix}$$

$$\mathbf{A} = \begin{bmatrix} 5 & \sqrt{2} \\ \sqrt{2} & 4 \end{bmatrix}$$

$$-d_1 = \text{tr } \mathbf{A} = 4 + 5 = 9$$

$$\mathbf{B}_1 = \mathbf{A} + d_1 \mathbf{I}$$

$$= \begin{bmatrix} 5 & \sqrt{2} \\ \sqrt{2} & 4 \end{bmatrix} - 9 \begin{bmatrix} 1 & 0 \\ 0 & 1 \end{bmatrix}$$

$$= \begin{bmatrix} -4 & \sqrt{2} \\ \sqrt{2} & -5 \end{bmatrix}$$

[†]V. N. Faddeeva, *Computational Methods of Linear Algebra*, Dover Publications, Inc., New York, 1959, pp. 177–182.

```
      SUBROUTINE FADEVA(A,B,C,D,L,IT,M,N)                       FDVA   1
C  PROGRAM FOR INVERTING MATRICES A AND (SI-A) USING FADEEVAS METHOD   FDVA   2
C  (SEE INTRODUCTION TO CONTROL SYSTEMS BY B.WATKINS 1969 MACMILLAN    FDVA   3
C  P260). INPUT REAL ELEMENTS A(I,J) ROW WISE AND ORDER N.OUTPUT B     FDVA   4
C  MATRICES A INVERSE AND CHARACTERISTIC EQUATION COEFFICIENTS D(M)    FDVA   5
C  IN DESCENDING ORDER.ORDER N MUST NOT BE OVER TEN.                   FDVA   6
C  MATRIX A IS DESTROYED                                              FDVA   7
C  N:   ORDER OF MATRIX A (NOT OVER TEN).                             FDVA   8
C  A:   A MATRIX ELEMENTS AND A INVERSE ELEMENTS                      FDVA   9
C  B:   (N-1) B MATRICES ELEMENTS                                    FDVA  10
C  C:   C MATRIX(TEMPORARY) AND CHECK MATRIX.                        FDVA  11
C  D:   CHARACTERISTIC EQUATION COEFFICIENTS.                        FDVA  12
      DIMENSION A(10,10),B(9,9,9),C(10,10),D(11)                     FDVA  13
      REAL*8A,B,C,D,ZAP                                              FDVA  14
      DO 10 I=1,N                                                    FDVA  15
      DO 10 J=1,N                                                    FDVA  16
   10 C(I,J)=A(I,J)                                                  FDVA  17
      D(1)=1.0                                                       FDVA  18
      DO 40 L=1,N                                                    FDVA  19
      IT=L-1                                                         FDVA  20
      M=L+1                                                          FDVA  21
C  COMPUTE CHARACTERISTIC EQUATION COEFFICIENTS D(M) FROM TRACE      FDVA  22
      D(M)=0                                                         FDVA  23
      DO 20 K=1,N                                                    FDVA  24
   20 D(M)=D(M)+C(K,K)                                               FDVA  25
      D(M)=-D(M)/L                                                   FDVA  26
      IF (L.EQ.N) GO TO 50                                          FDVA  27
C  COMPUTE B(L) MATRICES FOR (SI-A) INVERSE                         FDVA  28
      DO 30 I=1,N                                                    FDVA  29
      DO 30 J=1,N                                                    FDVA  30
      B(I,J,L)=C(I,J)                                                FDVA  31
   30 IF (I.EQ.J)B(I,J,L)=C(I,J)+D(M)                               FDVA  32
C  COMPUTE AB MATRIX                                                FDVA  33
      DO 40 I=1,N                                                    FDVA  34
      DO 40 J=1,N                                                    FDVA  35
      C(I,J)=0.0                                                     FDVA  36
      DO 40 IND=1,N                                                  FDVA  37
   40 C(I,J)=C(I,J)+A(I,IND)*B(IND,J,L)                             FDVA  38
C  COMPUTE A INVERSE                                                FDVA  39
   50 DO 60 I=1,N                                                    FDVA  40
      DO 60 J=1,N                                                    FDVA  41
      C(I,J)=-C(I,J)/D(M)                                            FDVA  42
   60 A(I,J)=-B(I,J,IT)/D(M)                                        FDVA  43
      RETURN                                                        FDVA  44
      END                                                           FDVA  45
```

Figure 10.2   Faddeeva algorithm for inverting a
matrix.

$$\mathbf{AB}_1 = \begin{bmatrix} 5 & \sqrt{2} \\ \sqrt{2} & 4 \end{bmatrix} \begin{bmatrix} -4 & \sqrt{2} \\ \sqrt{2} & -5 \end{bmatrix} = \begin{bmatrix} -18 & 0 \\ 0 & -18 \end{bmatrix}$$

$$-d_2 = \frac{1}{2}(-18 - 18) = -18$$

$$\mathbf{AB}_1 + d_2\mathbf{I} = 0 \qquad \text{(check)}$$

Then

$$(s\mathbf{I} - \mathbf{A})^{-1} = \frac{\begin{bmatrix} s & 0 \\ 0 & s \end{bmatrix} + \begin{bmatrix} -4 & \sqrt{2} \\ \sqrt{2} & -5 \end{bmatrix}}{s^2 - 9s + 18}$$

10.5   Inverse Laplace Transform

There is no need to invert $(s\mathbf{I} - \mathbf{A})^{-1}$ in the Laplace sense for analysis in the complex s-domain. However, it is necessary to apply the inverse Laplace transform to each element of the matrix resulting from summing the matrices $\mathbf{B}$ in Eq. (10.31) to find $\phi(t)$. We obtain the inverse transform of an element of this matrix by the following procedure:

1. Factor the characteristic equation.

2. Find a partial-fraction expansion of the element.

3. Take the inverse Laplace transform of each term of the partial-fraction expansion.

The characteristic equation may be factored by a polynomial root-solving routine such as the routine (POLYS) presented in Appendix B. As previously discussed, if the coefficients of the characteristic equation are randomly produced floating-point numbers, repeated roots seldom occur. If some do, a slight change in one or more of the coefficients will alter the roots (see the Problems). Thus, we can find

$$d_0 s^n + d_1 s^{n-1} + d_2 s^{n-2} + \cdots + d_n = d_0 (s - \lambda_1)(s - \lambda_2) \ldots (s - \lambda_n) \qquad (10.35)$$

where $\lambda_1, \lambda_2, \ldots, \lambda_n$ are the located roots, none of which are repeated. The λ's constitute the eigenvalues of $\mathbf{A}$ and the poles of the network.

A typical matrix element in the numerator of Eq. (10.31) is

$$a_0 s^{n-1} + a_2 s^{n-2} + \cdots + a_{n-2} s + a_{n-1}$$

where $a_0, a_1, \ldots, a_{n-1}$ are elements at corresponding locations in the matrices $\mathbf{B}_0$, $\mathbf{B}_1, \ldots, \mathbf{B}_{n-1}$. We must then find the following partial-fraction expansion

$$\frac{a_0 s^{n-1} + a_1 s^{n-2} + \cdots + a_{n-2} s + a_{n-1}}{d_0 s + d_1 s^{n-1} + d_2 s^{n-2} + \cdots + d_n} = \frac{K_1}{s - \lambda_1} + \frac{K_2}{s - \lambda_2} + \cdots + \frac{K_n}{s - \lambda_n} \qquad (10.36)$$

for each term in the matrix equation. For each complex root, a complex-conjugate root exists. If

$$\lambda_i = -\sigma + j\omega$$

then

$$\lambda_{i+1} = -\sigma - j\omega$$

The constants on the right-hand side of Eq. (10.36) will also be complex for this complex pair, so

$$K_i = w + jy$$

$$K_{i+1} = w - jy$$

We may then combine two terms of the right-hand side of Eq. (10.36) by writing

$$\frac{w + j\omega}{s + \sigma - j\omega} + \frac{w - jy}{s + \sigma + j\omega} = \frac{2[ws + (w\sigma - y\omega)]}{s^2 + 2\sigma s + (\sigma^2 + \omega^2)} = \frac{As + B}{(s + \sigma)^2 + \omega^2}$$

$$= \frac{As + B}{s^2 + ps + q}$$

Thus, if we let

$$N(s) = a_0 s^{n-1} + a_1 s^{n-2} + \cdots + a_n s + a_{n-1}$$

and

$$D(s) = d_0 s^n + d_1 s^{n-1} + d_2 s^{n-2} + \cdots + d_n$$

then Eq. (10.36) becomes

$$\frac{N(s)}{D(s)} = \sum_{i=1}^{r} \frac{K_k}{s + \alpha_i} + \sum_{i=r+1}^{n/2} \frac{A_i s + B_i}{s^2 + p_i s + q_i} \tag{10.37}$$

where $D(s)$ has r real roots and n total roots. The real roots are $\lambda_i = -\alpha_i$ and the used complex roots (one of each pair) are $\lambda_i = -\sigma_i + j\omega_i$. The inverse transform of the terms with real poles in Eq. (10.37) is given by

$$f_i(t) = K_i e^{-\alpha_i t} \tag{10.38a}$$

where

$$K_1 = \frac{N(-\alpha_i)}{D'(-\alpha_i)} \tag{10.38b}$$

where $D'(s) = dD(s)/ds$.† The inverse transform of the terms with complex poles in Eq. (10.37) is given by

$$f_i(t) = C_i e^{-\sigma_i t} \sin(\omega_i t + \theta_i) \tag{10.39a}$$

where

$$C_i = 2\sqrt{\frac{B(B - Ap) + A^2 q}{4q - p^2}} = 2\sqrt{w_i^2 + y_i^2} \tag{10.39b}$$

$$= \frac{1}{\omega}\sqrt{(B - A\sigma)^2 + (A\omega)^2} \tag{10.39c}$$

$$\theta_i = \tan^{-1}\frac{A\sqrt{4q - p^2}}{2B - Ap} = \tan^{-1}\frac{w_i}{-y_i} \tag{10.39d}$$

$$= \tan^{-1}\frac{A\omega}{B - A\sigma} \tag{10.39e}$$

and

$$w_1 + jy_i = \frac{N(-\sigma_i + j\omega_i)}{D'(-\sigma_i + j\omega_i)} \tag{10.39f}$$

where $p = 2\sigma$, $q = \sigma^2 + \omega^2$, $A = 2w$, $B = 2(w\sigma - y\omega)$.

The inverse transforms are determined once we know the K_i, A_i, B_i, and the corresponding factors $(s + \alpha_i)$ and $(s^2 + p_i s + q_i) = (s + \sigma_i)^2 + \omega_i^2$. We assume that the factors of the characteristic equation are known, and now present a technique to find the constants K_i, A_i, and B_i which avoids taking the derivative in Eqs. (10.38b) and (10.39d) and the complex arithmetic indicated in Eq. (10.39f).

---

†See Appendix A.

Following Henrici,[†] let

$$P(s) = a_0 s_n + a_1 s^{n-1} + \cdots + a_n$$
$$= (s^2 + ps + q)(b_0 s^{n-2} + b_1 s^{n-3} + \cdots + b_{n-2} \qquad (10.40)$$
$$+ b_{n-1}(s + p) + b_n$$

Multiplying out the right-hand side of Eq. (10.40) and comparing coefficients, we get

$$b_0 = a_0$$
$$b_1 = a_1 - pb_0$$
$$b_2 = a_2 - pb_1 - qb_0$$

or in general

$$\cdots \cdots \cdots \cdots \cdots$$
$$b_k = a_k - pb_{k-1} - qb_{k-2} \qquad (10.41)$$

Equation (10.41) holds for $k = 2, \ldots, n$, and also for $k = 0, 1$ if we let $b_{-1} = b_{-2} = 0$. Hence to divide out a factor $(s^2 + ps + q)$ from $P(s)$ find $b_k = a_k - pb_{k-1} - qb_{k-2}$, $k = 0, 1, \ldots, n$ with $b_{-2} = b_{-1} = 0$. The coefficients b_{n-1} and b_n are the numerator of the remainder of $P(s)/(s^2 + ps + q)$ if $(s^2 + ps + q)$ is not a factor, while $b_{n-1} = b_n = 0$ if the quadratic is a factor.

Then given

$$\frac{N(s)}{D(s)} = \frac{a_0 s^{n-1} + a_1 s^{n-2} + \cdots + a_{n-1}}{(s^2 + ps + q)(b_0 s^{n-2} + b_1 s^{n-3} + \cdots + b_{n-2})} \qquad (10.42a)$$

$$= \frac{N_1(s)}{D_1(s)} + \frac{As + B}{s^2 + ps + q} \qquad (10.42b)$$

where

$$b_0 = d_0$$

we wish to find A and B.

From Eq. (10.40),

$$N(s) = (s^2 + ps + q)(c_0 s^{n-3} + c_1 s^{n-4} + \cdots + c_{n-3}) + c_{n-2}(s + p) + c_{n-1} \qquad (10.43a)$$

and

$$D_1(s) = (s^2 + ps + q)(g_0 s^{n-4} + g_1 s^{n-5} + \cdots + g_{n-4}) + g_{n-3}(s + p) + g_{n-2} \qquad (10.43b)$$

$$sD_1(s) = (s^2 + ps + q)(g_0 s^{n-3} + g_1 s^{n-4} + \cdots + g_{n-3}) + g_{n-2}(s + p) + g_{n-1} \qquad (10.43c)$$

[†]Peter Henrici, *Elements of Numerical Analysis*, John Wiley & Sons, Inc., New York, 1964, p. 108.

Now

$$\frac{N_1(s)}{D_1(s)} = \frac{N(s)}{D(s)} - \frac{As + B}{s^2 + ps + q}$$

$$= \frac{N(s)(s^2 + ps + q) - D(s)(As + B)}{D(s)(s^2 + ps + q)}$$

or

$$\frac{N_1(s)}{D_1(s)} = \frac{N(s) - D_1(s)(As + B)}{D(s)} \tag{10.44}$$

If we divide the numerator and denominator of the right-hand side of Eq. (10.44) by $(s^2 + ps + q)$, then

$$\frac{N_1(s)}{D_1(s)} = \frac{[N(s) - D_1(s))(As + B)]/(s^2 + ps + q)}{D_1(s)}$$

or

$$N_1(s) = [N(s) - D_1(s)(A_1s + B_1)]/(s^2 + ps + q) \tag{10.45}$$

Since $N_1(s)$ is a polynomial of degree $n - 3$, the remainder in Eq. (10.45) must be zero. Substituting Eqs. (10.43) into Eq. (10.44), the numerator of the remainder is

$$c_{n-2}(s + p) + c_{n-1} = B[g_{n-3}(s + p) + g_{n-2}] - A[g_{n-2}(s + p) + g_{n-1}] \tag{10.46}$$

Setting the constant and linear terms of Eq. (10.46) independently to zero we get

$$c_{n-2}p + g_{n-1} - Bg_{n-3}p - Bg_{n-2} - Ag_{n-2}p - Ag_{n-1} = 0$$

and

$$c_{n-2} - Bg_{n-3} - Ag_{n-2} = 0$$

Now multiply the second equation by p and subtract from the first to obtain (using the second equation again)

$$g_{n-1}A + g_{n-2}B = c_{n-1}$$
$$g_{n-2}A + g_{n-3}B = c_{n-2}$$

These two equations give A and B or

$$A = \frac{g_{n-3}c_{n-1} - g_{n-2}c_{n-2}}{g_{n-1}g_{n-3} - g_{n-2}^2} \tag{10.47a}$$

$$B = \frac{g_{n-1}c_{n-2} - g_{n-2}c_{n-1}}{g_{n-1}g_{n-3} - g_{n-2}^2} \tag{10.47b}$$

From Eq. (10.41) the c and g coefficients in Eqs. (10.47) are determined by the recurrence relations

$$c_k = -pc_{k-1} - qc_{k-2} + a_k \qquad (10.48a)$$

$$k = 0, 1, \ldots, n-1 \qquad c_k = 0, \quad \text{for } k < 0$$

$$g_k = -pg_{k-1} - qg_{k-2} + b_k \qquad (10.48b)$$

$$k = 0, 1, \ldots, n-1 \qquad g_k = 0 \quad \text{for } k < 0 \qquad b_{n-1} = 0$$

Note that the algorithm operates on the polynomial $D_1(s)$, which is $D(s)$ reduced by the quadratic factor $(s^2 + ps + q)$. The reduced polynomial $D_1(s)$ may be obtained by using the algorithm on $D(s)$. Thus, the algorithm must be applied twice to $D(s)$; once to reduce it, and again to find the remainder.

EXAMPLE 10.5

Let

$$\frac{N(s)}{D(s)} = \frac{5s^3 + 9s^2 + 3s - 2}{(s^2 + s - 2)(s^2 + 2s + 2)}$$

$$= \frac{A_1 s + B_1}{s^2 + 2s + 2} + \frac{A_2 s + B_2}{s^2 + s - 2}$$

To find A_1 and B_1, $p = 2$, $q = 2$, $a_0 = 5$, $a_1 = 9$, $a_2 = 3$, $a_3 = -2$, $b_0 = 1$, $b_1 = 1$, $b_2 = -2$, $b_3 = 0$. Then

k			k	
0	$c_0 = 5$		0	$g_0 = 1$
1	$c_1 = -2(5) + 9 = -1$		1	$g_1 = -2(1) + 1 = -1$
2	$c_2 = -2(-1) - 2(5) + 3 = -5$		2	$g_2 = -2(-1) - 2(1) - 2 = -2$
3	$c_3 = -2(-5) - 2(-1) - 2 = 10$		3	$g_3 = -2(-2) - 2(-1) + 0 = 6$

Since $n = 4$, $n - 1 = 3$, $n - 2 = 2$, $n - 3 = 1$,

$$A_1 = \frac{-1(10) - (-2)(-5)}{6(-1) - (-2)^2}$$

$$= 2$$

$$B_1 = \frac{6(-5) - (-2)(10)}{-10}$$

$$= 1$$

To find A_2 and B_2, $p = 1$, $q = -2$, $b_0 = 1$, $b_1 = 2$, $b_2 = 2$, and the a_i are the same. Then

k			k	
0	$c_0 = 5$		0	$g_0 = 1$
1	$c_1 = -1(5) + 9 = 4$		1	$g_1 = -1(1) + 2 = 1$
2	$c_2 = -1(4) + 2(5) + 3 = 9$		2	$g_2 = -1(1) + 2(1) + 2 = 3$
3	$c_3 = -1(9) + 2(4) - 2 = -3$		3	$g_3 = -1(3) + 2(1) + 0 = -1$

$$A_2 = \frac{1(-3) - 3(9)}{-1(1) - 3^2} = 3$$

$$B_2 = \frac{-1(9) - 3(-3)}{-10} = 0$$

or finally,

$$\frac{5s^3 + 9s^2 + 3s - 2}{(s^2 + s - 2)(s^2 + 2s + 2)} = \frac{3s}{s^2 + s - 2} + \frac{2s + 1}{s^2 + 2s + 2}$$

We may proceed similarly to find the K_i values of Eq. (10.37); that is, given

$$\frac{N(s)}{D(s)} = \frac{N_1(s)}{D_1(s)} + \frac{K}{s + \alpha} \tag{10.49}$$

we wish to find K.

In this case let

$$\begin{aligned} P(s) &= a_0 s^n + a_1 s^{n-1} + \cdots + a_n \\ &= (s + \alpha)(b_0 s^{n-1} + b_1 s^{n-2} + \cdots + b_{n-1}) + b_n \end{aligned} \tag{10.50}$$

Multiplying out the right side of Eq. (10.50) and comparing coefficients as before, we find that

$$\begin{aligned} b_0 &= a_0 \\ b_1 &= a_1 - \alpha b_0 \\ b_2 &= a_2 - \alpha b_1 \\ &\cdot \ \cdot \ \cdot \ \cdot \ \cdot \ \cdot \ \cdot \\ b_k &= a_k - \alpha b_{k-1} \end{aligned} \tag{10.51}$$

Equation (10.51) holds for $k = 0, 1, \ldots, n$ if we let $b_{-1} = 0$. If $b_n = 0$, then $(s + \alpha)$ is a factor of $P(s)$; otherwise it is a remainder. From Eq. (10.49)

$$\begin{aligned} \frac{N_1(s)}{D_1(s)} &= \frac{N(s)}{D(s)} - \frac{K}{s + \alpha} \\ &= \frac{N(s)(s + \alpha) - D(s)K}{D(s)(s + \alpha)} \end{aligned}$$

or

$$\frac{N(s)}{D(s)} = \frac{N(s)D_1 - (s)K}{D(s)} \tag{10.52}$$

Following the same arguments as used previously, but now applying Eq. (10.50), the numerator of the remainder in Eq. (10.52) is

$$c_{n-1} = K g_{n-1}$$

Thus,

$$K = \frac{c_{n-1}}{g_{n-1}}$$

The c_k and g_k coefficients are determined by the recurrence relations

$$c_k = -\alpha c_{k-1} + a_k \tag{10.53}$$

$$k = 0, 1, \ldots, n - 1 \qquad c_k = 0 \quad \text{for } k < 0$$

$$g_k = -\alpha g_{k-1} + b_k \tag{10.54}$$

$$k = 0, 1, \ldots, n - 1 \qquad g_k = 0 \quad \text{for } k < 0$$

EXAMPLE 10.6

Let

$$\frac{N(s)}{D(s)} = \frac{5s^3 + 9s^2 + 3s - 2}{(s + 1)(s + 2)(s^2 + s2 + 2)}$$

$$= \frac{K_1}{s + 1} + \frac{K_2}{s + 2} + \frac{As + B}{s^2 + 2s + 2}$$

To find K_1, $\alpha = 1$, $a_0 = 5$, $a_1 = 9$, $a_2 = 3$, $a_3 = -2$, $b_0 = 1$, $b_1 = 4$, $b_2 = 6$, $b_3 = 4$. Then

k		k	
0	$c_0 = 5$	0	$g_0 = 1$
1	$c_1 = -1(5) + 9 = 4$	1	$g_1 = -1(1) + 4 = 3$
2	$c_2 = -1(4) + 3 = -1$	2	$g_2 = -1(3) + 6 = 3$
3	$c_3 = -1(-1) - 2 = -1$	3	$g_3 = -1(3) + 4 = 1$

$$K_1 = -\frac{1}{1} = -1$$

To find K_2, $\alpha = 2$, $b_0 = 1$, $b_1 = 3$, $b_2 = 4$, $b_3 = 2$, and the a_i are the same. Then

k		k	
0	$c_0 = 5$	0	$g_0 = 1$
1	$c_1 = -2(5) + 9 = -1$	1	$g_1 = -2(1) + 3 = 1$
2	$c_2 = -2(-1) + 3 = 5$	2	$g_2 = -2(1) + 4 = 2$
3	$c_3 = -2(5) - 2 = -12$	3	$g_3 = -2(2) + 2 = -2$

$$K_2 = \frac{-12}{-2} = 6$$

Using the previous technique, $A = 0$, $B = -5$, or

$$\frac{5s^3 + 9s^2 + 3s - 2}{(s + 1)(s + 2)(s^2 + 2s + 2)} = \frac{-1}{s + 1} + \frac{6}{s + 2} + \frac{-5}{s^2 + 2s + 2}$$

Note again that the algorithm given by Eq. (10.51) must be used twice on the denominator $D(s)$; once to remove the factor and obtain $D_1(s)$, and again on $D_1(s)$ to find the remainder.

A similar technique may be used to find partial-fraction expansions with higher-order denominator factors, but is of little interest in inverse-transformation work. A factor $(s + 0)$ is a special case where $\alpha = 0$.

Figure 10.3 presents a computer program to find a partial-fraction expansion following the technique shown above. Inverse transforms of any but the simplest ratio of polynomials are almost impossible to calculate by hand without error, and a computer program becomes almost a "must."†

---

†For a numerical inversion method, see C. D. Han, "A Note of Numerical Inversion of the Laplace Transform by Legendre Polynomials," *IEEE Transactions on Automatic Control*, Vol. AC-12, April 1967, pp. 230–231.

Handwritten at top:

1 FORMAT (//,2x, "PARTIAL FRACTION COEFFICIENT{K)=',F10.5,5x,'FOR
THE REAL ROOTS',F10.5)

```
      SUBROUTINE PARTL(DA,C,Z,ZIM,ZA,XK,A,B,P,Q,N,L,J)         PRTL   1
C THIS PROGRAM OBTAINS THE PARTIAL FRACTION EXPANSION OF THE RATIO OF   PRTL   2
C TWO POLYNOMIALS N(S)/D(S) WITH REAL COEFFICIENTS.THE DENOMINATOR D(S)PRTL   3
C ROOTS Z(I)+JZIM(I),Z(I)-JZIM(I),I=1,N MUST BE INPUTED.        PRTL   4
C ROOTS ARE ASSUMED TO NOT BE REPEATED.                         PRTL   5
C BASED ON AN ALGORITHM BY HENRICI AND WATKINS.                 PRTL   6
C DA : DENOMINATOR D(S) COEFFICIENTS IN DESCENDING ORDER.ORDER N LESS  PRTL   7
C      THAN 20.                                                 PRTL   8
C C  : NUMERATOR N(S) COEFFICIENTS IN DESCENDING ORDER.ORDER N-1.      PRTL   9
C Z  : REAL PART OF A ROOT OF D(S).                             PRTL  10
C ZIM: IMAGINARY PART OF A ROOT OF D(S).IF ZIM=0,IT MUST BE ENTERED AS PRTL  11
C      SUCH. COMPLEX ROOTS MUST BE ENTERED IN CONJUGATE PAIRS.  PRTL  12
C XK : CONSTANT IN FACTOR XK/(S+Z).                             PRTL  13
C P  : SECOND TERM IN QUADRATIC FACTOR S*S+PS+Q;=-2.*Z.         PRTL  14
C Q  : LAST TERM IN QUADRATIC FACTOR S*S+PS+Q;=Z*Z+ZIM*ZIM.     PRTL  15
C A  : CONSTANT.SEE NEXT LINE.                                  PRTL  16
C B  : CONSTANT IN FACTOR AS+B/(S*S+PS+Q).                      PRTL  17
C A RESULT OF ZERO FOR XK OR A AND B INDICATES AN INCORRECT ROOT WAS   PRTL  18
C ENTERED.                                                      PRTL  19
      DIMENSION DA(20),CA(20),Z(20),ZIM(20),D(20),C(20),BB(20),ZA(20),P(PRTL  20
     110),Q(10),A(10),B(10),XK(20)                              PRTL  21
      REAL*8DA,C,ZA,BB,D,CA,P,Q,Z,ZIM,DEN                       PRTL  22
C NEXT CARD IS USED FOR WATFIV COMPILER.REMOVE IF NOT NEEDED.   PRTL  23
      REAL*8DABS                                                PRTL  24
   10 FORMAT (1X,'REAL ROOT NUMBER',I2,'INCORRECT')             PRTL  25
   20 FORMAT (1X,'COMPLEX ROOT NUMBER',I2,'INCORRECT')          PRTL  26
      DO 30 K=1,20                                              PRTL  27
   30 XK(K)=0.                                                  PRTL  28
      DO 40 K=1,10                                              PRTL  29
      A(K)=0.                                                   PRTL  30
   40 B(K)=0.                                                   PRTL  31
      M=N+1                                                     PRTL  32
      J=0                                                       PRTL  33
      L=0                                                       PRTL  34
      BB(1)=DA(1)                                               PRTL  35
      D(1)=DA(1)                                                PRTL  36
      CA(1)=C(1)                                                PRTL  37
      DO 150 I=1,N                                              PRTL  38
C CHECK FOR LINEAR OR QUADRATIC FACTOR                          PRTL  39
      IF (ZIM(I).NE.0.) GO TO 90                                PRTL  40
C FIND CONSTANT XK FOR LINEAR FACTOR.                           PRTL  41
      L=L+1                                                     PRTL  42
      ZA(L)=-Z(I)                                               PRTL  43
C DIVIDE OUT LINEAR FACTOR FROM DENOMINATOR.                    PRTL  44
      DO 50 K=2,M                                               PRTL  45
   50 BB(K)=DA(K)-ZA(L)*BB(K-1)                                 PRTL  46
C CHECK REAL ROOT                                               PRTL  47
      IF (DABS(BB(M)).GT.1.D-03) GO TO 80                       PRTL  48
C FIND REMAINDER FOR DENOMINATOR.                               PRTL  49
      DO 60 K=2,N                                               PRTL  50
   60 D(K)=-ZA(L)*D(K-1)+BB(K)                                  PRTL  51
C FIND REMAINDER FOR NUMERATOR.                                 PRTL  52
      DO 70 K=2,N                                               PRTL  53
   70 CA(K)=-ZA(L)*CA(K-1)+C(K)                                 PRTL  54
C FIND CONSTANT XK(I).                                          PRTL  55
      XK(L)=CA(N)/D(N)                                          PRTL  56
      GO TO 150                                                 PRTL  57
   80 WRITE (6,10) I                                            PRTL  58
      GO TO 150                                                 PRTL  59
C FIND CONSTANTS A+B FOR QUADRATIC FACTOR.                      PRTL  60
   90 IF (I.LT.2) GO TO 100                                     PRTL  61
C BY PASS CONJUGATE ROOT.                                       PRTL  62
      IF (ZIM(I).EQ.-ZIM(I-1)) GO TO 150                        PRTL  63
  100 J=J+1                                                     PRTL  64
C FIND P,Q FROM Z,ZIM.                                          PRTL  65
      P(J)=-2.0*Z(I)                                            PRTL  66
      Q(J)=Z(I)*Z(I)+ZIM(I)*ZIM(I)                              PRTL  67
C DIVIDE OUT QUADRATIC FACTOR FROM DENOMINATOR.                 PRTL  68
      BB(2)=DA(2)-P(J)*DA(1)                                    PRTL  69
      DO 110 K=3,M                                              PRTL  70
```

Handwritten annotations in right margin and center:

DA,C

→ WRITE (1,1) Z(I),XK(L)   root =',

1 FORMAT (//,2x,'THE PARTIAL FRACTION
coefficient K=',F10.5)

Figure 10.3 Partial fraction algorithm.

```
      110 BB(K)=DA(K)-P(J)*BB(K-1)-Q(J)*BB(K-2)              PRTL  71
    C   CHECK COMPLEX ROOTS                                  PRTL  72
          IF (DABS(BB(M)).GT.1.D-03.OR.DABS(BB(N)).GT.1.D-03) GO TO 140   PRTL  73
    C   FIND REMAINDER FOR DENOMINATOR.                      PRTL  74
          D(2)=BB(2)-P(J)*D(1)                               PRTL  75
          DO 120 K=3,N                                       PRTL  76
      120 D(K)=BB(K)-P(J)*D(K-1)-Q(J)*D(K-2)                 PRTL  77
    C   FIND REMAINDER FOR NUMERATOR.                        PRTL  78
          CA(2)=C(2)-P(J)*C(1)                               PRTL  79
          DO 130 K=3,N                                       PRTL  80
      130 CA(K)=C(K)-P(J)*CA(K-1)-Q(J)*CA(K-2)               PRTL  81
    C   FIND CONSTANTS A(I)+B(I).                            PRTL  82
          DEN=D(N)*D(N-2)-D(N-1)*D(N-1)                      PRTL  83
          A(J)=(D(N-2)*CA(N)-D(N-1)*CA(N-1))/DEN             PRTL  84
          B(J)=(D(N)*CA(N-1)-D(N-1)*CA(N))/DEN               PRTL  85
          GO TO 150                                          PRTL  86
      140 WRITE (6,20) I                                     PRTL  87
      150 CONTINUE                                           PRTL  88
          RETURN                                             PRTL  89
          END                                                PRTL  90
```

$$C(J)=2*((B(J)*(B(J)-A(J)*P(J))$$
$$+(A(J)**2)*Q(J))/(4*Q(J)-(P(J)**2)))$$
$$WRITE(1,2) P(J),Q(J),A(J),B(J),C(J),Z(I),ZIM(I)$$

Figure 10.3—Cont.

10.6 Laplace-Transform Solutions, Driven Network

If $\mathbf{m} \neq 0$, then we need to solve the equation

$$\dot{\mathbf{x}} = \mathbf{A}\mathbf{x} + \mathbf{B}\mathbf{m} \qquad (10.55)$$

Again, taking the Laplace transform, if the network is fixed, we get

$$(s\mathbf{I} - \mathbf{A})\mathbf{X}(s) = \mathbf{x}_0 + \mathbf{B}\mathbf{M}(s)$$

or

$$\mathbf{X}(s) = (s\mathbf{I} - \mathbf{A})^{-1}\mathbf{x}_0 + (s\mathbf{I} - \mathbf{A})^{-1}\mathbf{B}\mathbf{M}(s) \qquad (10.56)$$

Then

$$\mathbf{x}(t) = \mathcal{L}^{-1}[\boldsymbol{\Phi}(s)\mathbf{x}_0 + \boldsymbol{\Phi}(s)\mathbf{B}\mathbf{M}(s)] \qquad (10.57)$$

where

$$\boldsymbol{\Phi}(s) = (s\mathbf{I} - \mathbf{A})^{-1}$$

Thus, the driven solution for the state variables may be found by taking the inverse Laplace transform of $\boldsymbol{\Phi}(s)\mathbf{B}\mathbf{M}(s)$. The output is

$$\mathbf{y} = \mathbf{C}\mathbf{x} + \mathbf{D}\mathbf{m}$$

so that

$$y(t) = \mathcal{L}^{-1}[\mathbf{C}\boldsymbol{\Phi}(s)\mathbf{x}_0 + \mathbf{C}\boldsymbol{\Phi}(s)\mathbf{B}\mathbf{M}(s) + \mathbf{D}\mathbf{M}(s)] \qquad (10.58)$$

If $\mathbf{D} = \mathbf{O}$, as it usually does, then the driven output for $\mathbf{x}_0 = 0$ becomes

$$\mathbf{y}(t) = \mathcal{L}^{-1}[\mathbf{C}\boldsymbol{\Phi}(s)\mathbf{B}\mathbf{M}(s)] \qquad (10.59)$$

The transient output is as found in Section 10.3. If we let

$$\mathbf{H}(s) = \mathbf{C}\boldsymbol{\Phi}(s)\mathbf{B} + \mathbf{D}$$

then the steady-state output neglecting initial conditions is

$$\mathbf{y}(t) = \mathcal{L}^{-1}[\mathbf{H}(s)\mathbf{M}(s)] \tag{10.60}$$

Using the convolution theorem in Laplace transforms, we get an explicit solution for the output as

$$\mathbf{y}(t) = \int_0^t \mathbf{h}(t - \tau)\mathbf{m}(\tau)\, d\tau \tag{10.61}$$

where $\mathbf{h}(t) = C\boldsymbol{\phi}(t)\mathbf{B} + \mathbf{D} = \mathcal{L}^{-1}\mathbf{H}(s)$. The matrix $\mathbf{H}(s)$ is called the *transfer function*, as it relates the output with the input in the *s*-domain.

From Eq. (10.60), if all entries in $\mathbf{M}(s)$ are unity, $\mathbf{y}(t)$ is the impulse response, or the output when all inputs are impulses. Thus, $\mathbf{h}(t)$ is also called the *impulse response matrix*. It may be determined by applying an impulse to each input in turn, determining the corresponding outputs, and finally summing these outputs. Since C, B, and D are known, this would be an indirect way of determining $\boldsymbol{\phi}(t)$. When using this method, it is assumed that all inputs and all ouputs are coupled to the network. With such a complete coupling we define the network as being controllable (all inputs affect at least one state variable) and observable (all state variables affect at least one output).

We may solve the driven network on the computer by either taking an inverse transform using Eq. (10.60), or by integration in time, using Eq. (10.61). The transition matrix $\boldsymbol{\phi}(t)$ [or its transform $\boldsymbol{\Phi}(s)$] must be known in either case. If $\mathbf{M}(s)$ is a ratio of polynomials, the preceding partial-fraction program may be used, but for many sources $\mathbf{M}(s)$ may not be such a ratio.

The direct solution of Eqs. (10.4) or (10.6) by utilizing an integration algorithm may be considerably less difficult and time consuming than applying the Laplace transform methods previously described. Such algorithms results in open-ended solutions which are subject to errors similar to those encountered in using a finite number of terms in an infinite series. In contrast, the Laplace methods give analytic, closed-form solutions. The direct-solution method may nevertheless be faster, more economical, and sufficiently accurate. We present some of these direct methods in Chapter 12.

Additional Reading

DeRusso, P. M., R. J. Roy, and C. M. Close. *State Variables for Engineers.* John Wiley & Sons, Inc., New York, 1965.

Faddeeva, V. N. *Computational Methods of Linear Algebra.* Dover Publications, Inc., New York, 1959.

Finkbeiner, D. T., II. *Introduction to Matrices and Linear Transformations.* W. H. Freeman and Company, Publishers, San Francisco, 2nd ed., 1966.

FRAME, J. S. "Matrix Functions and Applications." *IEEE Spectrum*, Vol. 1, 1964.

SCHWARZ, R. J., and B. FRIEDLAND. *Linear Systems*. McGraw-Hill Book Company, New York, 1965.

ZADEH, L. A., and C. A. DESOER. *Linear System Theory*. McGraw-Hill Book Company, New York, 1963.

Problems

10.1. (a) Set up the state equations for the differential equation

$$\frac{d^2y}{dt^2} + 5\frac{dy}{dt} + 2y = m(t)$$

where $y = y(t)$. Represent the equation by the vector-matrix state equation

$$\dot{\mathbf{x}} = \mathbf{Ax} + \mathbf{Bm}$$

and the output equation

$$\mathbf{y} = \mathbf{Cx} + \mathbf{Dm}$$

(b) Find the Laplace state transition matrix $\mathbf{\Phi}(s)$ and the state transition matrix $\boldsymbol{\phi}(t)$.

(c) Solve the state equations for $y(t)$ assuming $m(t) = 0$, $y(0) = 1$, $dy/dt(0) = 2$. Plot x_1 as a function of x_2, with t varying.

10.2. Solve the differential equation of Problem 10.1 assuming zero initial conditions and $m(t) = 2\delta(t)$. Repeat for $m(t) = 2u(t)$.

10.3. (a) Set up the state equations for the network characterized by the differential equation

$$2\frac{d^2y}{dt^2} + 3\frac{dy}{dt} + 18y = \frac{dm}{dt} + 2m$$

Represent the equation in matrix form as in Problem 10.1.

(b) Find the Laplace transition matrix $\mathbf{\Phi}(s)$.

(c) Solve the state equations for $y(t)$ assuming zero initial conditions and $m(t) = e^{-2t}$.

(d) Solve for $y(t)$ if $m(t) = 0$ and $y(0) = 1$, $dy/dt(0) = 2$. Plot x_2 versus x_1 as t varies. Compare with Problem 10.2.

10.4. (a) Set up the state equations for the network characterized by the differential equation

$$\frac{d^3y}{dt^3} + 3\frac{d^2y}{dt^2} + 2y = m$$

Represent the equation in matrix form as in Problem 10.1.

(b) Find the matrix $\mathbf{\Phi}(s)$.

(c) Solve the state equations for $y(t)$ assuming zero initial conditions with $m(t) = 2u(t)$.

10.5. (a) Set up the state equations for the network characterized by the second-order differential equation

$$\frac{d^2y}{dt^2} + 3\frac{dy}{dt} + 2y = 3\frac{d^2m}{dt^2} + 5\frac{dm}{dt} + m$$

Represent the equation in matrix form as in Problem 10.1.

(b) Find the matrix $\mathbf{\Phi}(s)$.

(c) Solve the state equations for $y(t)$ assuming zero initial conditions with $m(t) = 2e^{-2t}$.

10.6. (a) Set up the state equations $\dot{\mathbf{x}} = \mathbf{A}\mathbf{x} + \mathbf{B}m$ and $\mathbf{y} = \mathbf{C}\mathbf{x} + \mathbf{D}m$ related to the differential equation

$$\frac{d^3y}{dt^3} + 6\frac{d^2y}{dt^2} + 11\frac{dy}{dt} + 6y = 2\frac{d^2m}{dt^2} + 8\frac{dm}{dt} + 4m$$

(b) Find a diagonalizing matrix $\mathbf{V}$ such that

$$\mathbf{V}^{-1}\mathbf{A}\mathbf{V} = \begin{bmatrix} \lambda_1 & 0 & 0 \\ 0 & \lambda_2 & 0 \\ 0 & 0 & \lambda_3 \end{bmatrix} = [\lambda_{ii}]$$

(see Chapter 2). Let $\mathbf{x} = \mathbf{V}\mathbf{w}$ and convert the equation $\dot{\mathbf{x}} = \mathbf{A}\mathbf{x} + \mathbf{B}m$ to

$$\dot{\mathbf{w}} = [\lambda_{ii}]\mathbf{w} + \mathbf{V}^{-1}\mathbf{B}m$$

Solve for $\mathbf{w}$ if $\mathbf{m} = 0$.

(c) Find the initial conditions for $\mathbf{x}$ and $\mathbf{w}$ in terms of $y(0)$, $dy/dt(0)$, $d^2y/dt^2(0)$, $m(0)$ and $dm/dt(0)$.

10.7. Solve the following state equations and plot the state trajectories:

(a) $\qquad\qquad\qquad \dot{x}_1 = x_1 \qquad\qquad x_1(0) = 1$

$\qquad\qquad\qquad\qquad \dot{x}_2 = -x_1 + 2x_2 \qquad x_2(0) = 1$

(b) $\qquad\qquad\qquad \dot{x}_1 = x_1 - 2x_2 \qquad x_1(0) = 1$

$\qquad\qquad\qquad\qquad \dot{x}_2 = x_1 \qquad\qquad x_2(0) = -1$

10.8. The state equations of a linear network are given as

$$\begin{bmatrix} \dot{x}_1 \\ \dot{x}_2 \end{bmatrix} = \begin{bmatrix} 3 & 1 \\ 1 & -1 \end{bmatrix}\begin{bmatrix} x_1 \\ x_2 \end{bmatrix}$$

Determine the new state equations of the network if the state variables are redefined as

$$q_1 = x_1 + x_2 \qquad q_2 = x_1 - x_2$$

10.9. Using the Faddeeva algorithm find the inverse $\mathbf{A}^{-1}$ of the following matrices $\mathbf{A}$. Find $(s\mathbf{I} - \mathbf{A})^{-1}$ and the characteristic equation in each case. Show that $\mathbf{A}^{-1}\mathbf{A} = \mathbf{I}$.

(a) $$\begin{bmatrix} 0 & 1 & 6 \\ 0 & 0 & 1 \\ -6 & -11 & -6 \end{bmatrix}$$

(b) $$\begin{bmatrix} 1 & -2 & 3 & 5 & 6 \\ 3 & -6 & 4 & 9 & 6 \\ 2 & 5 & 2 & 4 & 3 \\ 4 & 3 & -6 & 8 & 5 \\ 8 & 10 & 3 & 2 & 6 \end{bmatrix}$$

10.10. Find the roots of the following equations:

(a) $s^3 + 7s^2 + 16s + 12 = 0$
(b) $s^3 + 7.007s^2 + 16s + 12 = 0$
(c) $s^3 + 7s^2 + 16.016s + 12 = 0$
(d) $s^3 + 7s^2 + 16s + 12.012 = 0$

Use the subroutine given in Appendix B for (b), (c), and (d).

10.11. Find the partial-fraction expansion of the following polynomial ratios:

(a) $\dfrac{s + 3}{(s + 1)(s + 2)}$

(b) $\dfrac{2s^2 + 3s + 1}{(s^2 + 4s + 8)(s + 2)}$

(c) $\dfrac{s + 3}{(s^2 + 4s + 8)(s^2 + 6s + 18)(s^2 + 8s + 32)}$

(d) $\dfrac{5s^4 + 4s^3 + 3s^2 + s + 6}{s^7 + 4.1s^6 + 3.8s^5 + 5.2s^4 + 8.7s^3 + 7.6s^2 + 9s + 31}$

Note: Use the root-finding subroutine in Appendix B to find the factors for the denominator in (d).

10.12. Find the partial-fraction expansion of the following polynomial ratios:

(a) $\dfrac{s + 3}{s(s + 2)}$

(b) $\dfrac{2s^2 + 3s + 1}{s(s + 2)(s^2 + 4s + 8)}$

(c) $\dfrac{s + 3}{s(s^2 + 8s + 32)(s^2 + 4s + 8)}$

Chapter 11

STATE EQUATIONS
OF
ELECTRICAL NETWORKS

*Doing easily what others find
difficult is talent; doing what is
impossible for talent is genius.*

HENRI-FRÉDÉRIC AMIEL, *Journal*

11.1 Introduction

From the material presented in Chapter 10, we observe a different approach for the state-variable method as compared to the methods discussed earlier in the text. In the mesh, node, loop, and cut-set analysis methods, the complete time-domain solution for the dependent circuit variables i and v requires a knowledge of the initial condition of these variables. Since we must derive these initial values from the initial values of the inductance currents and capacitance voltages, a process which becomes both tedious and difficult, we might as well use the inductance currents and capacitance voltages as dependent variables in the first instance.

This scheme provides two advantages over the classical methods. First, the initial conditions can be applied directly in obtaining a solution, and second, the use of capacitance voltages and inductance currents results in a system of first-order differential equations which can be solved directly using digital and/or analog computational methods. There are many integration techniques (some of which we discuss in Chapter 12) available for the digital computer which solve first-order systems of differential equations in the time domain.

The state-variable analysis of network problems is based upon the selection of capacitance voltages and inductance currents (or linear combinations of these) as the state variables of the network. Using these voltages and currents as state variables, the state equations depend upon expressing the voltages across the inductances $L\,di_L/dt$, and the current through the capacitances $C\,dv_c/dt$, as functions of the state variables and the inputs. For a given network† with l inductances (no mutual inductance), m capacitances, and n sources (voltage and current), the state equations of the network can be written as

$$L_i \frac{di_i}{dt} = \sum_{j=1}^{l} R_{ij} i_j + \sum_{j=l+1}^{l+m} G_{ij} v_j + \sum_{j=1}^{n} U_{ij} s_j \tag{11.1}$$

$$C_k \frac{dv_k}{dt} = \sum_{j=1}^{l} \mathcal{R}_{kj} i_j + \sum_{j=l+1}^{l+m} \mathcal{G}_{kj} v_j + \sum_{j=1}^{n} \mathcal{U}_{kj} s_j \tag{11.2}$$

where
$$i = 1, 2, 3, \ldots, l, \; k = l+1, l+2, \ldots, l+m$$
v_j = voltage across jth capacitance
i_j = current through jth inductance
s_j = jth source (voltage or current)

---

†We assume here that the network graph can be represented by a "proper" tree, as subsequently defined.

402

The elements of R are dimensionally resistance, while the elements of $\mathcal{G}$ are conductance; G and $\mathcal{R}$ are unitless, U elements are unitless or resistance depending on whether the source is voltage or current, respectively. Similarly, $\mathcal{U}$ elements are either conductance or unitless.

In this chapter we pursue two methods of writing state equations for electrical networks. The first is the inspection method. The second, more general, method utilizes network graph theory.

11.2 State Equations by Inspection

In this section we demonstrate a method for writing state equations which represent several simple network types by inspection. These network types fall into three primary categories as follows:

1. Networks containing no voltage source–capacitance loops, and no nodes connecting only inductances and current sources.

2. Networks containing voltage source–capacitance loops.

3. Networks containing inductance–current source cut-sets.

The best way to determine the category of the network to be analyzed consists of sketching the network graph. The branches† should be incorporated into the tree in the following order:

A. All branches containing voltage sources are included. If a loop with only voltage sources develops, the network is indeterminate.

B. Add branches containing capacitances to the tree. In certain networks, some of the capacitances cannot be incorporated into the tree without forming loops in the graph. These capacitances (referred to as excess or link capacitances) cause the network to fall into the second category. In either case, usually all the capacitances can be placed in the tree.

C. If the network tree is not complete (not all nodes connected), as many resistance branches as necessary are added to complete the tree. (Usually, part of the resistance elements can be designated as tree branches and part as links.)

D. If still unable to complete the tree using all resistance elements as branches, use inductive branches to finish the tree structure. This can only occur if a node forms an inductance–current source cut-set. The inductances in this cut-set (referred to as *excess inductances*) cause the network to fall in the third category. A network may contain both excess capacitances and inductances, in which case techniques applying to both the second and third categories must be used.

---

†A branch can contain a single R, L, C, E, or J element. Combinations, such as R and E, as used in Chapter 7, are not allowed.

The state equations are now written for each of the tree capacitances and link inductances. The system of equations we obtain from this procedure will be unique and independent. The following examples clarify the process.

EXAMPLE 11.1

As an example of a category 1 network type, consider the network of Fig. 11.1a. Using the previous rules in order, the associated graph of Fig. 11.1b emerges. Solid lines are used to represent the tree branches, and dotted lines are used to represent the links.

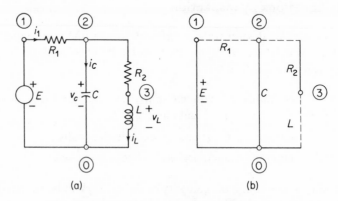

(a)   (b)

Figure 11.1 *RLC* electrical network.

Since the element C in this network is a tree capacitance and L is a link inductance, we apply KCL to node 2 to obtain

$$C\frac{dv_c}{dt} = i_1 - i_L$$
$$= \frac{E - v_c}{R_1} - i_L \tag{11.3}$$

Similarly, we apply KVL to the loop containing L and C to obtain

$$L\frac{di_L}{dt} = v_c - R_2 i_L \tag{11.4}$$

Rearranging, the equations become

$$\frac{dv_c}{dt} = \frac{E - v_c}{R_1 C} - \frac{i_L}{C} \tag{11.5}$$

$$\frac{di_L}{dt} = \frac{v_c}{L} - \frac{R_2}{L} i_L \tag{11.6}$$

In normal-state form, the equations are

$$\begin{bmatrix} \dfrac{dv_c}{dt} \\ \dfrac{di_L}{dt} \end{bmatrix} = \begin{bmatrix} -\dfrac{1}{R_1 C} & -\dfrac{1}{C} \\ \dfrac{1}{L} & -\dfrac{R_2}{L} \end{bmatrix} \begin{bmatrix} v_c \\ i_L \end{bmatrix} + \begin{bmatrix} \dfrac{1}{R_1 C} \\ 0 \end{bmatrix} E \tag{11.7}$$

Using the format specified in Eq. (10.9) we define

$$\mathbf{A} = \begin{bmatrix} -\dfrac{1}{R_1 C} & -\dfrac{1}{C} \\[2mm] \dfrac{1}{L} & -\dfrac{R_2}{L} \end{bmatrix} \qquad \mathbf{B} = \begin{bmatrix} \dfrac{1}{R_1 C} \\[2mm] 0 \end{bmatrix}$$

and

$$\mathbf{x} = \begin{bmatrix} v_c \\ i_L \end{bmatrix}$$

If we designate the output of the network as $\mathbf{Y} = \begin{bmatrix} v_{R_1} \\ v_{R_2} \end{bmatrix}$, the output equation is

$$\mathbf{Y} = \begin{bmatrix} -1 & 0 \\ 0 & R_2 \end{bmatrix} \begin{bmatrix} v_c \\ i_L \end{bmatrix} + \begin{bmatrix} 1 \\ 0 \end{bmatrix} E \tag{11.8}$$

or

$$\mathbf{C} = \begin{bmatrix} -1 & 0 \\ 0 & R_2 \end{bmatrix}$$

and

$$\mathbf{D} = \begin{bmatrix} 1 \\ 0 \end{bmatrix}$$

From the preceding example, we summarize the procedure for writing state equations as

1. Using KCL, express the current in each tree capacitance in terms of the link inductance currents, the tree capacitance voltages, and the source voltages and currents. The first two variables are the state variables and the remaining two variables are the driving functions.

2. Using KVL, express the voltage across each of the link inductances in terms of the state variables and the driving functions.

These two steps appear straightforward, but we sometimes encounter difficulties, as shown in the next example.

EXAMPLE 11.2

We complicate the network of Fig. 11.1 slightly by adding a third resistance R_3 in series with the capacitance, as shown in Fig. 11.2a. The resulting tree appears in Fig. 11.2b. The equations written for the modified network are

$$C \frac{dv_c}{dt} = \frac{E - v_c - R_3 C (dv_c/dt)}{R_1} - i_L$$
$$= \frac{E}{R_1} - \frac{v_c}{R_1} - \frac{R_3 C}{R_1} \frac{dv_c}{dt} - i_L \tag{11.9}$$

and

$$L \frac{di_L}{dt} = v_c + R_3 C \frac{dv_c}{dt} - R_2 i_L \tag{11.10}$$

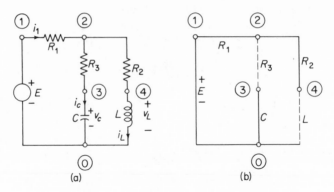

Figure 11.2 *RLC* network with R_3 addition.

These equations are not in the proper form since derivatives of the state variables appear on the right-hand side of Eqs. (11.9) and (11.10). Rearranging Eq. (11.9) to place it in the proper form, we obtain

$$\frac{dv_c}{dt} = -\frac{v_c}{C(R_1 + R_3)} - \frac{R_1 i_L}{C(R_1 + R_3)} + \frac{E}{C(R_1 + R_3)} \qquad (11.11)$$

Substituting Eq. (11.11) into Eq. (11.10) and rearranging, we find that

$$\frac{di_L}{dt} = \frac{R_1 v_c}{L(R_1 + R_3)} - \frac{(R_1 R_2 + R_1 R_3 + R_2 R_3)i_L}{L(R_1 + R_3)} + \frac{R_3 E}{L(R_1 + R_3)} \qquad (11.12)$$

In normal form, Eqs. (11.11) and (11.12) can now be written as

$$\begin{bmatrix} \dfrac{dv_c}{dt} \\[2ex] \dfrac{di_L}{dt} \end{bmatrix} = \begin{bmatrix} -\dfrac{1}{C(R_1 + R_3)} & -\dfrac{R_1}{C(R_1 + R_3)} \\[2ex] \dfrac{R_1}{L(R_1 + R_3)} & -\dfrac{(R_1 R_2 + R_1 R_3 + R_2 R_3)}{L(R_1 + R_3)} \end{bmatrix} \begin{bmatrix} v_c \\[2ex] i_L \end{bmatrix} + \begin{bmatrix} \dfrac{1}{C(R_1 + R_3)} \\[2ex] \dfrac{R_3}{L(R_1 + R_3)} \end{bmatrix} E$$

$$(11.13)$$

For an output of v_{R_3} the output equation is now

$$v_{R_3} = \begin{bmatrix} \dfrac{-R_3}{R_1 + R_3} & -\dfrac{R_1 R_3}{R_1 + R_3} \end{bmatrix} \begin{bmatrix} v_c \\[2ex] i_L \end{bmatrix} + \begin{bmatrix} \dfrac{R_3}{R_1 + R_3} \end{bmatrix} E \qquad (11.14)$$

EXAMPLE 11.3

Consider the network of Fig. 11.3 and the associated graph developed as before. In this example the network contains a voltage source which depends on the state voltage v_c. [Had the voltage been dependent upon the inductance voltage or another nonstate quantity, the voltage $E_2 = kv_L$ would have to be redefined in terms of state variables as $E_2 = k(E - v_c)$.] In the network given the capacitance current is

$$C\frac{dv_c}{dt} = i_L - \frac{v_c - kv_c}{R_1}$$

$$= -\frac{1 - k}{R_1} v_c + i_L \qquad (11.15)$$

The inductance voltage is

$$L\frac{di_L}{dt} = E - v_c \qquad (11.16)$$

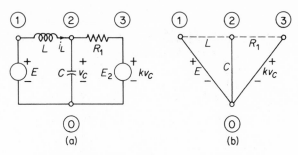

Figure 11.3 *RLC* network with dependent
source.

In normal form, the state equations for the network are

$$
\begin{bmatrix} \dfrac{dv_c}{dt} \\[2ex] \dfrac{di_L}{dt} \end{bmatrix} = \begin{bmatrix} -\dfrac{1-k}{R_1 C} & \dfrac{1}{C} \\[2ex] -\dfrac{1}{L} & 0 \end{bmatrix} \begin{bmatrix} v_c \\[1ex] i_L \end{bmatrix} + \begin{bmatrix} 0 \\[1ex] \dfrac{1}{L} \end{bmatrix} E \tag{11.17}
$$

As shown in this example, we insert dependent voltage sources in the network tree along with the independent sources. Dependent and independent current sources must be link elements. The dependent quantities, if not defined as functions of state variables, must be respecified in terms of the state variables and dependent quantities.

One of the more frightening network configurations encountered in network analysis consists of a network containing mutual inductance. The state-equation formulation can be handled using one of two methods:

1. The transformer can be replaced with its model as discussed in Chapter 2. This reduces the problem to a single *RLC* network with dependent sources.

2. The problem can be approached directly as in the following example. This approach involves the use of the equations defining the transformer action.

EXAMPLE 11.4

Consider the network of Fig. 11.4a. The appropriate tree is shown in Fig. 11.4b. Inductances L_1 and L_2 are mutually coupled. The derivatives of the current through the inductances [see Eq. (2.31)] are given by

$$
\frac{di_{L_1}}{dt} = \mathbf{\Gamma}_1 v_{L_1} - \mathbf{\Gamma}_M v_{L_2} \tag{11.18a}
$$

$$
\frac{di_{L_2}}{dt} = -\mathbf{\Gamma}_M v_{L_1} + \mathbf{\Gamma}_2 v_{L_2} \tag{11.18b}
$$

where

$$
\mathbf{\Gamma}_1 = \frac{L_2}{L_1 L_2 - M^2}
$$

$$
\mathbf{\Gamma}_2 = \frac{L_1}{L_1 L_2 - M^2}
$$

$$
\mathbf{\Gamma}_M = \frac{M}{L_1 L_2 - M^2}
$$

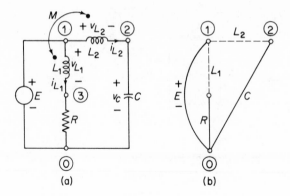

Figure 11.4 *RLC* network with coupled inductances.

By inspection, we see that

$$v_{L_1} = E - R_1 i_{L_1} \tag{11.19a}$$

and

$$v_{L_2} = E - v_c \tag{11.19b}$$

Combining Eqs. (11.19) and (11.18) we obtain

$$
\begin{aligned}
\frac{di_{L_1}}{dt} &= \mathbf{\Gamma}_1(E - R_1 i_{L_1}) - \mathbf{\Gamma}_M(E - v_c) \\
&= -\mathbf{\Gamma}_1 R_1 i_{L_1} + \mathbf{\Gamma}_M v_c + (\mathbf{\Gamma}_1 - \mathbf{\Gamma}_M)E
\end{aligned}
\tag{11.20a}
$$

and

$$
\begin{aligned}
\frac{di_{L_2}}{dt} &= -\mathbf{\Gamma}_M(E - R_1 i_{L_1}) + \mathbf{\Gamma}_2(E - v_c) \\
&= \mathbf{\Gamma}_M R_1 i_{L_1} - \mathbf{\Gamma}_2 v_c + (\mathbf{\Gamma}_2 - \mathbf{\Gamma}_M)E
\end{aligned}
\tag{11.20b}
$$

These are the first two equations required for the state formulation of the network. The final equation for the tree capacitance becomes

$$\frac{dv_c}{dt} = \frac{1}{C} i_{L_2} \tag{11.20c}$$

The resulting equation in normal form is

$$
\begin{bmatrix} \dfrac{dv_c}{dt} \\[2mm] \dfrac{di_{L_1}}{dt} \\[2mm] \dfrac{di_{L_2}}{dt} \end{bmatrix}
=
\begin{bmatrix} 0 & 0 & \dfrac{1}{C} \\[2mm] \mathbf{\Gamma}_M & -\mathbf{\Gamma}_1 R_1 & 0 \\[2mm] -\mathbf{\Gamma}_2 & \mathbf{\Gamma}_M R_1 & 0 \end{bmatrix}
\begin{bmatrix} v_c \\[2mm] i_{L_1} \\[2mm] i_{L_2} \end{bmatrix}
=
\begin{bmatrix} 0 \\[2mm] \mathbf{\Gamma}_1 - \mathbf{\Gamma}_M \\[2mm] \mathbf{\Gamma}_2 - \mathbf{\Gamma}_M \end{bmatrix} E
\tag{11.21}
$$

Category 2 and category 3 networks form special cases which we now consider in our discussion of state-equation formulation by inspection. The two categories consist of networks containing capacitance–voltage source loops, and networks containing inductance–current source cut-sets. These cases are frequently not allowed in computer methods of analysis and do not occur in nature; that is, all voltage sources, inductors, and capacitors contain a finite series resistance, and all current sources exhibit a finite shunt resistance at the terminals. The difficulties arise when we attempt to approximate the "real world" with a simple model which reasonably simulates the response of the world for a limited class of inputs. In practical terms, the judicious use of parasitic elements (capacitance, inductance, and resistance) in the model will usually prevent either of the special cases from arising. However, we now demonstrate the state-equation formulation for these special situations.

EXAMPLE 11.5 CAPACITIVE EXCESS BRANCH

A network with an excess capacitive branch is shown in Fig. 11.5a, and the associated graph is shown in Fig. 11.5b. In this example we arbitrarily construct the graph so that we designate C_3 as the excess branch. That is, C_3 is a link branch and the tree is not "proper."†

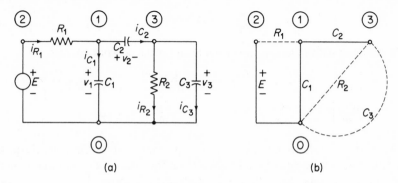

Figure 11.5 Network containing capacitive excess branch.

As before, we begin by writing the equations for the capacitive tree branches in terms of the state variables and the source voltages and currents. For this network at node 0 we see that

$$C_1 \frac{dv_1}{dt} = i_{C_1} = i_{R_1} - i_{R_2} - i_{C_3}$$

$$= \frac{E}{R_1} - v_1 \left(\frac{1}{R_1} + \frac{1}{R_2} \right) + \frac{v_2}{R_2} - C_3 \frac{dv_3}{dt}$$

but $v_3 = v_1 - v_2$, so

$$(C_1 + C_3) \frac{dv_1}{dt} = \frac{E}{R_1} - v_1 \left(\frac{1}{R_1} + \frac{1}{R_2} \right) + \frac{v_2}{R_1} + C_3 \frac{dv_2}{dt} \qquad (11.22)$$

Similarly,

$$C_2 \frac{dv_2}{dt} = i_{C_2}$$

$$= \frac{v_1 - v_2}{R_2} + C_3 \frac{dv_3}{dt}$$

---

†A proper tree of a network is defined as one whose branches contain every capacitive element of the network or every capacitive element plus resistive elements. The proper tree contains no inductances or current sources.

so

$$(C_2 + C_3)\frac{dv_2}{dt} = \frac{v_1}{R_2} - \frac{v_2}{R_2} + C_3\frac{dv_1}{dt} \tag{11.23}$$

$$\frac{dv_2}{dt} = \frac{v_1}{R_2(C_2 + C_3)} - \frac{v_2}{R_2(C_2 + C_3)} + \frac{C_3}{C_2 + C_3}\frac{dv_1}{dt} \tag{11.24}$$

Substituting Eq. (11.24) into Eq. (11.22) we obtain

$$\frac{C_T}{C_2 + C_3}\frac{dv_1}{dt} = \frac{E}{R_1} - v_1\frac{R_1C_2 + R_2C_2 + R_2C_3}{R_1R_2(C_2 + C_3)} + v_2\frac{C_2}{R_2(C_2 + C_3)}$$

or

$$\frac{dv_1}{dt} = \frac{-v_1}{C_T}\left(\frac{C_2}{R_2} + \frac{C_2 + C_3}{R_1}\right) + \frac{v_2}{C_T}\frac{C_2}{R_2} + \frac{C_2 + C_3}{C_TR_T}E \tag{11.25}$$

where $C_T = C_1C_2 + C_1C_3 + C_2C_3$. Substituting Eq. (11.25) into Eq. (11.24) we find

$$\frac{dv_2}{dt} = \frac{v_1}{C_T}\left(\frac{C_1}{R_2} - \frac{C_3}{R_1}\right) - \frac{v_2}{C_T}\frac{C_1}{R_2} + \frac{C_3}{C_TR_1}E \tag{11.26}$$

The resulting state equations in normal form for the network are then

$$\begin{bmatrix} \dfrac{dv_1}{dt} \\ \dfrac{dv_2}{dt} \end{bmatrix} = \begin{bmatrix} -\dfrac{C_2}{R_2C_T} - \dfrac{C_2 + C_3}{R_1C_T} & \dfrac{C_2}{R_2C_T} \\ \dfrac{C_1}{R_2C_T} - \dfrac{C_3}{R_1C_T} & -\dfrac{C_1}{R_2C_T} \end{bmatrix}\begin{bmatrix} v_1 \\ v_2 \end{bmatrix} + \begin{bmatrix} \dfrac{C_2 + C_3}{R_1C_T} \\ \dfrac{C_3}{R_1C_T} \end{bmatrix}E \tag{11.27}$$

The next example illustrates several important points. First, it illustrates the method of writing state equations by inspection for networks containing inductance–current source cut-sets. Second, the example demonstrates that the state variables cannot always be defined as the currents in the inductances and the voltages across the capacitances. It also illustrates a method of eliminating derivatives of source voltages and currents which may occur in equations of this type.

EXAMPLE 11.6 INDUCTIVE EXCESS BRANCHES

The category 3 network has an inductive branch in the tree of the network graph. This inductance branch is referred to as an inductive excess branch. A network with an inductive excess branch and its associated network graph is shown in Fig. 11.6.

Since node 1 has only inductances and a current source attached to it, a proper tree cannot be formed. In this example, we designate the branch containing L_2 as a tree branch. (See Fig. 11.6b.)

The first step in the analysis is to express the current through the capacitances as functions of the tree-branch voltages, link currents, and the voltage and current sources. Using node 0, the current i_{c_1} through C_1 is

$$C_1\frac{dv_1}{dt} = J - i_{c_2}$$

However,

$$i_{c_2} = i_{R_2} + i_1$$

Therefore,

$$C_1\frac{dv_1}{dt} = J - i_1 - \frac{v_1 - v_2}{R_2} \tag{11.28}$$

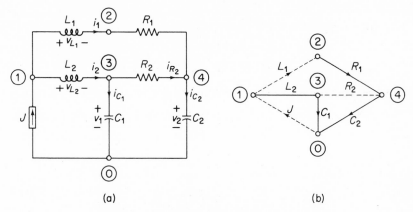

(a)   (b)

Figure 11.6 Network containing inductive excess branch.

The current through C_2 is

$$C_2\frac{dv_2}{dt} = i_1 + \frac{v_1 - v_2}{R_2} \tag{11.29}$$

The link inductance voltage v_{L_1} by KVL is

$$L_1\frac{di_1}{dt} = v_1 + v_{L_2} - v_2 - v_{R_3} \tag{11.30}$$

$$= v_1 - v_2 + L_2\frac{di_2}{dt} - i_1 R_1$$

However, $i_2 = J - i_1$, so

$$L_1\frac{di_1}{dt} = v_1 - v_2 + L_2\frac{dJ}{dt} - L_2\frac{di_1}{dt} - i_1 R_1$$

or

$$(L_1 + L_2)\frac{di_1}{dt} = v_1 - v_2 - i_1 R_1 + L_2\frac{dJ}{dt} \tag{11.31}$$

Combining Eqs. (11.28), (11.29), and (11.31) into normal form we have

$$\begin{bmatrix} \dfrac{dv_1}{dt} \\[2mm] \dfrac{dv_2}{dt} \\[2mm] \dfrac{di_1}{dt} \end{bmatrix} = \begin{bmatrix} -\dfrac{1}{R_2 C_1} & \dfrac{1}{R_2 C_1} & -\dfrac{1}{C_1} \\[2mm] \dfrac{1}{R_2 C_2} & -\dfrac{1}{R_2 C_2} & \dfrac{1}{C_2} \\[2mm] \dfrac{1}{L_1 + L_2} & -\dfrac{1}{L_1 + L_2} & -\dfrac{R_1}{L_1 + L_2} \end{bmatrix} \begin{bmatrix} v_1 \\[2mm] v_2 \\[2mm] i_1 \end{bmatrix} + \begin{bmatrix} \dfrac{1}{C_1} \\[2mm] 0 \\[2mm] \dfrac{L_2}{L_1 + L_2}\dfrac{d}{dt} \end{bmatrix} J \tag{11.32}$$

Equation (11.32) is not exactly in the proper normal form since the equation must appear as $\dot{\mathbf{x}} = f(\mathbf{x}, \mathbf{m}, t)$, which does not allow derivatives of the input signals. Therefore, the dJ/dt term should be eliminated before the equation can be considered a state equation. This can be achieved by redefining the state variables. In this case, let

$$z_1 = v_1$$

$$z_2 = v_2$$

$$z_3 = i_1 - \frac{L_2}{L_1 + L_2}J$$

In matrix form the transformation is written

$$
\begin{bmatrix} v_1 \\ v_2 \\ i_1 \end{bmatrix} = \begin{bmatrix} 1 & 0 & 0 \\ 0 & 1 & 0 \\ 0 & 0 & 1 \end{bmatrix} \begin{bmatrix} z_1 \\ z_2 \\ z_3 \end{bmatrix} + \begin{bmatrix} 0 \\ 0 \\ \dfrac{L_2}{L_1 + L_2} \end{bmatrix} J \tag{11.33}
$$

Substituting Eq. (11.33) into Eq. (11.32) and simplifying, the resulting state equations are

$$
\begin{bmatrix} \dfrac{dz_1}{dt} \\[2mm] \dfrac{dz_2}{dt} \\[2mm] \dfrac{dz_3}{dt} \end{bmatrix} = \begin{bmatrix} -\dfrac{1}{R_2 C_1} & \dfrac{1}{R_2 C_1} & -\dfrac{1}{C_1} \\[2mm] \dfrac{1}{R_2 C_2} & -\dfrac{1}{R_2 C_2} & \dfrac{1}{C_2} \\[2mm] \dfrac{1}{L_1 + L_2} & -\dfrac{1}{L_1 + L_2} & -\dfrac{R_1}{L_1 + L_2} \end{bmatrix} \begin{bmatrix} z_1 \\ z_2 \\ z_3 \end{bmatrix} + \begin{bmatrix} \dfrac{L_1}{C_1(L_1 + L_2)} \\[2mm] \dfrac{L_2}{C_2(L_1 + L_2)} \\[2mm] -\dfrac{R_1 L_2}{(L_1 + L_2)^2} \end{bmatrix} J \tag{11.34}
$$

11.3 Network Graph Method for Obtaining State Equations of Electrical Networks

RLC Networks

We observed several equations in Chapter 7 that related network quantities to link and tree voltages and currents. Some of these relationships are

$$
\mathbf{B}_f \mathbf{v}_b = \mathbf{0} \tag{7.92}
$$

and

$$
\mathbf{v}_b = \mathbf{Q}^\mathsf{T} \mathbf{v}_t \dagger \tag{7.109}
$$

which are expressions of Kirchhoff's voltage law and relate network branch voltages to tree branch voltages. The relationships

$$
\mathbf{i}_b = \mathbf{B}_f^\mathsf{T} \mathbf{i}_l \tag{7.95}
$$

and

$$
\mathbf{Q}\mathbf{i}_b = \mathbf{0} \tag{7.107}
$$

are expressions of Kirchhoff's current law and relate the network branch currents to the link branch currents.

We also observed that the fundamental cut-set matrix $\mathbf{Q}$ is related to the fundamental loop matrix $\mathbf{B}_f$ by the expressions

$$
\mathbf{Q} = [\mathbf{I} \mid -\mathbf{H}^\mathsf{T}] \tag{7.108}
$$

$$
\mathbf{B}_f = [\mathbf{H} \mid \mathbf{I}] \tag{7.106}
$$

where $\mathbf{H}$ is an $l \times n$ matrix, l is the number of link branches, and n is the number of tree branches in the network.

———————

†See footnote to Eq. (7.109), page 290.

Since we have equations which relate the tree and link quantities of a network from the network graph, it seems obvious that we should be able to formulate the network state equations using topological methods.

Let us assume that we have a network with m capacitances, n inductances, p tree resistances, q voltage sources, r link resistances, and s current sources. The network graph branches are numbered sequentially starting with the tree capacitances, followed by the tree resistances, voltage sources, link inductances, link resistances, and current sources. From the graph we can now construct the cut-set matrix:

$$
\mathbf{Q} = \begin{array}{c} C_t \\ R_t \\ E \end{array}\!\!
\begin{array}{c} C_t \quad R_t \quad E \qquad L_l \qquad R_l \qquad J \\
\left[\begin{array}{ccc:ccc}
 & & & \mathbf{Q}_{lCL} & \mathbf{Q}_{lCR} & \mathbf{Q}_{lCJ} \\
 & \mathbf{I} & & \mathbf{Q}_{lRL} & \mathbf{Q}_{lRR} & \mathbf{Q}_{lRJ} \\
 & & & \mathbf{Q}_{lEL} & \mathbf{Q}_{lER} & \mathbf{Q}_{lEJ}
\end{array}\right]
\end{array} \qquad (11.35)
$$

The portion of the cut-set matrix corresponding to the link branches is designated $\mathbf{Q}_{l--}$. The dimensions of the $\mathbf{Q}_l$ submatrices are

$$
\begin{array}{lll}
\mathbf{Q}_{lCL} \quad m \times n & \mathbf{Q}_{lCR} \quad m \times r & \mathbf{Q}_{lCJ} \quad m \times s \\
\mathbf{Q}_{lRL} \quad p \times n & \mathbf{Q}_{lRR} \quad p \times r & \mathbf{Q}_{lRJ} \quad p \times s \\
\mathbf{Q}_{lEL} \quad q \times n & \mathbf{Q}_{lER} \quad q \times r & \mathbf{Q}_{lEJ} \quad q \times s
\end{array}
$$

The tree-branch currents are given by

$$\mathbf{i}_t = -\mathbf{Q}_l \mathbf{i}_l \qquad (11.36)$$

or

$$
\begin{bmatrix} \mathbf{i}_{tC} \\ \mathbf{i}_{tR} \\ \mathbf{i}_{tE} \end{bmatrix} = -
\begin{bmatrix}
\mathbf{Q}_{lCL} & \mathbf{Q}_{lCR} & \mathbf{Q}_{lCJ} \\
\mathbf{Q}_{lRL} & \mathbf{Q}_{lRR} & \mathbf{Q}_{lRJ} \\
\mathbf{Q}_{lEL} & \mathbf{Q}_{lER} & \mathbf{Q}_{lEJ}
\end{bmatrix}
\begin{bmatrix} \mathbf{i}_{lL} \\ \mathbf{i}_{lR} \\ \mathbf{i}_{lJ} \end{bmatrix} \qquad (11.37)
$$

where $\mathbf{i}_{tC}$ is the vector of tree capacitance currents, $\mathbf{i}_{tR}$ the vector of tree resistance currents, $\mathbf{i}_t E$ the vector of voltage source currents, $\mathbf{i}_{lL}$ the vector of link inductance currents, $\mathbf{i}_{lR}$ the vector of link resistance currents, and $\mathbf{i}_{lJ}$ the vector of current sources. The tree currents $\mathbf{i}_{tC}$, $\mathbf{i}_{tR}$, and $\mathbf{i}_{tE}$ can be expressed as

$$\mathbf{i}_{tC} = -\mathbf{Q}_{lCL}\mathbf{i}_{lL} - \mathbf{Q}_{lCR}\mathbf{i}_{lR} - \mathbf{Q}_{lCJ}\mathbf{i}_{lJ} \qquad (11.38)$$

$$\mathbf{i}_{tR} = -\mathbf{Q}_{lRL}\mathbf{i}_{lL} - \mathbf{Q}_{lRR}\mathbf{i}_{lR} - \mathbf{Q}_{lRJ}\mathbf{i}_{lJ} \qquad (11.39)$$

Using the relationship between the cut-set matrix $\mathbf{Q}$ in Eq. (7.108) and the fundamental loop matrix $\mathbf{B}_f$ in Eq. (7.106), we can express $\mathbf{B}_f$ as

$$
\mathbf{B}_f = \begin{array}{c} L_l \\ R_l \\ J \end{array}\!\!
\begin{array}{c} C_t \qquad\quad R_t \qquad\quad E \qquad\; L_l \; R_l \; J \\
\left[\begin{array}{ccc:ccc}
-\mathbf{Q}^{\mathsf{T}}_{lCL} & -\mathbf{Q}^{\mathsf{T}}_{lRL} & -\mathbf{Q}^{\mathsf{T}}_{lEL} & & & \\
-\mathbf{Q}^{\mathsf{T}}_{lCR} & -\mathbf{Q}^{\mathsf{T}}_{lRR} & -\mathbf{Q}^{\mathsf{T}}_{lER} & & \mathbf{I} & \\
-\mathbf{Q}^{\mathsf{T}}_{lCJ} & -\mathbf{Q}^{\mathsf{T}}_{lRJ} & -\mathbf{Q}^{\mathsf{T}}_{lEJ} & & &
\end{array}\right]
\end{array} \qquad (11.40)
$$

We see from Eq. (7.92) that the link branch voltages can be expressed as

$$[-\mathbf{Q}_l^\mathsf{T} \mid \mathbf{I}]\begin{bmatrix} \mathbf{v}_t \\ \text{---} \\ \mathbf{v}_l \end{bmatrix} = \mathbf{0}$$

or

$$\mathbf{v}_l = \mathbf{Q}_l^\mathsf{T}\mathbf{v}_t \qquad\qquad (11.41)$$

Then

$$\begin{bmatrix} \mathbf{v}_{lL} \\ \mathbf{v}_{lR} \\ \mathbf{v}_{lJ} \end{bmatrix} = \begin{bmatrix} \mathbf{Q}_{lCL}^\mathsf{T} & \mathbf{Q}_{lRL}^\mathsf{T} & \mathbf{Q}_{lEL}^\mathsf{T} \\ \mathbf{Q}_{lCR}^\mathsf{T} & \mathbf{Q}_{lRR}^\mathsf{T} & \mathbf{Q}_{lER}^\mathsf{T} \\ \mathbf{Q}_{lCJ}^\mathsf{T} & \mathbf{Q}_{lRJ}^\mathsf{T} & \mathbf{Q}_{lEJ}^\mathsf{T} \end{bmatrix}\begin{bmatrix} \mathbf{v}_{tC} \\ \mathbf{v}_{tR} \\ \mathbf{v}_{tE} \end{bmatrix} \qquad (11.42)$$

From Eq. (11.42) $\mathbf{v}_{lL}$ and $\mathbf{v}_{lR}$ can be expressed as

$$\mathbf{v}_{lL} = \mathbf{Q}_{lCL}^\mathsf{T}\mathbf{v}_{tC} + \mathbf{Q}_{lRL}^\mathsf{T}\mathbf{v}_{tR} + \mathbf{Q}_{lEL}^\mathsf{T}\mathbf{v}_{tE} \qquad (11.43)$$

and

$$\mathbf{v}_{lR} = \mathbf{Q}_{lCR}^\mathsf{T}\mathbf{v}_{tC} + \mathbf{Q}_{lRR}^\mathsf{T}\mathbf{v}_{tR} + \mathbf{Q}_{lER}^\mathsf{T}\mathbf{v}_{tE} \qquad (11.44)$$

Equations (11.38) and (11.43) are now combined to give

$$\begin{bmatrix} \mathbf{i}_{tC} \\ \mathbf{v}_{lL} \end{bmatrix} = \begin{bmatrix} \mathbf{0} & -\mathbf{Q}_{lCL} \\ \mathbf{Q}_{lCL}^\mathsf{T} & \mathbf{0} \end{bmatrix}\begin{bmatrix} \mathbf{v}_{tC} \\ \mathbf{v}_{lL} \end{bmatrix} + \begin{bmatrix} \mathbf{0} & -\mathbf{Q}_{lCR} \\ \mathbf{Q}_{lRL}^\mathsf{T} & \mathbf{0} \end{bmatrix}\begin{bmatrix} \mathbf{v}_{tR} \\ \mathbf{i}_{lR} \end{bmatrix}$$
$$+ \begin{bmatrix} \mathbf{0} & -\mathbf{Q}_{lCJ} \\ \mathbf{Q}_{lEL}^\mathsf{T} & \mathbf{0} \end{bmatrix}\begin{bmatrix} \mathbf{v}_{tE} \\ \mathbf{i}_{lJ} \end{bmatrix} \qquad (11.45)$$

The left-hand side of Eq. (11.45) contains the capacitance currents and inductance voltages, which are proportional to the first derivatives of the capacitance voltages and inductance currents, respectively. These derivatives are expressed in terms of capacitance voltages and currents, and resistive tree voltages and link currents. In order to express the state equations in proper form, we must express the variables $\mathbf{v}_{tR}$ and $\mathbf{v}_{lR}$ in terms of $\mathbf{v}_{tC}$, $\mathbf{i}_{lL}$, $\mathbf{v}_{tE}$, and $\mathbf{v}_{lJ}$.

Combining Eqs. (11.39) and (11.44), we have

$$\begin{bmatrix} \mathbf{v}_{lR} \\ \mathbf{i}_{tR} \end{bmatrix} = \begin{bmatrix} \mathbf{Q}_{lCR}^\mathsf{T} & \mathbf{0} \\ \mathbf{0} & -\mathbf{Q}_{lRL} \end{bmatrix}\begin{bmatrix} \mathbf{v}_{tC} \\ \mathbf{i}_{lL} \end{bmatrix} + \begin{bmatrix} \mathbf{Q}_{lRR}^\mathsf{T} & \mathbf{0} \\ \mathbf{0} & -\mathbf{Q}_{lRR} \end{bmatrix}\begin{bmatrix} \mathbf{v}_{tR} \\ \mathbf{i}_{lR} \end{bmatrix}$$
$$+ \begin{bmatrix} \mathbf{Q}_{lER}^\mathsf{T} & \mathbf{0} \\ \mathbf{0} & \mathbf{Q}_{lRJ} \end{bmatrix}\begin{bmatrix} \mathbf{v}_{tE} \\ \mathbf{i}_{lJ} \end{bmatrix} \qquad (11.46)$$

The resistive voltages and currents are related by the expression

$$\begin{bmatrix} \mathbf{v}_{lR} \\ \mathbf{i}_{tR} \end{bmatrix} = \begin{bmatrix} \mathbf{0} & \mathbf{R}_l \\ \mathbf{G}_t & \mathbf{0} \end{bmatrix}\begin{bmatrix} \mathbf{v}_{tR} \\ \mathbf{i}_{lR} \end{bmatrix} \qquad (11.47)$$

where $\mathbf{R}_l$ is an $r \times r$ diagonal matrix whose elements are the link branch resistances and $\mathbf{G}_t$ is a $p \times p$ matrix whose diagonal elements are the tree-branch conductances.

Equations (11.46) and (11.47) are now combined to form

$$\begin{bmatrix} \mathbf{v}_{tR} \\ \mathbf{i}_{lR} \end{bmatrix} = \begin{bmatrix} -\mathbf{Q}_{lRR}^{\mathsf{T}} & \mathbf{R}_l \\ \mathbf{G}_t & \mathbf{Q}_{lRR} \end{bmatrix}^{-1} \begin{bmatrix} \mathbf{Q}_{lCR}^{\mathsf{T}} & \mathbf{0} \\ \mathbf{0} & -\mathbf{Q}_{lRL} \end{bmatrix} \begin{bmatrix} \mathbf{v}_{tC} \\ \mathbf{i}_{lL} \end{bmatrix}$$
$$+ \begin{bmatrix} -\mathbf{Q}_{lRR}^{\mathsf{T}} & \mathbf{R}_l \\ \mathbf{G}_t & \mathbf{Q}_{lRR} \end{bmatrix}^{-1} \begin{bmatrix} \mathbf{Q}_{lER}^{\mathsf{T}} & \mathbf{0} \\ \mathbf{0} & -\mathbf{Q}_{lRJ} \end{bmatrix} \begin{bmatrix} \mathbf{v}_{tE} \\ \mathbf{i}_{lL} \end{bmatrix} \qquad (11.48)$$

$$= [\mathcal{3C}] \begin{bmatrix} \mathbf{v}_{tC} \\ \mathbf{i}_{lL} \end{bmatrix} + [\hat{\mathcal{3C}}] \begin{bmatrix} \mathbf{v}_{tE} \\ \mathbf{i}_{lJ} \end{bmatrix} \qquad (11.49)$$

Combining Eqs. (11.46) and (11.49) to eliminate the unwanted variables $\mathbf{v}_{tR}$ and $\mathbf{i}_{lR}$, we obtain

$$\begin{bmatrix} \mathbf{i}_{tC} \\ \mathbf{v}_{lL} \end{bmatrix} = \left\{ \begin{bmatrix} \mathbf{0} & -\mathbf{Q}_{lCL} \\ \mathbf{Q}_{lCL}^{\mathsf{T}} & \mathbf{0} \end{bmatrix} + \left(\begin{bmatrix} \mathbf{0} & -\mathbf{Q}_{lCR} \\ \mathbf{Q}_{lRL}^{\mathsf{T}} & \mathbf{0} \end{bmatrix} [\mathcal{3C}] \right) \right\} \begin{bmatrix} \mathbf{v}_{tC} \\ \mathbf{i}_{lL} \end{bmatrix}$$
$$+ \left\{ \begin{bmatrix} \mathbf{0} & -\mathbf{Q}_{lCJ} \\ \mathbf{Q}_{lEL}^{\mathsf{T}} & \mathbf{0} \end{bmatrix} + \left(\begin{bmatrix} \mathbf{0} & -\mathbf{Q}_{lCR} \\ \mathbf{Q}_{lRL}^{\mathsf{T}} & \mathbf{0} \end{bmatrix} [\hat{\mathcal{3C}}] \right) \right\} \begin{bmatrix} \mathbf{v}_{tE} \\ \mathbf{i}_{lJ} \end{bmatrix} \qquad (11.50a)$$

$$= [\mathcal{G}] \begin{bmatrix} \mathbf{v}_{tC} \\ \mathbf{i}_{lL} \end{bmatrix} + [\hat{\mathcal{G}}] \begin{bmatrix} \mathbf{v}_{tE} \\ \mathbf{i}_{lJ} \end{bmatrix} \qquad (11.50b)$$

Before Eq. (11.50) is in the proper state form the left-hand side of the equation must be of the form

$$\frac{d}{dt} \begin{bmatrix} \mathbf{v}_{tC} \\ \mathbf{i}_{lL} \end{bmatrix} \qquad (11.51)$$

Notice that

$$\begin{bmatrix} \mathbf{i}_{tC} \\ \mathbf{v}_{lL} \end{bmatrix} = \begin{bmatrix} \mathbf{C}_t & \mathbf{0} \\ \mathbf{0} & \mathbf{L}_l \end{bmatrix} \frac{d}{dt} \begin{bmatrix} \mathbf{v}_{tC} \\ \mathbf{i}_{lL} \end{bmatrix} \qquad (11.52)$$

Substituting Eq. (11.52) into Eq. (11.50b) and solving for Eq. (11.51) completes the state-equation formulation as follows:

$$\frac{d}{dt} \begin{bmatrix} \mathbf{v}_{tC} \\ \mathbf{i}_{lL} \end{bmatrix} = \begin{bmatrix} \mathbf{C}_t & \mathbf{0} \\ \mathbf{0} & \mathbf{L}_l \end{bmatrix}^{-1} [\mathcal{G}] \begin{bmatrix} \mathbf{v}_{tC} \\ \mathbf{i}_{lL} \end{bmatrix} + \begin{bmatrix} \mathbf{C}_t & \mathbf{0} \\ \mathbf{0} & \mathbf{L}_l \end{bmatrix}^{-1} [\hat{\mathcal{G}}] \begin{bmatrix} \mathbf{v}_{tE} \\ \mathbf{i}_{lJ} \end{bmatrix} \qquad (11.53)$$

We observe that we require a discouragingly large number of submatrices for the derivation of the state equations. The major part of the state-equation formulation consists of the manipulation of these submatrices and a lot of bookkeeping. Fortunately, the computer does both well. We can thus use the computer to both set up the state equations and obtain a solution to them. Before we attempt an example of the topological method of formulating a set of state equations, let us review the procedure for obtaining the state equations:

1. Construct a proper tree for the network and number the branches as described.

2. Construct the topological matrix $\mathbf{Q}_l$. (The first two steps are relatively easy for us or the computer.)

3. Partition the $\mathbf{Q}_l$ matrix to provide the necessary set of submatrices.

4. Construct the matrices $\mathbf{R}_l$ and $\mathbf{G}_t$ and calculate the resistive tree voltages and link currents ($\mathbf{v}_{tR}$ and $\mathbf{i}_{lR}$).

5. Formulate the equations for the capacitive tree currents as in Eq. (11.50b).

6. Generate the submatrices $\mathbf{C}_t$ and $\mathbf{L}_l$ and formulate the state equations in Eq. (11.53) by further matrix manipulation.

EXAMPLE 11.7

To illustrate the preceding method for writing the network state equations by topological methods, consider the network of Fig. 11.7 and its associated oriented graph. The cut-sets for the network are selected to correspond with Eq. (11.35), so the cut-set matrix is

$$\mathbf{Q} = \begin{array}{c} \begin{array}{cccccc} 1 & 2 & 3 & 4 & 5 & 6 \end{array} \\ \begin{bmatrix} 1 & 0 & 0 & 0 & 1 & 1 \\ 0 & 1 & 0 & 0 & -1 & 0 \\ 0 & 0 & 1 & 0 & 1 & 1 \\ 0 & 0 & 0 & 1 & 0 & -1 \end{bmatrix} \end{array}$$

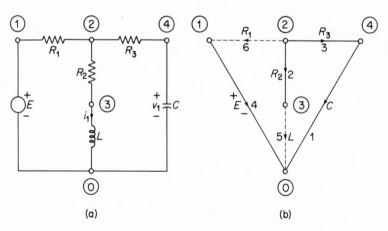

(a)   (b)

Figure 11.7 *RLC* network and oriented graph.

The link cut-set submatrix is

$$\mathbf{Q}_l = \begin{bmatrix} 1 & 1 \\ \hline -1 & 0 \\ \hline 1 & 1 \\ \hline 0 & -1 \end{bmatrix}$$

$$= \begin{bmatrix} \mathbf{Q}_{lCL} & \mathbf{Q}_{lCR} \\ \hline \mathbf{Q}_{lRL} & \mathbf{Q}_{lRR} \\ \hline \mathbf{Q}_{lEL} & \mathbf{Q}_{lER} \end{bmatrix}$$

The link resistance matrix $\mathbf{R}_l$ and tree conductance matrix $\mathbf{G}_t$ are

$$\mathbf{R}_l = [R_1] \qquad \mathbf{G}_t = \begin{bmatrix} G_2 & 0 \\ 0 & G_3 \end{bmatrix}$$

The resistive tree voltages and link currents are next calculated from Eq. (11.48):

$$\begin{bmatrix} -\mathbf{Q}_{lRR}^\mathsf{T} & \mathbf{R}_l \\ \mathbf{G}_t & \mathbf{Q}_{lRR} \end{bmatrix} = \begin{bmatrix} 0 & -1 & R_1 \\ G_2 & 0 & 0 \\ 0 & G_3 & 1 \end{bmatrix}$$

and

$$\begin{bmatrix} -\mathbf{Q}_{lRR}^\mathsf{T} & \mathbf{R}_l \\ \mathbf{G}_t & \mathbf{Q}_{lRR} \end{bmatrix}^{-1} = \begin{bmatrix} 0 & R_2 & 0 \\ \dfrac{-R}{R_1 + R_3} & 0 & \dfrac{R_1 R_3}{R_1 + R_3} \\ \dfrac{-1}{R_1 + R_3} & 0 & \dfrac{R_3}{R_1 + R_3} \end{bmatrix}$$

Then

$$\begin{bmatrix} \mathbf{v}_{tR} \\ \mathbf{i}_{lR} \end{bmatrix} = \begin{bmatrix} 0 & R_2 & 0 \\ -\dfrac{R_3}{R_1 + R_3} & 0 & \dfrac{R_1 R_3}{R_1 + R_3} \\ \dfrac{1}{R_1 + R_3} & 0 & \dfrac{R}{R_1 + R_3} \end{bmatrix} \begin{bmatrix} 1 & 0 \\ 0 & 1 \\ 0 & -1 \end{bmatrix} \begin{bmatrix} \mathbf{v}_{tC} \\ \mathbf{i}_{lL} \end{bmatrix}$$

$$+ \begin{bmatrix} 0 & R_2 & 0 \\ -\dfrac{R_3}{R_1 + R_3} & 0 & -\dfrac{R_1 R_3}{R_1 + R_3} \\ \dfrac{1}{R_1 + R_3} & 0 & -\dfrac{R_3}{R_1 + R_3} \end{bmatrix} \begin{bmatrix} -1 & 0 \\ 0 & 0 \\ 0 & 0 \end{bmatrix} v_{tE}$$

$$\begin{bmatrix} \mathbf{v}_{tR} \\ \mathbf{i}_{lR} \end{bmatrix} = \begin{bmatrix} 0 & R_2 \\ -\dfrac{R_3}{R_1 + R_3} & -\dfrac{R_1 R_3}{R_1 + R_3} \\ \dfrac{1}{R_1 + R_3} & -\dfrac{R_3}{R_1 + R_3} \end{bmatrix} \begin{bmatrix} \mathbf{v}_{tC} \\ \mathbf{i}_{lL} \end{bmatrix} + \begin{bmatrix} 0 \\ \dfrac{R_3}{R_1 + R_3} \\ -\dfrac{1}{R_1 + R_3} \end{bmatrix} v_{tE}$$

Next, the equations for the capacitive tree currents and inductive link voltages are formed according to Eq. (11.50):

$$\begin{bmatrix} \mathbf{i}_{tL} \\ \mathbf{v}_{lL} \end{bmatrix} = \left\{ \begin{bmatrix} 0 & -1 \\ \hline 1 & 0 \end{bmatrix} + \left(\begin{bmatrix} 0 & 0 & -1 \\ \hline -1 & 1 & 0 \end{bmatrix} \begin{bmatrix} 0 & R_2 \\ -\dfrac{R_3}{R_1 + R_3} & -\dfrac{R_1 R_3}{R_1 + R_3} \\ \dfrac{1}{R_1 + R_3} & -\dfrac{R_3}{R_1 + R_3} \end{bmatrix} \right) \right\} \begin{bmatrix} \mathbf{v}_{tC} \\ \mathbf{i}_{lL} \end{bmatrix}$$

$$+ \left\{ \begin{bmatrix} 0 & 0 \\ 0 & 0 \end{bmatrix} + \left(\begin{bmatrix} 0 & 0 & -1 \\ \hline -1 & 1 & 0 \end{bmatrix} \begin{bmatrix} 0 \\ \dfrac{R_3}{R_1 + R_3} \\ -\dfrac{1}{R_1 + R_3} \end{bmatrix} \right) \right\} v_{tE}$$

from which

$$
\begin{bmatrix} \mathbf{i}_{tC} \\ \mathbf{v}_{lL} \end{bmatrix} = \begin{bmatrix} -\dfrac{1}{R_1 + R_3} & -\dfrac{R_1}{R_1 + R_3} \\[2ex] \dfrac{R_1}{R_1 + R_3} & \dfrac{R_1 R_2 + R_2 R_3 + R_1 R_3}{R_1 + R_3} \end{bmatrix} \begin{bmatrix} \mathbf{v}_{tC} \\ \mathbf{i}_{lL} \end{bmatrix} + \begin{bmatrix} \dfrac{1}{R_1 + R_3} \\[2ex] \dfrac{R_3}{R_1 + R_3} \end{bmatrix} \mathbf{v}_{tE}
$$

Finally, the state equations for the network are

$$
\begin{bmatrix} \dfrac{dv_C}{dt} \\[2ex] \dfrac{di_L}{dt} \end{bmatrix} = \begin{bmatrix} -\dfrac{1}{C(R_1 + R_3)} & -\dfrac{R_1}{C(R_1 + R_3)} \\[2ex] \dfrac{R}{L(R_1 + R_3)} & -\dfrac{R_1 R_2 + R_2 R_3 + R_1 R_3}{L(R_1 + R_3)} \end{bmatrix} \begin{bmatrix} v_C \\ i_L \end{bmatrix} + \begin{bmatrix} \dfrac{1}{C(R_1 + R_3)} \\[2ex] \dfrac{R}{L(R_1 + R_3)} \end{bmatrix} E
$$

RLC Networks with Dependent Sources

The state-equation-formulation procedure for networks containing dependent sources is similar to the development in the last section except that another unwanted vector must be eliminated. The cut-set matrix for a network containing dependent sources is extended to

$$
\mathbf{Q} = \begin{matrix} & \begin{matrix} C_t & R_t & E & E_d & \quad L_l & R_l & J & J_d \end{matrix} \\ \begin{matrix} C_t \\ R_t \\ E \\ E_d \end{matrix} & \begin{bmatrix} & & & & \mathbf{Q}_{lCL} & \mathbf{Q}_{lCR} & \mathbf{Q}_{lCJ} & \mathbf{Q}_{lCD} \\ & & \mathbf{I} & & \mathbf{Q}_{lRL} & \mathbf{Q}_{lRR} & \mathbf{Q}_{lRJ} & \mathbf{Q}_{lRD} \\ & & & & \mathbf{Q}_{lEL} & \mathbf{Q}_{lER} & \mathbf{Q}_{lEJ} & \mathbf{Q}_{lED} \\ & & & & \mathbf{Q}_{lDL} & \mathbf{Q}_{lDR} & \mathbf{Q}_{lDJ} & \mathbf{Q}_{lDD} \end{bmatrix} \end{matrix} \qquad (11.54)
$$

The tree current vector can be partitioned into four subvectors relating to tree capacitance, tree resistance, voltage source, and dependent voltage source currents. The link current vector is likewise partitioned into inductance, link resistance, current source, and dependent current source currents. We can write

$$
\begin{bmatrix} \mathbf{i}_{tC} \\ \mathbf{i}_{tR} \\ \mathbf{i}_{tE} \\ \mathbf{i}_{tE_d} \end{bmatrix} = - \begin{bmatrix} \mathbf{Q}_{lCL} & \mathbf{Q}_{lCR} & \mathbf{Q}_{lCJ} & \mathbf{Q}_{lCD} \\ \mathbf{Q}_{lRL} & \mathbf{Q}_{lRR} & \mathbf{Q}_{lRJ} & \mathbf{Q}_{lRD} \\ \mathbf{Q}_{lEL} & \mathbf{Q}_{lER} & \mathbf{Q}_{lEJ} & \mathbf{Q}_{lED} \\ \mathbf{Q}_{lDL} & \mathbf{Q}_{lDR} & \mathbf{Q}_{lDJ} & \mathbf{Q}_{lDD} \end{bmatrix} \begin{bmatrix} \mathbf{i}_{lL} \\ \mathbf{i}_{lR} \\ \mathbf{i}_{lJ} \\ \mathbf{i}_{lJ_d} \end{bmatrix} \qquad (11.55)
$$

We will assume that there are u dependent voltage sources and w dependent current sources so that the new submatrix dimensions are

$$
\begin{array}{llll} \mathbf{Q}_{lCD} & m \times w & \mathbf{Q}_{lRD} \quad p \times w & \mathbf{Q}_{lED} \quad q \times w & \mathbf{Q}_{lDD} \quad u \times w \\[1ex] \mathbf{Q}_{lDL} & u \times n & \mathbf{Q}_{lDR} \quad u \times r & \mathbf{Q}_{lDJ} \quad u \times s & \end{array}
$$

The link branch voltages are expressed in terms of the tree-branch voltages as

$$
\begin{bmatrix} \mathbf{v}_{lL} \\ \mathbf{v}_{lR} \\ \mathbf{v}_{tE} \\ \mathbf{v}_{lJ_d} \end{bmatrix} =
\begin{bmatrix}
\mathbf{Q}^{\mathsf{T}}_{lCL} & \mathbf{Q}^{\mathsf{T}}_{lRL} & \mathbf{Q}^{\mathsf{T}}_{lEL} & \mathbf{Q}^{\mathsf{T}}_{lDL} \\
\mathbf{Q}^{\mathsf{T}}_{lCR} & \mathbf{Q}^{\mathsf{T}}_{lRR} & \mathbf{Q}^{\mathsf{T}}_{lER} & \mathbf{Q}^{\mathsf{T}}_{lDR} \\
\mathbf{Q}^{\mathsf{T}}_{lCJ} & \mathbf{Q}^{\mathsf{T}}_{lRJ} & \mathbf{Q}^{\mathsf{T}}_{lEJ} & \mathbf{Q}^{\mathsf{T}}_{lDJ} \\
\mathbf{Q}^{\mathsf{T}}_{lCD} & \mathbf{Q}^{\mathsf{T}}_{lRD} & \mathbf{Q}^{\mathsf{T}}_{lED} & \mathbf{Q}^{\mathsf{T}}_{lDD}
\end{bmatrix}
\begin{bmatrix} \mathbf{v}_{tC} \\ \mathbf{v}_{tR} \\ \mathbf{v}_{tE} \\ \mathbf{v}_{tE_d} \end{bmatrix}
\tag{11.56}
$$

The equations for the capacitance currents and inductance voltages can be written from Eqs. (11.55) and (11.56) as

$$
\begin{bmatrix} \mathbf{i}_{tC} \\ \mathbf{v}_{lL} \end{bmatrix} =
\begin{bmatrix} \mathbf{0} & -\mathbf{Q}_{lCL} \\ \mathbf{Q}^{\mathsf{T}}_{lCL} & \mathbf{0} \end{bmatrix}
\begin{bmatrix} \mathbf{v}_{tC} \\ \mathbf{i}_{lL} \end{bmatrix} +
\begin{bmatrix} \mathbf{0} & -\mathbf{Q}_{lCR} \\ \mathbf{Q}^{\mathsf{T}}_{lRL} & \mathbf{0} \end{bmatrix}
\begin{bmatrix} \mathbf{v}_{tR} \\ \mathbf{i}_{lR} \end{bmatrix}
$$
$$
+ \begin{bmatrix} \mathbf{0} & -\mathbf{Q}_{lCJ} \\ \mathbf{Q}^{\mathsf{T}}_{lEL} & \mathbf{0} \end{bmatrix}
\begin{bmatrix} \mathbf{v}_{tE} \\ \mathbf{i}_{lJ} \end{bmatrix} +
\begin{bmatrix} \mathbf{0} & -\mathbf{Q}_{lCD} \\ \mathbf{Q}^{\mathsf{T}}_{lDL} & \mathbf{0} \end{bmatrix}
\begin{bmatrix} \mathbf{v}_{tE_d} \\ \mathbf{i}_{lJ_d} \end{bmatrix}
\tag{11.57}
$$

The resistive tree-branch currents and resistive link voltages are given by

$$
\begin{bmatrix} \mathbf{v}_{lR} \\ \mathbf{i}_{tR} \end{bmatrix} =
\begin{bmatrix} \mathbf{Q}^{\mathsf{T}}_{lCR} & \mathbf{0} \\ \mathbf{0} & -\mathbf{Q}_{lRL} \end{bmatrix}
\begin{bmatrix} \mathbf{v}_{tC} \\ \mathbf{i}_{lL} \end{bmatrix} +
\begin{bmatrix} \mathbf{Q}^{\mathsf{T}}_{lRR} & \mathbf{0} \\ \mathbf{0} & -\mathbf{Q}_{lRR} \end{bmatrix}
\begin{bmatrix} \mathbf{v}_{tR} \\ \mathbf{i}_{lR} \end{bmatrix}
$$
$$
+ \begin{bmatrix} \mathbf{Q}^{\mathsf{T}}_{lCR} & \mathbf{0} \\ \mathbf{0} & -\mathbf{Q}_{lRJ} \end{bmatrix}
\begin{bmatrix} \mathbf{v}_{tE} \\ \mathbf{i}_{lJ} \end{bmatrix} +
\begin{bmatrix} \mathbf{Q}^{\mathsf{T}}_{lDR} & \mathbf{0} \\ \mathbf{0} & \mathbf{Q}_{lRD} \end{bmatrix}
\begin{bmatrix} \mathbf{v}_{tE_d} \\ \mathbf{i}_{lJ_d} \end{bmatrix}
\tag{11.58}
$$

The state equations can be obtained from Eq. (11.57) if the vectors relating the resistive quantities and the dependent sources are eliminated. Since the dependent voltage and current sources are functions of the other network variables, it is possible to represent the dependent quantities as

$$
\begin{bmatrix} \mathbf{v}_{tE_d} \\ \mathbf{i}_{lJ_d} \end{bmatrix} = [K_1]\begin{bmatrix} \mathbf{v}_{tC} \\ \mathbf{i}_{lL} \end{bmatrix} = [K_2]\begin{bmatrix} \mathbf{v}_{tR} \\ \mathbf{i}_{lR} \end{bmatrix} + [K_3]\begin{bmatrix} \mathbf{v}_{tE} \\ \mathbf{i}_{lJ} \end{bmatrix}
\tag{11.59}
$$

Equation (11.59) is now substituted into Eqs. (11.58) and (11.57) to eliminate the dependent source quantities. Next, Ohm's law is applied as in Eq. (11.47), the resistive quantities are eliminated, and the state equation can finally be expressed in terms of the tree capacitance voltages, the link inductance currents, and the independent voltage and current sources. The best way to outline the procedure with dependent sources is by means of a simple example.

Example 11.8

Write the state equations for the common-emitter transistor amplifier shown in Fig. 11.8.

The initial step in the analysis is to construct the network graph and number the branches to conform with the **Q** matrix in Eq. (11.54). The network graph is shown in Fig. 11.9.

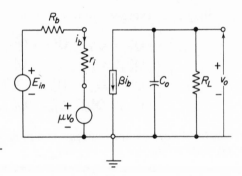

Figure 11.8 Transistor amplifier equivalent circuit.

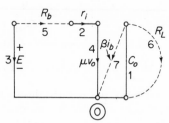

Figure 11.9 Transistor amplifier network graph.

The cut-set matrix for the network is

$$
\mathbf{Q} =
\begin{array}{c}
\quad\;\; 1 \;\; 2 \;\; 3 \;\; 4 \quad\; 5 \;\; 6 \;\; 7 \\
\left[
\begin{array}{cccc|ccc}
1 & 0 & 0 & 0 & 0 & 1 & 1 \\
0 & 1 & 0 & 0 & -1 & 0 & 0 \\
0 & 0 & 1 & 0 & 1 & 0 & 0 \\
0 & 0 & 0 & 1 & -1 & 0 & 0
\end{array}
\right]
\end{array}
$$

from which

$$
\mathbf{Q} =
\left[
\begin{array}{c|cc}
0 & 1 & 1 \\
\hline
-1 & 0 & 0 \\
\hline
1 & 0 & 0 \\
\hline
-1 & 0 & 0
\end{array}
\right]
=
\begin{bmatrix}
\mathbf{Q}_{ICR} & \mathbf{Q}_{ICD} \\
\mathbf{Q}_{IRR} & \mathbf{Q}_{IRD} \\
\mathbf{Q}_{IER} & \mathbf{Q}_{IED} \\
\mathbf{Q}_{IDR} & \mathbf{Q}_{IDD}
\end{bmatrix}
$$

The tree capacitance current can now be written as

$$
[\mathbf{i}_{tC}] = [0][\mathbf{v}_{tC}] + \begin{bmatrix} 0 & 0 & -1 \end{bmatrix}\begin{bmatrix} \mathbf{v}_{tR} \\ \mathbf{i}_{lR} \end{bmatrix} + [0][\mathbf{v}_{tE}] + \begin{bmatrix} 0 & -1 \end{bmatrix}\begin{bmatrix} \mathbf{v}_{tE_d} \\ \mathbf{i}_{lJ_d} \end{bmatrix}
$$

and

$$
\begin{bmatrix} \mathbf{v}_{lR} \\ \mathbf{i}_{tR} \end{bmatrix} = \begin{bmatrix} 0 \\ 1 \\ 0 \end{bmatrix}[\mathbf{v}_{tC}] + \begin{bmatrix} -1 & 0 & 0 \\ 0 & 0 & 0 \\ 0 & 1 & 0 \end{bmatrix}\begin{bmatrix} \mathbf{v}_{tR} \\ \mathbf{i}_{lR} \end{bmatrix} + \begin{bmatrix} 1 \\ 0 \\ 0 \end{bmatrix}[\mathbf{v}_{tE}] + \begin{bmatrix} -1 & 0 \\ 0 & 0 \\ 0 & 0 \end{bmatrix}\begin{bmatrix} \mathbf{v}_{tE_d} \\ \mathbf{i}_{lJ_d} \end{bmatrix}
$$

The dependent source equation is

$$\begin{bmatrix} \mathbf{v}_{tE_d} \\ \mathbf{i}_{lJ_d} \end{bmatrix} = \begin{bmatrix} \mu \\ 0 \end{bmatrix} [\mathbf{v}_{tC}] + \begin{bmatrix} 0 & 0 & 0 \\ 0 & \beta & 0 \end{bmatrix} \begin{bmatrix} \mathbf{v}_{tR} \\ \mathbf{i}_{lR} \end{bmatrix}$$

Notice that the dependent current source was made dependent upon the link resistance current. (Why?)

Substituting the dependent source equation into the two previous network equations we obtain

$$[\mathbf{i}_{tC}] = [0][\mathbf{v}_{tC}] + \begin{bmatrix} 0 & -\beta & -1 \end{bmatrix} \begin{bmatrix} \mathbf{v}_{tR} \\ \mathbf{i}_{lR} \end{bmatrix} + [0][\mathbf{v}_{tE}]$$

and

$$\begin{bmatrix} \mathbf{v}_{lR} \\ \mathbf{i}_{tR} \end{bmatrix} = \begin{bmatrix} -\mu \\ 1 \\ 0 \end{bmatrix} [\mathbf{v}_{tC}] + \begin{bmatrix} -1 & 0 & 0 \\ 0 & 0 & 0 \\ 0 & 1 & 0 \end{bmatrix} \begin{bmatrix} \mathbf{v}_{tR} \\ \mathbf{i}_{lR} \end{bmatrix} + \begin{bmatrix} 1 \\ 0 \\ 0 \end{bmatrix} [\mathbf{v}_{tE}]$$

The resistive voltages and currents are related by Ohm's law as

$$\begin{bmatrix} \mathbf{v}_{lR} \\ \mathbf{i}_{tR} \end{bmatrix} = \begin{bmatrix} 0 & R_b & 0 \\ 0 & 0 & R_L \\ g_i & 0 & 0 \end{bmatrix} \begin{bmatrix} \mathbf{v}_{tR} \\ \mathbf{i}_{lR} \end{bmatrix}$$

Substituting the relationship between the resistive voltages and currents into the equation defining the link resistance voltages and tree resistance currents we obtain

$$\begin{bmatrix} \mathbf{v}_{tR} \\ \mathbf{i}_{lR} \end{bmatrix} = \begin{bmatrix} \dfrac{r_i}{r_i + R_b} & 0 & \dfrac{r_i R_b}{R_i + R_b} \\ \dfrac{1}{r_i + R_b} & 0 & \dfrac{-r_i}{r_i + R_b} \\ 0 & \dfrac{1}{R_L} & 0 \end{bmatrix} \left\{ \begin{bmatrix} -\mu \\ 1 \\ 0 \end{bmatrix} [\mathbf{v}_{tC}] + \begin{bmatrix} 1 \\ 0 \\ 0 \end{bmatrix} [\mathbf{v}_{tE}] \right\}$$

$$= \begin{bmatrix} \dfrac{-\mu r_i}{r_i + R_b} \\ \dfrac{-\mu}{r_i + R_b} \\ \dfrac{1}{R_L} \end{bmatrix} [\mathbf{v}_{tC}] + \begin{bmatrix} \dfrac{r_i}{r_i + R_b} \\ \dfrac{1}{r_i + R_b} \\ 0 \end{bmatrix} [\mathbf{v}_{tE}]$$

Substituting this equation into the equation defining the capacitance current the result is

$$[\mathbf{i}_{tC}] = \left[\dfrac{\mu\beta}{r_i + R_b} - \dfrac{1}{R_L} \right][\mathbf{v}_{tC}] + \dfrac{-\beta}{r_i + R_b}[\mathbf{v}_{tE}]$$

At this point we note that $\mathbf{i}_{tC} = C_0(dv_0/dt)$, $\mathbf{v}_{tC} = v_0$, and $\mathbf{v}_{tE} = E$. The state equation now reduces to

$$\frac{d}{dt} v_0 = \frac{1}{C_0}\left(\frac{\mu\beta}{r_i + R_b} - \frac{1}{R_L} \right)v_0 - \frac{\beta}{C_0(r_i + R_b)} E$$

which is in the normal-state form.

In the previous development, we did not consider topological methods for writing state equation in which excesss capacitances and/or inductances could occur. This omission was intentional for the following reasons: (1) The methods for developing the matrix equations are quite similar to those used in the development of this section. (2) The derivation would add little to the value of the material in the text. (3) Finally, networks in the "real world" seldom, occur with excess elements. For a deeper study of the topological formulation of the state equations including excess elements, the texts listed in Additional Reading are suggested.

Additional Reading

BALABANIAN, N., and T. A. BICKART, *Electrical Network Theory*, John Wiley & Sons, Inc., New York, 1969.

BOWERS, J. C., and S. R. SEDORE, *SCEPTRE, A Computer Program for Circuit and Systems Analysis*, Prentice-Hall, Inc., Englewood Cliffs, N.J., 1971.

KUO, B. C., *Linear Networks and Systems*, McGraw-Hill Book Company, New York, 1967.

Problems

11.1. (a) Write the state equations in normal form for the network of Fig. P11.1 by inspection.

   (b) Replace C by L_1 and repeat.

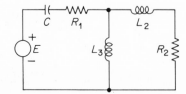

Figure P11.1

11.2. Write the state equations in normal form for the network shown in Fig. P11.2 using network graph methods.

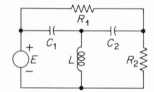

Figure P11.2

11.3. Repeat Problem 11.2 using inspection methods.

11.4. Write the state equations in normal form for the network of Fig. P11.3 using network graph methods. Write the output equation assuming that V_{R_3} is the desired
output.

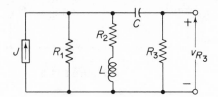

Figure P11.3

11.5. Repeat Problem 11.4 using inspection methods.

11.6. Write the state equations in normal form for the network of Fig. P11.4 using network graph methods. Write the output equation to output v_0 in terms of the state
variables and the sources.

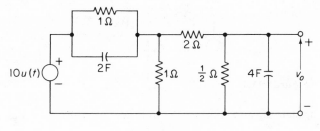

Figure P11.4

11.7. Repeat Problem 11.6 using inspection methods.

11.8. Write the state equations in normal form for the network of Fig. P11.5 using network
graph methods.

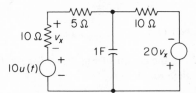

Figure P11.5

11.9. Repeat Problem 11.8 by inspection.

11.10. Write the state equations in normal form for the network of Fig. P11.6 using network graph methods assuming that $C_L = 0$.

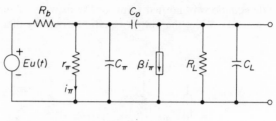

Figure P11.6

11.11.   Repeat Problem 11.10 by inspection.

11.12.   Repeat Problem 11.10 by inspection assuming that $C_L > 0$.

11.13.   Write the state equations in normal form for the network of Fig. P11.7 by any
method.

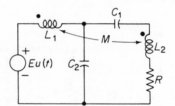

Figure P11.7

11.14.   Write the state equations in normal form for the network of Fig. P11.8 by any
method.

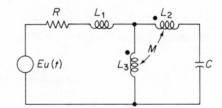

Figure P11.8

11.15.   Determine $i_c(t)$ in the circuit of Fig. P11.9 for $t \geq 0$ using state-variable techniques.
Solve the problem for
(a) $i_L(0) = 0$
(b) $i_L(0) = 1$ A

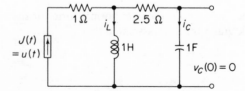

Figure P11.9

11.16.   Write a computer program to formulate the state equation for *RLC* networks. The
program must read the input branch data and sort the data to conform with the
cut-set matrix definition used in the chapter. The program output should include

(a) A listing of the branch data as rearranged by the program.

(b) The cut-set matrix **Q**.

(c) The normal-state equation matrices **A** and **B**.

The input-data format should be the same as specified for the ac analysis program of Chapter 7.

11.17. Repeat Problem 11.16 for *RLC* networks with dependent sources.

Chapter 12

DIFFERENTIAL EQUATIONS
Computer Solutions

*A journey of a thousand miles
must begin with a single step.*

LAO TZU

12.1 Introduction

Most of the time-varying problems in network analysis reduce to the problem of finding a solution to a differential equation or a series of such equations. In Chapter 10 we presented one method for getting time-domain results through the Laplace transform and its inversion. There are other techniques which obtain the transition matrix $\phi(t)$, thus providing a solution.[†] All these methods attack differential equations from an analytic standpoint and are normally restricted to linear differential equations with constant coefficients.

In this chapter we present methods which solve differential equations directly in the time domain by using numerical techniques and the digital computer. At least two advantages exist for this approach as follows: (1) large networks require many operations and much time for solution when we use analytic methods, (2) analytic solutions for nonlinear or time-varying networks are often difficult or impossible to find. Disadvantages of numerical methods include (1) the requirement to put the equations in a state-space form, and (2) the fact that results are obtained in a numerical format as opposed to an analytic formula.

Solutions by digital computer differ from analytic solutions in that the computer algorithm advances by taking finite steps in the independent variable; also the results are necessarily numerical, that is, not expressed by a formula. The value of the function obtained at any step is an approximation of the value which would be obtained analytically, whereas the analytical solution is exact. However, an analytical solution may be difficult or time consuming or even impossible to find. Thus, we seek ways to use the computer to perform a task which may not produce exact results but which will produce useful answers within a reasonable time.

In this chapter we consider only three basic methods for obtaining solutions to networks defined by systems of first-order ordinary differential equations. The methods included are Euler's, predictor–corrector, and Runge-Kutta. We will refer to the single ordinary differential equation and systems of such equations as ODE throughout the discussion.

The material presented is by no means intended to be an exhaustive study of numerical solutions, but a limited introduction. The references at the end of the chapter provide more complete presentations of the subject.

[†] B. O. Watkins, *Introduction to Control Systems*, The Macmillan Company, 1969, pp. 258–313.

12.2   Euler's Method

Suppose we are given the first-order differential equation

$$\dot{x} = f(x, t) \qquad x(t_0) = x_0 \qquad \qquad (12.1)$$

which has a unique solution $x(t)$ on the interval $t_0 \leq t \leq t_f$, where t_f is the final time of interest. Our problem consists of finding an approximate solution and is thus one of approximating $x(t)$ between t_0 and t_f.

We would have an impossible task to find a value $x(t)$ for every t in the interval, but this is not really necessary. We can find approximate values for $x(t)$ at enough points t_i ($i = 1, 2, \ldots, m$) between t_0 and t_f to clearly define the function. Although theory does not require these points to be equally spaced, we find it convenient computationally to do so, and this assumption also makes the discussion simpler. Thus, we wish to approximate the function at the discrete points $t_i = t_0 + ih$ ($i = 1, 2, \ldots, m$). The constant h is the step size of the integration, and the integer m is such that $t_0 + mh \geq t_f$.

In the differential-equation literature as well as the remainder of this chapter, an exact solution at t_i is denoted as $x(t_i)$, while the approximation to the solution is denoted as x_i. In other words, x_i is an approximation to $x(t_i)$.

Euler's method comprises the simplest of all the numerical integration methods. Although it is the least accurate of the methods, it serves to solve many types of problems. By studying this method in detail we will be able to grasp the basic ideas involved in numerical solutions of ODE and can more easily understand the more powerful, complex methods described in the following sections. Given Eq. (12.1), Euler's method states that

$$x_{i+1} = x_i + hf(x_i, t_i) \qquad \qquad (12.2a)$$

$$t_{i+1} = t_i + h \qquad \qquad (12.2b)$$

for $i = 0, 1, \ldots, m - 1$.

EXAMPLE 12.1

Apply Euler's method to the equation $\dot{x} = x$, $t_0 = 0$, $x(0) = 1$, with $h = 0.1$ and $t_f = 0.4$.

Solution

$$x_1 = x_0 + hf(x_0, t_0) = 1 + 0.1(1)$$
$$= 1.10 \qquad t_1 = 0.1$$
$$x_2 = x_1 + hf(x_1, t_1) = 1.1 + 0.1(1.1)$$
$$= 1.21 \qquad t_2 = 0.2$$
$$x_3 = x_2 + hf(x_2, t_2)$$
$$= 1.331 \qquad t_3 = 0.3$$
$$x_4 = x_3 + hf(x_3, t_3)$$
$$= 1.4641 \qquad t_4 = 0.4$$

The exact solution to this ODE is $x = e^t$. The exact values of the solution to four decimal places are given by

$$y(0.1) = 1.1052$$
$$y(0.2) = 1.2214$$
$$y(0.3) = 1.3499$$
$$y(0.4) = 1.4918$$

The results obtained in Example 12.1 indicate that Euler's method does indeed generate an approximation to the solution. The approximation may be considered poor, but we will defer a discussion of the errors.

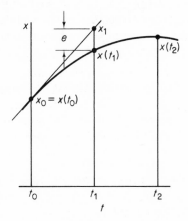

Figure 12.1 Euler's method for finding x, given x_0 and $f(x_0, t_0)$.

At this point a geometric interpretation of Euler's method might shed some light on our discussion. We are given the initial point $x(t_0)$ on the solution curve $x(t)$ in Fig. 12.1. The slope of the curve at t_0 is given by $f(x_0, t_0)$. Thus, we have the tangent line to the solution curve at the initial point. Euler's method estimates the solution at t_1 by assuming that it lies on this tangent line.

Mathematically, the tangent is given by

$$\dot{x} = f(x_0, t_0)$$
$$= \frac{1}{h}(x_1 - x_0)$$

Solving for x_1, we have

$$x_1 = x_0 + hf(x_0, t_0) \tag{12.3}$$

which is one step of Euler's method. The next step of the process is executed by assuming that x_1, t_1 is the initial point and repeating the process.

From Fig. 12.1 we can see that even if the calculations are exact (they aren't), the exact solution curve $x(t)$ will differ from the approximate curve unless the solution is a straight line. If we have an exact starting point $(x(t_0), t_0)$, the starting point for the second step (x_1, t_1) will be in error, so the derivative $f(x_1, t_1)$ will also be in error. We might gloomily summarize by observing that each subsequent step starts from an incorrect point with an incorrect slope under the incorrect assumption that the solution curve is linear. Fortunately, things are not as bad as they first appear. By using a reasonably small value of h we decrease the error between successive points. In the limit as $h \rightarrow 0$ the difference $|x_i - x(t_i)|$ tends to zero for the function $\dot{x} = x$.

In the attempt to improve accuracy, we can see that as the step size decreases, the number of steps must proportionally increase. This causes two problems which plague those who use any numerical method for the solution of ODE:

1. As we have stated before, the computer results, being approximations, are always slightly in error. In this type of problem the error tends to accumulate at each step, so an error which was insignificant at any one step may dominate the solution after 100,000 steps. In addition, computer roundoff error increases directly with the number of operations.

2. It takes time for the computer to execute each integration step. A very fine time of say $h = 10^{-5}$ may not provide any significant increase in accuracy over a step size $h = 10^{-3}$, but it will take 100 times as long to obtain the solution. Apparently, the step size must be selected small enough to obtain a reasonably accurate solution, but at the same time, large enough to be economical and avoid the numerical limitations of the computer.

In Fig. 12.2 we present a program utilizing Euler's method to solve the problem of Example 12.1. By following the program carefully, you can see how the algorithm proceeds. The results obtained by the program of Fig. 12.2 for $h = 0.05$, $t_0 = 0$, $x_0 = 1$, $t_f = 0.5$ to four decimal places is shown in Table 12.1 along with the exact values for $f(x) = e^t$. To show the effect of the step size on the accuracy of the solution, Table 12.2 contains the results of the calculations of $x_{0.5}$ for several step-size selections. The value of $x_{0.5}$ for $h = 0.2$ is obtained by interpolation.

We now extend Euler's method to a system of simultaneous first-order ODE. Suppose we are given the equations in the form

$$
\begin{aligned}
\dot{x}_1 &= f_1(t, x_1, x_2, \ldots, x_n) & x_1(t_0) &= x_{10} \\
\dot{x}_2 &= f_2(t, x_1, x_2, \ldots, x_n) & x_2(t_0) &= x_{20} \\
&\cdot \quad \cdot \quad \cdot \quad \cdot \quad \cdot \quad \cdot \quad \cdot \quad \cdot \quad \cdot \quad \cdot \quad \cdot \quad \cdot \\
\dot{x}_n &= f_n(t, x_1, x_2, \ldots, x_n) & x_n(t_0) &= x_{n0}
\end{aligned}
\tag{12.4}
$$

The problem is to find approximate values for n unknown functions $(x_1(t), x_2(t), \ldots, x_n(t))$ at the points $t_1, t_2, \ldots, t_m$ as we did for the single first-order ODE.

To visualize the problem, consider the point in n-space $(t_0, x_{10}, x_{20}, \ldots, x_{n0})$. We want to estimate the next point in the space from the point and the derivative in the direction $\mathbf{x}$.

Let $t_0, x_{10}, x_{20}, \ldots, x_{n0}, h$, and t_f be given. Calculate the values x_{ij} ($i = 1, 2,$

```
C       EULER INTEGRATION PROGRAM EXAMPLE                              EULR    1
C                                                                      EULR    2
C       PROGRAM TO DEMONSTRATE  EULER'S METHOD FOR THE EQUATION        EULR    3
C       X' = X                                                         EULR    4
C                                                                      EULR    5
C       DESCRIPTION OF VARIABLES                                       EULR    6
C           FUN - AN EXTERNAL FUNCTION TO GENERATE DERIVATIVE INFORMATION  EULR 7
C           H - INTEGRATION STEP SIZE                                  EULR    8
C           T - INDEPENDENT VARIABLE                                   EULR    9
C           TF - FINAL VALUE OF INDEPENDENT VARIABLE                   EULR   10
C           TO - INITIAL VALUE OF INDEPENDENT VARIABLE                 EULR   11
C           X - DEPENDENT VARIABLE                                     EULR   12
C           XO - INITIAL VALUE OF DEPENDENT VARIABLE                   EULR   13
C                                                                      EULR   14
C       STATEMENT FUNCTION DEFINITION OF X' = X                        EULR   15
        FUN(X,T)=X                                                     EULR   16
C       INPUT INITIAL DATA                                             EULR   17
        READ (5,10) TO,XO,H,TF                                         EULR   18
     10 FORMAT (4F20.0)                                                EULR   19
C       OUTPUT INITIAL VALUES                                          EULR   20
        WRITE (6,20) TO,XO                                             EULR   21
     20 FORMAT (5X,'T = ',1PE20.5,'   X = ',1PE20.5)                   EULR   22
C       INTEGRATION LOOP                                               EULR   23
C       INITIALIZE VALUES                                              EULR   24
        X=XO                                                           EULR   25
        T=TO                                                           EULR   26
C       ESTIMATE NEXT POINT                                            EULR   27
     30 X=X+H*FUN(X,T)                                                 EULR   28
        T=T+H                                                          EULR   29
        WRITE (6,20) T,X                                               EULR   30
C       TEST FOR FINAL STEP                                            EULR   31
        IF (T.LT.TF) GO TO 30                                          EULR   32
        STOP                                                           EULR   33
        END                                                            EULR   34
```

Figure 12.2 Program utilizing Euler's method.

TABLE 12.1

t_i	x_i	$x(t_i)$
0.00	1.0000	1.0000
0.05	1.0500	1.0513
0.10	1.1025	1.1052
0.15	1.1576	1.1618
0.20	1.2155	1.2214
0.25	1.2763	1.2840
0.30	1.3401	1.3499
0.35	1.4071	1.4191
0.40	1.4774	1.4918
0.45	1.5513	1.5683
0.50	1.6289	1.6487

TABLE 12.2

h	$x_{0.5}$
0.005	1.6466
0.01	1.6446
0.02	1.6408
0.05	1.6289
0.1	1.6105
0.2	~ 1.584

$\ldots, n; j = 1, 2, \ldots, m$), where $t_0 + mh \geq t_f$ according to

$$x_{1,i+1} = x_{1i} + hf_1(t_i, x_{1i}, x_{2i}, \ldots, x_{ni})$$
$$x_{2,i+1} = x_{2i} + hf_2(t_i, x_{1i}, x_{2i}, \ldots, x_{ni}) \qquad (12.5)$$
$$\cdots \cdots \cdots \cdots \cdots \cdots$$
$$x_{n,i+1} = x_{ni} + hf_n(t_i, x_{1i}, x_{2i}, \ldots, x_{ni})$$
$$t_{i+1} = t_i + h \qquad (12.6)$$

for $i = 0, \ldots, m - 1$.

EXAMPLE 12.2

Apply Euler's method to the system of first-order ODE defined by

$$\dot{x}_1 = x_1 + t \qquad x_1(0) = 0$$
$$\dot{x}_2 = x_1 + x_2 \qquad x_2(0) = 1.0$$

for $h = 0.1$ and $t_f = 0.2$.

$$x_{11} = x_{10} + hf(t, x_{10}, x_{20})$$
$$= x_{10} + h(x_{10} + t)$$
$$= 0 + 0.1(0 + 0) = 0.0$$
$$x_{21} = x_{20} + h(x_{10} + x_{20})$$
$$= 1.0 + 0.1(0 + 1.0) = 1.1$$
$$x_{12} = x_{11} + h(x_{11} + t) = 0 + 0.1(0.1) = 0.01$$
$$x_{22} = x_{21} + h(x_{11} + x_{21}) = 1.1 + 0.1(1.1) = 1.21$$

etc.

A Fortran subroutine which applies Euler's method to a system of first-order ODE can be written using the method illustrated in Example 12.2. We conclude this section by writting a subroutine EULINT built around the Euler algorithm which solves an nth-order system of equations ($n \leq 10$). The subroutine requires a function subprogram FUN (I, X, T) which evaluates $\dot{x}_i = f_i(x_1, x_2, \ldots, x_n, t)$ for each derivative $\dot{x}_1$. That is, FUN (2, X, T) $= f_2(x_1, x_2, \ldots, x_n, t)$. The subroutine is given in Fig. 12.3. The subroutine also requires an output routine to which the independent variable and the computed dependent variables are supplied for output processing. The form required in Fig. 12.3 is

SUBROUTINE OUTPUT (Y, X, N)

where Y is a vector of dependent variables and X is the independent variable, and N is the number of dependent variables.

If the equations describing the network are not in the form of Eq. (12.4), we must first put them in this form. One method for putting linear differential equations in the proper form is described in Chapter 10. If the given equations are nonlinear, the achievement of the proper form becomes a more difficult matter.

```
        SUBROUTINE EULINT(N,TO,STEP,ENDPT,X)                          EULS   1
C                                                                     EULS   2
C       PURPOSE                                                       EULS   3
C           TO SOLVE A SYSTEM OF FIRST-ORDER ODE WITH GIVEN INITIAL   EULS   4
C           CONDITIONS                                                EULS   5
C       DEFINITION OF VARIABLES                                       EULS   6
C           ENDPT - FINAL VALUE OF INDEPENDENT VARIABLE               EULS   7
C           FUN - AN EXTERNAL FUNCTION SUBPROGRAM TO SUPPLY DERIVATIVE EULS  8
C               INFORMATION.  THE ARGUEMENTS ARE I (EQUATION NO.),    EULS   9
C               X,AND TIME                                            EULS  10
C           N - NO. OF EQUATIONS IN SYSTEM                            EULS  11
C           OUTPUT - AN EXTERNAL SUBROUTINE SUBPROGRAM TO OUTPUT RESULTS EULS 12
C           STEP - INTEGRATION STEP SIZE H                            EULS  13
C           TEMP - A TEMPORARY WORKING VECTOR TO HOLD INTERMEDIATE    EULS  14
C               VALUES OF X                                           EULS  15
C           TIME- PRESENT VALUE OF INDEPENDENT VARIABLE               EULS  16
C           TO - INITIAL VALUE OF INDEPENDENT VARIABLE                EULS  17
C           X - NTH-ORDER VECTOR OF INITIAL VALUES.  LATER X IS THE   EULS  18
C               RESULTING VECTOR OF DEPENDENT VARIABLES COMPUTED      EULS  19
C               AT INTERMEDIATE POINTS                                EULS  20
C       NOTE ... X MUST BE DIMENSIONED IN CALLING PROGRAM             EULS  21
C                                                                     EULS  22
        DIMENSION X(10),TEMP(10)                                      EULS  23
C       INITIALIZE INDEPENDENT VARIABLE                               EULS  24
        TIME=TO                                                       EULS  25
C       INTEGRATION LOOP                                              EULS  26
C       COMPUTE NEXT VALUE OF DEPENDENT VARIABLE                      EULS  27
     10 DO 20 I=1,N                                                   EULS  28
     20 TEMP(I)=X(I)+STEP*FUN(I,X,TIME)                               EULS  29
        TIME=TIME+STEP                                                EULS  30
        DO 30 I=1,N                                                   EULS  31
     30 X(I)=TEMP(I)                                                  EULS  32
C       OUTPUT VALUES FOR INDEPENDENT AND DEPENDENT VARIABLES         EULS  33
        CALL OUTPUT(X,TIME,N)                                         EULS  34
C       TEST FOR FINAL VALUE OF INDEPENDENT VARIABLE                  EULS  35
        IF (TIME.LT.ENDPT) GO TO 10                                   EULS  36
        RETURN                                                        EULS  37
        END                                                           EULS  38
```

Figure 12.3 Program for a system of simultaneous first-order ordinary differential equations, Euler's method.

12.3 Function Approximation

As already intimated, the basis of finding a function which satisfies a given differential equation with given initial conditions consists of approximating the actual function by a more tractable function. Since polynomials have quite manageable features, very often the approximating function consists of a polynomial, although a polynomial may not always be the best approximating function.

Suppose we are given a tabulated function; that is, we have a table of function values corresponding to specific independent variable values, as in Table 12.3. Such a table is one way of defining $f(y)$ as a function of y. We can plot these points on a graph such as in Fig. 12.4. Given this functional relation, we might then ask: What value does $f(y)$ have if y lies between y_2 and y_3, or $y_2 < y < y_3$? Perhaps the simplest answer to this would involve drawing a straight line on Fig. 12.4 between the points at y_2 and y_3, and then taking $f(y)$, $y_2 < y < y_3$, as lying on this line. The line may be

TABLE 12.3

Variable		Function
y	Value	$f(y)$
y_0	-0.4	0.5
y_1	0	1.0
y_2	0.5	1.1
y_3	1.2	1.5
y_4	1.4	2.1
y_5	1.8	3.5
y_6	2.1	5.2
y_7	2.5	6.7
y_8	3.0	9.2

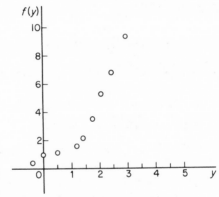

Figure 12.4 The points of Table 12.3 plotted on a graph.

expressed in analytic terms by the function

$$g_1(y) = \frac{f(y_3) - f(y_2)}{y_3 - y_2}(y - y_2) + f(y_2) \qquad (12.7)$$

Thus, if $y = 1.0$, using Table 12.3 and Eq. (12.7), we obtain (to three significant figures)

$$g_1(1) = \frac{1.5 - 1.1}{1.2 - 0.5}(1.0 - 0.5) + 1.1$$

$$= \frac{0.4}{0.7}(0.5) + 1.1$$

$$= 1.385$$

An alternate technique might consist of drawing a straight line between the points at y_1 and y_2 and extending this line to y. In this case,

$$g_2(y) = \frac{f(y_2) - f(y_1)}{y_2 - y_1}(y - y_1) + f(y_1) \qquad (12.8)$$

Again, if $y = 1.0$, we now obtain

$$g_2(1) = \frac{1.1 - 1.0}{0.5 - 0}(1.0 - 0) + 1.0$$

$$= \frac{0.1}{0.5}(1.0) + 1.0$$

$$= 1.200$$

These two examples illustrate interpolation and extrapolation, respectively. Equation (12.7) demonstrates a linear interpolation; Eq. (12.8) gives a linear extrapolation. Clearly the results are not the same.

In linear interpolation (or extrapolation) we apply the knowledge we have of two values of $f(x)$, whereas we may have many more values. Can we improve our estimate utilizing additional $f(y)$ values? We see that the functions $g_1(y)$ and $g_2(y)$ are really polynomials in y, such as

$$g(y) = ay + b \tag{12.9}$$

where a and b are constants. This makes us think that perhaps we could use a higher-order polynomial, such as

$$g(y) = ay^2 + by + c \tag{12.10}$$

We might find the coefficients a, b, and c in Eq. (12.10) by solving three equations, that is, by using the simultaneous equations

$$\begin{aligned} ay^2 + by + c &= f_1(y) \\ ay^2 + by + c &= f_2(y) \\ ay^2 + by + c &= f_3(y) \end{aligned} \tag{12.11}$$

By a different approach Lagrange developed the interpolating polynomial $P(y)$ such that

$$P(y) = a_1 + a_2 y + a_3 y^2 + \cdots + a_{n+1} y^n \tag{12.12}$$

with the property that

$$P(y_i) = f(y_i) \qquad i = 0, 1, 2, \ldots, n \tag{12.13}$$

The Lagrange interpolating polynomial is given by†

$$P(y) = \sum_{j=0}^{n} L_j(y) f(y_j) \tag{12.14a}$$

---

†D. G. Moursund and C. S. Duris, *Elementary Theory and Application of Numerical Analysis*, McGraw-Hill Book Company, New York, 1967.

where

$$L_j(y) = \prod_{\substack{k=0 \\ k \neq j}}^{n} \frac{y - y_k}{y_j - y_k} \tag{12.14b}$$

EXAMPLE 12.3

Find $f(y)$ in Table 12.3 if $y = 1.0$ using a Lagrange interpolating polynomial of degree $n = 2$, and y_1, y_2, and y_3 of that table. To use Eq. (12.14b), we must reduce the subscripts of y by 1. Then in Eq. (12.14b), $y_0 = 0$, $y_1 = 0.5$, $y_2 = 1.2$, and

$$L_0(y) = \frac{y - y_1}{y_0 - y_1} \frac{y - y_2}{y_0 - y_2}$$

$$= \frac{(y - 0.5)(y - 1.2)}{(0 - 0.5)(0 - 1.2)}$$

$$= \frac{1}{(0.5)(1.2)} (y^2 - 1.7y + 0.60)$$

$$L_1(y) = \frac{(y - y_0)(y - y_2)}{(y_1 - y_0)(y_1 - y_2)}$$

$$= \frac{(y - 0)(y - 1.2)}{(0.5 - 0)(0.5 - 1.2)}$$

$$= \frac{-1}{(0.5)(0.7)} (y^2 - 1.2y + 0)$$

$$L_2(y) = \frac{(y - y_0)(y - y_1)}{(y_2 - y_0)(y_2 - y_1)}$$

$$= \frac{(y - 0)(y - 0.5)}{(1.2 - 0)(1.2 - 0.5)}$$

$$= \frac{1}{(1.2)(0.7)} (y^2 - 0.5y + 0)$$

Then $P(y) = L_0(y)f(y_0) + L_1(y)f(y_1) + L_2(y)f(y_2)$, where $L_0(y)$, $L_1(y)$, and $L_2(y)$ have been previously determined. Thus,

$$P(y) = 1.667(y^2 - 1.7y + 0.60) - 2.857(y^2 - 1.2y)(1.1) + 1.190(y^2 - 0.5y)(1.5)$$

$$= 0.309y^2 + 0.044y + 1.0$$

$$P(1) = 1.35$$

Note that we really need about six-digit accuracy to obtain a result satisfactory to three digits. This example has been worked by slide rule and is accurate to only about two digits.

We could also find a $P(y)$ from the y_0, y_1, and y_2 table values, and extrapolate this; or use the y_2, y_3, y_4 values; and so on. Many possible combinations exist. Furthermore, we could with additional computations use a polynomial of third or higher degree. It would seem off-hand that polynomials of higher degree might be more accurate than those of lower degree, but this may not necessarily be true. Going back to our original tabular function, this function might also be known in analytical form (although in this case no such a priori attempt was made). In such a form, $f(y)$ is known for all y in some range. Now it should be clear that the approximation by some other function [such as $P(y)$] will not in general give the same value as $f(y)$. That is, except for special cases,

$$f(y) = P(y) + R(y) \tag{12.15}$$

where $R(y)$ is some remainder.

There are many other approximating polynomials; for example, the Taylor-series expansion and the least-squares fit. Also other nonpolynomial functions may provide better approximations to a given function in certain situations. One of the virtues of the polynomial is its simplicity, but it may not fit asymptotic curves, for example, as well as an exponential function. In numerical analysis, there is no one universal technique for any situation.

12.4 Integration

Our interest in the Lagrange interpolation formula must now be obvious, for we find it easy to integrate a polynomial. From Eqs. (12.15) and (12.14a)

$$f(y) = P(y) + R(y)$$

or

$$f(y) = \sum_{j=0}^{n} L_j(y)f(y_j) + R(y) \tag{12.16}$$

To integrate between two points a and b which lie in the set determined by the points y_i we integrate both sides of Eq. (12.16) to obtain

$$\int_a^b f(y)\, dy = \int_a^b \sum_{i=0}^{n} L_i(y)f(y_i)\, dy + \int_a^b R(y)\, dy$$

$$= \sum_{i=0}^{n} f(y_i) \int_a^b L_i(y)\, dy + \int_a^b R(y)\, dy$$

$$\int_a^b f(y)\, dy = \sum_{i=0}^{n} C_i f(y_i) + E(f) \tag{12.17}$$

where $C_i = \int_a^b L_i(y)\, dy$ and $E(f)$ is the error term. Equation (12.17), omitting the error term, is called a *quadrature formula*.

EXAMPLE 12.4

Find a formula for $\int_{y_0}^{y_1} f(y)\, dy$ in terms of $f(y_0)$ and $f(y_1)$.

Solution. The Lagrange interpolation formula is

$$f(y) = \frac{y - y_1}{y_0 - y_1} f(y_0) + \frac{y - y_0}{y_1 - y_0} f(y_1) + R(y)$$

Using Eq. (12.17)

$$C_0 = \frac{y_1 - y_0}{2} \qquad C_1 = \frac{y_1 - y_0}{2}$$

or

$$\int_{y_2}^{y_1} f(y)\,dy = \frac{y_1 - y_0}{2}[f(y_0) + f(y_1)] + E(f)$$

This is the simple trapezoidal rule of integration. We approximate the curve of $f(y)$ between points 0 and 1 by a straight line connecting the two points. Now if we examine Fig. 12.5 it is clear that when we divide $f(y)$ into sufficiently small intervals, the area under $f(y)$ can be approximated quite closely by the series of trapezoids formed by drawing straight lines between points on the curve $f(y)$. [We assume $f(y)$ to be reasonably well behaved.]

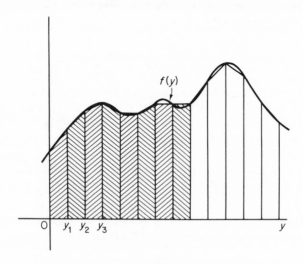

Figure 12.5 Approximating a function $f(y)$ by a series of straight lines (trapezoidal integration).

From our previous discussion of the Lagrange interpolation formula, we see that a large number of other possible quadrature formulas exist, depending on how we pick the interpolation points, the degree of the polynomial selected, and so on. In most work with digital computers it becomes convenient to select equal step sizes; that is, $y_0, y_1, \ldots, y_n$ are spaced equally. Thus, we let $y_1 = y_0 + h, y_2 = y_1 + h$, and so on, where h is the step size, to simplify the programming effort.

The points used to determine the approximating polynomial constitute some interval of the function. If the region of integration is the same as this interval, we obtain a *closed type* of quadrature formula. If it does not, and we integrate outside the interval (extrapolate), we obtain an *open type* of quadrature formula. Some closed-type formulas follow.

For two points, using a linear interpolation, the quadrature formula becomes

$$\int_{y_0}^{y_1} f(y)\,dy = \frac{h}{2}(f_0 + f_1) - \frac{h^3}{12}f''(\xi) \tag{12.18}$$

where $f_0 = f(y_0), f_1 = f(y_1)$, and $f''(\xi)$ is the second derivative at $y = \xi$, an unknown point in the interval of integration. This is the same as the trapezoidal rule derived earlier. For three points, using a quadratic polynomial approximation, we obtain

$$\int_{y_0}^{y_2} f(y)\, dy = \frac{h}{3}(f_0 + 4f_1 + f_2) - E(f) \tag{12.19}$$

while for four points and a cubic approximation

$$\int_{y_0}^{y_3} f(y)\, dx = \frac{3h}{8}(f_0 + 3f_1 + 3f_2 + f_3) - E(f) \tag{12.20}$$

Equations (12.18) and (12.19) are some of the formulas known as *Simpson's integration rules*. The error term for Eq. (12.19) is

$$E(f) = \frac{h^5}{90} f^{(4)}(\xi) \tag{12.21}$$

and for Eq. (12.20) is

$$E(f) = \frac{3h^5}{80} f^{(4)}(\xi) \tag{12.22}$$

where $f^{(4)}$ represents the fourth derivative of $f(y)$, and $y = \xi$ is an unknown point in the interval of integration. Simpson's rules become quite accurate as the interval h is decreased. Reducing the interval h results in more computer roundoff error and greater solution time, so we cannot decrease h indefinitely. Higher-order closed formulas can also be derived using the methods we have described.

Open-type quadrature formulas use points interior to the interval of integration, and allow us to extrapolate the integration to regions outside of the known points.

EXAMPLE 12.5

Show that

$$\int_{y_1}^{y_2} f(y)dy = \frac{h}{2}(-f_0 + 3f_1) + E(f)$$

where

$$h = y_1 - y_0 = y_2 - y_1$$

Solution.

$$L_0 = \frac{y - y_1}{y_0 - y_1} \qquad L_1 = \frac{y - y_0}{y_1 - y_0}$$

Then from Eq. (12.17)

$$C_0 = \frac{1}{y_0 - y_1}\left(\frac{y^2}{2} - yy_1\right)\Big|_{y_1}^{y_2} \qquad C_1 = \frac{1}{y_1 - y_0}\left(\frac{y^2}{2} - yy_0\right)\Big|_{y_1}^{y_2}$$

$$C_0 = -\frac{1}{2h}(y_2^2 - 2y_2 y_1 + y_1^2) \qquad C_1 = \frac{1}{2h}(y_2^2 - 2y_2 y_0 - y_1^2 + 2y_1 y_0)$$

$$C_0 = -\frac{1}{2h}(y_2 - y_1)^2 \qquad C_1 = \frac{1}{2h}[(y_2 - y_0)^2 - (y_1 - y_0)^2]$$

$$C_0 = -\frac{h}{2} \qquad C_1 = \frac{1}{2h}[4h^2 - h^2]$$

$$= \frac{3h}{2}$$

Hence

$$\int_{y_1}^{y_2} f(y)\,dy = \frac{h}{2}(-f_0 + 3f_1) + E(f) \tag{12.23}$$

Equation (12.23) has use as a "predictor" in predictor–corrector methods for solving ODE. Another open-type formula is

$$\int_{y_0}^{y_1} f(y)\,dy = hf_0 + \frac{h^2}{2} f'(\xi) \tag{12.24}$$

The equation

$$\int_{y_3}^{y_4} f(y)\,dy = \frac{h}{24}(-9f_0 + 37f_1 - 59f_2 + 55f_3) + \frac{251h^5}{720} f^{(4)}(\xi) \tag{12.25}$$

is a fourth-order open-integration formula used in the Adams predictor–corrector integration method.

12.5 Predictor-Corrector Integration Methods

The integration methods presented in this section are based upon the quadrature formulas obtained by the Lagrange-function approximations discussed in the preceding section.

We begin by considering the ordinary differential equation, Eq. (12.1), or

$$\dot{x} = f(x, t) \qquad x(t_0) = x_0$$

We will assume a unique solution exists on the interval $[t_0 \le t \le b]$ and is sufficiently differentiable to allow us to develop the following integration methods. Let $t_i = t_0 + ih, i = 1, 2, \ldots, m$. Then from Eq. (12.1) we obtain

$$\int_{t_i}^{t_{i+1}} \dot{x}(t)\,dt = \int_{t_i}^{t_{i+1}} f(x(t), t)\,dt$$

The left-hand is simply

$$x(t_{i+1}) - x(t_i) = \int_{t_i}^{t_{i+1}} f(x(t), t)\,dt \tag{12.26}$$

We now replace the integral on the right-hand side of Eq. (12.26) with a quadrature formula $Q(f(x), t_i, t_{i+1})$ and its associated error term, respectively, on the interval $[t_i, t_{i+1}]$. Then Eq. (12.26) can be written as

$$x(t_{i+1}) = x(t_i) + Q(f, t_i, T_{i+1}) + E(f, t_i, t_{i+1}) \tag{12.27}$$

Initially, we will demonstrate by using Eq. (12.24),

$$\int_{t_i}^{t_{i+1}} g(t)\,dt = hg(t_i) + \frac{h^2}{2} f'(\xi_i)$$

where $t_i < \xi_i < t_{i+1}$. The result is $x(t_{i+1}) = x(t_i) + hf(x(t_i), t_i) + (h^2/2)f'(\xi_i)$. You notice that this reduces to Euler's method if the error term $(h^2/2)f'(\xi_i)$ is truncated. Thus,

$$x_{i+1} = x_i + hf(x_i, t_i) \tag{12.28}$$

We thus predict the value $x(t_{i+1})$ from a given value at t_i.

To continue, we try another quadrature formula such as the trapezoidal rule in Eq. (12.18), or

$$\int_{t_i}^{t_{i+1}} g(t)\, dt = \frac{h}{2}\left[g(t_i) + g(t_{i+1})\right] - \frac{h^3}{12} g''(\xi_i)$$

Using this formula in Eq. (12.27), we have

$$x_{i+1} = x_i + \frac{h}{2}\left[f(x(t_i), t_i) + f(x(t_{i+1}), t_{i+1})\right] - \frac{h^3}{12} f^{(3)}(\xi_i) \tag{12.29}$$

Proceeding as before, we drop the remainder to get

$$x_{i+1} = x_i + \frac{h}{2}\left[f(x_i, t_i) + f(x_{i+1}, t_{i+1})\right] \tag{12.30}$$

Equation (12.30) introduces the dilemma of having x_{i+1} appearing on both sides of the formula. However, we might consider the possibility of solving Eq. (12.30) for x_{i+1} by successive substitution. Under suitable conditions, which exist in most network problems, x_{i+1} can by computed by fixed-point iteration, as illustrated subsequently. (See the Additional Reading for further information.)

The solution can then be advanced from t_i to t_{i+1} by using a predictor (extrapolation) to approximate the solution and a corrector (interpolation) to perform one or more fixed-point iterations to improve the approximation. The number of iterations to be performed can be varied to obtain a reasonably accurate solution for x_{i+1}.

EXAMPLE 12.6

Given the differential equation $\dot{x} = x + t$, $x(0) = 1$. Find solutions for $x(0.1)$ and $x(0.2)$ using a predictor–corrector scheme discussed above with $h = 0.1$.

Solution. One predictor formula (Euler's method) is

$$x_1^0 = x_0 + hf(x_0, t_0)$$

Hence,

$$x_1^0 = x_0 + h(x_0 + t_0)$$
$$= 1 + 0.1(1 + 0) = 1.1$$

We refine the approximation using the trapezoidal rule as follows:

$$x_1^{(J+1)} = x_0 + \frac{h}{2}\left[f(x_0, t_0) + f(x_1^{(J)}, t_1)\right]$$

where the superscript refers to the iteration number.

Hence

$$x_1^{(1)} = 1 + 0.05[1 + (1.1 + 0.1)] = 1.11$$

$$x_1^{(2)} = 1 + 0.05[1 + (1.11 + 0.1)] = 1.1105$$

$$x_1^{(3)} = 1 + 0.05(1 + 1.2105) = 1.110525$$

$$x_1^{(4)} = 1 + 0.05(1 + 1.210525) = 1.110526$$

$$x_1 = 1.110526$$

The exact solution in this case at $t = 0.1$ is 1.110342. Continuing, $x_2^{(0)}$ using the same predictor is

$$x_2^{(0)} = x_1 + hf(x_1, t_1)$$

$$= 1.110526 + 0.1(1.110526 + 0.1) = 1.231579$$

The trapezoidal-rule correction formula for x_2 is

$$x_2^{(j+1)} = x_1 + \frac{h}{2}[f(x_1, t_1) + f(x_2^{(j)}, t_2)]$$

Then

$$x_2^{(1)} = 1.110526 + 0.05(1.210526 + 1.431579) = 1.242631$$

$$x_2^{(2)} = 1.110526 + 0.05(1.210526 + 1.442631) = 1.243184$$

$$x_2^{(3)} = 1.110526 + 0.05(1.210526 + 1.443184) = 1.243212$$

$$x_2^{(4)} = 1.110526 + 0.05(1.210526 + 1.443212) = 1.243213$$

$$x_2 = 1.243213.$$

The exact solution is $x(0.2) = 1.242806$.

The algorithm described is a second-order corrector method with iteration. The results obtained are not extremely accurate. The significant advantages of the predictor–corrector methods are not too apparent in the lower-order methods. However, the results obtained are more accurate than those obtained directly from Euler's method.

We can easily extend the above procedure to higher-order methods of integration utilizing higher-order polynomial approximations. In general, the predictor formula will be of the form

$$x_{i+1} = x_i + a_i f(x_i, t_i) + \cdots + a_{i-n} f(x_{i-n}, t_{i-n}) \tag{12.31}$$

where all x_j, $i - n, \leq j \leq i$, are known values. The corrector formula will be of the form

$$x_{i+1} = x_i + b_{i+1} f(x_{i+1}, t_{i+1}) + b_i f(x_i, t_i) + \cdots$$
$$+ a_{i-m} f(x_{i-m}, t_{i-m}) \tag{12.32}$$

The subscripts n and m are related to the order of the predictor and the corrector, respectively. For example, $n = 3$ and $m = 2$ in the fourth-order Adams method. Equation (12.32) may be solved for x_{i+1} by fixed-point iteration. One of the most common forms of the predictor–corrector integration approach is the fourth-order Adams method. The Adams method is not self-starting, in that it requires the initial value x_0 and approximate values for three additional points, x_1, x_2, and x_3, before the integration process can begin. Usually, the starting values are obtained by using another integration technique, such as the fourth-order Runge-Kutta algorithm described later in this chapter.

Adams Method

Let the integration step h, the upper integration limit m, and the ordinary differential equation

$$\dot{x} = f(x, t) \qquad x(t_0) = x_0$$
$$x(t_1) = x_1$$
$$x(t_2) = x_2$$
$$x(t_3) = x_3$$

be given. The function values x_1, x_2, and x_3 are obtained using a one-step integration method. The values of x_{i+1} for $3 \leq i \leq m - 1$ are computed using

$$
\begin{aligned}
x_{i+1}^{(0)} = x_i + \frac{h}{24} [-9f(x_{i-3}, t_{i-3}) + 37f(x_{i-2}, t_{i-2}) \\
- 59f(x_{i-1}, x_{i-1}) + 55f(x_i, t_i)]
\end{aligned}
\tag{12.33}
$$

as a predictor, and

$$
\begin{aligned}
x_{i+1}^{(j+1)} = x_i + \frac{h}{24} [f(x_{i-2}, t_{i-2}) - 5f(x_{i-1}, t_{i-1}) \\
+ 19f(x_i, t_i) + 9f(x_{i+1}^{(j)}, t_{i+1})] \qquad j = 0, 1, \ldots
\end{aligned}
\tag{12.34}
$$

iteratively as a corrector. The final value of x_{i+1} is $x_{i+1}^{(k)}$, where k is specified or where $x_{i+1}^{(j)}, j = 0, 1, \ldots$, satisfies a specified convergence criteria.

Since the Adams method uses a fourth-order predictor, the difference between the predicted and actual values of x_{i+1} is quite small assuming that a reasonable step size h is chosen. The value obtained for x_{i+1} by the corrector is needed. In many cases, the fourth-order predictor–corrector needs only two evaluations of $f(x, t)$ per integration step compared to four evaluations of other fourth-order techniques.

We conclude this section by extending the Adams integration algorithm to a system of n simultaneous first-order ordinary differential equations. Let

$$\dot{\mathbf{x}} = f(\mathbf{x}, t) \qquad \mathbf{x}(t_0) = \mathbf{x}_0 \tag{12.35}$$

describe the system

$$
\begin{aligned}
\dot{x}_1 &= f_1(x_1, x_2, \ldots, x_n, t) & x_1(t_0) &= x_1, 0 \\
\dot{x}_2 &= f_2(x_1, x_2, \ldots, x_n, t) & x_2(t_0) &= x_2, 0 \\
& \cdot \quad \cdot \quad \cdot \quad \cdot \quad \cdot \quad \cdot \quad \cdot \quad \cdot \quad \cdot \quad \cdot \\
\dot{x}_n &= f_n(x_1, x_2, \ldots, x_n, t) & x_n(t_0) &= x_n, 0
\end{aligned}
$$

Then the predictor is given by

$$\mathbf{x}_{i+1}^{(0)} = \mathbf{x}_i + \frac{h}{24}[-9f(\mathbf{x}_{i-3}, t_{i-3}) + 37f(\mathbf{x}_{i-2}, t_{i-2})$$
$$- 59f(\mathbf{x}_{i-1}, t_{i-1}) + 55f(\mathbf{x}_i, t_i)] \tag{12.36}$$

The corrector is given by

$$\mathbf{x}_{i+1}^{(j+1)} = \mathbf{x}_i + \frac{h}{24}[f(\mathbf{x}_{i-2}, t_{i-2}) - 5f(\mathbf{x}_{i-1}, t_{i-1})$$
$$+ 19f(\mathbf{x}_i, t_i) + 9f(\mathbf{x}_{i+1}^{(j)}, t_{i+1})] \qquad j = 0, 1, \ldots \tag{12.37}$$

We can easily write a Fortran subroutine subprogram PCINT for Adams algorithm for a system of ordinary differential equations. We must supply an output subroutine identical to the OUTPUT routine required in Fig. 12.3 and a function subprogram FUN(K, X, T) defined as follows:

$$\text{FUN}(1, X, T) = f_1(x_1, x_2, \ldots, x_n, t)$$
$$\text{FUN}(2, X, T) = f_2(x_1, x_2, \ldots, x_n, t)$$
$$\cdot \quad \cdot \quad \cdot \quad \cdot \quad \cdot \quad \cdot \quad \cdot \quad \cdot \quad \cdot \quad \cdot \quad \cdot$$
$$\text{FUN}(N, X, T) = f_n(x_1, x_2, \ldots, x_n, t)$$

All starting values must have been previously computed and available to the subroutine PCINT. The arguments supplied to the subroutine are N (the number of equations in the system), TO (the initial point), STEP (step size), ENDPT (the end point of the integration), ITER8 (the number of the iterations to be used by the corrector), X [the values of $x_{1,3}, x_{2,3}, x_{3,3}, \ldots, x_{n,3}$, the numerical values of $x_1(t_3), x_2(t_3), \ldots, x_n(t_3)$], and a matrix F [containing in its columns the vectors $f(\mathbf{x}_0, t_0)$, $f(\mathbf{x}_1, t_1)$, $f(\mathbf{x}_2, t_2)$, and $f(\mathbf{x}_3, t_3)$]. The subroutine is given in Fig. 12.6.

12.6   Runge-Kutta Methods

There are many Runge-Kutta formulas, but we will limit our discussion to the second- and fourth-order methods. Again we are given the differential equation

$$\dot{x} = \frac{dx}{dt} = f(x, t) \qquad x(t_0) = x_0 \tag{12.38}$$

where $f(x, t)$ is any reasonably well-behaved function of x and t and $x = x(t)$. If the value of the function x at some time $t = t_m$ is x_m, then we might estimate the value at t_{m+1}, where $(t_{m+1} - t_m) = h$, a time interval, by using the slope at t_m given by Eq. (12.38). As we saw in Section 12.2, this results in the Euler formula

$$x_{m+1} = x_m + hf(x_m, t_m) \tag{12.39}$$

which is a first-order Runge-Kutta method.

```
C      ADAM'S 4TH-ORDER PREDICTOR-CORRECTOR INTEGRATION METHOD       ADAM   1
       SUBROUTINE PCINT(N,TO,STEP,ENDPT,ITER8,Y,F)                   ADAM   2
C                                                                    ADAM   3
C      PURPOSE                                                       ADAM   4
C      TO SOLVE A SYSTEM OF FIRST-ORDER ODE WITH GIVEN               ADAM   5
C         INITIAL CONDITIONS                                         ADAM   6
C      DEFINITION OF VARIABLES                                       ADAM   7
C         ENDPT - FINAL VALUE OF INDEPENDENT VARIABLE                ADAM   8
C         F - NX4 INPUT MATRIX OF STARTING FUNCTION VALUES FOR       ADAM   9
C             THE INTEGRATION ROUTINE.  FOUR INITIAL PCINTS MUST     ADAM  10
C                THE INTEGRATION ROUTINE.  FOUR INITIAL POINTS       ADAM  11
C                MUST BE KNOWN TO START ADAM'S METHOD.  MATRIX USED   ADAM  12
C             BE KNOWN TO START ADAM'S METHOD.  MATRIX USED LATER    ADAM  13
C             TO UPDATE F(Y,T) VALUES AT INTERMEDIATE POINTS         ADAM  14
C         FUN - AN EXTERNAL FUNCTICN SUBPROGRAM TO SUPPLY DERIVATIVE ADAM  15
C               INFORMATION.  THE ARGUEMENTS ARE I (EQUATION NO.),   ADAM  16
C               Y1 ( Y(I+1) ), AND TIME                              ADAM  17
C         ITER8 - NO. OF ITERATIONS IN CORRECTOR PHASE OF            ADAM  18
C                 INTEGRATION STEP                                   ADAM  19
C         N - NO. OF EQUATIONS IN SYSTEM                             ADAM  20
C         OUTPUT - AN EXTERNAL SUBROUTINE SUBPROGRAM TO OUTPUT RESULTS ADAM 21
C         STEP - INTEGRATION STEP SIZE H                             ADAM  22
C         TO - INITIAL VALUE OF INDEPENDENT VARIABLE                 ADAM  23
C         TIME - PRESENT VALUE OF INDEPENDENT VARIABLE               ADAM  24
C         Y - NTH-ORDER INPUT VECTOR OF INITAL VALUES OF DEPENDENT   ADAM  25
C             VARIABLE.  LATER Y IS THE RESULTING VECTOR OF VALUES   ADAM  26
C             COMPUTED AT INTERMEDIATE POINTS.                       ADAM  27
C         Y1 - NTH-ORDER VECTOR CONTAINING NEXT VALUES OF            ADAM  28
C              DEPENDENT VARIABLE ( Y(I+1) )                         ADAM  29
C      NOTE ... VARIABLES F AND Y MUST BE DIMENSIONED IN CALLING PROGRAM ADAM 30
C                                                                    ADAM  31
       DIMENSION F(10,4),Y(10),Y1(10)                               ADAM  32
C      INITIALIZE ROUTINE PARAMETERS                                 ADAM  33
       TIME=TO+3.*STEP                                               ADAM  34
       S=STEP/24.0                                                   ADAM  35
C      INTEGRATION LOOP                                              ADAM  36
   10 TIME=TIME+STEP                                                 ADAM  37
C      PREDICT VALUE FOR Y(I+1)                                      ADAM  38
       DO 20 I=1,N                                                   ADAM  39
       Y1(I)=Y(I)+S*(-9.*F(I,1)+37.*F(I,2)-59.*F(I,3)+55.*F(I,4))    ADAM  40
C      SHIFT F VALUES FOR CORRECTOR STEP                             ADAM  41
       F(I,1)=F(I,2)                                                 ADAM  42
       F(I,2)=F(I,3)                                                 ADAM  43
   20 F(I,3)=F(I,4)                                                  ADAM  44
C      CORRECT PREDICTED VALUE OF Y(I+1)                             ADAM  45
       J=0                                                           ADAM  46
   30 DO 40 I=1,N                                                    ADAM  47
   40 F(I,4)=FUN(I,Y1,TIME)                                          ADAM  48
       IF (J.GE.ITER8) GO TO 60                                      ADAM  49
       DO 50 I=1,N                                                   ADAM  50
   50 Y1(I)=Y(I)+S*(F(I,1)-5.*F(I,2)+19.*F(I,3)+9.*F(I,4))           ADAM  51
       J=J+1                                                         ADAM  52
       GO TO 30                                                      ADAM  53
C      UPDATE FUNCTION VALUE                                         ADAM  54
   60 DO 70 I=1,N                                                    ADAM  55
   70 Y(I)=Y1(I)                                                     ADAM  56
C      OUTPUT X-Y VARIABLES FOR LATEST INTEGRATION STEP             ADAM  57
       CALL OUTPUT(Y,TIME,N)                                         ADAM  58
       IF (TIME.LT.ENDPT) GO TO 10                                   ADAM  59
       RETURN                                                        ADAM  60
       END                                                           ADAM  61
```

Figure 12.6 PCINT—fourth-order predictor-
corrector integration subroutine.

Since Euler's method is inaccurate, we might do better by using a slope which averages $\dot{x}_m$ and $\dot{x}_{m+1}$. Thus, we compute a temporary x_{m+1} as before, but we now find a new slope $\dot{x}_{m+1}$ from Eq. (12.38). We next average the slopes $\dot{x}_m$ and $\dot{x}_{m+1}$. Using this average slope, we recompute x_{m+1}. The average slope is

$$\dot{x}_{ma} = \tfrac{1}{2}[f(x_m, t_m) + f(x_m + h\dot{x}_m, t_m + h)] \tag{12.40}$$

Then

$$x_{m+1} = x_m + h\dot{x}_{ma} \tag{12.41}$$

We can now get some idea as to the accuracy of Eq. (12.41) by comparing it with a Taylor-series expression of the function. An expansion of $f(x, t)$ about the point (x_m, t_m) gives

$$f(x, t) = f(x_m, t_m) + \frac{\partial f}{\partial t}(t - t_m) + \frac{\partial f}{\partial x}(x - x_m) + \cdots \tag{12.42}$$

where $\partial f/\partial x$ and $\partial f/\partial t$ are evaluated at (x_m, t_m). Let $x = x_m + h\dot{x}_m$ and $t = t_m + h$. Since $\dot{x}_m = f(x_m, t_m)$, we get

$$f(x_m + h\dot{x}_m, t_m + h) = f + hf_t + hff_x + O(h^2) \tag{12.43}$$

where $f = f(x_m, t_m)$, $f_t = \partial f/\partial t$, and $f_x = \partial f/\partial x$, all evaluated at (x_m, t_m), and $O(h^2)$ means to the order of h^2. Substituting Eqs. (12.43) into Eq. (12.42), we obtain

$$\dot{x}_m = f + \frac{h}{2}(f_t + ff_x) + O(h^2) \tag{12.44}$$

which substituted in Eq. (12.41) results in the equation

$$x_{m+1} = x_m + hf + \frac{h^2}{2}(f_t + ff_x) + O(h_3) \tag{12.45}$$

The Taylor-series expansion of $x(t)$ at a point x_m is

$$x(t) = x_m + \dot{x}_m(t - t_m) + \tfrac{1}{2}\ddot{x}_m(t - t_m)^2 + \cdots \tag{12.46}$$

Assume that somehow we have a (approximate) solution for times $t_0, t_1, \ldots, t_m$, where $t_m = t_0 + mh$, m being an integer. Then the next point $x(t_{m+1}) = x_{m+1}$ may be estimated by substituting t_{m+1} for t in Eq. (12.46), giving

$$x_{m+1} = x_m + h\dot{x}_m + \frac{h^2}{2}\ddot{x}_m + \frac{h^3}{6}x_m^{(3)} + \cdots \tag{12.47}$$

From Eq. (12.38)

$$\dot{x}_m = f(x_m, t_m)$$

and differentiating with respect to t,

$$\ddot{x} = \frac{\partial f(x, t)}{\partial t} + f(x, t)\frac{\partial f(x, t)}{\partial x}$$

or

$$\ddot{x}_m = f_t + f f_x$$

Substituting this in Eq. (12.47), we get

$$x_{m+1} = x_m + hf + \frac{h^2}{2}(f_t + f f_x) + O(h^3) \qquad (12.48)$$

Thus, Eq. (12.44) agrees with a Taylor-series expansion through terms in h^2, and hence we can say that Eq. (12.41) is accurate to $O(h^3)$ in this sense. The Taylor-series expansion of Eq. (12.48) requires three function evaluations, f, f_x, and f_t, but Eq. (12.41) requires only two function evaluations, $f(x_m, t_m)$ and $f(x_m + h\dot{x}_m, t_m + h)$. Thus, Eq. (12.41) is computationally faster.

Rather than averaging slopes as we did in obtaining Eq. (12.41), we might develop an alternate basis, as indicated by Fig. 12.7. In this figure we propose proceeding

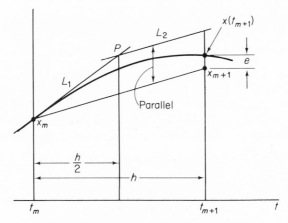

Figure 12.7 Average slopes in a second-order Runge-Kutta method.

along the line L_1 at the known slope $\dot{x}_m$ of the function at x_m to a point P halfway in the interval, and calculating the slope at this point, giving the line L_2 as in Fig. 12.7. Now we return to x_m and calculate x_{m+1} using this last slope.

Hence, let the slope at the point P be $\dot{x}_{m+h/2}$; then

$$\dot{x}_{m+h/2} = f(x_m + (h/2)\dot{x}_m, t_{m+h/2}) \qquad (12.49)$$

where $\dot{x}_m = f(x_m, t_m)$. Then

$$x_{m+1} = x_m + h\dot{x}_{m+h/2} \qquad (12.50)$$

Both Eq. (12.41) and (12.50) agree with the Taylor series through terms in h^2 and are known as second-order Runge-Kutta methods.

EXAMPLE 12.7

Use the Runge-Kutta method of Eq. (12.50) to solve the equation of Example 12.6, that is, $\dot{x} = x + t$, $x(0) = 1$ with $h = 0.1$.

Solution

$$x_1 = x_0 + h\dot{x}_{h/2}$$
$$= 1.0 + 0.1[f(x_0 + (h/2)\dot{x}_0, 0.05]$$
$$= 1.0 + 0.1[f(x_0 + 0.05(1.0 + 0)), 0.05]$$
$$= 1.0 + 0.1(1.0 + 0.05 + 0.05)$$
$$= 1.0 + 0.1(1.10) = 1.110$$
$$x_2 = x_1 + h\dot{x}_{1 + h/2}$$
$$= 1.110 + 0.1[f(x_1 + (h/2)\dot{x}_1), 0.1 + 0.05]$$
$$= 1.110 + 0.1[f(x_1 + 0.05(1.11)), 0.15]$$
$$= 1.110 + 0.1(1.110 + 0.0555 + 0.15)$$
$$= 1.110 + 0.1(1.3155)$$
$$= 1.2416$$

We may compare these answers with the results of Example 12.6.

If you review the previous alternatives, you see that there are a number of possible ways to approximate x_{m+1} given x_m and $\dot{x}_m$. Continuing with similar but more sophisticated averaging methods, there are an infinite number of Runge-Kutta formulas of higher order. Perhaps the most used formula consists of the one-third rule, which is of fourth order, as follows:

$$x_{m+1} = x_m + \Delta x \tag{12.51}$$
$$\Delta x = \tfrac{1}{6}(a_1 + 2a_2 + 2a_3 + a_4)$$

and

$$a_1 = f(x, t)h$$
$$a_2 = f(x + \tfrac{1}{2}a_1, t + \tfrac{1}{2}h)h$$
$$a_3 = f(x + \tfrac{1}{2}a_2, t + \tfrac{1}{2}h)h$$
$$a_4 = f(x + a_3, t + h)h$$

The derivation follows the lines previously explored and may be found in Kunz.† The error (comparing to the Taylor-series expansion) is to $O(h^5)$, and hence the formula is quite accurate. Interestingly, if f is a function of t only, then Eq. (12.51) is analogous to Simpson's one-third rule.

---

† W. S. Kunz, *Numerical Analysis*, McGraw-Hill Book Company, New York, 1957, pp. 167–189.

EXAMPLE 12.8

Solve $\dot{x} = x + t$, $x(0) = 1$ using Eq. (12.51) and $h = 0.1$.

Solution. To find x_1:

$$a_1 = (0 + 1)0.1 = 0.1$$
$$a_2 = (0.05 + 1.05)0.1 = 0.11$$
$$a_3 = (0.05 + 1.055)0.1 = 0.1105$$
$$a_4 = (0.1 + 1.1105)0.1 = 0.12105$$
$$\Delta x = \tfrac{1}{6}(0.1 + 0.22 + 0.2210 + 0.12105)$$
$$= 0.11034$$
$$x_1 = 1.11034$$

To find x_2:

$$a_1 = (0.1 + 1.11034)0.1 = 0.121034$$
$$a_2 = (0.15 + 1.17086)0.1 = 0.132086$$
$$a_3 = (0.15 + 1.17638)0.1 = 0.132638$$
$$a_4 = (0.2 + 1.24298)0.1 = 0.144298$$
$$\Delta x = \tfrac{1}{6}(0.121034 + 0.264172 + 0.265276 + 0.144298)$$
$$= 0.132463$$
$$x_2 = 1.24280$$

The Runge-Kutta formula, Eq. (12.51), is slow since four function evaluations must be made for each step. Moreover, no test exists to determine the optimum size of h as the computation proceeds. One technique might be to repeat the solution with varying h, which might improve the confidence in the results. Another would be to integrate forward to a desired point and then integrate back. The difference between the initial condition and the final condition after integrating back represents some error measure. Both methods are time consuming. Decreasing h beyond a certain point (not known in advance) does not improve accuracy, owing to larger computer roundoff errors. Since the fourth-order Runge-Kutta method is slow, we may use it to find starting values for the faster predictor–corrector methods previously discussed. For example, with the Adams method described in Section 12.4 we require four values, including the initial value $\mathbf{x}(0)$. Three additional values may be calculated with the Runge-Kutta method, and then we may apply the Adams predictor–corrector technique.

The extension from a single equation to a series of equations follows the same lines as followed in Sections 12.2 and 12.4. We give in Fig. 12.8 a Fortran subroutine RUNGA utilizing Eq. (12.51) for a set of first-order ODE. The user must supply an output routine identical to the one required in Fig. 12.3 and a function subprogram FUN(K, X, T) which computes the functions $f_1(x_1, x_2, \ldots, x_n, t), f_2(x_1, x_2, \ldots, x_n, t),$ $\ldots, f_n(x_1, x_2, \ldots, x_n, t)$. This subprogram is the same as that required by subroutine PCINT, as previously described. RUNGA may be used to integrate over the entire range, or to supply the first four function values to PCINT, which will then complete

```
      SUBROUTINE RUNGA(N,X,TO,ENDT,ITER8,H,KP)                  RUNG   1
C     PURPOSE                                                   RUNG   2
C     TO SOLVE A SYSTEM OF FIRST-ORDER ODE WITH GIVEN           RUNG   3
C        INITIAL CONDITIONS.  THE SUBROUTINE IS BASED UPON THE  RUNG   4
C        FOURTH-ORDER RUNGA-KUTTA INTEGRATION ALGORITHM, BUT    RUNG   5
C        CAN BE USED AS A DRIVING ROUTINE FOR THE PREDICTOR-    RUNG   6
C        CORRECTOR ROUTINE PCINT BY SETTING THE FLAG KP=2.      RUNG   7
C     DEFINITION OF VARIABLES                                   RUNG   8
C        DELX(K) - X(J+1) = X(J) + DELX FOR ALL X(K),K=1,N      RUNG   9
C        ENDT - FINAL VALUE OF INDEPENDENT VARIABLE             RUNG  10
C        F - NX4 INPUT MATRIX OF STARTING VALUES OF PCINT       RUNG  11
C        F1 - NX4 MATRIX STORING FUNCTION EVALUATIONS FOR RUNGA RUNG  12
C        H - INTEGRATION STEP SIZE                              RUNG  13
C        ITER8 - NO. OF ITERATIONS IN CORRECTOR PHASE OF        RUNG  14
C                INTEGRATION STEP IN PCINT.  (INPUT ZERO IF PCINT RUNG 15
C                IS NOT USED.)                                  RUNG  16
C        J - ITERATION COUNT                                    RUNG  17
C        KP - =1 USE ONLY RUNGA FOR INTEGRATION.                RUNG  18
C             =2 USE RUNGA TO SUPPLY THE STARTING VALUES FOR    RUNG  19
C                THE PREDICTOR-CORRECTOR ROUTINE (PCINT).       RUNG  20
C        N - NO. OF EQUATIONS IN SYSTEM                         RUNG  21
C        OUTPUT - AN EXTERNAL SUBROUTINE SUBPROGRAM TO OUTPUT RESULTS RUNG 22
C        T - TEMPORARY VALUE OF INDEPENDENT VARIABLE            RUNG  23
C        TO - INITIAL TIME                                      RUNG  24
C        T1 - TO+J*H WHERE J IS THE ITERATION                   RUNG  25
C        X - NTH-ORDER INPUT VECTOR OF INITIAL VALUES OF DEPENDENT RUNG 26
C            VARIABLE.  LATER X IS THE RESULTING VECTOR OF VALUES RUNG 27
C            COMPUTED AT INTERMEDIATE POINTS.                   RUNG  28
C        Z - TEMPORARY VALUE FOR X                              RUNG  29
C                                                               RUNG  30
      DIMENSION X(10),F(10,4),Z(10),F1(10,4),DELX(10)           RUNG  31
      J=0                                                       RUNG  32
      T1=TO                                                     RUNG  33
C     OUTPUT INITIAL CONDITIONS                                 RUNG  34
      CALL OUTPUT(X,T1)                                         RUNG  35
      DO 10 K=1,N                                               RUNG  36
   10 F(K,1)=FUN(K,X,T1)                                        RUNG  37
C     START ITERATION                                           RUNG  38
   20 J=J+1                                                     RUNG  39
C     TEST FOR COMPLETION                                       RUNG  40
      IF (T1.LT.ENDT) GO TO 30                                  RUNG  41
      RETURN                                                    RUNG  42
   30 DO 40 I=1,N                                               RUNG  43
   40 Z(I)=X(I)                                                 RUNG  44
      T=T1                                                      RUNG  45
C     EVALUATE F1(K,1) FOR RUNGA                                RUNG  46
      DO 50 K=1,N                                               RUNG  47
   50 F1(K,1)=FUN(K,Z,T)                                        RUNG  48
      T=T+0.5*H                                                 RUNG  49
C     FIND TEMPORARY X(K) FOR RUNGA                             RUNG  50
      DO 60 K=1,N                                               RUNG  51
   60 Z(K)=X(K)+0.5*F1(K,1)*H                                   RUNG  52
C     EVALUATE F1(K,2) FOR RUNGA                                RUNG  53
      DO 70 K=1,N                                               RUNG  54
   70 F1(K,2)=FUN(K,Z,T)                                        RUNG  55
C     FIND TEMPORARY X(K) FOR RUNGA                             RUNG  56
      DO 80 K=1,N                                               RUNG  57
   80 Z(K)=X(K)+0.5*F1(K,2)*H                                   RUNG  58
C     EVALUATE F1(K,3) FOR RUNGA                                RUNG  59
      DO 90 K=1,N                                               RUNG  60
   90 F1(K,3)=FUN(K,Z,T)                                        RUNG  61
      T=T+0.5*H                                                 RUNG  62
C     FIND TEMPORARY X(K) FOR RUNGA                             RUNG  63
      DO 100 K=1,N                                              RUNG  64
  100 Z(K)=X(K)+F1(K,3)*H                                       RUNG  65
C     EVALUATE F1(K,4) FOR RUNGA                                RUNG  66
      DO 110 K=1,N                                              RUNG  67
```

Figure 12.8 Runge-Kutta integration sub-routine.

```
  110 F1(K,4)=FUN(K,Z,T)                                    RUNG  68
C         APPLY RUNGA-KUTTA FORMULA                         RUNG  69
          DO 120 K=1,N                                      RUNG  70
          DELX(K)=(F1(K,1)+2.0*(F1(K,2)+F1(K,3))+F1(K,4))*H/6.0  RUNG 71
C         INCREMENT X(I) = X(I)+DELX(I)                     RUNG  72
  120 X(K)=X(K)+DELX(K)                                     RUNG  73
C         INCREMENT INDEPENDENT VARIABLE                    RUNG  74
          T1=T1+H                                           RUNG  75
C         OUTPUT RESULTS AT ITERATION J                     RUNG  76
          CALL OUTPUT(X,T)                                  RUNG  77
C         RETURN TO RUNGA                                   RUNG  78
          IF (J.GT.3) GO TO 20                              RUNG  79
          DO 130 K=1,N                                      RUNG  80
  130 F(K,J+1)=FUN(K,X,T1)                                  RUNG  81
C         TEST FOR TRANSFER TO PCINT                        RUNG  82
          IF (J.EQ.3) GO TO (20,140),KP                     RUNG  83
          GO TO 20                                          RUNG  84
  140 CALL PCINT(N,TO,H,ENDT,ITER8,X,F)                     RUNG  85
          RETURN                                            RUNG  86
          END                                               RUNG  87
```

Figure 12.8—*Cont.*

the integration. The arguments to be supplied to RUNGA are N, the number of variables, TO (initial time), H (step size), ENDT (final time), and KP. If KP = 1, RUNGA is used for the entire range, while if KP = 2, a transfer is made to PCINT after three iterations. If PCINT is to be used, it must be supplied, and a value of ITER8 $\neq$ 0 must be supplied with RUNGA to be used by PCINT. If PCINT is not to be used, subroutine PCINT must either still be supplied, or as a minimum, the statements SUBROUTINE PCINT and END must be supplied. [See Fig. 12.8. The initial values of x (X(I), I = 1, N) must be supplied in either case.] The step size H is transferred to PCINT as STEP. By adding a statement just before the transfer, STEP may be changed to a value different than H.

12.7 Time-Domain Solutions

To complete this chapter, we demonstrate the application of state-variable techniques and numerical integration in obtaining the time-domain response of networks to arbitrary inputs. As an example of the analysis procedure, we consider the capacitance voltage $v_c(t)$ for $t \geq 0$ in the series RLC network shown in Fig. 12.9 to a unit-step-input voltage applied at $t = 0$.

Rather than selecting the state equations by graph-theory methods, we see by inspection that two possible state variables are

$$x_1 = v_c$$

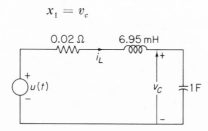

Figure 12.9 *RLC* network for time-domain solution example.

and

$$x_2 = i_L$$
$$= C\dot{x}_1$$

Then

$$\dot{x}_1 = \frac{1}{C} x_2$$

$$\dot{x}_2 = \frac{E}{L} - \frac{R}{L} x_2 - \frac{x_1}{L}$$

The state equations in matrix form are

$$
\begin{bmatrix} \dot{x}_1 \\ \dot{x}_2 \end{bmatrix}
=
\begin{bmatrix} 0 & \dfrac{1}{C} \\ -\dfrac{1}{L} & -\dfrac{R}{L} \end{bmatrix}
\begin{bmatrix} x_1 \\ x_2 \end{bmatrix}
+
\begin{bmatrix} 0 \\ \dfrac{1}{L} \end{bmatrix} E
\tag{12.52a}
$$

$$
=
\begin{bmatrix} 0 & 1 \\ -144 & -7.2 \end{bmatrix}
\begin{bmatrix} x_1 \\ x_2 \end{bmatrix}
+
\begin{bmatrix} 0 \\ 144 \end{bmatrix} E
$$

The output equation is

$$
y = \begin{bmatrix} 1 & 0 \end{bmatrix}
\begin{bmatrix} x_1 \\ x_2 \end{bmatrix}
\tag{12.52b}
$$

Equations (12.52) are in the proper form to be supplied to an integration routine such as those given in Figs. 12.3, 12.6, and 12.8. The functions for the EULINT, PCINT, or RUNGA integration routines are evaluated by the function subprogram FUN shown in Fig. 12.10. The integration routine evaluates the state variables x_1 and x_2 but does not print or plot the output variable, which, in this case, is simply x_1. The output equation is evaluated in the subroutine OUTPUT.

```
      FUNCTION FUN(K,X,T)                         FUN    1
      DIMENSION X(10)                             FUN    2
      GO TO (10,20),K                             FUN    3
   10 FUN=X(2)                                    FUN    4
      RETURN                                      FUN    5
   20 FUN=-144.*X(1)-7.2*X(2)+144.                FUN    6
      RETURN                                      FUN    7
      END                                         FUN    8
```

Figure 12.10 Function subprogram for PCINT
or RUNGA network of Fig. 12.9.

Figure 12.11 shows the response obtained using the integration routine RUNGA and PCINT (KP = 2) with a step size H of 0.05 s. The solution may also be plotted in the state-space coordinates x_2 versus x_1 (i_L versus v_c), with time as a parameter, as in Fig. 12.12. The solution in this form is known as a state-space trajectory, and provides considerable insight into network operation, particularly in nonlinear cases.

The value of ITER8 used with PCINT in obtaining the data for Figs. 12.11 and 12.12 was 2, or there were two iterations of the corrector on each step.

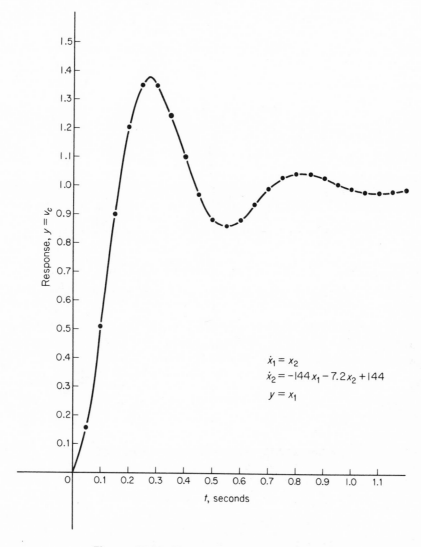

$$\dot{x}_1 = x_2$$
$$\dot{x}_2 = -144\, x_1 - 7.2\, x_2 + 144$$
$$y = x_1$$

Figure 12.11 Time response of network in
in Fig. 12.9.

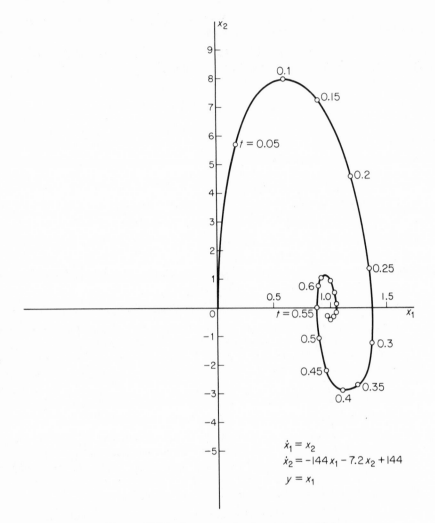

Figure 12.12 Response of network in Fig. 12.9
plotted in state-space form.

Additional Reading

HAMMING, R. W., *Numerical Methods for Scientists and Engineers*, McGraw-Hill Book Company, New York, 1962.

HENRICI, P., *Elements of Numerical Analysis*, John Wiley & Sons, Inc., New York, 1964.

KUNZ, K. S., *Numerical Analysis*, McGraw-Hill Book Company, New York, 1957.

McCRACKEN, D. D., and W. S. DORN, *Numerical Methods and Fortran Programming*, John Wiley & Sons, Inc., New York, 1964.

MOURSUND, D. G., and C. S. DURIS, *Elementary Theory and Application of Numerical Analysis*, McGraw-Hill Book Company, New York, 1967.

Problems

12.1. Given $\dot{x} = 0.05x$, $x(0) = 1000$.

   (a) Compute $x(1)$ with Euler's method with $h = 0.05$ and $h = 0.2$.

   (b) Compare the results of part (a) with the exact result $x = 1000e^{0.05t}$.

12.2. Repeat Problem 12.1 for the equation $\dot{x} = 0.1x$, $x(0) = 1000$. The exact result is $x = 1000e^{0.1t}$.

12.3. Compute the percentage error in Problem 12.1 (using the exact values) of $x(1)$ and $x(10)$ using $h = 0.1$ and $h = 0.2$. Compare.

12.4. Repeat Problem 12.3 using the fourth-order Runge-Kutta formula of Eq. (13.51).

12.5. A rocket with an initial mass of 500 kg has a constant thrust of 10,000 newtons. The propellant mass decreases at a rate of 10 kg/s for 30 seconds until the propellant is exhausted. If the rocket is at rest at $t = 0$ and there are no other forces on it, find the velocity at $t = 30$. The acceleration equation is $\dot{v} = 10,000/(500 - 10t)$, for $0 \leq t \leq 30$. Compare the results obtained by the Euler method with the Runge-Kutta method for $h = 5$ and $h = 1$. Compare the computed results with the exact solution $v = 1000 \log [500/(500 - 10t)]$.

12.6. If the rocket of Problem 12.5 encounters an air resistance of $5v$, the acceleration equation is $\dot{v} = (10,000 - 5v)/(500 - 10t)$. Find the velocity at $t = 30$ using the subroutine RUNGA to establish the first four points, and then the subroutine PCINT for the remaining points, with $h = 1$. Compare the computation time required with that required by the subroutine RUNGA.

12.7. Given $\dot{x} = 2t$, $x(0) = 0$. Show that since $f(x, t) = f(t) = 2t$ is independent of x, that in (Eq. 12.51) $a_1 = f(t)h$, $a_2 = f(t + h/2)h$, $a_3 = f(t + h/2)h$, and $a_4 = f(t + h)h$; therefore, $\Delta x = (h/6)[f(t) + 4f(t + h/2) + f(t + h)]$, and $x_{m+1} = x_m + (h/6)(f_m + 4f_{m+1/2} + f_{m+1})$, where $f_m = f(t_m)$. Using $f(t) = 2t$, show that we obtain the exact solution values as given by $x = t^2$.

12.8. Given the set of differential equations

$$\dot{x}_1 = x_2$$
$$\dot{x}_2 = -x_1$$
$$x_1(0) = 1 \qquad x_2(0) = 0$$

Solve these using subroutine RUNGA with $h = 0.1$ from $t = 0$ to $t = \pi/2$. Taking the derivative of the first equation above and substituting in the second equation we obtain the ODE $\ddot{x} + x = 0$. Hence, the solution is $x = \sin t$ with the given initial conditions. Plot $x_2(t)$ versus $x_1(t)$ for various times in a rectangular coordinate system.

12.9. Given the ODE $\dddot{x} + 4\ddot{x} + 3\dot{x} + 49x = 49u(t)$, let $y_1 = x$, $y_2 = \dot{x}$, and $y_3 = \ddot{x}$. Then for $t > 0$ we may write the equations

$$\dot{y}_1 = y_2$$
$$\dot{y}_2 = y_3$$
$$\dot{y}_3 = -49y_1 - 3y_2 - 4y_3 + 49$$
$$x = y_1$$

Show that this set of simultaneous equations is indeed equivalent to the original third-order ODE. Solve this with all initial conditions zero using subroutines RUNGA and PCINT with an appropriate step size for $0 \leq t \leq 2$.

12.10. Solve the ODE

$$\dot{x} = t + |x^{1/2}| \qquad x(0) = 1$$

in the range $0 \leq t \leq 1$ with $h = 0.1$ using any appropriate method.

12.11. The ODE $\ddot{x} = (1 + t^2)x$ may be represented by the set of equations

$$\dot{y}_1 = y_2$$
$$\dot{y}_2 = (1 + t^2)y_1$$
$$x = y_1$$

Let $x(0) = y_1(0) = 0$, $\dot{x}(0) = y_2(0) = 1$. Using different step sizes, solve for the range $0 \leq t \leq 2$.

12.12. Using a small step size h in a numerical integration scheme involves a large number of mathematical operations, and as a consequence, a large number of roundoff errors. The truncation error in this case should be negligible compared to the round-off error. In order to investigate this effect, write programs that use the Euler, Adams, and Runge-Kutta integration formulas to compute $y(t)$ from the ODE

$$\dot{y} = 2t \quad y(0) = 0 \qquad \text{for } 0 \leq t \leq 1$$

The exact solution is $y = t^2$. Use $h = 0.5, 0.2, 0.1, 0.01, 0.001,$ and 0.0001. Plot the error and the execution time as a function of h and comment on the resulting plots.

12.13. Assuming zero initial conditions, use Euler's method to solve for the currents in each of the networks shown in Fig. P12.1.

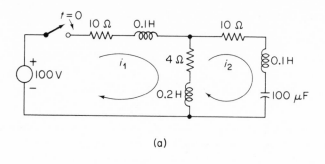

(a)

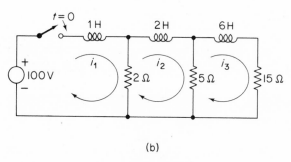

(b)

Figure P12.1

12.14. Write a Fortran subroutine PCINT2 which is based upon the predictor formula given in Eq. (12.28) and the corrector formula given in Eq. (12.30) capable of solving a system of equations. ($N \leq 10$.)

12.15. Modify subroutine RUNGA to drive the predictor–corrector integration routine PCINT2 developed in Problem 12.14.

12.16. Write a Fortran subroutine RUNGA2 based upon a second-order Runge-Kutta integration formula. (This formula is sometimes called a modified Euler formula.) The subroutine is not to be used as a driver for another integration routine. Use the subroutine to solve Problem 12.4. Compare the results obtained with the fourth-order formula.

APPENDICES

*Nothing is more conducive to
arriving nowhere, than to be going
nowhere. That's the spot where,
with no guide posts, and little urge,
one arrives with utmost certainty.*

(AUTHOR UNKNOWN)

Appendix A

COMPLEX VARIABLES

Complex numbers probably first arose in the solution of quadratic polynomials. Thus, given

$$az^2 + bz + c = 0 \qquad \text{(A.1)}$$

the quadratic formula prescribes that

$$z = \frac{-b \pm \sqrt{b^2 - 4ac}}{2a} \qquad \text{(A.2)}$$

If $b^2 - 4ac < 0$, then we may write Eq. (A.2) as

$$z = \frac{-b \pm j\sqrt{4ac - b^2}}{2a} \qquad \text{(A.3)}$$

where $j = \sqrt{-1}$ and is called the *imaginary unit*. From Eq. (A.3),

$$z_1 = \frac{-b + j\sqrt{4ac - b^2}}{2a}$$

$$z_2 = \frac{-b - j\sqrt{4ac - b^2}}{2a}$$

or

$$z_1 = x + jy \qquad \text{(A.4a)}$$

$$z_2 = x - jy \qquad \text{(A.4b)}$$

The quantities z_1 and z_2 are called *complex numbers* if x and y are real numbers. Since by definition, if $z = x + jy$, and the complex conjugate z^* is the number

$$z^* = x - jy \qquad \text{(A.5)}$$

then z_1 and z_2 in Eqs. (A.4) are complex conjugates.

Any polynomial in z can be formed of factors of the form given by the left-hand side of Eq. (A.1), and factors of the form $(az + b)$. Since polynomials continually arise in mathematical work, complex numbers become very important in engineering.

By definition, the real part of a complex number $z = x + jy$, designated $\Re(z)$, is

$$\Re(z) = \Re(x + jy)$$
$$= x \tag{A.6}$$

and the imaginary part, designated $\mathcal{I}(z)$, is

$$\mathcal{I}(z) = \mathcal{I}(x + jy)$$
$$= y \tag{A.7}$$

Complex numbers may be represented as a point in a complex plane with a real axis x and a perpendicular imaginary axis y, as in Fig. A.1. In this figure we attach the quantity j to the imaginary axis to emphasize the difference between the two axes, although by our previous definitions this is not required. From Fig. A.1,

$$z = x + jy$$

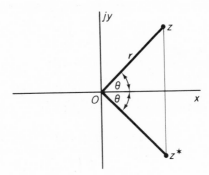

Figure A.1 Representation of a complex number.

or

$$z = r(\cos \theta + j \sin \theta) \tag{A.8}$$

Equation (A.8) defines the polar form of z. From Fig. A.1,

$$r = \sqrt{x^2 + y^2}$$
$$= |z| \tag{A.9}$$

where $|z|$ is called the *magnitude* of z. Also from Fig. A.1,

$$\theta = \tan^{-1} \frac{y}{x} \tag{A.10}$$

where θ is called the *argument* of z.

Complex numbers may be added, subtracted, multiplied, and divided. The sum of two complex numbers is the complex number whose real part is the sum of their real components and whose imaginary part is the sum of their imaginary components.

Thus, if

$$z_1 = x_1 + jy_1$$

and

$$z_2 = x_2 + jy_2$$

then

$$z_1 + z_2 = (x_1 + x_2) + j(y_1 + y_2) \tag{A.11}$$

Using Eq. (A.11) and Fig. A.2 it is clear that complex numbers may be considered as vectors in the complex plane, and that they are added by adding components in a similar way to adding real vectors in a two-dimensional space.

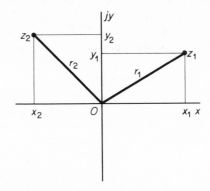

Figure A.2 Two complex numbers shown in the complex plane.

The product of z_1 and z_2 is defined as

$$
\begin{aligned}
z_1 z_2 &= (x_1 + jy_1)(x_2 + jy_2) \\
&= (x_1 x_2 - y_1 y_2) + j(x_1 y_2 + x_2 y_1)
\end{aligned} \tag{A.12}
$$

Since $j^2 = -1$, the product is found by multiplying the two factors in the usual way. Division is the inverse of multiplication, or

$$z_3 = \frac{z_1}{z_2}$$

implies that

$$z_2 z_3 = z_1 \tag{A.13}$$

Thus,

$$
\begin{aligned}
z_3 &= \frac{z_1}{z_2} = \frac{x_1 + jy_1}{x_2 + jy_2} \\
&= \frac{(x_1 + jy_1)(x_2 - jy_2)}{(x_2 + jy_2)(x_2 - jy_2)}
\end{aligned}
$$

or

$$z_3 = \frac{(x_1 x_2 + y_1 y_2) + j(x_2 y_1 - x_1 y_2)}{x_2^2 + y_2^2} \tag{A.14}$$

Substitution of Eq. (A.14) in Eq. (A.13) satisfies the later equation.

By using the previous definition, you may easily show the following relations:

$$|z| = zz^* \tag{A.15}$$

$$|z_1 z_2| = |z_1||z_2| \tag{A.16}$$

$$\left|\frac{z_1}{z_2}\right| = \frac{|z_1|}{|z_2|} \tag{A.17}$$

$$|z_1 + z_2 + z_3| \leq |z_1| + |z_2| + |z_3| \tag{A.18}$$

By expanding $e^{j\theta}$, $\cos\theta$, and $\sin\theta$ in a Taylor series, we may show that

$$e^{j\theta} = \cos\theta + j\sin\theta \tag{A.19a}$$

and

$$e^{-j\theta} = \cos\theta - j\sin\theta \tag{A.19b}$$

or

$$\cos\theta = \frac{e^{j\theta} + e^{-j\theta}}{2} \tag{A.20a}$$

$$\sin\theta = \frac{e^{j\theta} - e^{-j\theta}}{2j} \tag{A.20b}$$

Thus, the polar form of (A.18) may be represented as

$$z = |z|e^{j\theta} \tag{A.21}$$

where $|z| = \sqrt{x^2 + y^2}$ and $\tan\theta = y/x$. The form (A.21) is convenient when multiplying and dividing, since, following the laws of exponents,

$$z_1 z_2 = (|z_1|e^{j\theta_1})(|z_2|e^{j\theta_2})$$
$$= |z_1||z_2|e^{j(\theta_1 + \theta_2)}$$
$$= |z_1||z_2|[\cos(\theta_1 + \theta_2) + j\sin(\theta_1 + \theta_2)]$$

and

$$\frac{z_1}{z_2} = \frac{|z_1|e^{j\theta_1}}{|z_2|e^{j\theta_2}}$$
$$= \frac{|z_1|}{|z_2|}e^{j(\theta_1 - \theta_2)}$$
$$= \frac{|z_1|}{|z_2|}[\cos(\theta_1 - \theta_2) + j(\sin\theta_1 - \theta_2)]$$

If several sequential multiplications and divisions are to be performed, the polar form becomes much better from a computational standpoint than the rectangular forms of Eqs. (A.12) and (A.14), since we only need to divide and multiply magnitudes and add and subtract arguments. This becomes even more evident in finding powers, since obviously

$$z^n = |z|^n e^{jn\theta}$$
$$= |z|^n(\cos n\theta + j\sin n\theta) \tag{A.22}$$

and

$$z^{-n} = \frac{1}{|z|^n} \left[\cos\left(\frac{\theta + 2k\pi}{n}\right) + j \sin\left(\frac{\theta + 2k\pi}{n}\right) \right] \qquad \text{(A.23)}$$

where

$$k = 0, 1, 2, \ldots, n - 1$$

The multiplicity of solutions in Eq. (A.23) comes about because $n\theta = \theta + 2k\pi$, where k is any integer. Thus,

$$(1)^{1/3} = e^{j0°}, e^{j120°}, e^{-j120°}$$

This equation may be verified multiplying each exponent by 3, or rotating each vector in Fig. A.3 by three times its angle. Also, by examining this figure, we see that jz simply rotates z through an angle of 90°. By convention, positive angles are measured counterclockwise from the positive real axis ($+x$ in Fig. A.3).

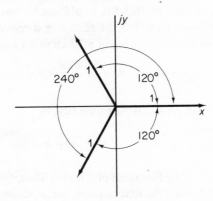

Figure A.3 Representation of the cube root of unity.

Example A.1

Given $z_1 = 3 + j4$ and $z_2 = 4 + j3$, find $z_1 + z_2, z_1 - z_2, z_1 z_2, z_1/z_2, |z_1|, |z_2|,$ arg z_1, arg z_2, z_1^5, $z_2^{1/2}$, and $z_1^* z_2$.

Solution

$$
\begin{array}{cc}
3 + j4 & 3 + j4 \\
4 + j3 & -4 - j3 \\
\hline
z_1 + z_2 = 7 + j7 \qquad & z_1 - z_2 = -1 + j1
\end{array}
$$

$$z_1 z_2 = (3 + j4)(4 + j3) = 12 - 12 + j25 = 0 + j25$$

$$|z_1| = \sqrt{3^2 + 4^2} = 5 \qquad |z_2| = \sqrt{4^2 + 3^2} = 5$$

$$\arg z_1 = \theta_1 = \tan^{-1}\tfrac{4}{3} = 53.2° \qquad \arg z_2 = \theta_2 = \tan^{-1}\tfrac{3}{4} = 36.8°$$

$$z_1 z_2 = |z_1||z_2| e^{j(53.2° + 36.8°)} = 25 e^{j90°} = 0 + j25$$

$$\frac{z_1}{z_2} = \frac{(3 + j4)(4 - j3)}{16 + 9} = \frac{(24 + j7)}{25} = 1 e^{j16.4°}$$

$$\frac{z_1}{z_2} = \frac{z_1}{z_2} e^{j(53.2° - 36.8°)} = 1 e^{+j16.4°} = 0.96 + j0.28$$

$$z_1^* = 3 - j4 = 5e^{-j53.2°}$$

$$z_1^* z_2 = 5e^{-j53.2°} 5e^{j36.8°} = 25e^{-j16.4°} = 24 - j7$$

$$z_1^5 = 5^5 e^{j5(53.2°)} = 5^5 e^{j266°} = 5^5 e^{-j94°}$$

$$z_2^{1/2} = \sqrt{5} \, e^{j36.8°/2} = \sqrt{5} \, e^{j18.4°}$$

$$= \sqrt{5} \, e^{j396.8°/2} = \sqrt{5} \, e^{j198.4°}$$

If z represents any point in the complex plane, then we term it a complex variable. Turning back to the factors of a polynomial with real coefficients, since the polynomial may be broken into factors of first and second order, but no higher, and since all roots, real or complex, thus may be represented in the complex plane, it seems reasonable that if the polynomial coefficients are random, most roots of a high-order polynomial must be complex (a much higher probability exists that a root will be off the real axis than it will be on the real axis). If the polynomial is odd, it will of course have at least one real root. For each complex root there must be a complex conjugate, as otherwise the polynomial coefficients would not be real.

If w is related to z such that for each z in a region of the complex plane there corresponds a value or set of values of w, then w is a function of z, or

$$w = f(z)$$

In general, w also will be a complex quantity, or

$$w = u + jv$$

where both u and v will be functions of x and y. Thus, to graph a complex function, we need two complex planes: the first to show the excursions of z and the second to show the corresponding excursions of w. In some cases, the plane showing the excursions of z is omitted, but the excursions are described or understood.

$w = f(z)$ is said to be analytic if $f(z)$ has a derivative at the point z. To make sense, this derivative must have the same value no matter in which direction in the complex z plane we make the change in z (to determine the derivative). This leads to the Cauchy–Riemann conditions for analyticity, which are

$$\frac{\partial u}{\partial x} = \frac{\partial v}{\partial y}$$

$$\frac{\partial u}{\partial y} = -\frac{\partial v}{\partial x}$$

(A.24)

where $z = x + jy$, and $w = f(z) = f(x + jy) = u + jv$.

Example A.2

See where $z^2 = (x + jy)^2$ is analytic.

$$z^2 = x^2 - y^2 + j2xy$$

or $u = x^2 - y^2$, $v = 2xy$,

$$\frac{\partial u}{\partial x} = 2x \qquad \frac{\partial v}{\partial y} = 2x$$

$$\frac{\partial u}{\partial y} = -2y \qquad \frac{\partial v}{\partial x} = 2y$$

Hence z^2 is analytic for all $z < \infty$.

If a function is analytic in the neighborhood of z_0, but not at z_0, then the function is said to have an isolated singularity at z_0. We will be dealing largely with functions consisting of the ratios of polynomials with real coefficients such as

$$f(z) = \frac{N(z)}{D(z)}$$

where $N(z)$ and $D(z)$ are polynomials in z. These functions are analytic at every point $z < \infty$ in the complex z plane except where $f(z) \to \infty$. This will occur when $D(z) = 0$, or when z approaches z_1, a root of $D(z)$. These singularities are called *poles*.

From Green's theorem, a theorem known as Cauchy's theorem may be shown to be true. This states that if $f(z)$ is analytic inside and on a closed contour c, then

$$\int_c f(z)\, dz = 0 \qquad (A.25)$$

where the symbol c indicates integration counterclockwise around the contour c. Using Cauchy's theorem, it may immediately be shown that if $f(z)$ is analytic on or within a closed contour c, and if z_0 is interior to c, then†

$$f(z_0) = \frac{1}{2\pi j} \int_c \frac{f(z)}{z - z_0}\, dz \qquad (A.26)$$

EXAMPLE A.3

Find $\int_c f(z)/(z - z_0)\, dz$, where the path is a circle $|z| = 2$, $f(z) = z/(9 - z^2)$ and $z_0 = j$.

Solution. Since $f(z)$ is analytic inside a circle of radius 3, we can apply Eq. (A.26). Thus,

$$\int_c \frac{f(z)}{z - z_0}\, dz = 2\pi j f(z_0)$$

$$= 2\pi j \frac{z}{9 - z^2}\bigg|_{z=z_0}$$

$$= 2\pi j \frac{j}{10} = -\frac{\pi}{5}$$

---

†R. V. Churchill, *Complex Variables*, McGraw-Hill Book Company, New York, 1948, pp. 90–95.

Since the theorem makes no restrictions on the contour shape, we would obtain the same result as long as the contour c remained in the region of analyticity.

By continued differentiation of Eq. (A.26),

$$\frac{d^n f(z_0)}{dz^n} = \frac{1}{2\pi j} \int_c^{\cdot} \frac{f(z)\,dz}{(z - z_0)^{n+1}} \tag{A.27}$$

We may now expand $f(z)$ in a series. If we expand near a singular point z_s, we obtain

$$f(z) = b_0 + b_1(z - z_s) + b_2(z - z_s)^2 + \cdots$$
$$+ b_{-1}(z - z_s)^{-1} + b_{-2}(z - z_s)^{-2} + \cdots$$

or

$$f(z) = \sum_{-\infty}^{\infty} b_n (z - z_s)^n \tag{A.28}$$

where

$$b_n = \frac{1}{2\pi j} \int_c \frac{f(z)\,dz}{(z - z_s)^{n+1}}$$

The closed contour c is taken counterclockwise. The region of validity for Eq. (A.28) is between circular contours c_1 and c_2, where c_2 is the largest circle around z_s not touching another singularity and c_1 is a small circle not touching z_s.

The series of Eq. (A.28) is called the *Laurent series* and the coefficient b_{-1} is called the *residue*. If $b_{-n} = 0$ for all values of $n > m$, then z_s is called a *pole of order m*. For ratios of polynomials, or $f(z) = N(z)/D(z)$, the series will always terminate $(b_{-n} = 0)$ on the low end. The order of a pole at z_s will be the order of repetition of the root of $D(z)$ at z_s.

EXAMPLE A.4

$$\frac{N(z)}{D(z)} = \frac{N(z)}{(z + 6)^3 (z + 3)^2 (z + 1)}$$

has a third-order pole at $z = -6$, a second-order pole at $z = -3$, and a simple (first-order) pole at $z = -1$. The Laurent expansion about $z = -6$ will terminate for all $n < -3$ with the term $b_{-3}/(z + 6)^3$.

EXAMPLE A.5

Expand e^z/z^2 about $z = 0$.

Solution

$$\frac{e^z}{z^2} = \frac{1}{z^2}\left(1 + \frac{z^2}{2!} + \frac{z^3}{3!} + \cdots\right)$$
$$= \frac{1}{z^2} + \frac{1}{z} + \frac{1}{2!} + \frac{z}{3!} + \cdots$$

We note from Example A.5 that since the series is unique, if we can find a valid series by any means, it will be the proper series.

We now get to the point of our contour-integration discussion. We see from Eq.

(A.28) that for a closed contour c around a single isolated pole,

$$b_{-1} = \frac{1}{2\pi j} \int_c f(z)\, dz \qquad (A.29)$$

Thus, the integral on the right-hand side of Eq. (A.29) is known if we know the residue b_{-1}.

EXAMPLE A.6

Find

$$\frac{1}{2\pi j} \int_c \frac{\cos z}{z^3}\, dz$$

where c encloses the pole at the origin.

Solution. Since

$$\frac{\cos z}{z^3} = \frac{1}{z^3} - \frac{1}{2z} + \frac{1}{4!}z - \frac{1}{6!}z^3 + \cdots$$

then

$$\frac{1}{2\pi j} \int_c \frac{\cos z}{z^3}\, dz = -\frac{1}{2}$$

We then state the residue theorem as follows: If c is a closed curve within and on which $f(z)$ is analytic except for a finite number of singular points $z_1, z_2, \ldots, z_k, \ldots, z_n$ inside of c, then

$$\int_c f(z)\, dz = 2\pi j(b_{-11} + b_{-12} + \cdots + b_{-1k} + \cdots + b_{-1n})$$

where b_{-1k} represents the residue at the kth pole inside c. The integration is taken counterclockwise.

Let us now define $\phi(z)$ as

$$\phi(z) = (z - z_s)^m f(z) \qquad (A.30)$$

where m is the order of an isolated pole at $z = z_s$. Since

$$f(z) = \sum_{n=0}^{\infty} b_n(z - z_s)^n + \frac{b_{-1}}{z - z_s} + \frac{b_{-2}}{z - z_s} + \cdots + \frac{b_{-m}}{z - z_s}$$

then

$$\phi(z) = b_{-1}(z - z_s)^{m-1} + b_{-2}(z - z_s)^{m-2} + \cdots + b_{-m} + \sum_{n=0}^{\infty} b_n(z - z_s)^n \qquad (A.31)$$

is analytic at $z = z_s$. Thus,

$$b_{-m} = \lim_{z \to z_s} \phi(z)$$

or

$$b_{-m} = \lim_{z \to z_s} (z - z_s)^m f(z) \qquad (A.32)$$

and from the Taylor series,

$$b_{-1} = \frac{\phi^{(m-1)}(z_s)}{(m-1)!}$$ (A.33)

where the superscript denotes the $(m-1)$th derivative and the argument indicates evaluation at $z = z_s$. If $m = 1$, obviously

$$b_{-1} = \lim_{z \to z_s} (z - z_s) f(z)$$

If

$$f(z) = \frac{p(z)}{q(z)}$$

then we may show that for a simple pole, b_{-1} is also given by

$$b_{-1} = \lim_{z \to z_s} \frac{p(z)}{q'(z)}$$ (A.34)

where $q'(z)$ is the first derivative of $q(z)$ with respect to z.

The inverse Laplace transform is defined as

$$\mathcal{L}^{-1}[F(s)] = \frac{1}{2\pi j} \int_{\sigma-\infty}^{\sigma+\infty} F(s) e^{st}\, dt$$ (A.35)

The contour of the integral is the straight line c_1 in the complex s-plane as shown in Fig. A.4. We may now evaluate the right-hand side of Eq. (A.35) by finding the residues of $F(s)e^{st}$ at all the poles of this function lying inside the contours c_1 and c_2 (c_2 is usually taken as a circle with infinite radius), and subtracting the integral along c_2. The value of σ is adjusted such that all poles are enclosed. For many functions (but

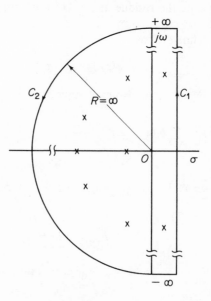

Figure A.4 Path of integration for the inverse Laplace transform.

not all), the contour integral along c_2 is zero, or

$$\frac{1}{2\pi j}\int_{c_2} F(s)e^{st}\, dt = 0 \tag{A.36}$$

In particular, if $\lim_{s\to\infty}|F(s)| \leq K|1/s|$, then Eq. (A.36) holds† and

$$\mathcal{L}^{-1}[F(s)] = \sum \text{ residues of poles of } F(s)e^{st} \tag{A.37}$$

An example of an exception to Eq. (A.37) occurs in using the inverse z transform, where $F(z)$ contains functions of the form e^{az}. For ratios of polynomials such as $F(s) = N(s)/D(s)$, Eq. (A.36) applies if $N(s)$ is lower in degree than $D(s)$. Applying Eq. (A.33), then

$$
\begin{aligned}
f(t) &= \mathcal{L}^{-1}[F(s)] \\
&= \sum_k \frac{1}{(n-1)!}\left\{\frac{d^{n-1}}{ds^{n-1}}\left[(s-s_k)^n F(s)e^{st}\right]\right\}_{s=s_k}
\end{aligned}
\tag{A.38}
$$

where there are k total poles, each of order n.

EXAMPLE A.7

Find the inverse transform of the function

$$
\begin{aligned}
F(s) &= \frac{7s+2}{s^3 + 2s^2 + 2s} \\
&= \frac{7s+2}{s(s+1)(s+2)}
\end{aligned}
\tag{A.39}
$$

Solution. The residue of $F(s)e^{st}$ at $s = 0$ is

$$
\begin{aligned}
R_0 &= \left[\frac{7s+2}{(s+1)(s+2)}\, e^{st}\right]_{s=0} \\
&= \frac{2}{2}\, e^0 = 1
\end{aligned}
$$

Similarly,

$$
\begin{aligned}
R_{-1} &= \left[\frac{7s+2}{s(s+2)}\, e^{st}\right]_{s=-1} \\
&= 5e^{-t} \\
R_{-2} &= \left[\frac{7s+2}{s(s+1)}\, e^{st}\right]_{s=-2} \\
&= -6e^{-2t}
\end{aligned}
$$

Thus,

$$
\begin{aligned}
f(t) &= \sum_{\text{all poles}} \text{ residues of } F(s)e^{st} \\
&= 1 + 5e^{-t} - 6e^{-2t}
\end{aligned}
\tag{A.40}
$$

---

†G. C. Newton, Jr., L. A. Gould, and J. F. Kaiser, *Analytic Design of Linear Feedback Controls*, John Wiley & Sons, Inc., New York, 1957, pp. 295–298.

EXAMPLE A.8

Consider the function

$$F(s) = \frac{1}{(s + 3)^4} \tag{A.41}$$

The inverse transform of this function by the residue method is

$$
\begin{aligned}
\mathcal{L}^{-1}[F(s)] &= \mathcal{L}^{-1}\left[\frac{1}{(s + 3)^4}\right] \\
&= \frac{1}{3!}\left[\frac{d^3}{ds^3} e^{st}\right]_{s=-3} \\
&= \frac{t^3}{3!} e^{-3t}
\end{aligned} \tag{A.42}
$$

The partial-fraction technique discussed in Chapter 8 produces the same results, but with a slightly different approach.

Appendix B

COMPUTER PROGRAMS

B.1 Program BODE

This program finds the magnitude and phase angle of $G(s) = KP(s)/Q(s)$ when $s = j\omega$ and ω varies. $Q(s)$ degree ND and $P(s)$ degree NN must be less than 20. The basis of the program is to find the gain and phase of each linear or quadratic factor of $P(s)$ and $Q(s)$ and to add these together properly. The starting ω value OMGI must be inputted by the user, and must not be zero. The program iterates by augmenting OMG1 at each iteration until it reaches the value OMGX. OMG1 may be augmented in one of two ways: (1) If the user inputs a value for OMGML, the previous value of OMG1 is multiplied by this factor; (2) if the user inputs a value for OMGAD $\neq 0$, the value OMGAD is added to OMG1 at each iteration (OMGML is ignored). The user may input $P(s)$ and $Q(s)$ either as (1) polynomial coefficients AQ(K) for the denominator $Q(s)$ in descending order, and polynomial coefficients AP(K) for the numerator $KP(s)$ in descending order, or (2) polynomial factors $AY(1)s^2 + AY(2)s + AY(3)$ for the denominator and $BY(1)s^2 + BY(2)s + BY(3)$ for the numerator where $s = j\omega$. The user must provide sufficient factors to correlate with the polynomial degrees ND and NN, respectively. If a linear factor $(s + z)$ is to be entered, $BY(1) = 0$, $BY(2) = 1$, $BY(3) = z$, etc. If the numerator $P(\omega) = 1$, then $BY(1) = BY(2) = 0$, $BY(3) = 1.0$. If quadratic factors are entered, the roots must be complex. That is, a quadratic with real roots must be factored into two linear factors and entered in this form. If the variable IN $= 1$, polynomial coefficients are entered; if IN $= 2$, polynomial factors are entered. K is entered as AK, and must have a value.

If polynomial coefficients are entered, $P(s)$ and $Q(s)$ are first factored by using subroutine POLYS.

The program outputs the polynomial coefficients in descending order, or the factors, as the case may be. It then outputs the current ω value OMG1 and the associated magnitude and phase angle at each iteration. The program may be repeated by making IDEX $\neq 0$ on the data card subsequent to the first data deck, and inputting new data, or terminated by making IDEX $= 0$.

The program description is as follows using the card number to the right (abbreviated C). C 63 inputs IDEX, ND, the order of $Q(s)$, NN, the order of $P(s)$, and IN. If IN $= 1$, polynomial coefficients are to be entered; if IN $= 2$, factors are to be entered. C 66 enters OMG1, the starting value of ω, OMGML, OMGAD, and OMGMX, the stopping value of ω. If OMGAD $= 0$ (no entry on the data card),

OMGML is to be used. C 67 enters the gain K in KP(s)/Q(s) as the program variable AK. The magnitude is expressed as $20 \log |G(j\omega)|$ in decibels. C71 starts this process. C 73 tests IN for the type of entry. If coefficients are to be entered, a branch is made to C 77, where the coefficients AQ(K), K = 1, ND + 1 are entered. C 77 reads AP(K), K = 1, NN + 1, while C 79–81 writes out the AQ(K) and AP(K). After C 82-85 finds the magnitude of AP(1)/AQ(1), C 86–88 finds and stores the roots of the denominator ZD(K) $\pm j$ ZID(K), K = 1, ND + 1 and ZN(K) $\pm j$ ZIN(K),K = 1, NN + 1 by using the subroutine POLYS. C 93–94 calculates the magnitude of all numerator factors (as GAIN) and adds to the preceding magnitude; C 95–96 calculates the magnitude of all denominator factors and subtracts from the preceding magnitude. C 99–106 similarly finds the phase angle, first for the numerator factors and then for the denominator factors. (C 97 writes out GAIN, the magnitude, in decibels.) C 107 writes out the phase angle, and C 109 tests for termination. If OMG1 is greater than OMGMX, the program goes back to C 63 to read new data; if it is not, C 112–114 augments OMG1 using OMGML or OMGAD, as required, and the program returns to C 89 to reiterate.

At C 73, if polynomial factors are to be entered, the program branches to C 118. After some initialization, C 127–128 reads and writes the first denominator factor AY(1)s^2 + AY(2)s + AY(3). This factor is analyzed to see if AY(1) or AY(2) = 0. If AY(1) = 0, the AY(2)s + AY(3) factor is linear, and C 129 sends the program to C 144, and if also AY(2) = 0, C 144 sends the program to C 172. At C 172 if AY(2) = 0, the root of the linear factor is calculated and stored as ZD(LD) + j0. If AY(1) $\neq$ 0 at C 129, then AY(1)s^2 + AY(2) + AY(3) is factored and the factors stored as ZD(LD) $\pm j$ZID(LD). The roots are assumed to be complex, or quadratics with real roots are entered as linear factors. If a quadratic with real roots is entered, it will fail at C 158. (Additional programming could provide for this eventuality.) A root count LD is kept and tested by C 142, and if LD is less than ND, the program returns to C 127 to read additional factors AY(1)s^2 + AY(2)s + AY(3): otherwise, the program goes to C 152, where the numerator factors BY(1)s^2 + BY(2)s + BY(3) are read in. We deal with the numerator factors in the same manner as those of the denominator, storing linear roots as ZN(LD) + j0, and complex roots as ZN(LD) + j ZIN(LD). Again quadratic factors are assumed to have complex roots. After all roots have been found and stored, the program goes to C 89, where the magnitude and phase angle are computed as before and iteration on ω occurs, as before.

```
C     PROGRAM BODE                                                   BODE   1
C     THIS PROGRAM FINDS THE MAGNITUDE AND PHASE OF G(OMEGA)=KP/Q AS OMEGA BODE  2
C     VARIES.Q DEGREE ND AND P DEGREE NN MUST BE LESS THAN 20. IF POLYNOM-BODE   3
C     IAL COEFFICIENTS ARE ENTERED,THE POLYNOMIALS ARE FACTORED BY SUB-  BODE   4
C     ROUTINE POLYS.                                                     BODE   5
C     IDEX : IF 0,RETURN.IF NOT 0 PROCEED.IDEX=0 ON LAST DATA CARD.      BODE   6
C     ND   : ORDER OF DENOMINATOR POLYNOMIAL Q(S).                       BODE   7
C     NN   : ORDER OF NUMERATOR PCLYNOMIAL P(S).                         BODE   8
C     IN   : IN=1 IF POYNOMIAL COEFFICIENTS TO BE ENTERED.IF N=2 POLYNOM- BODE  9
C            IAL FACTORS TO BE ENTERED.                                  BODE  10
C     OMG1 : ENTERED AS STARTING VALUE OF PARAMETER  .ALSO USED AS SUCEED-BODE 11
C            ING VALUES.                                                 BODE  12
C     OMGML: MULTIPLIES PREDEEDING VALUE OF OMG1  FOR NEW ITERATION.     BODE  13
```

Figure B.1   Program BODE.

```
C    OMGAD: ADDED TO PRECEEDING VALUE OF OMG1 FOR NEW ITERATION IF       BODE 14
C           OMGAD NOT =0.                                                BODE 15
C    OMGMX: MAXIMUM VALUE OF OMG1.                                       BODE 16
C    AK   : CONSTANT MULTIPLIER OF NUMERATOR P(S): (K).                  BODE 17
C    AQ(K): DENOMINATOR Q(S) COEFFICIENTS IN DESCENDING ORDER.           BODE 18
C    AP(K): NUMERATOR P(S) COEFFICIENTS IN DESCENDING ORDER.AT LEAST ONE BODE 19
C           AP(K) MUST HAVE A VALUE OTHER THAN ZERO                      BODE 20
C    AY(K): (AY(1)S*S+AY(2)S+AY(3))FACTOR OF DENOMINATOR Q(S). ANY ONE   BODE 21
·C           OR TWO AY MAY BE ZERO.                                      BODE 22
C    BY(K): (BY(1)S*S+BY(2)S+BY(3)) FACTOR OF NUMERATOR P(S).SEE ABOVE.  BODE 23
C           AT LEAST ONE BY(K) MUST HAVE A VALUE OTHER THAN ZERO         BODE 24
C    GAIN : MAGNITUDE IN DB.                                             BODE 25
C    PHI  : PHASE ANGLE IN DEGREES.                                      BODE 26
C    AK,BK: TEMPORARY STORAGE FOR GAIN.                                  BODE 27
C    THE PROGRAM OUTPUTS ENTERED PLYNOMIALS P(S),Q(S),OR FACTORS AY,BY,  BODE 28
C    PARAMETER OMG1,GAIN IN DB AND PHASE IN DEGREES FROM OMG1 INITIAL TO BODE 29
C    OMGMX.                                                              BODE 30
C    DATA CARD ORDER:                                                    BODE 31
C    CARD 1 FORMAT(4I3) IDEX,ND,NN,IN                                    BODE 32
C    CARD 2 FORMAT(4F10.2) OMG1,CMGML,OMGAD,OMGMX                        BODE 33
C    CARD 3 FORMAT (1E14.6) AK                                           BODE 34
C  IF IN=1                                                               BODE 35
C    CARD 4 ETC. FORMAT(5E14.6) AQ(K),K=1,ND+1                           BODE 36
C    NEXT CARDS FORMAT (5E14.6) AP(K),K=1,NN+1                           BODE 37
C  IF IN=2                                                               BODE 38
C    CARD 4 ETC. FORMAT(3E14.6) AY(K),K=1,3                              BODE 39
C           IF ALL THREE AY ARE ENTERED,ROOTS OF THIS FACTOR MUST BE COM-BODE 40
C           PLEX.REAL ROOTS SHOULD BE ENTERED WITH AY(1)=0.             BODE 41
C    NEXT CARDS FORMAT(3E14.6) BY(K),K=1,3. SEE NOTE ON AY ABOVE.        BODE 42
C    LAST CARD SAME AS CARD 1. IF IDEX NOT 0,REPEAT DATA DECK AS ABOVE.  BODE 43
C    DIMENSION AY(3),AQ(20),AP(20),ZD(20),ZN(20),ZIN(20),ZID(20),BY(3)   BODE 44
          REAL*8AQ,AP                                                    BODE 45
C    REMOVE THE FOLLOWING CARD IF WATFIV PROCESSOR NOT USED              BODE 46
          REAL*8DABS                                                     BODE 47
       10 FORMAT (4I3)                                                   BODE 48
       20 FORMAT (4F10.2)                                                BODE 49
       30 FORMAT (5E14.6)                                                BODE 50
       40 FORMAT (1H0'NEW PROBLEM')                                      BODE 51
       50 FORMAT (1H0'DENOMINATOR COEFFICIENTS IN DESCENDING ORDER')     BODE 52
       60 FORMAT (1H ,8E15.6)                                            BODE 53
       70 FORMAT (1H0,'NUMERATOR COEFFICIENTS IN DESCENDING ORDER')      BODE 54
       80 FORMAT (1H0,'DENOMINATOR FACTORS')                             BODE 55
       90 FORMAT (1H0,'NUMERATOR FACTORS')                               BODE 56
      100 FORMAT (1H ,'NUMERIC GAIN',1PE14.5)                            BODE 57
      110 FORMAT (1H ,E13.6,'S*S+',E13.6,'S+',E13.6)                     BODE 58
      120 FORMAT (1H0,'ANGULAR VELOCITY=',1PE12.3)                       BODE 59
      130 FORMAT (1H ,'GAIN IN DB=',1PE12.3)                             BODE 60
      140 FORMAT (1H+,30X,'PHASE ANGLE IN DEGREES=',1PE12.3)             BODE 61
      150 FORMAT (1H0,'ERROR;QUADRATIC HAS REAL ROOTS')                  BODE 62
      160 READ (5,10) IDEX,ND,NN,IN                                      BODE 63
          IF (IDEX.EQ.0) STOP                                           BODE 64
          WRITE (6,40)                                                   BODE 65
          READ (5,20) OMG1,OMGML,OMGAD,OMGMX                            BODE 66
          READ (5,30) AK                                                 BODE 67
          WRITE (6,100) AK                                               BODE 68
          MD=ND+1                                                        BODE 69
          MN=NN+1                                                        BODE 70
          CK=20.*ALOG10(AK)                                             BODE 71
C    BRANCH DEPENDING ON ENTERING POLYNOMIAL COEFFICIENTS OR FACTORS.   BODE 72
          GO TO (170,270),IN                                            BODE 73
C    POLYNOMIAL COEFFICIENTS ENTERED.                                    BODE 74
      170 READ (5,30) (AQ(K),K=1,MD)                                     BODE 75
C    IF NN=0, ENTER AP(1)=1.0                                           BODE 76
          READ (5,30) (AP(K),K=1,MN)                                     BODE 77
          WRITE (6,50)                                                   BODE 78
          WRITE (6,60) (AQ(K),K=1,MD)                                    BODE 79
          WRITE (6,70)                                                   BODE 80
          WRITE (6,60) (AP(K),K=1,MN)                                    BODE 81
          TEMP=DABS(AP(1)/AQ(1))                                         BODE 82
```

Figure B.1—*Cont.*

```
        BK=20.*ALOG10(TEMP)                                  BODE  83
        AK=BK+CK                                             BODE  84
C     USE POLYNOMIAL FACTORING SUBROUTINE.                   BODE  85
        CALL POLYS(ND,AQ,ZD,ZID,KC1,K11)                     BODE  86
        IF (NN.EQ.0) GO TO 180                               BODE  87
        CALL POLYS(NN,AP,ZN,ZIN,KC,K1)                       BODE  88
  180 WRITE (6,120) OMG1                                     BODE  89
C     FIND MAGNITUDE.                                        BODE  90
        GAIN=AK                                              BODE  91
        IF (NN.EQ.0) GO TO 200                               BODE  92
        DO 190 K=1,NN                                        BODE  93
  190 GAIN=GAIN+10.*ALOG10(ZN(K)*ZN(K)+(OMG1+ZIN(K))**2)     BODE  94
  200 DO 210 K=1,ND                                          BODE  95
  210 GAIN=GAIN-10.*ALOG10(ZD(K)*ZD(K)+(OMG1+ZID(K))**2)     BODE  96
        WRITE (6,130) GAIN                                   BODE  97
C     FIND PHASE.                                            BODE  98
        PHD=0.                                               BODE  99
        PHN=0.                                               BODE 100
        IF (NN.EQ.0) GO TO 230                               BODE 101
        DO 220 K=1,NN                                        BODE 102
  220 PHN=PHN+ATAN2(OMG1+ZIN(K),-ZN(K))                      BODE 103
  230 DO 240 K=1,ND                                          BODE 104
  240 PHD=PHD+ATAN2(OMG1+ZID(K),-ZD(K))                      BODE 105
        PHI=(180./3.14159)*(PHN-PHD)                         BODE 106
        WRITE (6,140) PHI                                    BODE 107
C     TEST FOR TERMINATION.                                  BODE 108
        IF (OMG1.GT.OMGMX) GO TO 160                         BODE 109
C     USE OMGAD IF NOT ZERO                                  BODE 110
        IF (OMGAD.NE.0.) GO TO 250                           BODE 111
        OMG1=OMG1*OMGML                                      BODE 112
        GO TO 260                                            BODE 113
  250 OMG1=OMG1+OMGAD                                        BODE 114
C     NEW ITERATION.                                         BODE 115
  260 GO TO 180                                              BODE 116
C     ENTER POLYNOMIAL FACTORS.                              BODE 117
  270 LD=0                                                   BODE 118
        LN=0                                                 BODE 119
        AK=CK                                                BODE 120
        DO 280 K=1,MD                                        BODE 121
  280 ZID(K)=0.                                              BODE 122
        DO 290 K=1,MN                                        BODE 123
  290 ZIN(K)=0.                                              BODE 124
        WRITE (6,80)                                         BODE 125
C     ENTER DENOMINATOR FACTORS.                             BODE 126
  300 READ (5,30) (AY(K),K=1,3)                              BODE 127
        WRITE (6,110) (AY(K),K=1,3)                          BODE 128
        IF (AY(1).EQ.0.) GO TO 320                           BODE 129
        AK=-20.*ALOG10(AY(1))+AK                             BODE 130
C   WARNING:QUADRATIC FACTOR ROOTS ARE ASSUMED TO BE COMPLEX.IF FACTORS  BODE 131
C   ARE REAL,THE QUADRATIC SHOULD BE FACTORED AND ENTERED AS REAL FACT-  BODE 132
C   ORS; THAT IS WITH AY(1)=0.                               BODE 133
        LD=LD+1                                              BODE 134
        ZD(LD)=-AY(2)/(2.0*AY(1))                            BODE 135
        DISC=4.0*AY(1)*AY(3)-AY(2)*AY(2)                     BODE 136
        IF (DISC.LT.0.) WRITE (6,150)                        BODE 137
        ZID(LD)=SQRT(DISC)/(2.0*AY(1))                       BODE 138
        LD=LD+1                                              BODE 139
        ZD(LD)=ZD(LD-1)                                      BODE 140
        ZID(LD)=-ZID(LD-1)                                   BODE 141
  310 IF (LD.LT.ND) GO TO 300                                BODE 142
        GO TO 330                                            BODE 143
  320 IF (AY(2).EQ.0.) GO TO 370                             BODE 144
        LD=LD+1                                              BODE 145
        AK=-20.*ALOG10(AY(2))+AK                             BODE 146
        ZD(LD)=-AY(3)/AY(2)                                  BODE 147
        GO TO 310                                            BODE 148
  330 WRITE (6,90)                                           BODE 149
C     ENTER NUMERATOR FACTORS. IF NN=0, ENTER BY(3)=1.0.     BODE 150
  340 READ (5,30) (BY(K),K=1,3)                              BODE 151
        WRITE (6,110) (BY(K),K=1,3)                          BODE 152
```

Handwritten annotations:

N=ND
DO 15 L=1,MD
15 A(L) = AQ(L)
ZD = ZA
ZID=ZM3
KC1= KC
K11=K1

ZA=ZD
ZID= ZID
KC1= KC

Figure B.1—*Cont.*

```
        IF (BY(1).EQ.0.) GO TO 360                          BODE 153
        AK=20.*ALOG10(BY(1))+AK                             BODE 154
        LN=LN+1                                             BODE 155
        ZN(LN)=-BY(2)/(2.0*BY(1))                           BODE 156
        DISC=4.*BY(1)*BY(3)-BY(2)*BY(2)                     BODE 157
        IF (DISC.LT.0.) WRITE (6,150)                       BODE 158
        ZIN(LN)=SQRT(DISC)/(2.0*AY(1))                      BODE 159
        LN=LN+1                                             BODE 160
        ZN(LN)=ZN(LN-1)                                     BODE 161
        ZIN(LN)=-ZIN(LN-1)                                  BODE 162
C   ENTER PREVIOUS ROUTINE.                                 BODE 163
  350 IF (LN.LT.NN) GO TO 340                               BODE 164
        GO TO 180                                           BODE 165
  360 IF (BY(2).EQ.0.) GO TO 380                            BODE 166
        LN=LN+1                                             BODE 167
        AK=20.0*ALOG10(BY(2))+AK                            BODE 168
        ZN(LN)=-BY(3)/BY(2)                                 BODE 169
        GO TO 350                                           BODE 170
  370 AK=AK-20.*ALOG10(AY(3))                               BODE 171
        GO TO 310                                           BODE 172
  380 AK=AK+20.*ALOG10(BY(3))                               BODE 173
        GO TO 350                                           BODE 174
        END                                                 BODE 175


C    DATA CARD EXAMPLE FOR BODE
  1   5   2   2
        1.00        2.00              100.00
    .110800E+04
                .100000E+01
                .100000E+01     .333000E+00
                .100000E+01     .667000E+00
                .100000E+01     .333000E+02
                .100000E+01     .200000E+03
                .100000E+01     .2000000E+03
                .100000E+01     .2000000Q+01
                .100000E+01     .2000000E+01
                .100000E+01     .667000E+01
  0
```

Card 1:

IDEX $\neq$ 0

ND, order of denominator = 5

NN, order of numerator = 2

IN = 2, or polynomial factors are to be entered.

Card 2:

OMG1—starting value of omega (ω) = 1.00

OMGML—multiplying factor for each iteration of omega = 2.00

OMGAD = 0, hence not considered.

OMGMX—maximum value of omega (ω) = 100.00

Card 3:

AK = 1108.0, or K = 1108 in $KP(j\omega)/Q(j\omega)$.

Cards 4, 5, 6, 7, 8:

The denominator factors of $Q(s)$ are as follows: $(s + 0.0)$, $(s + 0.333)$, $(s + 0.667)$, $(s + 33.3)$, $(s + 200)$

Cards 9, 10:

The numerator factors of $P(s)$ are $(s + 2.0)$, $(s + 6.67)$

Card 11:

IDEX = 0, terminate computations

Figure B.1—*Cont.*

B.2 Program LOCUS

This program finds the root locus for $G(s) = KP(s)/Q(s)$, where $P(s)$ and $Q(s)$ are polynomials in s and K is the parameter. The basis of the program is to form the polynomial $Q(s) + KP(s)$ and to find the roots of this polynomial using subroutine POLYS. $P(s)$ and $Q(s)$ may each be of any degree less than 20. The degree ND of the denominator $Q(s)$ and NN or the numerator $P(s)$ must be entered. The starting value of K must be inputted by the user as CON1. The program iterates with an augmented value of CON1 [K] until reaching the maximum value, which must be inputted as CONMX. CON1 [K] may be augmented in one of two ways: (1) If the user inputs a value for CONML, the value of CON1 is multiplied on each iteration by this factor; (2) if the user inputs a value for CONAL, CONML is ignored and CONAD is added to CON1 on each iteration. The user may enter the polynomials in either of two forms: (1) polynomial coefficients AQ(K) in descending order for $Q(s)$ and AP(K) similarly for $P(s)$, or (2) polynomial factors $AY(1)s^2 + AY(2)s + AY(3)$ for the denominator $Q(s)$ and $BY(1)s^2 + BY(2)s + BY(3)$ for the numerator $P(s)$. The user must provide sufficient factors for the denominator and numerator to correlate with the polynomial degrees ND and NN, respectively, previously entered. If a linear factor $(s + z)$ is to be entered, $BY(1) = 0$, $BY(2) = 1$, $BY(3) = z, \ldots$. If the numerator $P(s) = 1$, then $BY(1) = BY(2) = 0$, $BY(3) = 1.0$. If a factor $s + 0$ is to be entered in the denominator, $AY(1) = 0$, $AY(2) = 1$, $AY(3) = 0, \ldots$. If the variable $IN = 1$, P and Q are to be entered in coefficient form, while if $IN = 2$, P and Q are to be entered in factored form. The user must provide the appropriate data cards.

The program outputs the polynomial coefficients in descending order, or the factors, as the case may be. It then outputs the current values of CON1 and the associated roots as $Z + j$ZIM, $Z - j$ZIM for each iteration. The program may be repeated by making IDEX $\neq 0$ in the first card subsequent to the data deck and inputting new data or terminated by making IDEX $= 0$ on the last data card. CON1 [K], CONAD, and CONMX may be given negative values if desired.

The program description is as follows: card (abbreviated C) refers to the right-hand card number. C 55 reads in IDEX, ND, NN, IN. C 58 reads in CON1, CONML, CONAD, CONMX. If CONAD $\neq 0$, then this value is added at each iteration; otherwise CONML is used. Because $P(s)$ may be greater or lesser in degree than $Q(s)$, C 63–68 finds the maximum degree, whichever is greater, and sets JA $= $ ND $-$ NN. C 70 causes a branch depending on whether the user enters in coefficient or factored form. If $IN = 1$, polynomial coefficients are entered and a branch is made to C 72. C 72–78 reads and writes $Q(s)$ and $P(s)$ coefficients. If NN $= 0$, AP(1) may be entered as 1.0 (any constant is included in CON1), or as some other real number, reducing CON1 accordingly. C 80 multiplies the $P(s)$ coefficients by CON1, or gives $KP(s)$. C 82–96 shifts the lesser degree polynomial coefficients so that the two polynomials $KP(s)$ and $Q(s)$ may be added, with the coefficients properly aligned. C 97–99 adds the two polynomial coefficients to form the sum polynomial coefficients AD(K), K $= 1$,

NN + 1 or K = 1, ND + 1, whichever is the greater. C 101 applies the factoring subroutine POLYS, after which C 102–103 writes out the results and C 105 tests for termination by comparing CON1 with CONMX. On termination, the program returns to C 55. C 108–110 augments CON1 by adding CONAD if CONAD $\neq$ 0, or by multiplying by CONML if CONAD = 0. (CONAD = 0 is entered by leaving this data entry blank.) The program now returns to C 80 to reiterate.

If IN = 2 at C 70, a branch is made to C 115, where factors are to be entered. After some initializations, C 123–125 reads and writes the first denominator factor. It is now necessary to prepare for multiplying the next factor by the one just entered. To easily do this, the order of the factors is reversed by C 127–129. (The order is now ascending, not descending.) C 130 tests the highest-order term. If it is not zero (all terms present), a branch occurs to C 141 to read in the numerator factors. If it is zero, more terms are needed, and the second factor is read and written by C 131–132. C 135–137 multiplies the reversed polynomial (originally the first factor) by the incoming factor. C 140 tests for all factors in, and if not, the program branches back to C 131 to read a new factor. Iteration occurs until all denominator factors are read in, or we now have Q(s). The program now goes to C 142 to repeat the entire process for the numerator. Thus, C 142–144 reads and writes the first numerator factor (there must be at least one term). C 145–148 reverses the order, C 149 makes a completion test, C 150–151 reads and writes a new factor, C 153–154 multiplies the new factor by the reversed polynomial, C 155 returns the program to C 150 to read more factors or continues to C 163–164, where the numerator factors are multiplied by CON1, or we now have KP(s). C 166-167 forms KP(s) + Q(s). C 168–170 changes the reversed order of KP(s) + Q(s) to the original order, and the program goes to C 101 to call POLYS as before. Iteration continues on CON1 as before, with C 111 sending the factor entering program to the correct place, C 160.

```
C    PROGRAM LOCUS                                                      LOCI    1
C    THIS PROGRAM FINDS ROOT LOCUS FOR KP(S)/Q(S) WHERE K IS THE PARAMET-LOCI    2
C    ER.THE CHARACTERISTIC EQUATION IS FORMED AND FACTORED USING SUBROUT-LOCI    3
C    INE POLYS.K IS VARIED AND THIS IS REPEATED.Q(S) OF ND DEGREE + P(S) LOCI    4
C    OF NN DEGREE MUST EACH BE LESS THAN 20. NO RESTRICTIONS ON RELATION LOCI    5
C    BETWEEN ND AND NN.                                                  LOCI    6
C    IDEX : IF 0,RETURN.IF NOT 0 PROCEED.IDEX=0 ON LAST DATA CARD.       LOCI    7
C    ND   : ORDER OF DENOMINATOR POLYNOMIAL Q(S).                        LOCI    8
C    NN   : ORDER OF NUMERATOR POLYNOMIAL P(S).                          LOCI    9
C    IN   : IN=1 IF POYNOMIAL COEFFICIENTS TO BE ENTERED.IF N=2 POLYNOM- LOCI   10
C           IAL FACTORS TO BE ENTERED.                                   LOCI   11
C    CON1 : ENTERED AS STARTING VALUE OF PARAMETER K.ALSO USED AS SUCEED-LOCI   12
C           ING VALUES.                                                  LOCI   13
C    CONML: MULTIPLIES PREDEEDING VALUE OF CON1 (K) FOR NEW ITERATION.   LOCI   14
C    CONAD: ADDED TO PRECEEDING VALUE OF CON1 (K) FOR NEW ITERATION IF   LOCI   15
C           CONAD NOT =0.                                                LOCI   16
C    CONMX: MAXIMUM VALUE OF CON1 (K).                                   LOCI   17
C    AQ(K): DENOMINATOR Q(S) COEFFICIENTS IN DESCENDING ORDER.           LOCI   18
C    AP(K): NUMERATOR P(S) COEFFICIENTS IN DESCENDING ORDER.AT LEAST ONE LOCI   19
C           AP(K) MUST HAVE A VALUE.                                     LOCI   20
C    AY(K): (AY(1)S*S+AY(2)S+AY(3))FACTOR OF DENOMINATOR Q(S). ANY ONE   LOCI   21
C           OR TWO AY MAY BE ZERO.                                       LOCI   22
C    BY(K): (BY(1)S*S+BY(2)S+BY(3)) FACTOR OF NUMERATOR P(S).SEE ABOVE.  LOCI   23
C           AT LEAST ONE BY(K) MUST HAVE A VALUE.                        LOCI   24
```

Figure B.2 Program LOCUS.

```
C    JA   : ND-NN                                                    LOCI 25
C    Z,ZIM: REAL AND IMAGINARY PARTS OF ROOTS OF CHARACTERISTIC EQUATION LOCI 26
C         Q(S)+KP(S)=0.                                              LOCI 27
C    THE PROGRAM OUTPUTS ENTERED PLYNOMIALS P(S),Q(S),OR FACTORS AY,BY, LOCI 28
C    PARAMETER CON1 (K),AND ROOTS Z+J ZIM FROM CON1 INITIAL TO CONMX. LOCI 29
C    DATA CARD ORDER:                                                LOCI 30
C    CARD 1 FORMAT(4I3) IDEX,ND,NN,IN                                LOCI 31
C    CARD 2 FORMAT(4F10.2) CON1,CONML,CONAD,CONMX                    LOCI 32
C    IF IN=1                                                         LOCI 33
C    CARD 3 ETC. FORMAT(5E14.6) AQ(K),K=1,ND+1                       LOCI 34
C    NEXT CARDS FORMAT (5E14.6) AP(K),K=1,NN+1                       LOCI 35
C    IF IN=2                                                         LOCI 36
C    CARD 3 ETC. FORMAT(3E14.6) AY(K),K=1,3                          LOCI 37
C    NEXT CARDS FORMAT(3E14.6) BY(K),K=1,3                           LOCI 38
C    LAST CARD SAME AS CARD 1. IF IDEX NOT 0,REPEAT DATA DECK AS ABOVE. LOCI 39
     DIMENSION AY(3),AQ(22),AP(22),CP(22),AD(22),BP(22),Z(20),ZIM(20),B LOCI 40
    1Y(3)                                                            LOCI 41
     REAL*8AQ,AP,AD,CP,BP                                            LOCI 42
  10 FORMAT (4I3)                                                    LOCI 43
  20 FORMAT (4F10.2)                                                 LOCI 44
  30 FORMAT (5E14.6)                                                 LOCI 45
  40 FORMAT (1H0'NEW PROBLEM,RCOT LOCUS')                            LOCI 46
  50 FORMAT (1H0'DENOMINATOR CCEFFICIENTS IN DESCENDING ORDER')      LOCI 47
  60 FORMAT (1H ,8E15.6)                                             LOCI 48
  70 FORMAT (1H0,'NUMERATOR COEFFICIENTS IN DESCENDING ORDER')       LOCI 49
  80 FORMAT (1H0,'GAIN=',F10.2)                                      LOCI 50
  90 FORMAT (3(3H Z=,E14.6,3X,5H ZIM=,E14.6))                        LOCI 51
 100 FORMAT (1H0,'DENOMINATOR FACTORS')                              LOCI 52
 110 FORMAT (1H ,E13.6,'S*S+',E13.6,'S+',E13.6)                      LOCI 53
 120 FORMAT (1H0,'NUMERATOR FACTORS')                                LOCI 54
 130 READ (5,10) IDEX,ND,NN,IN                                       LOCI 55
     IF (IDEX.EQ.0) STOP                                             LOCI 56
     WRITE (6,40)                                                    LOCI 57
     READ (5,20) CON1,CONML,CONAD,CONMX                              LOCI 58
     MD=ND+1                                                         LOCI 59
     MN=NN+1                                                         LOCI 60
C    NN MAY BE LESS OR GREATER THAN ND.FIND WHICH IS GREATER.CHARACTER- LOCI 61
C    ISTIC POLYNOMIAL WILL BE OF THIS DEGREE N.                      LOCI 62
     IF (MD.GE.MN) GO TO 140                                         LOCI 63
     MX=MN                                                           LOCI 64
     GO TO 150                                                       LOCI 65
 140 MX=MD                                                           LOCI 66
 150 N=MX-1                                                          LOCI 67
     JA=MD-MN                                                        LOCI 68
C    BRANCH DEPENDING ON ENTERING POLYNOMIAL COEFFICIENTS OR FACTORS. LOCI 69
     GO TO (160,300),IN                                              LOCI 70
C    POLYNOMIAL COEFFICIENTS ENTERED.                                LOCI 71
 160 READ (5,30) (AQ(I),I=1,MD)                                      LOCI 72
C    IF NN=0, ENTER AP(1)=1.0                                        LOCI 73
     READ (5,30) (AP(I),I=1,MN)                                      LOCI 74
     WRITE (6,50)                                                    LOCI 75
     WRITE (6,60) (AQ(I),I=1,MD)                                     LOCI 76
     WRITE (6,70)                                                    LOCI 77
     WRITE (6,60) (AP(I),I=1,MN)                                     LOCI 78
C    FORM CHARACTERISTIC EQUATION Q(S)+KP(S)=0.                      LOCI 79
 170 DO 180 K=1,MN                                                   LOCI 80
 180 BP(K)=AP(K)*CON1                                                LOCI 81
     IF (JA)  190,250,220                                            LOCI 82
 190 JA=-JA                                                          LOCI 83
     DO 200 K=JA,N                                                   LOCI 84
     KK=MX-K                                                         LOCI 85
     KJ=KK+JA                                                        LOCI 86
 200 AQ(KJ)=AQ(KK)                                                   LOCI 87
     DO 210 K=1,JA                                                   LOCI 88
 210 AQ(K)=0.                                                        LOCI 89
     GO TO 250                                                       LOCI 90
 220 DO 230 K=JA,N                                                   LOCI 91
     KK=MX-K                                                         LOCI 92
     KJ=KK+JA                                                        LOCI 93
 230 BP(KJ)=BP(KK)                                                   LOCI 94
```

Figure B.2—Cont.

```
        DO 240 K=1,JA                                             LOCI  95
240 BP(K)=0.                                                      LOCI  96
250 DO 260 K=1,MX                                                 LOCI  97
C   COEFFICIENTS OF CHARACTERISTIC POLYNOMIAL ARE AD(K).          LOCI  98
260 AD(K)=AQ(K)+BP(K)                                             LOCI  99
C   USE POLYNOMIAL FACTORING SUBROUTINE.                          LOCI 100
270 CALL POLYS(N,AD,Z,ZIM,KC,K1)                                  LOCI 101
        WRITE (6,80) CON1                                         LOCI 102
        WRITE (6,90) (Z(K),ZIM(K),K=1,KC)                         LOCI 103
C   TEST FOR TERMINATION.                                         LOCI 104
        IF (ABS(CON1).GT.ABS(CONMX)) GO TO 130                    LOCI 105
C   USE CONAD IF NOT ZERO.                                        LOCI 106
        IF (CONAD.NE.0) GO TO 280                                 LOCI 107
        CON1=CON1*CONML                                           LOCI 108
        GO TO 290                                                 LOCI 109
280 CON1=CON1+CONAD                                               LOCI 110
290 IF (IN.EQ.2) GO TO 430                                        LOCI 111
C   NEW ITERATION.                                                LOCI 112
        GO TO 170                                                 LOCI 113
C   POLYNOMIAL FACTORS ENTERED.                                   LOCI 114
300 MDM=MD+2                                                      LOCI 115
    INM=MN+2                                                      LOCI 116
    MAX=MX+2                                                      LOCI 117
    DO 310 K=1,MAX                                                LOCI 118
    AD(K)=0.                                                      LOCI 119
    AQ(K)=0.                                                      LOCI 120
310 AP(K)=0.                                                      LOCI 121
C   ENTER DENOMINATOR FACTORS.                                    LOCI 122
    READ (5,30) (AY(K),K=1,3)                                     LOCI 123
    WRITE (6,100)                                                 LOCI 124
    WRITE (6,110) (AY(K),K=1,3)                                   LOCI 125
C   REVERSE ORDER OF COEFFICIENTS SO THAT NEW FACTORS MAY BE MUTIPLIED. LOCI 126
    DO 320 K=1,3                                                  LOCI 127
    IT=6-K                                                        LOCI 128
320 AQ(IT)=AY(K)                                                  LOCI 129
    IF (AQ(MDM).NE.0.) GO TO 370                                  LOCI 130
330 READ (5,30) (AY(K),K=1,3)                                     LOCI 131
    WRITE (6,110) (AY(K),K=1,3)                                   LOCI 132
C   FORM DENOMINATOR POLYNOMIAL.                                  LOCI 133
    DO 340 K=3,MDM                                                LOCI 134
340 AD(K)=AQ(K-2)*AY(1)+AQ(K-1)*AY(2)+AQ(K)*AY(3)                 LOCI 135
    DO 350 K=3,MDM                                                LOCI 136
350 AQ(K)=AD(K)                                                   LOCI 137
    DO 360 K=1,MAX                                                LOCI 138
360 AD(K)=0.                                                      LOCI 139
    IF (AQ(MDM).EQ.0.) GO TO 330                                  LOCI 140
C   ENTER NUMERATOR FACTORS. IF NN=0, ENTER BY(3)=1.0.            LOCI 141
370 READ (5,30) (BY(K),K=1,3)                                     LOCI 142
    WRITE (6,120)                                                 LOCI 143
    WRITE (6,110) (BY(K),K=1,3)                                   LOCI 144
C   REVERSE ORDER OF COEFICIENTS.                                 LOCI 145
    DO 380 K=1,3                                                  LOCI 146
    IT=6-K                                                        LOCI 147
380 AP(IT)=BY(K)                                                  LOCI 148
    IF (AP(INM).NE.0.) GO TO 430                                  LOCI 149
390 READ (5,30) (BY(K),K=1,3)                                     LOCI 150
    WRITE (6,110) (BY(K),K=1,3)                                   LOCI 151
C   FORM NUMERATOR POLYNOMIAL                                     LOCI 152
    DO 400 K=3,INM                                                LOCI 153
400 AD(K)=AP(K-2)*BY(1)+AP(K-1)*BY(2)+AP(K)*BY(3)                 LOCI 154
    DO 410 K=3,INM                                                LOCI 155
410 AP(K)=AD(K)                                                   LOCI 156
    DO 420 K=1,MAX                                                LOCI 157
420 AD(K)=0.                                                      LOCI 158
    IF (AP(INM).EQ.0.) GO TO 390                                  LOCI 159
430 DO 440 K=1,MAX                                                LOCI 160
440 AD(K)=0.                                                      LOCI 161
C   FORM CHARACTERISTIC EQUATION Q(S)+KP(S)=0.                    LOCI 162
    DO 450 K=1,INM                                                LOCI 163
```

Figure B.2—*Cont.*

```
  450 AD(K)=AP(K)*CON1                                        LOCI 164
C   BRING ORDER OF COEFFICIENTS BACK TO ORIGINAL.            LOCI 165
      DO 460 K=3,MAX                                          LOCI 166
  460 CP(K)=AQ(K)+AD(K)                                       LOCI 167
      DO 470 K=1,MX                                           LOCI 168
      MNO=MX+3-K                                              LOCI 169
  470 AD(K)=CP(MNO)                                           LOCI 170
C   GO TO PREVIOUS ROUTINE BY FACTORING POLYNOMIAL.          LOCI 171
      GO TO 270                                               LOCI 172
      END                                                     LOCI 173


C     DATA CARD EXAMPLE FOR LOCUS.
   1  3   0   1
       1.00        2.00               100.00
     .100000E+01    .600000E+01    .800000E+01    .000000E+00
     .100000E+01
   0
```

Card 1:
  IDEX $\neq 0$
  ND, order of denominator = 3
  NN, order of numerator = 0
  IN = 1, or coefficients are to be entered

Card 2:
  CON1—starting value of K = 1.00 in KP(s)/Q(s)
  CONML—multiplying factor for each iteration of K = 2.00
  CONAD = 0, hence not considered
  CONMX—maximum value of K = 100.00

Cards 2, 3, 4:
  The denominator polynomial is $(s^3 + 6.0s^2 + 8.0s + 0.0)$

Card 5:
  The numerator polynomial is $(0.0s + 1.0)$

Card 6:
  IDEX = 0, terminate computations

Figure B.2—Cont.

B.3 Subroutine POLYS

This program finds the roots of a polynomial P(s) with real coefficients and of order N less than 20. The program utilizes the Q-D algorithm to find starting values for all roots. (See P. Henrici, *Elements of Numerical Analysis*, John Wiley & Sons, Inc., New York, 1964, pp. 162–179.) Because of this feature, the program is quite fast compared to routines using search techniques to obtain starting values. The starting values are refined using the Newton or Newton–Bairstow method. As each root is found, it is divided out, which speeds up the process. Since the Q-D algorithm depends on the difference of the root magnitudes, nearly equal magnitude roots may cause failure. The program circumvents this problem by changing σ, the real part of the variable $s = \sigma + j\omega$, and reiterating overall if all roots have not been found. The maximum number of such iterations is 10. Identical or nearly identical roots cause the algorithm to fail. With real floating-point of fixed-point coefficients somewhat randomly selected, such roots will be extremely rare, and usually computer roundoff errors will be sufficient to separate these roots. With integer coefficients, identical roots may be expected more frequently. (See P. Henrici and B. O. Watkins, "Finding Zeros of a Polynomial by the Q-D Algorithm," *Communications of the ACM*, Vol. 8, No. 9, Sept. 1965, pp 570–574.)

The detailed description of the program follows, where C refers to the card number in the right-hand column. The (dummy) variable N, the order of the polynomial, and $A(I)$, $I=1$, $N+1$, the polynomial coefficients in descending order, are entered from the calling program when POLYS is called. The returned roots are $ZA(I) + jZM3(I)$, $I=1$, N; KC is the root count; and K1 is an indicator set to 0 if all roots are found, or 2 if otherwise. H2 is the amount of the real part of the variable change if a zero polynomial coefficient is found, or if a reiteration occurs when all roots are not found, and H4 is the total sum of such changes. The value H4 at any iteration must be applied to the real part $ZA(I)$ of a root to correct for the shift of the real part of the variable x. $AC(I)$, $I=1$, $N+1$ are the shifted coefficients. After some initialization, C 51 bypasses most of the program if $N = 1$, while C 52 also bypasses most of the program if $N = 2$. C 53 sets H2 to 1.0E-1., and can be altered if desired. C 56–58 tests for a zero coefficient by comparing each coefficient with $AC(1)$. $AC(1) = A(1)$, the coefficient of the highest power, must NOT $= 0$. If any other coefficient is zero, the variable is shifted by C 61–66. The shifting process can occur 10 times as per C 69, and $H2 = -1.5H2$ each time. (The factor 1.5 in C 68 may be altered if desired.) Normally any zero coefficients are eliminated by one or two such shifts. After elimination of zero coefficients, the program enters the Q-D process at C 73, where it continues through C 130. The Q-D algorithm is done in single-precision arithmetic. L counts the number of Q-D iterations. (See above reference describing the Q-D algorithm.) The order of the reduced polynomial is NA, where NA=N on the first overall iteration. (NA $<$ N after found roots are divided out.) Roots are found in the Q-D algorithm by setting indicators $IND(K)$, $K = 2$, NA. Starting of indicator settings occurs in C 93 after $5 + NA/2$ iterations of the Q-D algorithm, and stopping of the indicator settings occurs after $20 + NA/2$ iterations of the Q-D algorithm. Q-D iteration occurs between C 86 and C 105. An indicator $IND(K)$ is increased by 1 if the test in C 100 is passed. The test value 1.0E-01 in C 100 may be changed if desired. Experience shows that very small or large numbers often occur, leading to overflow or underflow. C 88 prevents too small a value of any $Q(K)$ and C 91 prevents too small a value of $E(K)$, where $Q(K)$ and $E(K)$ are the variables used in the Q-D algorithm. Although the value inserted (1.E-10) is obviously not correct for the following iterations, some roots will be located. C 107–109 checks the $IND(K)$ settings. If these are 5 or greater, the indicator is considered set (the test in C 100 has been passed 5 times during 20 iterations of the Q-D algorithm). If any $IND(K) < 5$, it is reset to 0. Testing of the indicators $IND(K)$ starts in C 114 and continues to C 130. Starting values of real roots for linear factors $(s–z)$ are found and stored from C 114 to C 122. Starting values of quadratic factors $(s^2 - Rs + S)$ are found and stored from C 123 to C 130. L2 is the count of real roots found and L4 is the count of quadratic factors found.

Starting with C 131 the starting real roots are refined by the Newton method. All arithmetic is now done in double precision. The maximum number of possible Newton iterations is 14(C 138). B and C are temporary storage for the coefficients in the Newton method and ZTEMP is the temporary root. Convergence is tested by comparing the current ZTEMP, or the root from the previous iteration and Z(L1), the latest root value in C 146. The comparison (1×10^{-4}) may be altered if desired. B(MA) represents the remainder if the root is divided out (see Henrici; *Elements of Numerical Analysis*). This remainder is tested in C 147. The test value (1×10^{-3}) may be altered if desired.

If the root passes, it is stored as ZA(KC), where KC is the root count. The found root is now divided out from C 154–156 (see Henrici). This process is continued for all real roots found by the Q-D algorithm.

Starting with C 160 the quadratic factor is refined by the Newton–Bairstow method (see Henrici). The maximum number of iterations here is 30 (C 165). C 171–176 prevents B(K) and C(K) from getting over 1×10^{30}, which might lead to overflow or underflow on the next iteration. C 179 checks DELT to similarly prevent later overflow or underflow. The quadratic factor coefficients are checked with the previous iteration values in C 186 and C 187. The test number 1×10^{-4} may be altered, if desired. The polynomial remainder if the factor is divided out is tested in C 188–189. The test number 1×10^{-3} may be altered if desired. If the quadratic factor passes these tests, it is divided out from C 195–201. L is used here as a count of quadratic factors passing, whose coefficients are stored as RA(L) and SA(L) in C 192–193. The quadratic factors are factored and the complex roots obtained from C 205–229. Again the root count is KC.

A test to see if all roots are found is made in C 232. If all but two roots are found, the remaining quadratic factor is returned to C 205 and factored. If all but one root is found, the remaining linear factor root is stored in C 240. If more than two roots remain, the program returns to C 61 for a new overall iteration. The real part of the variable is shifted with the reduced polynomial, and the entire process repeated. If all roots are not found after 10 shifts, the program writes out a message to this effect, and sets K1 = 2.

```
      SUBROUTINE POLYS(N,A,ZA,ZM3,KC,K1)                              POLY0001
C   PROGRAM FINDS THE ROOTS OF A POLYNOMIAL OF ORDER N LESS THAN 20   POLY0002
C   WITH REAL COEFFICIENTS USING QD ALGORITHM TO GIVE STARTING VALUES POLY0003
C   SEE "ELEMENTS OF NUMERICAL ANALYSIS" P.HENRICI,WILEY 1964 PP 162 179 POLY0004
C   REFINED BY NEWTON BAIRSTOW METHOD.INPUT COEF A(K) IN DESCENDING ORDERPOLY0005
C   AND N.OUTPUT ZA(KC),REAL,AND ZM3(KC),IMAGINARY.NC.ROOTS FOUND=KC.  POLY0006
C   N:    ORDER OF POLYNOMIAL                                         POLY0007
C   A:    POLYNOMIAL COEFFICIENTS IN DESCENDING ORDER                 POLY0008
C   ZA:   ROOT REAL PART                                              POLY0009
C   ZM3:  ROOT IMAGINARY PART                                         POLY0010
C   KC:   ROOT COUNT                                                  POLY0011
C   K1:   INDICATES 0 FOR ALL ROOTS FOUND,2 OTHERWISE                 POLY0012
C   AC:   POLYNOMIAL COEFFICIENTS IN DESCENDING ORDER OF REDUCED POLYNOM- POLY0013
C         IAL OR PCLYNOMIAL WITH VARIABLE CHANGE.                     POLY0014
C   NA:   ORDER OF POLYNOMIAL AND REDUCED POLYNOMIAL.ON FIRST OVERALL POLY0015
C         ITERATION NA=N.                                            POLY0016
C   MA:   NA+1                                                        POLY0017
C   H2:   AMOUNT OF VARIABLE CHANGE OF REAL PART OF VARIABLE TO FIND MORE POLY0018
C         ROOTS ON NEW OVERALL ITERATION,OR TO ELIMINATE ZERO COEFFICIENTSPOLY0019
```

Figure B.3 Subroutine POLYS.

```
C  H4:   TOTAL AMOUNT OF VARIABLE CHANGE.                          POLY0020
C  Q:    QUOTIENT IN Q-D ALGORITHM.                                POLY0021
C  E:    DIFFERENCE IN Q-D ALGORITHM.                              POLY0022
C  L2:   COUNT OF LINEAR FACTOR ROOTS FOUND BY Q-D.                POLY0023
C  L4:   COUNT OF QUADRATIC FACTORS FOUND BY Q-D.                  POLY0024
C  IND:  INDICATORS SET TO LOCATE ROOTS IN Q-D ALGORITHM.          POLY0025
C  Z:    TEMPORARY REAL ROOT.                                      POLY0026
C  R,S:  COEFFICIENTS IN QUADRATIC FACTOR X*X-R*X+S.               POLY0027
C  B,C:  TEMPORARY COEFFICIENTS USED IN NEWTON REFINEMENT OF ROOTS. POLY0028
C  DELT:FACTOR IN NEWTON-BAIRSTOW REFINEMENT                       POLY0029
C  EPT:  AS ABOVE.                                                 POLY0030
C  RA:   TEMPORARY R.                                              POLY0031
C  SA:   TEMPORARY S.                                              POLY0032
C                                                                  POLY0033
      DIMENSION AC(20),Z(20),R(20),S(20),C(20),ZA(20),B(20),ZM3(20),ET(2POLY0034
     10),IND(20),Q(20),E(20),A(20),RA(20),SA(20)                   POLY0035
      REAL*8A,AC,Z,R,S,C,ZTEMP,STEMP,B,RTEMP,EPT,DELT,H2           POLY0036
C  THE NEXT CARD IS USED WITH THE WATFIVE COMPILER.REMOVE IF NOT NEEDED POLY0037
      REAL*8DABS,DSQRT                                             POLY0038
      EQUIVALENCE (Q,B),(E,C),(ET,R),(IND,S)                       POLY0039
      M=N+1                                                        POLY0040
      DO 10 K=1,M                                                  POLY0041
      ZA(K)=0.0                                                    POLY0042
      ZM3(K)=0.0                                                   POLY0043
      Z(K)=0.0                                                     POLY0044
   10 AC(K)=A(K)                                                   POLY0045
      MA=M                                                         POLY0046
      NA=N                                                         POLY0047
      KC=0                                                         POLY0048
      K1=0                                                         POLY0049
      H4=0.0                                                       POLY0050
      IF (N.EQ.1) GO TO 710                                        POLY0051
      IF (N.EQ.2) GO TO 690                                        POLY0052
      H2=1.0E-1                                                    POLY0053
      L=0                                                          POLY0054
C  TEST FOR ZERO COEF.                                            POLY0055
   30 DO 40 K=1,MA                                                 POLY0056
      IF (DABS(AC(K)/AC(1))-1.0E-20)  50,50,40                     POLY0057
   40 CONTINUE                                                     POLY0058
      GO TO 90                                                     POLY0059
C  MAKE COEF.A=A+H2 FOR VARIABLE CHANGE                            POLY0060
   50 H4=H4+H2                                                     POLY0061
      MZ=MA                                                        POLY0062
   60 DO 70 K=2,MZ                                                 POLY0063
   70 AC(K)=AC(K)+H2*AC(K-1)                                       POLY0064
      MZ=MZ-1                                                      POLY0065
      IF (MZ-2)  80,60,60                                          POLY0066
C  ALTER H2. VARIABLE CHANGE MAXIMUM OF TEN TIMES                  POLY0067
   80 H2=-1.5*H2                                                   POLY0068
      IF (ABS(H4).LT.2.0) GO TO 30                                 POLY0069
      K1=2                                                         POLY0070
      GO TO 720                                                    POLY0071
C  START QD ROUTINE                                                POLY0072
   90 L2=0                                                         POLY0073
      L4=0                                                         POLY0074
      Q(2)=-AC(2)/AC(1)                                           POLY0075
      DO 100 K=3,MA                                                POLY0076
  100 Q(K)=0.0                                                     POLY0077
      DO 110 K=2,NA                                                POLY0078
  110 E(K)=AC(K+1)/AC(K)                                          POLY0079
      DO 120 K=2,NA                                                POLY0080
  120 IND(K)=0                                                     POLY0081
      IND(1)=5                                                     POLY0082
      IND(MA)=5                                                    POLY0083
      E(MA)=0.0                                                    POLY0084
      E(1)=0.0                                                     POLY0085
  130 DO 140 K=2,MA                                                POLY0086
      Q(K)=Q(K)+E(K)-E(K-1)                                        POLY0087
      IF (ABS(Q(K)).LT.1.E-10)Q(K)=1.0E-10                         POLY0088
  140 CONTINUE                                                     POLY0089
```

Figure B.3—*Cont.*

```
        DO 150 K=2,NA                                              POLY0090
        IF (ABS(E(K)).LT.1.0E-10)E(K)=1.0E-10                      POLY0091
150 E(K)=E(K)*Q(K+1)/Q(K)                                          POLY0092
C   START CONVERGENCE TEST AFTER 5+NA/2 ITERATIONS                 POLY0093
        IF (L-5-NA/2)   210,160,180                                POLY0094
160 DO 170 K=2,NA                                                  POLY0095
170 ET(K)=ABS(E(K))                                                POLY0096
C   CHECK CONVERGENCE                                              POLY0097
180 DO 200 K=2,NA                                                  POLY0098
        IF (ABS(E(K)).LT.1.0E-20) GO TO 220                        POLY0099
        IF (ABS(E(K))/ET(K)-0.10)   190,190,200                    POLY0100
190 IND(K)=IND(K)+1                                                POLY0101
200 CONTINUE                                                       POLY0102
210 L=L+1                                                          POLY0103
C   QD GOES 20+NA/2 ITERATIONS                                     POLY0104
        IF (L-20-NA/2)   130,220,220                               POLY0105
C   SET INDICATORS                                                 POLY0106
220 DO 250 K=2,NA                                                  POLY0107
        IF (IND(K))   250,250,230                                  POLY0108
230 IF (IND(K)-5)   240,25C,250                                    POLY0109
240 IND(K)=0                                                       POLY0110
250 CONTINUE                                                       POLY0111
        L=0                                                        POLY0112
C   FIND START REAL ROOT                                           POLY0113
        DO 280 K=2,MA                                              POLY0114
        IF (IND(K)-5)   280,26C,260                                POLY0115
260 IF (IND(K-1)-5)   280,270,270                                  POLY0116
270 L2=L2+1                                                        POLY0117
        Z(L2)=Q(K)                                                 POLY0118
280 CONTINUE                                                       POLY0119
        DO 290 K=2,MA                                              POLY0120
290 Q(K)=Q(K)+E(K)-E(K-1)                                          POLY0121
        M1=NA-1                                                    POLY0122
C   FIND START QUADRATIC FACTOR                                    POLY0123
        DO 320 K=1,M1                                              POLY0124
        IF (IND(K+1))   320,300,320                                POLY0125
300 IF (IND(K)-5)   320,310,31C                                    POLY0126
310 L4=L4+1                                                        POLY0127
        R(L4)=Q(K+1)+Q(K+2)                                        POLY0128
        S(L4)=-(Q(K+1)-E(K+1)+E(K))*Q(K+2)                         POLY0129
320 CONTINUE                                                       POLY0130
        B(1)=AC(1)                                                 POLY0131
        C(1)=B(1)                                                  POLY0132
C   START NEWTON REFINE REAL ROOT                                  POLY0133
        IF (L2-1)   420,330,330                                    POLY0134
330 DO 410 L1=1,L2                                                 POLY0135
        KIP=1                                                      POLY0136
340 KIP=KIP+1                                                      POLY0137
        IF (KIP-14)   350,410,410                                  POLY0138
350 ZTEMP=Z(L1)                                                    POLY0139
        DO 360 K=2,MA                                              POLY0140
360 B(K)=AC(K)+Z(L1)*B(K-1)                                        POLY0141
        DO 370 K=2,NA                                              POLY0142
370 C(K)=B(K)+Z(L1)*C(K-1)                                         POLY0143
        Z(L1)=Z(L1)-B(MA)/C(NA)                                    POLY0144
C   TEST CONVERGENCE                                               POLY0145
        IF (DABS((ZTEMP-Z(L1))/Z(L1))-1.0E-04)   380,380,340       POLY0146
380 IF (DABS(B(MA)/Z(L1))-1.0E-03)   390,390,340                   POLY0147
390 KC=KC+1                                                        POLY0148
C   OBTAIN REAL ROOT                                               POLY0149
        ZA(KC)=Z(L1)+H4                                            POLY0150
        MA=MA-1                                                    POLY0151
        NA=NA-1                                                    POLY0152
C   DIVIDE OUT REAL ROOT                                           POLY0153
        DO 400 K=2,MA                                              POLY0154
        B(K)=AC(K)+Z(L1)*B(K-1)                                    POLY0155
400 AC(K)=B(K)                                                     POLY0156
        IF (NA.EQ.1) GO TO 710                                     POLY0157
410 CONTINUE                                                       POLY0158
C   START NEWTON BAIRSTOW REFINE ON QUADRATIC FACTOR               POLY0159
```

Figure B.3—*Cont.*

```
       420 IF (L4-1)  680,430,430                                    POLY0160
       430 DO 550 L3=1,L4                                            POLY0161
           IF (NA.LE.2) GO TO 560                                    POLY0162
           KIM=1                                                     POLY0163
       450 KIM=KIM+1                                                 POLY0164
           IF (KIM-30)   460,550,550                                 POLY0165
       460 RTEMP=R(L3)                                               POLY0166
           STEMP=S(L3)                                               POLY0167
           B(2)=AC(2)+R(L3)*B(1)                                     POLY0168
           DO 470 K=3,MA                                             POLY0169
           B(K)=AC(K)+R(L3)*B(K-1)+S(L3)*B(K-2)                      POLY0170
           IF (DABS(B(K)).GE.1.E30) GO TO 550                        POLY0171
       470 CONTINUE                                                  POLY0172
           C(2)=B(2)+R(L3)*C(1)                                      POLY0173
           DO 480 K=3,NA                                             POLY0174
           C(K)=B(K)+R(L3)*C(K-1)+S(L3)*C(K-2)                       POLY0175
           IF (DABS(C(K)).GE.1.E30) GO TO 550                        POLY0176
       480 CONTINUE                                                  POLY0177
           DELT=C(NA)*C(NA-2)-C(NA-1)*C(NA-1)                        POLY0178
           IF (DABS(DELT).GE.1.E35.OR.DABS(DELT).LE.1.E-30) GO TO 550 POLY0179
           EPT=(C(NA)*(-B(NA))-C(NA-1)*(-B(MA)))/DELT               POLY0180
           DELT=(-B(NA)-C(NA-2)*EPT)/C(NA-1)                         POLY0181
           R(L3)=R(L3)+DELT                                          POLY0182
           S(L3)=S(L3)+EPT                                           POLY0183
           ZZ=DSQRT(DABS(S(L3)))                                     POLY0184
    C   TEST CONVERGENCE                                             POLY0185
           IF (DABS((STEMP-S(L3))/S(L3))-1.0E-04)   490,490,450      POLY0186
       490 IF (DABS((RTEMP-R(L3))/R(L3))-1.0E-04)   500,500,450      POLY0187
       500 IF (DABS(B(MA)/ZZ)-1.0E-03)   510,510,450                 POLY0188
       510 IF (DABS(B(MA-1)/ZZ)-1.0E-03)   520,520,450               POLY0189
       520 L=L+1                                                     POLY0190
    C  C  OBTAIN QUADRATIC FACTOR                                    POLY0191
           RA(L)=R(L3)                                               POLY0192
           SA(L)=S(L3)                                               POLY0193
    C   CIVIDE OUT QUADRATIC FACTOR                                  POLY0194
           MA=MA-2                                                   POLY0195
           NA=NA-2                                                   POLY0196
           B(2)=AC(2)+R(L3)*AC(1)                                    POLY0197
           DO 530 K=3,MA                                             POLY0198
       530 B(K)=AC(K)+R(L3)*B(K-1)+S(L3)*B(K-2)                      POLY0199
           DO 540 K=2,MA                                             POLY0200
       540 AC(K)=B(K)                                                POLY0201
       550 CONTINUE                                                  POLY0202
       560 IF (L.EQ.0) GO TO 680                                     POLY0203
    C   FACTOR QUADRATIC FACTORS                                     POLY0204
       570 DO 670 K=1,L                                              POLY0205
           ZIMA=0.                                                   POLY0206
           ZIMB=0.                                                   POLY0207
           DISC=(RA(K)**2/4.0+SA(K)                                  POLY0208
           IF (ABS(DISC)-1.0E-20)   630,630,580                      POLY0209
       580 IF (DISC-0.)  640,630,590                                 POLY0210
       590 FACT=SQRT(DISC)                                           POLY0211
           IF (RA(K)-0.0)  610,600,600                               POLY0212
       600 ZTMPA=RA(K)/2.0+FACT                                      POLY0213
           GO TO 620                                                 POLY0214
       610 ZTMPA=RA(K)/2.0-FACT                                      POLY0215
       620 ZTMPB=-SA(K)/ZTMPA                                        POLY0216
           GO TO 660                                                 POLY0217
       630 ZTMPA=RA(K)/2.0                                           POLY0218
           GO TO 650                                                 POLY0219
       640 ZTMPA=RA(K)/2.0                                           POLY0220
           ZIMA=SQRT(-DISC)                                          POLY0221
       650 ZTMPB=ZTMPA                                               POLY0222
    C OUTPUT COMPLEX ROOTS                                           POLY0223
       660 KC=KC+1                                                   POLY0224
           ZA(KC)=ZTMPA+H4                                           POLY0225
           ZM3(KC)=ZIMA                                              POLY0226
           KC=KC+1                                                   POLY0227
           ZA(KC)=ZTMPB+H4                                           POLY0228
       670 ZM3(KC)=-ZIMA                                             POLY0229
```

Figure B.3—*Cont.*

```
C   TEST IF ALL ROOTS FOUND AND OBTAIN SINGLE QUADRATIC FACTOR          POLY0230
C   IF NOT ALL ROOTS FOUND MAKE COEF A=A+H2 AND REPEAT                   POLY0231
    680 IF (KC-N+2)  50,690,700                                          POLY0232
    690 RA(1)=-AC(2)/AC(1)                                               POLY0233
        SA(1)=-AC(3)/AC(1)                                              POLY0234
        L=1                                                             POLY0235
        GO TO 570                                                       POLY0236
C   TEST IF ALL ROOTS FOUND AND OBTAIN SINGLE REAL ROOT                 POLY0237
    700 IF (KC-N+1)  50,710,740                                         POLY0238
    710 KC=KC+1                                                         POLY0239
        ZA(KC)=-AC(2)/AC(1)+H4                                          POLY0240
        RETURN                                                          POLY0241
    720 WRITE (6,730) KC                                                POLY0242
    730 FORMAT (30H NOT ALL ROOTS FOUND NO.ROOTS=I3)                    POLY0243
    740 RETURN                                                          POLY0244
        END                                                            POLY0245
```

Figure. B.3—*Cont.*

INDEX